1987

NUCLEAR POWER
REACTOR SAFETY

Nuclear Power Reactor Safety

E. E. Lewis

Nuclear Engineering Program
The Technological Institute
Northwestern University

A WILEY–INTERSCIENCE PUBLICATION

JOHN WILEY & SONS

NEW YORK / CHICHESTER / BRISBANE / TORONTO

Library of Congress Cataloging in Publication Data:

Lewis, Elmer Eugene, 1938–
  Nuclear power reactor safety.

  "A Wiley–Interscience publication."
  Includes index.
  1. Nuclear reactors—Safety measures. 2. Nuclear reactors—Accidents. I. Title.

TK9152.L48   621.48'35   77-21360
ISBN 0-471-53335-1

Printed in the United States of America

10 9 8 7 6

to my nuclear family

Ann, Elizabeth, and Paul

Preface

In recent years the effort associated with assuring the safety of large nuclear power reactors has accelerated rapidly. The work is being documented in increasing numbers of research reports, journal articles, conference proceedings, and monographs. Nevertheless, it is difficult to recommend a reference to which a student, faculty member, or practicing engineer can turn for a unified overview of reactor safety. The objective of this text is to fill this void by providing a coherent treatment of the primary facets of reactor safety within a single volume of reasonable length.

The scope of the text is confined to nuclear safety as it pertains to power reactor accidents that may lead to releases of radioactive materials to the environment. Moreover, primary emphasis is given to the discussion of those more serious situations with the potential for causing significant public health problems. Although the setting of operating margins, the accommodation of less consequential accidents, and other related topics also have important safety implications, their treatment in comparable detail would necessitate a greatly expanded volume in which much of the material would closely parallel that contained in standard reactor engineering texts. This volume does not include the safety issues raised as a result of reactor fuel transport or reprocessing, or by other phases of the nuclear fuel cycle. Also not treated are problems related to radioactive waste disposal, routine radioactive emission, or diversion of plutonium for weapons manufacture. Although much of the material has wider application, only power reactors with solid, stationary fuel are discussed explicitly.

For the most part the text is organized along generic lines, rather than being divided according to reactor concept. It was found that this organization not only eliminated a great deal of redundancy that would otherwise be necessary, but also facilitated the comparison of safety characteristics of the various reactor concepts. Thus unified treatments of reactor kinetics, fuel behavior, coolant transients, and so on, form the basic structure of the text, and the specifics of particular reactor types are incorporated into the discussions. Only where the behavior of the various reactor concepts is fundamentally different, as, for example, in loss-of-coolant accidents, are subdivisions made according to reactor type.

Because the study of reactor safety must necessarily cut across many scientific and engineering disciplines, it is impossible to cover any one aspect of the subject with the depth available in more specialized publications. Thus, instead of emphasizing computational refinements, I have attempted to utilize the simplest mathematical models that will provide a quantitative understanding of the more important phenomena: Neutronics is treated with one-group diffusion theory, heat transfer, and fluid flow by lumped parameter models, and so on. Where these models are inadequate, emphasis is placed on the graphical presentation of results obtained from more sophisticated analysis or from experimental data.

The text has evolved, in part, from the lecture notes for a graduate-level course on reactor safety that I have taught a number of times within the Nuclear Engineering program at Northwestern University. The students in the course have been previously exposed at least to an elementary course in reactor engineering, and have a sound undergraduate physics and mathematics background. I expect that the primary use for this book will be in courses of a similar nature. The text is, however, self-contained, and I hope that it will also be found useful as supplementary reading in reactor engineering or related courses, and as a reference for the practicing engineer.

The comments and discussion of my students and colleagues have been invaluable in preparing the manuscript. I am most indebted to L. L. Briggs and D. T. Eggen of Northwestern University; to W. F. Miller, Jr., and J. Bolstad of Los Alamos Scientific Laboratory; and to H. Wider, R. W. Weeks, and E. U. Khan of Argonne National Laboratory. The thoughtful review provided by William B. Cottrell of Oak Ridge National Laboratory was very helpful in preparing the final manuscript.

I am particularly grateful to Mrs. Deana Rottner for her persevering efforts in translating my illegible scrawl and meandering equations to typed copy for each of the manuscript's several drafts. Finally, the completion of this work would have been impossible without the patience and encouragement of my wife Ann, who, with our children Elizabeth and Paul, has endured the long gestation period of this text.

E. E. LEWIS

Evanston, Illinois
May 1977

Contents

NUCLEAR POWER
REACTOR SAFETY

NUCLEAR POWER REACTOR CHARACTERISTICS

Chapter 1

1-1 INTRODUCTION

In recent years the difficulties in relying on fossil fuels for the long-term supply of electricity to industrialized society have become apparent. The reserves of crude oil and natural gas are limited, and it is argued that their use should be restricted for higher-priority applications. The economical use of the more extensive coal reserves is hindered by a number of unresolved problems of environmental degradation. At the same time the nuclear reactor based on the fission reaction has become economically competitive, and there appears to be a nearly unlimited supply of fuel, provided breeder reactor concepts are developed over the long term. As a result, the nuclear power reactor is rapidly becoming a major power source. As of 1976 there were 60 licensed power reactors with a capacity of some 42,000 MW(e) in the United States. Although this amounts to only about 8% of the total U.S. capacity, there is a growing commitment to nuclear power; if power reactors in the construction and planning stages also are included, the total U.S. commitment to nuclear power amounts to 226 reactors with a capacity of 226,000 MW(e) (Atomic Industrial Forum, 1976).

In considering the safety of nuclear power plants, the possibility of a catastrophic nuclear explosion can be disregarded. By now it is common knowledge that a nuclear power reactor cannot behave like a nuclear bomb. The fuel in a reactor does not approach the purity required in weapons grade uranium or plutonium. Neither is a mechanism present for the implosive compaction that is required for a successful nuclear detonation. There are, however, a number of potentially serious problems that must be addressed if nuclear reactors are to be used successfully as a large-scale source of power for the foreseeable future. These include the diversion of plutonium from reactors for use in the production of nuclear weapons, the long-term disposal of radioactive wastes from nuclear power plants, and the potential for catastrophic accidents in which a significant part of the large inventory of

1

radioactive material in a reactor core is released to the environment. This text is addressed to the problem of reactor accidents that may result in the release of radioactive materials to the environment.

The inventory of radioactive materials in a reactor core is an increasing function both of the power level at which the reactor is operated and of the fuel exposure—or total energy produced by the fuel in the core. In early reactors, which operated at relatively low power and fuel exposure levels, the inventories of radioactive materials were not large. Therefore sufficient protection for the public was afforded by locating the reactors in remote areas where atmospheric dispersion would reduce the concentration of radioactivity to acceptable levels, even in the event of a major accident. As power levels increased, however, the distance criteria for remote siting became inconvenient, if not prohibitive; as a result, containment vessels came into use to prevent radioactivity from escaping to the atmosphere in the event of a major accident. The first of these was built in the early 1950s for the Submarine Intermediate Reactor Mark A-SIR in West Milton, New York.

The high power levels and fuel exposures of more modern nuclear power plants have resulted in core inventories that may exceed 10 billion curies (Ci) of radioactive material. A major accident that allowed the unimpeded release of a significant fraction of the radioactivity would create an intolerable public health problem. The consequences first were forcefully pointed out in the now famous report WASH-740 (U.S. Atomic Energy Commission, 1957), which indicated that the release of the radioactive fission products from a 500 MW(e) reactor in a populated area might result in up to 43,000 fatalities, 43,000 injuries, and the contamination of between 18 and 150,000 square miles of land. Although these results were based on highly pessimistic assumptions, and indeed the authors stressed the exceedingly small probability that such an event could occur, the report set a grave tone.

Because of the potentially severe consequences of the release of a major part of the radioactive inventory of present-day power reactors, remote siting, even in conjunction with the use of a single containment shell, is not considered adequate for the protection of the public. Instead a rigorous scrutiny of the power plant designs, construction, and operational procedures is required to eliminate to the greatest extent possible mechanisms by which major accidents might arise. Multiple barriers to the release of radioactive material to the atmosphere are required, and the proximity of nuclear power plants to population centers is limited by minimum distance requirements. In addition, elaborate safety systems are incorporated into the plant to ameliorate the effects of an accident, should one occur, and, in particular, to prevent accidents from propagating to the point that significant amounts of radioactive materials are released into the atmosphere. As a result of this effort, it can be argued that the risks to the public from accidents in large nuclear power plants are far less than those resulting from other causes, both natural and man-made, in an industrial society.* In this text we examine the nature of reactor accidents, strive to gain some insight as to how they can be prevented, and investigate the means for minimizing their consequences.

1-2 RADIOACTIVE MATERIALS

We begin our study of reactor safety by examining the nature of the radioactive materials found in nuclear reactors and of the health hazards they may cause.

Inventory of Radioactive Materials

We consider first a fission reaction for uranium-235 as shown in Fig. 1-1. The products of such reactions are two or three neutrons, two lighter nuclei—called fission fragments—and gamma rays and neutrinos. A neutron chain reaction is perpetuated by the fission neutrons, the properties of which are discussed in detail in subsequent sections. Most of the approximately 200 MeV of energy produced per fission first appears as kinetic energy of the fission fragments, neutrons, and other particles. This kinetic energy is dissipated to heat nearly instantaneously as the reaction products interact with the surrounding media, producing ionization and molecular and atomic

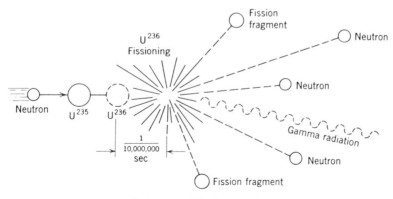

FIGURE 1-1   A fission reaction.

* See Section 2-2.

excitation. The fission fragments account for about 83% of the energy production, their kinetic energy dissipating within a distance of several microns in the surrounding medium. The two fission fragments typically have unequal masses. For example,

$$_0n^1 + _{92}U^{235} \rightarrow _{54}X^{140} + _{38}Sr^{94} + 2_0n^1. \tag{1-1}$$

The fission fragments are radioactive because they have neutron to proton ratios that are larger than those for stable nuclei. This is illustrated in Fig. 1-2. Less than 1% of these neutron-rich nuclei decay by direct neutron emission. The predominant mode of decay is through beta emission along with one or more gamma rays. Moreover, a chain of several beta decays is most often necessary before the fission fragments are transmuted to stable

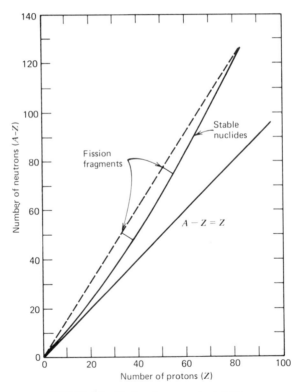

FIGURE 1-2 Fission fragment instability.

nuclei. For the example of Eq. (1-1),

$$_{54}Xe^{140} \rightarrow {}_{55}Cs^{140} + \beta + \gamma$$
$$\rightarrow {}_{56}Ba^{140} + \beta + \gamma$$
$$\rightarrow {}_{57}La^{140} + \beta + \gamma \tag{1-2}$$
$$\rightarrow {}_{58}Ce^{140} + \beta + \gamma,$$

and

$$_{38}Sr^{94} \rightarrow {}_{39}Y^{94} + \beta + \gamma$$
$$\rightarrow {}_{40}Zr^{94} + \beta + \gamma. \tag{1-3}$$

Equation (1-1) shows only one example of the more than 40 different fragment pairs that result from fission. Fission fragments may have mass numbers between 72 and 160. The mass frequency distribution is shown in Fig. 1-3 for the three nuclides fissionable by neutrons of all energies. For fissions caused by neutrons having energies of a few ev or less, nearly all of the fission fragments fall into two broad groups. The light group has mass numbers between 80 and 110, and the heavy group has mass numbers between 125 and 155. The probability of symmetrical fission increases with increasing incident neutron energy, and the valley in the curves disappears for fission caused by neutrons with incident energies in the tens of

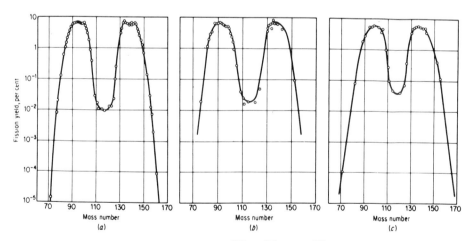

FIGURE 1-3 Fission product yields for U^{235}, Pu^{239}, and U^{233}. Adapted from National Bureau of Standards "Nuclear Data," Circular 449, 1950.

megaelectron volts. Because nearly all the 80 fission fragments lead to characteristic chains of radioactive decay products resulting from successive beta particle emission, more than 200 different fission products exist in a nuclear reactor. The general term *fission products* is used to refer to nuclides that are either fission fragments or their decay products.

Yields and Half-Lives.   Each fission product is characterized by a yield and a half-life. If we have simple decay of a radioisotope, with no production taking place, then the number of nuclei $N(t)$ is governed by the equation

$$\frac{d}{dt}N(t) = -\lambda N(t), \tag{1-4}$$

where the decay constant λ is related to the half-life by

$$\lambda = \frac{0.693}{T_{1/2}}. \tag{1-5}$$

The solution of Eq. (1-4) is

$$N(t) = N(0) e^{-\lambda t}. \tag{1-6}$$

It is often more convenient to express radioactivity in terms of the rate at which the nuclei decay: $\lambda N(t)$. This decay rate is often expressed as activity in curies: One curie is the quantity of any radioactive species decaying at a rate of 3.70×10^{10} disintegrations $\sec^{-1}$. Thus, instead of using the number of nuclei, the decay of fission products is often expressed in curies of activity:

$$A(t) = \frac{\lambda N(t)}{3.7 \times 10^{10}}, \tag{1-7}$$

where λ is in $\sec^{-1}$.

Both the yields and the half-lives of the fission products are important in reactor safety. To illustrate, first consider a simple chain in which a fission fragment undergoes only one decay. While the reactor is in operation fission fragments are produced from fission and are lost by decay. If P is the thermal reactor power, γ is the thermal energy produced per fission, and y_1 is the yield of the fission fragment per fission, the number of fission fragments is determined from

$$\frac{dN_1(t)}{dt} = \frac{y_1}{\gamma}P - \lambda_1 N_1(t), \tag{1-8}$$

where it is assumed that the fraction of fission fragments consumed by neutron capture is negligible. If the reactor is started up at $t = 0$ and remains at a fixed power P, then

$$N_1(t) = \frac{y_1}{\gamma\lambda_1}P[1 - e^{-\lambda_1 t}]. \tag{1-9}$$

Taking $\gamma = 3.3 \times 10^{-11}$ W-sec per fission, Eqs. (1-7) and (1-9) may be combined to give the reactor inventory of the fragment in kilocuries as

$$A_1(t) = 0.82 y_1 P[1 - e^{-\lambda_1 t}], \tag{1-10}$$

where P is the reactor power in megawatts (thermal). If the reactor has operated for many half-lives, the inventory reaches an equilibrium that is proportional to the reactor power:

$$A_1(\infty) = 0.82 y_1 P. \tag{1-11}$$

Conversely, when the half-life is very long compared to the irradiation time, the activity is proportional to the total energy that has been produced by the reactor:

$$A_1(t) \simeq 0.82 y_1 \lambda P t. \tag{1-12}$$

Because fuel in present-day reactors typically has a useful life of a few years, Eq. (1-11) is often a reasonable approximation for fission fragments with half-lives of less than about a year; Eq. (1-12) is appropriate for half-lives in the tens of years or more.

The yield and half-life are two important parameters in determining the significance of a particular fission fragment as a health hazard. Obviously, if the yield is sufficiently small, the inventory will be insignificant. If the half-life is small, the hazard is also decreased. To illustrate this we consider the following example: Assume that there is an accident after a reactor has been operating at some fixed, thermal power P for a time t^*, and, as a result of the accident, the reactor is shut down at $t = t^*$. The fission fragment inventory following the accident will obey Eq. (1-6) with the axis shifted by t^*. Thus determining $A(t^*)$ from Eq. (1-10), we have

$$A_1(t) = 0.82 y_1 P[1 - e^{-\lambda_1 t^*}] e^{-\lambda_1(t-t^*)}. \tag{1-13}$$

where $t - t^*$ is the time elapsed since the accident. For short half-lived fragments the activity is at the equilibrium value, $0.82 y_1 P$, at the time of the accident. However, if the half-life is sufficiently small—say substantially less than an hour—the exponential on the far right of the equation causes the activity to become insignificantly small before the fragment can penetrate the reactor containment and enter the environment under conceivable accident conditions.

Fission Product Decay Chains. Not only the fission fragments themselves, but their decay chains must be considered in evaluating the hazards associated with fission product inventories. General treatment of chains containing any number of radioactive decay products is given elsewhere (Kaplan, 1963). We illustrate with a chain consisting only of a fission fragment and the product formed by a single beta decay. The two nuclides are denoted by subscripts 1 and 2, respectively. During reactor operation the behavior of the first nuclide in the chain is determined by Eq. (1-9). The inventory of the second nuclide is given by

$$\frac{dN_2(t)}{dt} = \frac{y_2}{\gamma}P + \lambda_1 N_1(t) - \lambda_2 N_2(t). \tag{1-14}$$

The first term on the right accounts for the fact that the second nuclide may be produced directly by fission with a yield y_2. The second term is due to production by decay of the first nuclide. The last term is the decay rate of the second nuclide.

Equation (1-14) may be solved by first replacing $N_1(t)$ by Eq. (1-9) and then applying an integrating factor of $\exp(-\lambda_2 t)$. The result is

$$N_2(t) = \frac{P}{\gamma}\left[\frac{(y_1+y_2)}{\lambda_2}(1-e^{-\lambda_2 t}) + \frac{y_1}{\lambda_2-\lambda_1}e^{-\lambda_2 t} - e^{-\lambda_1 t}\right], \tag{1-15}$$

where the initial condition is $N_2(0) = 0$. After an irradiation time t^* the inventory in kilocuries is

$$A_2(t^*) = 0.82P\left[(y_1+y_2)(1-e^{-\lambda_2 t^*}) + \frac{y_1\lambda_2}{\lambda_2-\lambda_1}(e^{-\lambda_2 t^*} - e^{-\lambda_1 t^*})\right]. \tag{1-16}$$

A useful approximation is often obtained by noting that many fission product decay chains contain half-lives that differ greatly. In these cases the fission product inventory tends to be determined predominantly by the largest half-life isotopes (i.e., the smallest λ_i). Thus it is sometimes appropriate to approximate the remaining nuclides by zero half-lives (infinite λ_i). For $\lambda_1 \gg \lambda_2$ Eq. (1-16) becomes

$$A_2(t^*) \simeq 0.82P(y_1+y_2)(1-e^{-\lambda_2 t^*}), \tag{1-17}$$

whereas for $\lambda_1 \ll \lambda_2$

$$A_2(t^*) \simeq 0.82P(y_1+y_2)\left(1 - \frac{y_1}{y_1+y_2}e^{-\lambda_1 t^*}\right). \tag{1-18}$$

Following a reactor shutdown at $t = t^*$ similar methods may be used to determine the fission product activity as a function of time. For example, in the preceding two-nuclide chain the activity of the first nuclide again is given by Eq. (1-7). The activity of the second is determined from Eq. (1-14) with the reactor power set equal to zero:

$$\frac{dN_2(t)}{dt} = \lambda_1 N_1(t) - \lambda_2 N_2(t). \tag{1-19}$$

The solution in kilocuries may be shown to be

$$A_2(t) = A_2(t^*)e^{-\lambda_2(t-t^*)} + \frac{\lambda_2}{\lambda_2 - \lambda_1} A_1(t^*)[e^{-\lambda_1(t-t^*)} - e^{-\lambda_2(t-t^*)}], \qquad t > t^*, \tag{1-20}$$

where $A_1(t^*)$ and $A_2(t^*)$ can be obtained from Eqs. (1-13) and (1-16).

From these expressions it may be seen that even if the half-life of a fission fragment is small, its decay chain must be considered if the chain contains a longer half-lived nuclide. If in our two-nuclide example we take $\lambda_2 \ll \lambda_1$, then even though the first nuclide has decayed to negligible concentrations, we have from Eqs. (1-12) and (1-17)

$$A_2(t) \simeq 0.82P(y_1 + y_2)(1 - e^{-\lambda_2 t^*}) e^{-\lambda_2(t-t^*)}, \qquad t > t^*. \tag{1-21}$$

The behavior thus is determined by the longer-lived nuclide. Conversely, if the half-life of the decay product is short, but $\lambda_1 \ll \lambda_2$, the activity may be shown to be governed by the parent nuclide, $A_2(t) = A_1(t)$, for $t > t^*$.

The detailed description of fission product inventories requires that the decay chains of many different fission fragments be followed (Blomke and Todd, 1958). Moreover, it is often necessary to modify the nuclide balance equations to account for consumption by neutron capture and to account for changes in yield with incident neutron energy and with fissionable nuclide. The decay chains encompass a wide variety of half-lives and yields, including some combinations of half-lives in successive decay products that lead to activities that increase with time following reactor shutdown.

A list of the significant fission products is given in Tables 1-1 and 1-2. They are categorized as either short or long half-lived. An estimate of the fission product inventory may be made by assuming that after a long period of steady-state operation, the short half-lived fission products have reached equilibrium and therefore their activity per unit of reactor power is constant as shown in Table 1-1. The activity of the long half-lived fission products, listed in Table 1-2, grows with the total energy generated by the reactor.

The boiling points and volatility for the various fission products are also listed in Tables 1-1 and 1-2. These and other physical and chemical properties are important in determining the likelihood that a particular nuclide will escape to the environment under accident conditions. Gaseous

TABLE 1-1   Characteristics of Important Short-Half-Life Fission-Product Isotopes

Isotope	Half-Life	Activity in Kilocuries per Megawatt of Thermal Power		Boiling Point (°C)	Volatility	Health Physics Properties
		Shutdown	1 Day after Shutdown			
Br-83	2.3 h	3	0	59	Highly volatile	External whole-body
-84	32 m	6	0	59	Highly volatile	radiation, moderate
-85	3 m	8	0	59	Highly volatile	health hazard
-87	56 s	15	0	59	Highly volatile	
Kr-83m	114 m	3	0	−153	Gaseous	External radiation,
-85m	4.4 h	8	0.2	−153	Gaseous	slight health
-87	78 m	15	0	−153	Gaseous	hazard
-88	2.8 h	23	0.1	−153	Gaseous	
-89	3 m	31	0	−153	Gaseous	
-90	33 s	38	0	−153	Gaseous	
I-131	8 d	25	23	185	Highly volatile	External radiation,
-132[1]	2.3 h	38	0	185	Highly volatile	internal irradiation
-133	21 h	54	25	185	Highly volatile	of thyroid, high
-134	52 m	63	0	185	Highly volatile	radiotoxicity
-135	6.7 h	55	4.4	185	Highly volatile	
-136	86 s	53	0	185	Highly volatile	
Xe-131m	12 d	0.3	0.3	−108	Gaseous	External radiation,
-133m	2.3 d	1	0.7	−108	Gaseous	slight health
-133	5.3 d	54	47	−108	Gaseous	hazard
-135m	15.6 m	16	0	−108	Gaseous	
-135	9.2 h	25	4	−108	Gaseous	
-137	3.9 m	48	0	−108	Gaseous	
-138	17 m	53	0	−108	Gaseous	
-139	41 s	61	0	−108	Gaseous	
Te-127m	105 d	0.5	0.5	Released from oxidizing uranium		External radiation,
-127	9.4 h	2.9	0.5	Released from oxidizing uranium		moderate health
-129m	34 d	2.3	2.3	Released from oxidizing uranium		hazard
-129	72 m	9.5	0	Released from oxidizing uranium		
-131m	30 h	3.9	2.2	Released from oxidizing uranium		
-131	25 m	26	0	Released from oxidizing uranium		Health hazard from
-132	77 h	38	31	Released from oxidizing uranium		I-132 daughter[1]
-133m	63 m	54	0	Released from oxidizing uranium		External radiation,
-133	2 m	54	0	Released from oxidizing uranium		moderate health
-134	44 m	63	0	Released from oxidizing uranium		hazard
-135	2 m	55	0	Released from oxidizing uranium		

[1] Thirty-eight kilocuries of I-132 per megawatt of thermal power are generated in the reactor by decay of Te-132. Analyses that follow also consider the I-132 formed outside the reactor by decay of Te-132 released from a reactor accident.

From Beattie, 1961.

or highly volatile fission products, for example, escape from overheated or molten reactor fuel more easily than refractory metals or other immobile substances. The health physics properties listed in the last column of Tables 1-1 and 1-2 are taken up in the next subsection.

TABLE 1-2   Characteristics of Important Long-Half-Life Fission-Product Isotopes

Isotope	Half-Life	Activity in Kilocuries per Megawatt of Thermal Power		Boiling Point (°C)	Volatility	Health Physics Properties
		After 1 Yr of Irradiation	After 5 Yr of Irradiation			
Kr-85	10.4 y	0.12	0.62	−153	Gaseous	Slight health hazard
Sr-89	54 d	39	39	1366	Moderately volatile	Internal hazard to bone and lung
-90	28 y	1.2	6.0	1366	Moderately volatile	
Ru-106	1.0 y	5	10	4080	Highly volatile oxides, RuO_3 and RuO_4	Internal hazard to kidney and GI tract
Cs-137	33 y	1.1	5.3	670	Highly volatile	Internal hazard to whole body
Ce-144	282 d	30	50	3470	Slightly volatile	Internal hazard to bone, liver, and lung
Ba-140	12.8 d	53	53	1640	Moderately volatile	Internal hazard to bone and lung

From Beattie, 1961.

Decay Heat.   The net energy carried by the radiation emitted by fission products plays an important part in accident assessment, because approximately 8% of the energy produced in fission is due to the decay of fission products. Moreover, even though the neutron chain reaction is abruptly terminated, decay energy continues to be produced in copious quantities. The energy carried by the beta and gamma radiation is rapidly converted to thermal energy as the radiation interacts with the surrounding medium. This thermal energy, called decay heat, amounts to approximately 8% of full power immediately following the shutdown of the reactor. Even after several hours the decay heat is produced at a rate of about 1% of full power.

If adequate cooling is not provided following reactor shutdown, the decay heat causes overheating and eventual melting of the reactor fuel. This in turn leads to the release of the more volatile fission products.

A quantitative estimate of the decay heat was first given by the Wigner–Way formula (Way and Wigner, 1948)

$$P_d(t) = 0.0622 P_0 [t^{-0.2} - (t_0 + t)^{-0.2}] \qquad (1\text{-}22)$$

where

$P_d(t)$ = power generation due to beta and gamma rays,

P_0 = reactor power before shutdown,

t_0 = time, in seconds, of power operation before shutdown,

t = time, in seconds, elapsed since shutdown.

The result is correct to within a factor of 2 between 10 seconds and 100 days. More recent empirical formulas may be used to gain additional accuracy (Untermyer and Weills, 1962; Shure, 1961). In Fig. 1-4 the time dependence of the decay heat is shown following an infinitely long period of reactor operation. The decay heat has approximately equal contributions from beta particles and from gamma rays.

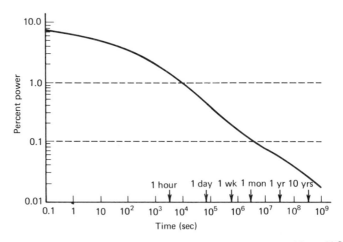

FIGURE 1-4 Heat production by decay of fission products. Adapted from U.S. Atomic Energy Commission "Reactor Safety Study," WASH-1400 (Draft), 1974.

Capture Products. Because of neutron capture reactions, radioactive materials other than fission products also appear in nuclear reactors. The most important of these involves the production of actinides as a result of neutron capture in uranium-238, thorium-232, or other heavy fuel isotopes. Plutonium is produced in large quantities in many present-day power reactors through the reaction

$$_0n^1 + _{92}U^{238} \rightarrow _{92}U^{239} + \gamma$$
$$\searrow _{93}Np^{239} + \beta \qquad (1\text{-}23)$$
$$\searrow _{94}Pu^{239} + \beta$$

Plutonium-239 is readily fissionable and therefore much of it may be consumed in the reactor by subsequent fission. However, higher isotopes of plutonium are also added to the inventory of radioactive material by successive neutron captures. For example,

$$_0n^1 + _{92}Pu^{239} \rightarrow _{92}Pu^{240} + \gamma. \qquad (1\text{-}24)$$

Significant amounts of plutonium-241 and plutonium-242 may also be produced in this way. Plutonium-239 is typical of many actinides in that it is an alpha particle emitter with a very long half-life. The decay reaction

$$_{94}Pu^{239} \rightarrow _{92}U^{235} + _2He^4 \qquad (1\text{-}25)$$

has a half-life of 24.4×10^3 years. Because of their characteristically long half-lives, plutonium and other actinides do not reach equilibrium concentration in a nuclear reactor. Their concentrations, rather, tend to increase with total energy produced by the reactor fuel. Following shutdown, actinide activity appears to be nearly time independent because of the long half-lives involved. In reactors where large amounts of plutonium or other actinides are present, it may be necessary to incorporate the heat produced by the decay of these nuclides into a composite decay heat curve.

In addition to fission products and actinides, radioactive isotopes are also produced by neutron capture in coolant, structural, shielding, and other materials. In specific instances it may be important to consider the effects of one or more such capture products in detail. By and large, however, they play a lesser role in hazards evaluation from reactor accidents than do the fission products and actinides.

Biological Effects of Radiation

The beta and gamma radiation produced by fission products causes biological damage by ionizing atoms and molecules of living tissue. Because the energy required to ionize a molecule has a relatively constant value, the total number of ionized molecules is roughly proportional to the radiation energy absorbed. As a result a reasonable measure of biological effects of radiation is the energy absorbed per unit mass of tissue. Thus the standard unit used in health physics is the radiation absorbed dose, or rad:

$$1 \text{ rad} = 100 \text{ ergs/g.}$$

This unit may also be used to assess damage to inorganic materials, since the mass does not necessarily have to be tissue.

In order to relate the rad to common health criteria—death, illness, or other detrimental effects—it is necessary to specify what organs are irradiated and in most cases the type of radiation. In this regard the gamma and beta radiation from fission products may cause damage by quite different mechanisms. Having no mass or electrical charge, gamma rays penetrate matter to a much greater extent than beta particles. In fact, for gamma ray energies in the MeV range, the collision mean free path is several centimeters in tissue, comparable to the lateral dimensions of the human body. The penetrability of gamma rays of even moderate energy has the effect of producing a radiation absorbed dose that is approximately uniform over the whole body. This is true whether the source is external (e.g., from the atmosphere or soil) or internal (through inhalation or ingestion).

In contrast, the charged beta particles do not penetrate matter to a great extent. The energy from external beta radiation is absorbed by the atmosphere, unless the source is within a few centimeters of the skin. Even then, only skin burns result, because beta particles will penetrate only the surface tissue. The situation is somewhat similar for low-energy or soft gamma rays, but for these the penetration increases very sharply with energy.

For beta particles or other nonpenetrating radiation, external sources are relatively unimportant. Rather, the predominant biological hazard results if the radioisotope is inhaled or ingested to where the localized energy deposition can do damage to the lining of the lungs, to the digestive tract, or to other organs. As might be expected, the physical and biochemical properties of the isotope or its compound that is inhaled or ingested to a great extent determine the damage that is done. Depending on its chemical properties, a given isotope tends to concentrate selectively in one or more organs. The concentration is usually specified in terms of the fraction of the ingested isotope that reaches the organ and a biological half-life, the latter being a measure of the average time the isotope remains in the organ before being passed off by metabolic processes. Taking into account these factors and the radiosensitivity of the various organs, there is normally one critical organ in which the biological damage of a particular radioisotope is concentrated. For example, the critical organ for radioiodine is the thyroid; for strontium, the bone marrow; and for insoluble compounds or aerosols containing most alpha or beta emitters, the lining of the lungs (U.S. Nuclear Regulatory Commission, 1975).

When radiation from neutrons, alphas, or other heavy particles is present, it is necessary to define a more general dose unit because the biological damage resulting from the same energy deposition in rads is greater than that caused by beta particles or gamma rays (Cember, 1969). For this purpose the radiation equivalent man, or rem, is defined:

$$rem = rad \times Q.F.$$

The quality factor, or Q.F., is a measure of the relative effects of the particles in producing damage for a given energy deposition. Typical values of the quality factor are given in Table 1-3. Note that for most fission product beta and gamma radiation, rad and rem can be used interchangeably.

TABLE 1-3 Quality Factor, Various Radiations

Radiation	QF
Gamma rays	1
X rays	1
Beta rays and electrons of energy >0.03 MeV	1
Beta rays and electrons of energy <0.03 MeV	1.7
Thermal neutrons	3
Fast neutrons	10
Protons	10
Alpha rays	10
Heavy ions	20

Acute whole-body radiation exposure—exposure to a high dose of radiation over a few days or less—affects all the organs of the body, with the response depending on the magnitude of the dose. Above 25 rems, changes in the blood begin to appear; above 100 rems, nausea and vomiting may appear, with the incidence increasing with dose. Above 200 rems, hemorrhaging and delayed infection appear, and the probability of death within a matter of months becomes significant, increasing to 50% at about 400 rems. As the whole-body dose approaches 1000 rems, death within a matter of weeks becomes certain. Only in situations where exposure is due to external sources of gamma rays is it likely that the dose will be uniform over the whole body. If the dose is due primarily to the inhalation or ingestion of alpha or beta emitters, different organs may receive widely varying doses, depending on the biochemical properties of the radionuclides. Generally, when only a

single organ is irradiated, substantially larger doses are tolerated before
acute effects appear. In Table 1-4, for example, the internal doses to the
lungs, bone, and thyroid required to produce effects comparable to a given
external whole-body dose are compared.

TABLE 1-4 Radiation Hazards

Relative Hazard	External Radiation	Internal Radiation
Lethal dose	400 rems	5000 rads to lungs, 7050 rads to thyroid, 3350 rads to bone
Illness	100 rems	700 rads to lungs, 900 rads to thyroid, 470 rads to bone
Emergency maximum permissible dose	25 rems	100 rads to lungs, 141 rads to thyroid, 67 rads to bone

The preceding symptoms are classified as early somatic effects, since they
appear either immediately or within weeks following acute radiation
exposure. Following nonlethal exposure to radiation, late somatic and
genetic effects may also appear. Late somatic effects are manifest as an
increased incidence of cancer; the genetic damage is caused by radiation-
induced mutations. It is not certain whether there is a threshold dose below
which radiation does not induce cancers. Thus for evaluating radiation risk it
is most often assumed that the probability of inducing cancer is proportional
to the exposure, no matter how small the dose (National Academy of
Sciences, 1972). For example, it is estimated that there would be 122×10^{-6}
cancers induced per year per man-rem of whole-body external radiation
(U.S. Nuclear Regulatory Commission, 1975). According to this linear
hypothesis, the same number of cancers would be caused per year whether
100,000 people received 10 rems or 10 million people received 0.1 rem,
since in both cases the total population dose is 1 million man-rems. A more
detailed discussion of both somatic and genetic effects appears in Section
10-5.

1-3 NEUTRON CHAIN REACTORS

As previously stated, power in a nuclear reactor is produced in the form of
heat generated by the approximately 200 MeV of energy released each time

a neutron causes a nucleus to fission. The rate at which this energy is released is determined by the state of the neutron chain reaction in which the neutrons produced by fission proceed to cause other fissions, which in turn result in other fissions, and so on, ad infinitum. The behavior of such neutron chain reactions is probably most easily understood in terms of the populations of neutrons in successive generations. Birth takes place as neutrons are produced as a result of fission. During a neutron lifetime a number of scattering collisions take place, each collision altering the energy and direction of travel as the neutron diffuses through the material medium. The ultimate fate of each neutron is either in absorption by a nucleus or in leakage out of the reactor. If the absorption results in a fissioning of the nucleus, then new neutrons are born, giving rise to the next generation. If no fission takes place the neutron is said to be captured, and no progeny result.

The multiplication of a reactor, denoted by k, is defined in terms of the neutron population of successive generations:

$$k = \frac{\text{no. of neutrons in } (n+1)\text{th generation}}{\text{no. of neutrons in the } n\text{th generation}}.$$

Suppose that we have a system in which the only mechanism for neutron production is fission. We then say that the system has no external sources. If there are N_1 neutrons in such a system, then in the nth generation there are

$$N_n = N_1 k^{n-1} \tag{1-26}$$

neutrons. An approximate equation can be derived for the time-dependent behavior of the neutron population by loosely defining l to be the effective neutron lifetime between birth and death. Hence the time between the birth of the first and the nth neutron generation is $t = (n-1)l$. Substituting this expression into Eq. (1-26) then yields $N(t) = k^{t/l} N(0)$. The right-hand side of this expression may also be written as $\exp\{(t/l)\ln k\}$. Normally k is very close to one, and hence $\ln k \simeq k - 1$. We therefore have

$$N(t) \simeq N(0) \exp\left(\frac{k-1}{l} t\right). \tag{1-27}$$

When treated precisely, the kinetic behavior of a neutron chain reaction is more complex than indicated by this equation. The complication arises from the fact that a small fraction of the neutrons are not produced instantaneously on fission, but result from the decay of certain fission products. For present purposes, however, Eq. (1-27) adequately represents the time variation of the neutron population. It indicates steady-state behavior when $k = 1$, and an increasing or decreasing neutron population when $k > 1$ or $k < 1$, respectively. A chain reaction is said to be subcritical, critical, or supercritical, depending on whether $k < 1$, $k = 1$, or $k > 1$, respectively.

Often it is more convenient to deal with the reactivity ρ, which is defined in terms of the multiplication by

$$\rho = \frac{k-1}{k}. \tag{1-28}$$

Obviously, the subcritical, critical, and supercritical states correspond to $\rho < 0$, $\rho = 0$, and $\rho > 0$, respectively.

Neutron Reactions

Before proceeding with the treatment of neutron multiplication, it is necessary to quantify the reactions of neutrons with matter through the definition of the neutron cross sections and the neutron flux. Let $\sigma(E)$ be the crosssection area of a nucleus at rest as it appears to a neutron with kinetic energy E. The probability that the neutron will undergo a collision with a nucleus in traveling a distance of 1 cm is then

$$\Sigma(E) = \bar{N}\sigma(E), \tag{1-29}$$

where $\bar{N}$ is the atomic density in nuclei per cubic centimeter. The quantities $\sigma(E)$ (in cm^2) and $\Sigma(E)$ (in cm^{-1}) are referred to as the microscopic and macroscopic cross sections, respectively.

To determine the number of neutrons colliding with the nuclei per second it is necessary to evaluate the neutron density distribution; $n(E)\,dE$ is defined as the number of neutrons per cubic centimeter with energies between E and $E + dE$. To determine the number of collisions occurring we need to know the total distance traveled by all the neutrons in 1 cm^3 of material in 1 sec. This will be just $v(E)n(E)\,dE$, where v is the neutron speed in centimeters per second. Thus defining the neutron flux as

$$\varphi(E) = v(E)n(E), \tag{1-30}$$

we see that $\varphi(E)\,dE$ is the total distance traveled by all neutrons located in 1 cm^3 with energies between E and $E + dE$ during one second. The number of collisions made by these neutrons per cubic centimeter per second is $\Sigma(E)\varphi(E)\,dE$. Integrating over all neutron energies, we obtain the total number of collisions per cubic centimeter per second

$$R = \int_0^\infty \Sigma(E)\varphi(E)\,dE. \tag{1-31}$$

This quantity is normally referred to as the collision density or reaction rate.

The cross sections $\sigma(E)$ and $\Sigma(E)$ refer to all collisions, regardless of the fate of the neutron. In subsequent discussions, however, it is necessary to distinguish between types of reaction rates. A neutron that undergoes collision is scattered, is captured, or causes a fission. Therefore $\sigma(E)$, referred to as the total cross section and often written as $\sigma_t(E)$, may be expressed as a superposition of the three possible reactions

$$\sigma_t(E) = \sigma_s(E) + \sigma_c(t) + \sigma_f(E), \tag{1-32}$$

where the subscripts t, s, c, f refer to *total, scattering, capture,* and *fission,* respectively. At times it is convenient to group fission and capture cross sections together. Thus an absorption cross section is also defined by

$$\sigma_a(E) = \sigma_c(E) + \sigma_f(E). \tag{1-33}$$

For each reaction type a macroscopic cross section is defined as

$$\Sigma_x(E) = \bar{N}\sigma_x(E), \tag{1-34}$$

where, hereafter, we use the subscript x to refer to any of the reaction types. The reaction rate of type x is seen to be

$$R_x = \int_0^\infty \Sigma_x(E)\varphi(E)\,dE. \tag{1-35}$$

Thus far we have assumed that only one nuclide is present, whereas in actuality reactor configurations consist of several materials, each of which is likely to have more than one isotope. Hence we define the microscopic cross section for the reaction x for a nuclide designated as the ith type as σ_x^i. The contribution to the macroscopic cross section from the ith nuclide is then

$$\Sigma_x^i(E) = \bar{N}_i\sigma_x^i(E). \tag{1-36}$$

where $\bar{N}_i$ is the atom density of the ith nuclide. The macroscopic cross section for all the species of nuclei is then

$$\Sigma_x(E) = \sum_i \Sigma_x^i(E). \tag{1-37}$$

Infinite Medium Reactors

With the cross-section notation in hand, we are prepared to examine the multiplication k of an infinite medium system. Neutrons can be born into a new generation only by fission. Because there is no spatial loss of neutrons in an infinite medium, they can be lost only by absorption (i.e., either by capture or fission), since scattering only changes their energy. The multiplication, defined as the ratio of the birth to death reaction rate, is thus

$$k_\infty = \frac{\int_0^\infty \nu\,\Sigma_f(E)\varphi(E)\,dE}{\int_0^\infty \Sigma_a(E)\varphi(E)\,dE}, \tag{1-38}$$

where ν is the average number of neutrons produced per fission and is here assumed to be energy independent.

The energy of the neutrons of interest in reactor physics ranges from about 10 MeV to less than 0.01 eV. The upper limit is determined by the energy spectrum of neutrons produced by fission. As shown in Fig. 1-5,

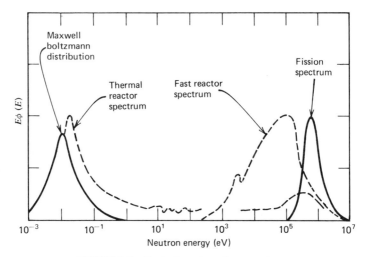

FIGURE 1-5 Typical neutron flux spectra.

fission neutrons are produced in the megaelectron volt energy range with an average energy of about 2 MeV and a most probable energy of about 1 MeV. All scattering collisions result in a degradation of neutron energy, until the neutrons—if they are not absorbed first—finally come to equilibrium with thermal motions of the atoms of the surrounding medium. In the latter event the neutrons form a Maxwell-Boltzmann distribution, as shown schematically in Fig. 1-5. If the medium is at room temperature, the average energy of the thermal neutron distribution is 0.025 eV, and the most probable energy is 0.01 eV.

The neutron energy spectrum in a reactor lies between the extremes of the fission and thermal equilibrium spectra. It is largely determined by the competition between scattering and absorption reactions. For neutrons with energies significantly above the thermal range, a scattering collision results in a degradation of neutron energy, whereas neutrons in thermal equilibrium would have equal probabilities of gaining or losing energy when interacting with the thermal motions of the nuclei in the surrounding media. Energy degradation (referred to as neutron slowing down) is produced by scattering collisions. In media for which the average energy loss per collision and the ratio of scattering to absorption cross sections are large, a neutron spectrum $\varphi(E)$, which is near thermal equilibrium, results. The spectrum is then referred to as soft or thermal. Conversely, in systems with small ratios of neutron degradation to absorption, neutrons are absorbed before significant slowing down takes place. The neutron spectrum then lies much closer to the fission spectrum and is said to be hard or fast.

The nature of the slowing down or energy degradation differs for elastic and for inelastic scattering. For elastic scattering events, in which kinetic energy is conserved between the neutron and the scattering nucleus, the average energy degradation increases with decreased atomic weight A of the scattering nucleus. The energy loss due to elastic scattering is most often measured in terms of the average value of the logarithm of the neutron energy before and after the scattering event:

$$\xi = \overline{\ln(E'/E'')}, \tag{1-39}$$

where E' and E'' are the neutron energy before and after the collision. It may be shown that (Lamarsh, 1966)

$$\xi = 1 - \frac{(A-1)^2}{2A} \ln \frac{A+1}{A-1}, \tag{1-40}$$

which for $A > 1$ can be approximated by

$$\xi \approx \frac{2}{A+2/3}. \tag{1-41}$$

If a nuclide is to have a substantial effect in slowing down neutrons, it must have not only a large energy loss per collision, but also a large scattering cross section for there to be large numbers of collisions. For this reason the moderating power, defined as $\xi \Sigma_s$, is a frequent measure of the ability of a nuclide to degrade the neutron energy spectrum by elastic collisions.

Neutron slowing down can also take place through inelastic scattering, in which some of the kinetic energy of the neutron is lost to the internal excitation of the nucleus. In order for a neutron to scatter inelastically, however, it must possess sufficient energy to excite a nucleus. Thus, although large amounts of energy are lost in inelastic collisions, the scattering cross section vanishes below some energy threshold. Typically, these thresholds range from the megaelectron volt range for light nuclei to the kiloelectron volt range for heavy nuclei. Thus inelastic scattering tends to have a significant effect only on neutrons in the kiloelectron volt through megaelectron volt range, and then primarily for heavy nuclei that have larger inelastic scattering cross sections.

Reactors are classified as fast or thermal according to their neutron energy spectrum $\varphi(E)$. To understand the neutronic significance of the energy spectrum, the cross sections of fissionable materials must be examined. Fissionable nuclides are classified as either fissile or fertile materials, depending on the behavior of the fission cross section. Fissile nuclides, such as U^{235}, Pu^{239}, and U^{233}, have a nonzero fission cross section over the entire range of neutron energies; fission is energetically possible in fertile nuclides, such as U^{238}, Th^{232}, and Pu^{240}, only for neutrons with energies above some threshold value in the high kiloelectron volt or low megaelectron volt range. The fission cross sections for the common fissile and fertile isotopes are shown in Fig. 1-6.

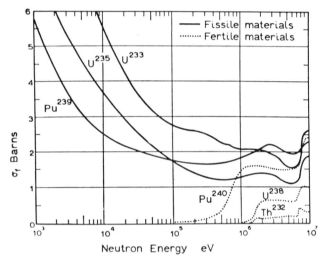

FIGURE 1-6 Fission cross sections of fissile and fertile materials at high energies. From D. Jakeman, *Physics of Nuclear Reactors*, American Elsevier, New York, 1966. Used by permission.

The only naturally occurring fissile isotope is U^{235}, which is 0.7% of natural uranium, with the remainder being U^{238} except for trace amounts of U^{234}. A reactor cannot be built out of pure natural uranium for reasons that can be seen by examining the capture (n, γ), fission (n, f), and scattering (n, n) cross sections of U^{235} and U^{238}. These are given in Figs. 1-7 and 1-8. Fission neutrons produced in the megaelectron volt range would be quickly slowed down past the fission threshold in the U^{238} by both elastic and inelastic scattering. Because of the preponderance of U^{238} in a pure natural uranium system, the neutrons would be captured primarily by the U^{238}, particularly in the energy range between 1 eV and 1 keV, where large resonance peaks exist in the capture cross section of U^{238}. To achieve a critical chain reaction, either additional material, called a moderator, must

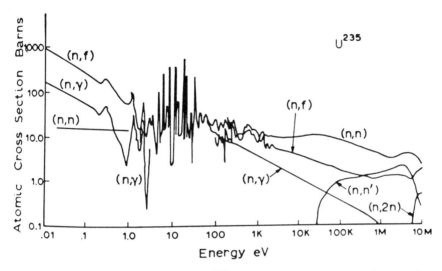

FIGURE 1-7 Neutron cross sections of U^{235}. Here (n, n) refers to the elastic scattering collisions, (n, n') to inelastic scattering collisions, (n, γ) to capture, and (n, f) to fission. From D. Jakeman, *Physics of Nuclear Reactors*, American Elsevier, New York, 1966. Used by permission.

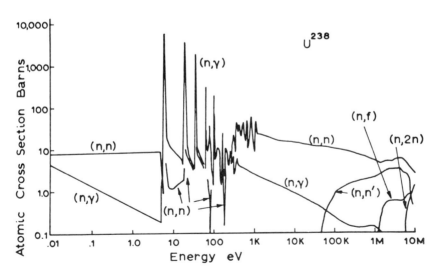

FIGURE 1-8 Neutron cross sections of U^{238}. From D. Jakeman, *Physics of Nuclear Reactors*, American Elsevier, New York, 1966. Used by permission.

be added to prevent excessive absorption in the capture resonances or the ratio of fissile to fertile material must be increased by enriching the fuel. Often both steps are taken.

A quantity that illustrates the neutron economy of a reactor fuel is the number of fission neutrons produced per neutron absorbed in the fuel:

$$\eta(E) = \frac{\nu \sigma_f(E)}{\sigma_a(E)}, \tag{1-42}$$

where

$$\sigma_a(E) = \sigma_c(E) + \sigma_f(E). \tag{1-43}$$

Obviously, if a chain reaction is to be achieved, on the average there must be more than one fission neutron produced for each neutron absorbed in the fuel, since there will also be neutrons absorbed in nonfuel components of the system. It is important to concentrate $\varphi(E)$ in energy ranges where $\eta(E)$ is large in order to maximize the flux-weighted energy-averaged value $\bar{\eta}$. Although ν, the number of neutrons per fission, is almost energy independent, the cross sections, and thus $\eta(E)$, vary substantially. Plots of $\eta(E)$ for fissile isotopes are shown in Fig. 1-9. Obviously, it is undesirable to have

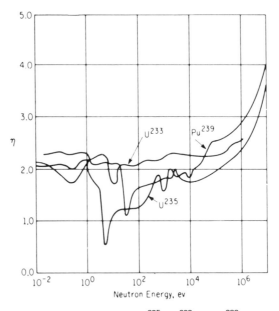

FIGURE 1-9 η versus neutron energy for U^{235}, Pu^{239} and U^{233}. From M. M. El-Wakil, *Nuclear Energy Conversion*, International Publishers, Scranton, Pa., 1971. Used by permission.

many neutrons in the energy-range from a few electron volts to a few kiloelectron volts, because too large a fraction of the neutrons in this range are lost to the large resonances in the capture cross sections.

The situation is more pronounced for a typical reactor fuel that consists of a mixture of fissile and fertile material. Suppose we define the enrichment $\tilde{e}$ of a fuel as the atom ratio of fissile to fertile plus fissile nuclei,

$$\tilde{e} = \frac{\bar{N}_{fi}}{\bar{N}_{fe} + \bar{N}_{fi}}, \tag{1-44}$$

where fi and fe denote fissile and fertile, respectively. The number of fission neutrons produced per neutron of energy E absorbed in the fuel may be shown to be

$$\eta(E) = \frac{\nu^{fi}\sigma_f^{fi}(E) + (\tilde{e}^{-1} - 1)\nu^{fe}\sigma_f^{fe}(E)}{\sigma_a^{fi}(E) + (\tilde{e}^{-1} - 1)\sigma_a^{fe}(E)}. \tag{1-45}$$

The presence of the fertile material is most noticeable in the resonance region between a few electron volts and a few kiloelectron volts, for it contributes a large capture cross section, as shown, for example, in Fig. 1-8. This further deepens the valley in the intermediate energy range of $\eta(E)$. At the same time, the fertile material contributes no fissions in this energy range, since $\sigma_f^{fe}(E)$ is zero, becoming significant only above its threshold in the low megaelectron volt range.

In order to avoid the sharp valley in $\eta(E)$ exhibited by nuclear fuels in the intermediate energy range, reactors are designed to concentrate the neutron flux $\varphi(E)$ either in the thermal or the fast energy range. In thermal reactors substantial amounts of a so-called moderator material are present. The function of the moderator is to cause elastic scattering collisions that will rapidly slow down the fission neutrons through the intermediate energy range. To be effective a moderator must have a low atomic mass number for the average energy loss per collision to be large as indicated by Eq. (1-41). It must have a large scattering cross section to promote collisions but a small capture cross section, so that neutrons are not lost as a result of the presence of the moderator. The most common moderator materials are water, heavy water, graphite, and beryllium. With sufficient moderation a large fraction of the neutrons escape capture in the intermediate energy range and arrive in the high $\bar{\eta}$, low-energy region to form a flux distribution that is nearly in thermal equilibrium with the surrounding medium.

The neutron flux distribution in a typical thermal reactor is sketched in Fig. 1-5. Most thermal power reactor fuels require enrichments of only a few percent; natural uranium fuels can be used in carefully designed fuel-moderator lattices using C, Be, or D_2O. In fast reactors lightweight material is eliminated to the greatest degree practicable, and the enrichment is increased sufficiently (from 10 to 20%) that neutrons cause fission in the energy range above a few kiloelectron volts, before they are slowed down into the high capture resonance range. A typical fast reactor spectrum is also sketched in Fig. 1-5.

Although fertile materials tend to decrease the multiplication in either a fast or thermal reactor, their presence is nevertheless desirable in order to create new fuel by converting the fertile nuclei to fissile nuclei through neutron capture and radioactive decay. The two most important such reactions are the production of plutonium,

$$
\begin{aligned}
n + {}_{92}U^{238} \rightarrow {}_{92}U^{239} &+ \gamma \\
&\searrow {}_{93}Np^{239} + \beta \\
&\qquad \searrow {}_{94}Pu^{239} + \beta,
\end{aligned}
\tag{1-46}
$$

and the production of uranium-233,

$$
\begin{aligned}
n + {}_{90}Th^{232} \rightarrow {}_{90}Th^{233} &+ \gamma \\
&\searrow {}_{91}Pa^{233} + \beta \\
&\qquad \searrow {}_{92}U^{233} + \beta.
\end{aligned}
\tag{1-47}
$$

An important achievement is that of breeding, in which more fertile material is converted to fissile material than fissile material is destroyed. In order for breeding to be achieved, however, the average value of $\eta(E)$ must be greater than 2, since for each neutron absorption in the fuel, one fission neutron is required to sustain the reaction and a second is required to replace the destroyed fissile nuclide by converting a fertile to a fissile nuclide. Fast reactors based on the U^{238}, Pu^{239} chain are particularly attractive in this regard, as can be observed from the large value of $\eta(E)$ for Pu^{239} in the megaelectron volt range.

The One-Group Diffusion Model

We turn now to the treatment of the spatial effects, an important part in any real reactor system, which necessarily is of finite dimensions. In a reactor the neutron flux is a function of space as well as energy:

$$
\Phi = \Phi(\mathbf{r}, E),
\tag{1-48}
$$

where $\mathbf{r}$ is a position vector. Most often the composition of a reactor is not completely uniform; therefore the macroscopic cross sections are also

spatially dependent:

$$\Sigma_x = \Sigma_x(\mathbf{r}, E). \qquad (1\text{-}49)$$

As a first approximation, the simplest method for handling both space and energy effects is to assume that their effects are approximately separable. Thus we take

$$\Phi(\mathbf{r}, E) \simeq \varphi(E)\phi(\mathbf{r}), \qquad (1\text{-}50)$$

with the normalization

$$\int_0^\infty \varphi(E)\, dE = 1. \qquad (1\text{-}51)$$

The formulas of the preceding section now may be simply modified. At a particular point in space $\mathbf{r}$, the reaction rates become

$$R_x(\mathbf{r}) = \int_0^\infty \Sigma_x(\mathbf{r}, E)\Phi(\mathbf{r}, E)\, dE, \qquad (1\text{-}52)$$

which with the separability assumption reduce to

$$R_x(\mathbf{r}) = \Sigma_x(\mathbf{r})\phi(\mathbf{r}), \qquad (1\text{-}53)$$

where the energy spectrum averaged cross section is defined as

$$\Sigma_x(\mathbf{r}) = \int_0^\infty \Sigma_x(\mathbf{r}, E)\varphi(E)\, dE. \qquad (1\text{-}54)$$

Although the separability is never completely valid, often the approximation of Eq. (1-50) is sufficient for a first-order quantitative understanding of many physical effects, particularly if different flux spectra $\varphi(E)$ are used in performing the preceding energy integrals for each distinctive region of material composition in a reactor.

With these preliminaries the simple one-energy-group treatment may be used to incorporate spatial diffusion with the physical effects described in the preceding section. Suppose we consider a case of an infinitesimal volume centered at the point $\mathbf{r} = \{x, y, z\}$ as shown in Fig. 1-10. Under steady-state conditions neutron conservation requires that

$$\begin{pmatrix} \text{source neutrons} \\ \text{emitted in } dx\,dy\,dz \end{pmatrix} + \begin{pmatrix} \text{fission neutrons} \\ \text{produced in } dx\,dy\,dz \end{pmatrix} - \begin{pmatrix} \text{neutrons absorbed} \\ \text{in } dx\,dy\,dz \end{pmatrix}$$
$$= \begin{pmatrix} \text{neutrons leaking} \\ \text{out of } dx\,dy\,dz \end{pmatrix}.$$

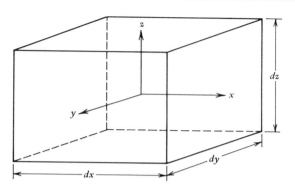

FIGURE 1-10   Volume element.

Let $S_e(r)\,dx\,dy\,dz$ be the number of neutrons produced per second in $dx\,dy\,dz$ by neutron emitting sources. The number of reactions taking place per cm^3/sec is given by Eq. (1-53). Therefore we may rewrite the neutron conservation condition as

$$S_e(r)\,dx\,dy\,dz + [\nu\,\Sigma_f(r) - \Sigma_a(r)]\phi(r)\,dx\,dy\,dz = \mathscr{L}, \qquad (1\text{-}55)$$

where $\mathscr{L}$ signifies the leakage from the cube. The leakage results from the net number of neutrons passing out through each of the six faces of the cube and may be expressed in terms of the neutron current components as follows. Let $J_x(x, y, z)$ be the net number of neutrons per cm^2/sec passing through the $y-z$ plane in the positive x direction at (x, y, z). Similarly, $J_y(x, y, z)$ and $J_z(x, y, z)$ are the net numbers of neutrons per cm^2/sec passing through the $x-z$ and $x-y$ planes in the positive y and z directions, respectively. For the cube in Fig. 1-10 the net number of neutrons passing out of the cube through the right-hand, front, or top faces is $J_x[x + (dx/2), y, z]\,dy\,dz$, $J_y[x, y + (dy/2), z]\,dx\,dz$, and $J_z[x, y, z + (dz/2)]\,dx\,dy$, respectively. Similarly, the net number of neutrons passing out of the cube through the left, back, or bottom faces is, respectively, $-J_x[x - (dx/2), y, z]\,dy\,dz$, $-J_y[x, y - (dy/2), z]\,dx\,dz$ and $-J_z[x, y, z - (dz/2)]\,dx\,dy$. Thus the net leakage from the cube is

$$\mathscr{L} = \left[J_x\left(x + \frac{dx}{2}, y, z\right) - J_x\left(x - \frac{dx}{2}, y, z\right)\right] dy\,dz$$

$$+ \left[J_y\left(x, y + \frac{dy}{2}, z\right) - J_y\left(x, y - \frac{dy}{2}, z\right)\right] dx\,dz \qquad (1\text{-}56)$$

$$+ \left[J_z\left(x, y, z + \frac{dz}{2}\right) - J_z\left(x, y, z - \frac{dz}{2}\right)\right] dx\,dy.$$

We now combine Eqs. (1-55) and (1-56) and divide by the cube volume.

Then as $dx, dy, dz \to 0$ we have

$$S_e(\mathbf{r}) + [\nu \Sigma_f(\mathbf{r}) - \Sigma_a(\mathbf{r})]\phi(\mathbf{r}) = \frac{d}{dx}J_x(\mathbf{r}) + \frac{d}{dy}J_y(\mathbf{r}) + \frac{d}{dz}J_z(\mathbf{r}). \quad (1\text{-}57)$$

The definition of the current vector as

$$\mathbf{J}(\mathbf{r}) = \hat{\imath}J_x(\mathbf{r}) + \hat{\jmath}J_y(\mathbf{r}) + \hat{k}J_z(\mathbf{r}) \quad (1\text{-}58)$$

permits Eq. (1-57) to be written more compactly as

$$S_e(\mathbf{r}) + [\nu \Sigma_f(\mathbf{r}) - \Sigma_a(\mathbf{r})]\phi(\mathbf{r}) = \nabla \cdot \mathbf{J}(\mathbf{r}). \quad (1\text{-}59)$$

By far the most common approximation in neutronics calculation is that of diffusion theory, which assumes that current and scalar flux may be related by Fick's law:

$$\mathbf{J}(\mathbf{r}) = -D(\mathbf{r})\nabla\phi(\mathbf{r}). \quad (1\text{-}60)$$

The more detailed derivations in standard textbooks (e.g., Henry, 1975) give D, the diffusion coefficient, in terms of the total cross section as

$$D(\mathbf{r}) = \int_0^\infty \frac{\varphi(E)}{3\Sigma(\mathbf{r}, E)} dE. \quad (1\text{-}61)$$

The combination of Eqs. (1-59) and (1-60) leads to

$$\nabla D(\mathbf{r}) \nabla\phi(\mathbf{r}) + [\nu \Sigma_f(\mathbf{r}) - \Sigma_a(\mathbf{r})]\phi(\mathbf{r}) = -S_e(\mathbf{r}). \quad (1\text{-}62)$$

This is referred to as the one-group diffusion equation because the averaging over the entire energy range permits treatment of the energy variable as a single group. In subsequent discussions emphasis is placed on understanding the effects of composition, temperature, and other changes on the state of neutron chain reactions. For this purpose it is often more convenient to deal not with the cross sections directly but rather with the following local core parameters. In analogy to Eq. (1-38) we may define a local multiplication as

$$k_\infty(\mathbf{r}) = \frac{\nu \Sigma_f(\mathbf{r})}{\Sigma_a(\mathbf{r})}, \quad (1\text{-}63)$$

which is the value of the multiplication that would result if the material composition at point $\mathbf{r}$ were to extend to infinity in all directions. Similarly, the local migration area

$$M^2(\mathbf{r}) = \frac{D(\mathbf{r})}{\Sigma_a(\mathbf{r})}, \quad (1\text{-}64)$$

or the corresponding migration length $M(\mathbf{r})$, is a measure of how far a neutron diffuses between birth and death; in an infinite medium the root-mean-squared travel distance between birth and death may be shown to be

$\sqrt{6}M$. The local value of the prompt neutron lifetime $l_\infty(\mathbf{r})$ is the mean lifetime of a neutron between emission by fission and absorption in an infinite medium with the same properties as those at $\mathbf{r}$. It is given by

$$l_\infty(\mathbf{r}) = \frac{1}{v \sum_a (\mathbf{r})},\tag{1-65}$$

where v is the mean neutron speed whose spatial variations are neglected. The one-group equation may now be expressed in terms of k_∞, M^2, and l_∞. Dividing Eq. (1-62) by the absorption cross section and using Eqs. (1-63) through (1-65), it may be shown, after some algebra, that

$$\nabla M^2(\mathbf{r})\nabla\phi(\mathbf{r}) - M^2(\mathbf{r})[\nabla \ln l_\infty(\mathbf{r})]\nabla\phi(r) + [k_\infty(\mathbf{r}) - 1]\phi(\mathbf{r}) = -vl_\infty(\mathbf{r})S_e(\mathbf{r}).\tag{1-66}$$

If V is defined as the volume of the reactor and $\mathscr{S}$ as the outermost surface, then the conditions associated with the neutron diffusion equation are

$$\Phi(\mathbf{r}) > 0, \quad \mathbf{r} \in V,\tag{1-67}$$

$$\Phi(\mathbf{r}) = 0, \quad \mathbf{r} \in \mathscr{S}.\tag{1-68}$$

In order for a reactor to operate at steady state, a time-independent neutron chain reaction must be sustained in the absence of external neutron sources. The reactor is said to be critical. In our one-group model, criticality requires the existence of a solution to Eq. (1-66) with $S_e = 0$:

$$\nabla M^2(\mathbf{r})\nabla\phi_0(\mathbf{r}) - M^2(\mathbf{r})[\nabla \ln l_\infty(\mathbf{r})]\nabla\phi_0(\mathbf{r}) + [k_\infty(\mathbf{r}) - 1]\phi_0(\mathbf{r}) = 0,\tag{1-69}$$

where $\phi_0(\mathbf{r})$ must meet the nonnegative conditions of Eqs. (1-67) and (1-68). Note that the second term in this equation vanishes if the material properties, and therefore l_∞, are space independent. In a typical situation in which a reactor consists of several uniform regions, the second term appears only at the regional interfaces. In such situations it may be eliminated by treating the equation separately in each region and applying the interface conditions of continuity of flux or normal current.

Physically, criticality is achieved by adjusting either the size or the composition of the reactor so that a steady-state neutron balance (as expressed in the one-group approximation by Eq. [1-69]) is maintained. In power reactors the volume is invariably fixed by thermal considerations, so criticality is maintained by changes in the composition of the core and therefore by changes in the material properties $k_\infty(\mathbf{r})$, $M^2(\mathbf{r})$, and $l_\infty(\mathbf{r})$. Most often this is accomplished by control rods that consist of nonfissionable neutron-absorbing materials that can be moved in or out of the core. The net effect of the insertion of control rods is to reduce $k_\infty(\mathbf{r})$ and therefore reduce the overall multiplication of the reactor. The same effect may be accomplished by varying the concentration of a neutron-absorbing material, such as boron, in a liquid coolant. Control rods, soluble boron, or other neutron-absorbing materials introduced into the reactor for controlling the chain reaction are often referred to as poisons, because they reduce the multiplication and therefore "poison" the chain reaction.

The adjustable control poison is used to maintain the reactor in a critical state by compensating for several other effects that cause changes in the multiplication. As a power reactor operates over a long period of time, the fuel concentration is depleted and the concentration of neutron-absorbing fission products increases. The changes in the material properties l_∞, M^2, k_∞ resulting from these fuel burnup effects must be compensated for by decreasing the poison concentration with time. The temperatures within the reactor core increase with power level and change with other operating conditions. Therefore neutron poison adjustments also must be made to compensate for the temperature dependence of the material properties l_∞, M^2, and k_∞. Finally, poison control is used to bring the reactor supercritical during reactor start-up or power increases and subcritical for power decreases or shutdown. In these situations where the reactor is not critical, Eq. (1-69) is not applicable, since the flux is then time dependent. Explicit relationships to describe this time dependence are taken up in Chapter 3. For the remainder of this chapter attention is focused on the behavior of critical, source-free systems.

The Uniform, Bare Reactor Core

The concepts necessary to the understanding of reactor safety are most easily introduced by treating a simple and highly idealized reactor core. We consider a reactor whose composition and temperature are uniform, and therefore k_∞, M^2, and l_∞ are space independent. This is referred to as a *bare, uniform reactor*. The criticality equation then reduces to

$$\nabla^2 \phi_0(\mathbf{r}) + \left\{ \frac{k_\infty - 1}{M^2} \right\} \phi_0(\mathbf{r}) = 0. \qquad (1\text{-}70)$$

For this equation to have a solution, $\nabla^2 \phi_0$ must be proportional to ϕ_0. Thus if the proportionality constant is taken as B_g^2, we have the equivalent equation

$$\nabla^2 \phi_0(\mathbf{r}) + B_g^2 \phi_0(\mathbf{r}) = 0. \qquad (1\text{-}71)$$

For a fixed shape and size reactor core this equation has a unique solution and a single value of B_g^2 for which $\phi_0(\mathbf{r})$ is nonnegative within the reactor and zero on the surface; B_g^2 is a function of the core shape and dimensions and is referred to as the *geometrical buckling*. Eliminating the ∇^2 term between Eqs. (1-70) and (1-71), we have

$$\frac{k_\infty - 1}{M^2} = B_g^2,\tag{1-72}$$

or

$$1 = \frac{k_\infty}{1 + M^2 B_g^2}.\tag{1-73}$$

This is the criticality condition for a bare, uniform reactor. It is a relationship between the composition of the core, as contained in k_∞ and M^2, and the size and shape of the core, as contained in B_g^2. It must be satisfied if a steady-state chain reaction is to be sustained.

Power reactors are almost always built in the form of right circular cylinders to facilitate heat transport from the core. Thus we consider the solution of Eq. (1-71) for a cylinder of radius $\tilde{R}$ and height $\tilde{H}$, with the origin of the (r, z) coordinates located at the center of the reactor. In cylindrical geometry Eq. (1-71) becomes

$$\frac{1}{r}\frac{d}{dr}r\frac{d}{dr}\phi_0(r, z) + \frac{d^2}{dz^2}\phi_0(r, z) + B_g^2\phi_0(r, z) = 0,\tag{1-74}$$

with $\Phi(r, z) > 0$ for $0 \leq r < \tilde{R}$ and $|z| < \tilde{H}/2$ and $\Phi(\tilde{R}, z), \Phi(r, \pm\tilde{H}/2) = 0$. The solution may be shown to be (Lamarsh, 1966)

$$B_g^2 = \left(\frac{2.405}{\tilde{R}}\right)^2 + \left(\frac{\pi}{\tilde{H}}\right)^2,\tag{1-75}$$

$$\phi_0(r, z) = \phi_{max}J_0\left(\frac{2.405r}{\tilde{R}}\right)\cos\left(\frac{\pi z}{\tilde{H}}\right),\tag{1-76}$$

where J_0 is the zero-order Bessel function of the first kind. The radial and axial flux distributions are shown as the solid reference lines in Fig. 1-11a. The buckling B_g^2 decreases with increased reactor volume. If we further specify that the diameter $\tilde{D}$ be equal to the height $\tilde{H}$, then the buckling is

$$B_g^2 = \frac{33.0}{\tilde{D}^2},\tag{1-77}$$

illustrating that for a fixed shape, the geometrical buckling varies as the inverse square of the characteristic core dimension.

The form of the criticality condition given in Eq. (1-73) is useful in assessing the relative importance of fission, absorption, and leakage in a reactor. Recall that the neutron multiplication may be defined as the ratio of

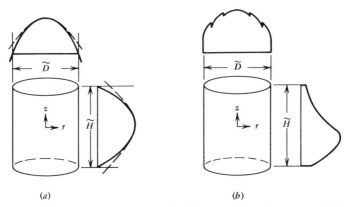

FIGURE 1-11 Core power distributions. (a) Uniform core: ——) bare; – – –) reflected. (b) Nonuniform core with reflector.

the neutron birth rate to neutron death rate. Thus for a finite system,

$$k = \frac{\text{fission neutron production}}{\text{neutron absorption} + \text{neutron leakage}}, \tag{1-78}$$

or rearranging,

$$k = \left[\frac{\text{fission neutron production}}{\text{neutron absorption}}\right]$$
$$\times \left[\frac{\text{neutron absorption}}{\text{neutron absorption} + \text{neutron leakage}}\right]. \tag{1-79}$$

For a critical system, $k = 1$. Noting that the first term on the right is just k_∞ as defined in Eq. (1-38), the second term must be just $(1 + M^2 B^2)^{-1}$ for a bare uniform reactor if Eq. (1-73) is to hold. Since all neutrons are either absorbed in the core or leak out of it, the ratio of absorption to absorption plus leakage is just the probability that a neutron will not escape from the system. Denoting this nonescape probability by P_N, we have

$$P_N = \frac{1}{1 + M^2 B_g^2}. \tag{1-80}$$

Rearranging terms in Eq. (1-79), we also see that

$$P_N = 1 \Big/ \left(1 + \frac{\text{neutron leakage}}{\text{neutron absorption}}\right), \tag{1-81}$$

or

$$\frac{\text{neutron leakage}}{\text{neutron absorption}} = M^2 B_g^2. \tag{1-82}$$

As the ratio of the migration length to the dimensions of a reactor becomes small, the ratio of leakage to absorption also diminishes. Using the expression for buckling given in Eq. (1-77), we see that this ratio goes as $\sim(M/\tilde{D})^2$ for our example core. When this ratio is small the reactor is said to be neutronically loosely coupled. Because M is a measure of the neutron diffusion distance between birth and death, a small value of $(M/\tilde{D})$ implies that a perturbation occurring at one end of the core will not be felt at the other end until many generations of neutrons later. As a result, loosely coupled cores are more susceptible to severe distortions in the spatial power distribution when subjected to localized perturbations, such as control rod ejection, coolant boiling, or fuel meltdown. The relative looseness of neutronic coupling for several power reactor types is indicated by the magnitudes of $M/\tilde{D}$ given in Table 1-5.

TABLE 1-5   Properties for 3000 MW(t) Power Reactors

Type	Description	Core-Averaged Power Density $\bar{P}'''$ (W/cm^3)	Migration Length M (cm)	Core Volume $V(m^3)$	Diameter in Migration Lengths $\tilde{D}/M$	Nonescape Probability (Eq. 1-80)
HTGR	Graphite-moderated He-cooled reactor	7.0	12.0	428.6	62.8	0.9917
BWR	Boiling-H$_2$O reactor	50.0	2.2	60.0	178.0	0.9990
PWR	Pressurized-H$_2$O reactor	75.0	1.8	40.0	190.0	0.9991
GCFR	He-cooled fast breeder reactor	280.0	6.6	10.7	33.4	0.9713
LMFBR	Na-Cooled fast breeder reactor	530.0	5.0	5.7	35.6	0.9746

Based on Fröehlich, 1973.

From the definition of k_∞ and P_N it is apparent that the right-hand side of Eq. (1-73) is just the ratio of total neutron births to total neutron deaths, and this in turn is just the definition of the multiplication given earlier. Thus we may write more generally,

$$k = \frac{k_\infty}{1 + M^2 B_g^2}. \qquad (1-83)$$

This is a useful expression for the multiplication, since if we have a reactor of a given size and composition, it allows us to determine whether the reactor is subcritical or supercritical and by how much. Moreover, this general form is applicable to nonuniform reactors if k_∞, M_g^2, and B_g^2 are interpreted as core-averaged values. Averaging procedures are treated in Chapter 3, and Eq. (1-83) is subsequently used to understand the perturbations of various accident effects on the multiplication.

If a reactor is operated at a constant power level we must have $k = 1$. During reactor operation, however, there are a number of mechanisms that tend to cause the multiplication to deviate from one. Among the more important of these are changes in the core composition due to fuel depletion and fission product buildup, and changes in the fuel temperatures and coolant density as operating conditions are varied. A control mechanism must be incorporated into the reactor core to compensate for these mechanisms and maintain $k = 1$ for steady-state operation, as well as to bring the reactor super-or subcritical for start-up, shutdown, or changes in power level. For this purpose, nonfissionable nuclides with large neutron absorption cross sections, referred to as control poisons, are placed in the reactor core. These usually take the form of either movable control rods, which may be inserted or withdrawn from the core, or a soluble poison. The worth of the poison is defined as the decrease in the reactivity ρ, defined by Eq. (1-28), due to the insertion of the poison into the core. Thus, for example, the control rod worth is the decrease in reactivity due to the insertion of the control rod, and the incremental control rod worth is the decrease in reactivity per unit length of control rod insertion.

Returning to the critical state where $k = 1$, we note that because Eq. (1-71) is homogeneous, the neutron flux normalization is yet unspecified. It is proportional to the power at which the reactor is operated. We recall that the fission rate per unit volume is given by $\Sigma_f(\mathbf{r})\phi(\mathbf{r})$. Then if γ is the energy produced per fission, we can define the power density P''' (or power produced per unit volume) as

$$P'''(\mathbf{r}) = \gamma \Sigma_f(\mathbf{r})\phi(\mathbf{r}) \tag{1-84}$$

where $\gamma = 3.2 \times 10^{-11}$ W-sec per fission. The reactor power is then

$$P = \gamma \langle \Sigma_f(\mathbf{r})\phi_0(\mathbf{r}) \rangle, \tag{1-85}$$

where $\langle \cdot \rangle$ is the integral over the reactor volume. For our example of a uniform cylindrical core, this reduces to

$$P = 2\pi\gamma \Sigma_f \int_{-\tilde{H}/2}^{\tilde{H}/2} \int_0^{\tilde{R}} \phi_0(r, z)r\, dr\, dz. \tag{1-86}$$

Inserting Eq. (1-76) and evaluating the integrals, we obtain

$$\phi_{max} = \frac{3.64P}{\gamma \Sigma_f V}. \tag{1-87}$$

Although it is a useful idealization, the bare uniform reactor neglects the fact that power reactors are invariably not bare, but are surrounded by a reflector or a blanket. Reflectors, used in thermal reactors, are constructed of low-absorption cross-section moderators to deflect neutrons back into the core. Blankets, used in fast reactors, serve as reflectors in that they also deflect neutrons back into the core, but they are constructed of depleted uranium or other fertile material, in order to increase the production of fissile material and to avoid degrading the neutron energy spectrum.

Because a reflector or blanket returns to the reactor core many of the neutrons that would have been lost from a bare system by leakage, a core of a given composition goes critical when the dimensions are appreciably smaller if it is surrounded by a reflector. In addition to the decreased core volume, a reflector or blanket results in a decrease in the ratio of the maximum-to-average neutron flux because the flux no longer goes to zero at the core surface. In the one-group approximation, reflector or blanket effects may be taken into account approximately simply by retaining the bare reactor flux distribution but replacing the core dimensions by extrapolated dimensions lying at some distance outside the core, the distance being determined by the reflector or blanket composition and thickness. For example, in the uniform cylindrical core treated earlier, the core dimensions would remain $\tilde{R}$ and $\tilde{H}$, but in determining the flux distribution, larger values $\tilde{R}'$, $\tilde{H}'$ would be used in Eqs. (1-75) and (1-76). The differences $\tilde{R}' - \tilde{R}$ and $\tilde{H}' - \tilde{H}$ are referred to as the reflector savings. The axial and radial distributions are plotted for a uniform reflected core in Fig. 1-11a to illustrate the decrease in the peak-to-average flux ratio.

1-4 THE POWER REACTOR CORE

The neutronics problems discussed in the preceding section treat only one of the many interrelated aspects of the design of a power reactor core. Heat transport, material compatibility, mechanical design, and radiation damage are among the other important considerations (Glasstone and Sesonske, 1963; El-Wakil, 1971). In particular, many of the features of a reactor core reflect not only the necessity of maintaining a critical chain reaction but also the requirement that large amounts of energy must be produced and transported out of a limited core volume in an orderly fashion.

Core Composition

A power reactor most often consists of an array of cylindrical cells, each of which extends the axial length of the core. Each cell includes a fuel element, consisting of fuel and cladding regions; a coolant channel (or channels); and in some cases a separate moderator region. Cross-section sketches of several typical power reactor lattice cells are shown in Fig. 1-12. The heat is produced in the fuel and transported to the coolant channel by conduction. It is then transported by coolant convection along the channel and out of the reactor core. The cladding is necessary to maintain the structural integrity of the fuel element configuration, to prevent gaseous fission products from escaping to the coolant channels, and in some cases to prevent fuel-coolant chemical reactions.

The fuel element has a small cross-section area for the following reason. Economic core performance requires a high specific power, that is, a high power per unit mass of fuel. Thus a large fuel element surface-to-mass ratio is necessary to transport the heat into the coolant channel without overheating the interior of the fuel or causing an excessive heat flux across the cladding-coolant interface. The outer dimension or pitch of the lattice cell is also limited, since economic core performance also requires that a high power density or power per unit volume of core be achieved. This in turn requires that a large number of fuel elements of a given design be present per cross-section area of the core. Restrictions are placed on the minimum fuel element and lattice dimensions by mechanical design, fabrication costs, and neutronics considerations as well as the requirements for an adequate coolant channel flow area.

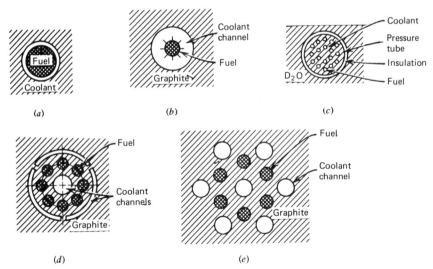

FIGURE 1-12 Reactor lattice cell cross sections (not to scale). (*a*) Light-water cooled or fast reactor. (*b*) Magnox gas-cooled reactor. (*c*) Pressure tube reactor. (*d*) and (*e*) High-temperature gas-cooled reactors.

In fast reactors, where there is no moderator, or in light-water-cooled reactors where the coolant also serves as the moderator, the basic fuel-cladding-coolant configuration shown in Fig. 1-12a is used. The fuel element, consisting of the fuel and cladding, is most often fabricated in the pin geometry shown, although fuel elements have also been formed as thin plates and in annular configurations. At present, most of these power reactor fuels are ceramics, with oxides and carbides of the fissionable material being the most popular. The cladding in Fig. 1-12a is usually a metal alloy.

In situations where the use of a solid moderator (e.g., graphite) or of a liquid moderator (e.g., heavy water) that is not also used as a coolant requires a separate moderator region in the lattice cell, a greater number of geometrical configurations appear in practice; four are shown in Fig. 1-12. In the first two the coolant region surrounds the fuel pin or pins with the moderator on the periphery. Figure 1-12b is typical of older gas-cooled reactors, where fins are added to the cladding to facilitate heat transfer. Figure 1-12c is most often used in heavy-water-moderated systems; a cluster of fuel pins is enclosed in a pressure tube through which the coolant flows. The heavy water is in a low-pressure tank, thermally insulated from the coolant. The remaining lattice cells are typical of new high-temperature helium-cooled systems. In these the heat must pass through some or all of the graphite moderator before reaching the coolant channel. The fuel in these helium-cooled systems consists of small particles of uranium carbide or oxide coated with graphite and silicon carbide. The coatings prevent the fission products from escaping from the fuel. The graphite moderator serves as structural support for the fuel region, thus eliminating the need for metal cladding and permitting the coolant to operate at substantially higher temperatures.

Typically, there may be several thousand fuel elements in a large power reactor. Because removing these individually for refueling would be an inordinately lengthy task, the fuel pins are grouped together to form fuel assemblies. The fuel assembly is designed to be moved as a whole in an out of the reactor.* Some typical fuel assemblies are shown in Fig. 1-13.

Provisions must also be made in the reactor core for the location of control poisons that are varied to compensate for fuel depletion and temperature effects as well as to execute changes in power level and to shut down the reactor. Control rods may be designated into subgroups of regulating, shim, and safety rods, depending on their function. Regulating rods are used to maintain fine control on reactor power level and compensate for the effects

---

* The nomenclature is not standardized. For some systems what we call fuel assemblies may be referred to as fuel elements, or fuel subassemblies. We shall reserve fuel element to refer to the fuel-cladding configuration within a single-lattice cell, and fuel assembly to refer to the unit that is moved as a whole during fueling.

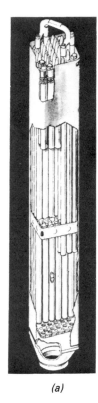
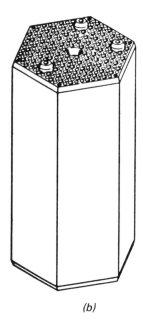

(a) (b)

FIGURE 1-13 Fuel assemblies. (a) Boiling water reactor (b) High-temperature gas-cooled reactor. Adapted from J. George Wills, *Nuclear Power Plant Technology,* Wiley, New York, 1967. Used by permission.

of changes in temperature and fuel depletion. In some systems compensation for fuel depletion and other slow-acting effects may be accomplished by a soluble neutron poison of variable composition in the coolant stream. Solid nonmovable burnable poisons are also used for this purpose. Shim rods are used to bring the reactor to critical and for coarse power level control; they are normally completely out of the core when the reactor is at full power. Sometimes separate safety rods are provided to shut the reactor down in an emergency. These rods are kept in a cocked position outside the core while it is critical; with their insertion it should be impossible for the reactor to become critical. Often one set of rods is designed to operate both for shim and safety. In addition to control rods, secondary backup shutdown devices are provided. Typical of these is the rapid injection of soluble poison into the coolant stream or small spheres of solid poison into designated channels. In the pressure tube reactor the heavy water moderator can be dumped to shut down the chain reaction.

The effect of a control rod is felt only within a few migration lengths from its location in a reactor core. Therefore in large, neutronically loosely coupled cores with dimensions of many migration lengths it is necessary to have a substantial number of control rods distributed uniformly through the core. The periodic array formed by these control rods is sometimes said to consist of supercells. Such an array is shown for a large boiling-water reactor in Fig. 1-14a. These rods are operated in uniform checkerboard banks or patterns, with the banks sometimes being used interchangeably for safety, shim, and regulating purposes. In a loosely coupled core it is extremely important not to let pathological patterns develop in which this uniformity of control rod distribution is grossly violated, since severe flux distortions and single rods that have an unacceptably large effect on the core multiplication may then occur. In a more tightly coupled core a smaller number of control rods may suffice, and their distribution is less critical. As an example, the control rod location in a fast gas-cooled reactor of comparable power is shown in Fig. 1-14b.

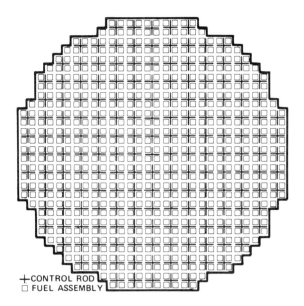

FIGURE 1-14   Control rod arrays. (a) Boiling water reactor. From U.S. Atomic Energy Commission, "The Safety of Nuclear Power Reactors (Light Water-Cooled) and Related Facilities," WASH-1250, 1973. (b) Gas cooled fast breeder reactor. From Gulf General Atomic, "Gas Cooled Fast Breeder Reactor Preliminary Safety Information Document," GA-10298, 1971. Used by permission.

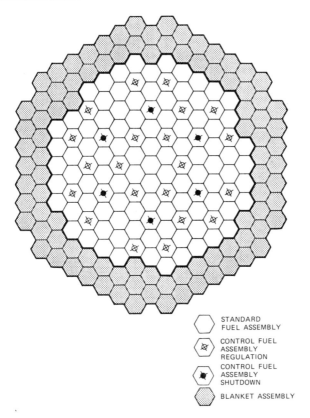

STANDARD
FUEL ASSEMBLY

CONTROL FUEL
ASSEMBLY
REGULATION

CONTROL FUEL
ASSEMBLY
SHUTDOWN

BLANKET ASSEMBLY

FIGURE 1-14(b)

Core Power Distribution

The uniform, bare reactor treated in Section 1-3 provides a convenient starting point to illustrate some power distribution considerations. Before proceeding, however, it is necessary to point out that in speaking of a uniform core—or, for that matter, a uniform core region—we do not mean to imply that the core consists of a homogeneous mixture of fuel, coolant, cladding, and so on. Rather, it consists of an array of identical cells, each containing fuel element, coolant channel, and possibly moderator or poison volumes. The space-independent material properties k_∞, M, $^2\Sigma_f$, and l_∞ used in neutronics calculations for the uniform region are determined by averaging cross sections over a lattice cell volume. In fast reactors where the flux may be expected to be nearly uniform over the cross-sectional area of a cell, calculations of the averages can usually be performed simply by homogenizing or smearing the atom density of each constituent uniformly over the cell volume. In thermal reactors, local flux depression in the fuel elements causes more elaborate averaging techniques to be required. These techniques are discussed in Chapter 4.

Consider, then, a uniform, bare reactor. There is nothing in the foregoing treatment of neutron chain reactions that limits the power at which the reactor can operate. The limit, rather, is set by the maximum power density P'''_{max} that can be achieved for a particular lattice configuration without overheating fuel, coolant, or other constituents. For a uniform, bare, cylindrical reactor the maximum power density is obtained by combining Eqs. (1-84) and (1-87):

$$P'''_{max} = 3.64 \frac{P}{V}. \qquad (1\text{-}88)$$

Thus if P'''_{max} is fixed from thermal considerations the only way to build a uniform reactor of higher power is to increase the volume. Increases in volume, however, imply decreases in B_g^2, the geometric buckling. Therefore to satisfy the criticality condition given by Eq. (1-73), material properties must be modified to lower k_∞ or increase M^2. Most often this is done by decreasing the enrichment of the fuel. In general, k_∞ is substantially more sensitive to fuel enrichment than M^2. Therefore the net effect of going to larger volumes to increase the power rating is to make the core more loosely coupled by decreasing $M^2 B_g^2$. In Table 1-5 estimates of achievable power densities, core volumes, and degree of neutronic coupling are presented for 3000 MW(t) cores of several types of power reactors. It is readily seen that high power densities and large migration lengths lead to more tightly coupled cores.

The foregoing core approximation is frequently inappropriate even after the reflector or blanket region is taken into account. In a power reactor several regions containing fuel of different enrichments may be used to optimize fuel management economics. Spatial nonuniformities in material properties are also introduced by the nonuniform fuel depletion and temperature distributions. The presence of control rods in the core during operation also creates nonuniform distributions of k_∞, M^2, and l_∞. Finally, during accident transients, coolant boiling, fuel melting, or other localized distortions may destroy the uniform composition of the core.

Whether a reactor core is uniform or nonuniform, the determination of the power distribution, and, in particular the maximum power density, is an important part of many safety assessment procedures, since in addition to core performance being limited by the maximum power density, core damage is likely to depend on the maximum rather than the average power density in the event of a major accident. Although the power density cannot be calculated directly from Eq. (1-76) for a nonuniform core, it is useful to retain the radial-axial separability of effects and express the power density as

$$P'''(r, z) = \bar{P}''' f_r(r) f_z(z). \qquad (1\text{-}89)$$

Since $\bar{P}'''$ is taken as the core average power density,

$$\bar{P}''' = \frac{P}{V}; \qquad (1\text{-}90)$$

$f_r(r)$ and $f_z(z)$ must be normalized to mean values of one:

$$\bar{f}_r \equiv \frac{2}{R^2} \int_0^{\tilde{R}} f_r(r) r \, dr = 1, \qquad (1\text{-}91)$$

$$\bar{f}_z = \frac{1}{H} \int_{-\tilde{H}/2}^{\tilde{H}/2} f_z(z) \, dz = 1. \qquad (1\text{-}92)$$

An important measure of the condition of the core is the nuclear hot spot, or peaking factor F_q, defined as the peak-to-average power density:

$$F_q = \frac{P'''(r, z)|_{\max}}{\bar{P}'''} \qquad (1\text{-}93)$$

The separability approximation yields

$$F_q = f_r(r)|_{\max} f_z(z)|_{\max}, \qquad (1\text{-}94)$$

provided the conservative assumption that the radial and axial maximum occur at the same point is used. The quantities on the right are defined as the radial and axial hot spot or peaking factors:

$$F_r = f_r(r)|_{\max}, \qquad (1\text{-}95)$$

$$F_z = f_z(z)|_{\max}, \qquad (1\text{-}96)$$

In addition, there is usually a local hot spot factor F_l to account for fuel element manufacturing tolerances, local control and instrumentation perturbations, and so on. Including this factor, we then have

$$F_q = F_l F_r F_z. \qquad (1\text{-}97)$$

In a typical power reactor core, two or more radial zones containing fuels of different enrichments may be used to decrease the radial peaking F_r. The radial power profile might then look as shown in Fig. 1-11b, where the discontinuities in the power distribution are due to the discontinuities in the fission cross section. In the axial direction, zoned fuel is also sometimes used. More commonly, the axial power is distorted because of the presence of partially inserted control rod banks entering from one end of the core. As shown in Fig. 1-11b, where a bank enters at $+\tilde{H}/2$, the flux is depressed in the upper end of the core, where the control rod bank is located; therefore a peak occurs in the lower half of the core, causing an increase in F_z.

In general, the more neutronically loosely coupled a core is, the more susceptible it becomes to large power-peaking factors due to changes in control rod bank positions or other, similar phenomena. These effects become more pronounced under transient conditions and become an important concern to the safety analyst. In large loosely coupled cores, local accidents involving control rod ejections, flow blockages, or fuel melting may cause very large values of F_q during the ensuing transient. The severity and extent of core damage is likely to depend very strongly on the local conditions where the power peak occurs, and only mildly on the core-averaged behavior. At the same time, such local conditions are normally much more difficult to predict accurately than the spatially averaged behavior of the reactor.

Core Heat Transport

The processes by which the heat produced in the reactor fuel is transferred to the coolant channels and then carried by the coolant out of the reactor play a prominent role in determining the behavior of a reactor under both steady-state and transient conditions. In Chapters 6 through 8 the heat transport from the reactor core is treated in some detail in conjunction with cooling failure accidents. In this subsection some of the more basic steady-state relationships relating to the power density, coolant flow rates, thermal and hydraulic resistances, and other parameters are presented.

The heat transport in a reactor core is closely associated with the thermohydraulic behavior within the individual lattice cells. Suppose we designate the properties at an axial distance z along a lattice cell whose centerline is a radial distance r from the core center by the arguments (r, z). We again assume that the reactor is a cylinder of height $\tilde{H}$ and radius $\tilde{R}$, with the lattice cells axially aligned in the z direction. We assume, for simplicity, that the power distribution is separable into r and z components as indicated by Eq. (1-89). The individual fuel element behavior is related to the power density as follows: Let $q'(r, z)$ be the thermal power produced per unit length of the fuel element located in the lattice cell at (r, z); q' is normally referred to as the linear heat rate. The thermal power in the entire fuel element must then be

$$q(r) = \int_{-\tilde{H}/2}^{\tilde{H}/2} q'(r, z)\, dz. \tag{1-98}$$

If A_{cl} is the cross-section area of the lattice cell associated with the fuel element, the thermal power produced per unit volume of the core is just $q'(r, z)/A_{cl}$. But this is just the definition of power density. Hence

$$P'''(r, z) = \frac{q'(r, z)}{A_{cl}}. \tag{1-99}$$

We may now define some approximate relationships between core temperatures, linear heat rates, and coolant flow. As derived in Chapter 6, the temperature drop between a fuel element interior and the coolant is proportional to the linear heat rate. Thus we may write

$$\tilde{T}_{fe}(r, z) - T_c(r, z) = R'_{fe} q'(r, z), \tag{1-100}$$

where $\tilde{T}_{fe}(r, z)$ is the fuel element temperature averaged by the specific heats over the element cross-section area, and $T_c(r, z)$ is the coolant temperature averaged over the coolant channel flow area. The proportionality constant R'_{fe} is the fuel element thermal resistance per unit length. For cylindrical fuel elements, R'_{fe} is nearly independent of the element diameter but varies inversely with fuel and cladding thermal conductivity and with the cladding-coolant heat transfer coefficient.

The axial dependence of coolant and fuel element temperatures along the length of a lattice cell may be determined from a heat balance for the coolant. Assume that the coolant flows upward through the core with a mass flow rate of W_{ch}, which is associated with the lattice cell coolant channel. The internal energy increase in the coolant over a distance dz is just

$$W_{ch} c_P \, dT_c(r, z) = q'(r, z) \, dz, \tag{1-101}$$

where c_P is the coolant specific heat at constant pressure. Integrating this energy balance yields

$$T_c(r, z) = \frac{1}{W_{ch} c_P} \int_{-\tilde{H}/2}^{z} q'(r, z') \, dz' + \bar{T}_i, \tag{1-102}$$

where $\bar{T}_i$ is the coolant inlet temperature, which we assume to be uniform over the entire core inlet. Combining Eqs. (1-100) and (1-102) yields the fuel element average temperature at any point in the core

$$\tilde{T}_{fe}(r, z) = R'_{fe} q'(r, z) + \frac{1}{W_{ch} c_P} \int_{-\tilde{H}/2}^{z} q'(r, z') \, dz' + \bar{T}_i. \tag{1-103}$$

Finally, the coolant temperature at the core outlet $T_o(r)$ may be evaluated by setting $z = \tilde{H}/2$ in Eq. (1-102). Using Eq. (1-98), we have

$$T_o(r) = \frac{q(r)}{W_{ch}c_P} + \bar{T}_i. \tag{1-104}$$

At this point we have calculated the distribution of fuel element and coolant temperatures throughout the reactor core. Often changes in the reactor power level and power distribution must be assessed to determine the impact on the fuel element and coolant temperatures. We first relate the core-averaged values $\tilde{T}_{fe}$ and T_c to the reactor power and then determine the maximum values in terms of the hot spot factors derived earlier.

We first express the linear heat rate in terms of the reactor power by combining Eqs. (1-89), (1-90), and (1-99):

$$q'(r, z) = \frac{A_{cl}}{\pi \tilde{R}^2 \tilde{H}} P f_r(r) f_z(z), \tag{1-105}$$

where we have expressed the core volume as $V = \pi \tilde{R}^2 \tilde{H}$. If the core consists of N lattice cells, each with cross-section area A_{cl}, then $\pi \tilde{R}^2 = N A_{cl}$, and we may write

$$q'(r, z) = \frac{P}{N\tilde{H}} f_r(r) f_z(z). \tag{1-106}$$

This relationship may be inserted into Eq. (1-100) to yield

$$\tilde{T}_{fe}(r, z) - T_c(r, z) = R P f_r(r) f_z(z), \tag{1-107}$$

where the core thermal resistance is defined by

$$R = \frac{R'_{fe}}{N\tilde{H}}. \tag{1-108}$$

Physically, it is the thermal resistance per unit length of fuel pin divided by the ratio of the whole-core heat transfer area to the heat transfer area per unit length of a fuel pin. If Eqs. (1-102) and (1-106) are combined, we have

$$T_c(r, z) = \frac{P f_r(r)}{W c_p} \frac{1}{\tilde{H}} \int_{-\tilde{H}/2}^{z} f_z(z') \, dz' + \bar{T}_i, \tag{1-109}$$

where

$$W = N W_{ch} \tag{1-110}$$

is just the total mass flow rate of coolant through the core.

For the cylindrical reactor under consideration, the core-volume-averaged temperatures are obtained from the relationship

$$\bar{T}_x = \frac{2}{\check{R}^2 \check{H}} \int_{-\check{H}/2}^{\check{H}/2} \int_0^{\check{R}} T_x(r, z)r\, dr\, dz, \tag{1-111}$$

where $\bar{T}_x$ is taken either for the fuel element or the coolant. Applying this definition of the average to Eq. (1-107) and using the normalization conditions, Eqs. (1-91) and (1-92), on f_r and f_z, we have

$$\bar{T}_{fe} = RP + \bar{T}_c. \tag{1-112}$$

Applying Eq. (1-111) to the coolant temperature as given by Eq. (1-109), we have

$$\bar{T}_c = \frac{P}{Wc_P} \frac{1}{\check{H}^2} \int_{-\check{H}/2}^{\check{H}/2} \int_{-\check{H}/2}^{z} f_z(z')\, dz'\, dz + \bar{T}_i. \tag{1-113}$$

If it is assumed that the power density is symmetric about the core midplane—a condition that is approximately met in many situations—the double integral reduces to simply $(\frac{1}{2})\check{H}^2$ and therefore

$$\bar{T}_c = \frac{P}{2Wc_P} + \bar{T}_i. \tag{1-114}$$

If we combine this expression with Eq. (1-112) the average fuel temperature may also be written in terms of P and $\bar{T}_i$:

$$\bar{T}_{fe} = \left\{ R + \frac{1}{2Wc_P} \right\} P + \bar{T}_i. \tag{1-115}$$

Similarly, the core outlet temperature, given by Eq. (1-104), may be expressed in terms of the power as

$$T_0(r) = \frac{P}{Wc_P} f_r(r) + \bar{T}_i, \tag{1-116}$$

and the average core outlet temperature is then

$$\bar{T}_o = \frac{P}{Wc_P} + \bar{T}_i. \tag{1-117}$$

Finally, Eqs. (1-114) and (1-117) provide a simple approximation for the average coolant temperature in terms of the inlet and outlet conditions:

$$\bar{T}_c = \frac{\bar{T}_i + \bar{T}_o}{2}. \tag{1-118}$$

In addition to the preceding average fuel element and coolant temperatures, it is often necessary to know the maximum temperature for purposes of evaluating safety margins and making damage estimates. These maximum temperatures are estimated by using the hot spot or peaking factors defined in Eqs. (1-95) and (1-96). From Eq. (1-116) we observe that the maximum coolant outlet temperature is

$$T_o(r)|_{\max} = \frac{F_r P}{W c_p} + \bar{T}_i.$$
(1-119)

Similarly, from Eq. (1-107) the maximum temperature drop between fuel and coolant is

$$[\tilde{T}_{fe}(r, z) - T_c(r, z)]_{\max} = F_r F_z R P.$$
(1-120)

It is difficult to predict the maximum fuel temperature precisely from this relationship, since the maximum temperature drop and the maximum fuel temperature may not occur at the same point. For cores in which the average fuel element-coolant temperature drop is much larger than the average coolant temperature rise (i.e., cores for which $R W c_p \gg 1$), a reasonable assumption is to replace $T_c(r, z)$ by $\bar{T}_c$ in Eq. (1-120). Using Eq. (1-114), we then have

$$\tilde{T}_{fe}(r, z)|_{\max} \simeq \left(R F_q + \frac{1}{2 W c_p} \right) P + \bar{T}_i,$$
(1-121)

where the local peaking factor F_l is included in F_q to account for local anomalies. For purposes of core damage analysis it is usually the maximum value of the fuel centerline temperature $T_{\mathfrak{C}}(r, z)$, instead of the fuel element cross-sectional average temperature $\tilde{T}_{fe}(r, z)$ that is needed. To determine this temperature, Eq. (1-108) or (1-120) is modified by replacing the core thermal resistance by $R'_{\mathfrak{C}}/(NH)$, where $R'_{\mathfrak{C}}$ is the fuel element thermal resistance per unit length, but calculated, as in Chapter 6, between the fuel element centerline and the coolant.

Finally, it is often required to know the maximum value of the heat flux $q''_c(r, z)$ across the cladding-coolant interface. If all the fuel elements are cylinders of radius b we have

$$q''_c(r, z) = \frac{q'(r, z)}{2 \pi b}$$
(1-122)

or using Eq. (1-106)

$$q''_c(r, z) = \frac{P}{A_{ht}} f_r(r) f_z(z),$$
(1-123)

where $A_{ht} = 2\pi b \tilde{H} N$ is the total cladding-coolant heat transfer area for the core. Taking the maximum, we have, using the peaking-factor definitions,

$$q_c''(r, z)_{\max} = \frac{P}{A_{ht}} F_q, \tag{1-124}$$

where F_l has again been included to account for local peaking anomalies.

In the foregoing equations the mass flow rate W plays a prominent role in determining the temperature distributions. For the turbulent flow conditions existing in power reactor cores, the flow rate may be related to the pressure drop across the core by the approximate relationship

$$\Delta p = \frac{\mathcal{K}_c}{2\bar{m}} \left(\frac{W}{A_T} \right)^2, \tag{1-125}$$

where

$\bar{m}$ = coolant density;

A_T = total core flow area;

$\mathcal{K}_c$ = core flow loss coefficient.

The coefficient $\mathcal{K}_c$ depends on the geometry of the flow channels, the viscosity of the coolant, and the roughness of the cladding surfaces. The flow is driven by the pressure drop across the core, which is caused in turn by the pressure head created by the coolant pumps.

The discussion in this section focuses on the determination of the reactor fuel and coolant temperatures. Both the average and maximum temperatures are needed to analyze the safety of a reactor. The effects of changing fuel and coolant temperature on the critical state of the reactor are closely associated with the average values, whereas damage criteria are associated with the maximum values that are reached. These temperatures are given in terms of P, $\bar{T}_i$, and W—the reactor power, inlet temperature, and mass flow rate, respectively. Very roughly, the effect of several accident types can be seen through these relationships. In an overpower transient P becomes too large, causing the core to overheat. In a loss-of-heat-sink accident, heat is no longer effectively removed from the coolant before it returns to the core, causing $\bar{T}_i$ to increase with time. And, in a loss-of-flow accident W deteriorates, either because of an increase in $\mathcal{K}_c$—caused by a flow obstruction in the core—or because of a decrease in Δp caused by a pump failure. Finally, a loss-of-coolant accident would decrease the coolant density, and therefore W, very rapidly.

The preceding equations are strictly applicable only to quasi-static changes in the state of the reactor. If accidents are considered that evolve more rapidly, the dynamic behavior of the heat transport processes must be modeled to understand the consequences of the accidents. Such modeling is addressed in Chapters 5 and 6 for the heat conduction out of the fuel pins, in Chapter 7 for the mass flow and energy transport in the coolant, and in Chapter 8 for the particularly difficult case where the coolant is lost. The quasi-static equations nevertheless offer a first approximation for relatively slow transients, and extensive use can be made of them in understanding the behavior of a reactor core under accident conditions. Often even though the behavior of the core-averaged parameters is not adequately described by the static equations, the peak-to-average ratios are still approximately correct, at least in the situations where the accident is not of a localized nature, and it affects the whole core more or less uniformly. It then may be helpful to investigate the extent of core damage by first determining the core-averaged temperature behavior and then multiplying by a peaking factor to determine what the situation is at the worst point in the core. Even when the peaking factors change abruptly because of shifting in power flow distribution during the course of the accident, an approximate analysis may be carried out using peaking factors that conservatively overpredict the adverse effects of the power redistribution.

1-5 NUCLEAR POWER PLANT CLASSIFICATION

Many different power reactor concepts have been proposed and/or built in the last three decades. These include not only the solid fuel element systems, to which the discussions of the preceding section are most applicable, but also a number of molten fuel, fluidized bed, and other concepts. Discussions of most of these systems may be found elsewhere (El-Wakil, 1971b). Although some of the material in the following chapters may be more widely applicable, the intent of this text is to treat generically only those systems that employ solid fuels. Moreover, where discussions of specific power reactor systems are presented, they are limited for the most part to the following reactor classes that it appears will constitute the preponderance of the world's nuclear power capacity for the foreseeable future: (1) pressurized-water and boiling-water reactors that utilize light water (H_2O) as both coolant and moderator; (2) heavy-water (D_2O) moderated, pressure tube reactors that may use light or heavy water or a number of other materials as a coolant; (3) gas-cooled graphite moderated reactors; (4) fast reactors cooled by sodium or by helium. In this section a brief description is first given of the three types of energy conversion systems by which heat is transported from the reactor core through the turbine-condenser system. Then, as background for the subsequent discussions of reactor safety, the distinguishing characteristics of each of the preceding four classes of reactors are outlined in the following subsections.

The energy transport and conversion systems associated with nuclear power plants may vary substantially with the reactor concept, and particularly with the nature of the coolant. The reactor core is joined to the turbine-condenser system by two or more parallel loops in one of the three configurations illustrated in Fig. 1-15. A flow diagram for the most common configuration is shown in Fig. 1-15a. The plant is divided into a primary and a secondary system. The primary system consists of the reactor, a heat exchanger, coolant pump, and connecting piping. The secondary system consists of the working fluid loop containing the turbine, condenser, feedwater pump, and secondary side of the heat exchanger. This configuration is used with either liquid or gas primary coolant. To date, the working fluid in the secondary system has invariably been water. The water enters the heat exchanger, called a steam generator, where it is vaporized. It then passes through the steam line to the turbine. In the turbine the thermal energy of the high-pressure steam is converted to mechanical energy, and the steam is exhausted to the condenser at subatmospheric pressure. The steam is then condensed by exchange of heat to the cooling water. The waste heat picked up by the cooling water is discharged either to a river or cooling pond or to the atmosphere, through cooling towers. The water in the secondary system is then returned to the steam generator by the feedwater pump, and the cycle is repeated. Not shown in the diagram are the reheating and feed water-heating systems, which are used to increase the thermodynamic efficiency of the working fluid cycle.

If the reactor coolant is a gas or a boiling liquid, the possibility exists of eliminating the steam generators and using a direct cycle in which the primary coolant and working fluid are the same. A schematic diagram for this configuration is shown in Fig. 1-15b. In the boiling-water system, water enters the reactor vessel, where it is circulated through the reactor and converted to steam. The boundary between primary and secondary systems now consists of sets of valves in the steam and feedwater lines, which may be closed to isolate the primary system. In proposed direct cycle gas-cooled systems the steam turbine and feedwater pump are replaced by a gas turbine and compressor. The entire working fluid cycle then lies within the primary system.

If the primary coolant and working fluid are chemically reactive, the rupture of a tube in the steam generator could cause a violent reaction. To eliminate the possibility of such an event directly affecting the primary system, an intermediate coolant loop is added. Shown in Fig. 1-15c is a schematic diagram for a sodium-cooled system. This configuration permits the physical isolation of the affected steam generator from the primary system.

The layouts of the steam cycle or secondary side of nuclear power plants tend to have more in common than the primary systems do. The secondary

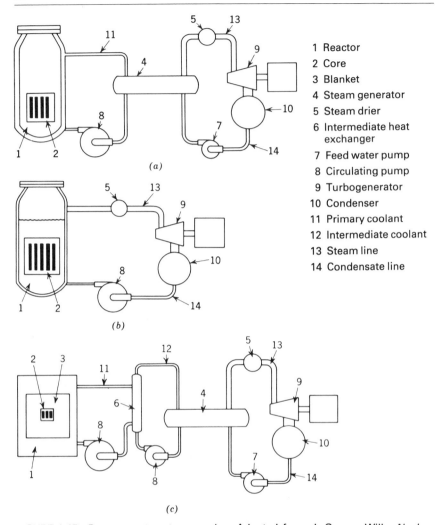

1 Reactor
2 Core
3 Blanket
4 Steam generator
5 Steam drier
6 Intermediate heat exchanger
7 Feed water pump
8 Circulating pump
9 Turbogenerator
10 Condenser
11 Primary coolant
12 Intermediate coolant
13 Steam line
14 Condensate line

(a)

(b)

(c)

FIGURE 1-15 Power reactor steam cycles. Adapted from J. George Wills, *Nuclear Power Plant Technology*, Wiley, New York, 1967. Used by permission.

system is discussed further in the last section of Chapter 7. For purposes of later comparisons of the behavior of different reactor concepts under accident conditions, it is useful to point out some of the differing characteristics of the primary systems. What follows is a brief description of the features of some of the more important power reactor systems either in use or under development (El-Wakil, 1971; Nuclear Engineering Int., 1973). Diagrams of the systems are shown in Fig. 1-16.

Light-Water Reactors

The pressurized-water reactor, or PWR, shown in Fig. 1-16d, evolved from the naval propulsion program in the United States to become widely used around the world. The substantial thermal capture cross section of the light water that is used for both coolant and moderator precludes the use of natural uranium fuel. However, when enriched fuel is used, the large slowing-down power ($\xi\Sigma_s$) of water results in optimized reactor lattices that are quite compact; fuel-to-coolant volume rates of one-to-one are typical. The fuel elements typically consist of 2 to 4% enriched UO_2 fuel pins, clad in zirconium alloys known as zircoloy. The pins are approximately 1 cm in diameter and are arranged in a square lattice with up to 200 pins per fuel assembly. As a result of the compact lattice configuration, PWR's have the highest power densities, and therefore the smallest core volumes of any thermal power reactor, as indicated in Table 1-5. The migration lengths in these tightly packed lattices are short, however; as a result, PWR cores, which may consist of 200 or more fuel assemblies, are neutronically very loosely coupled.

Pressurized-water reactors typically are operated from 12 to 18 months between partial refuelings. The refueling is carried out by depressurizing the system and removing the head of the reactor vessel. The reactivity over the core life is partially controlled by burnable poisons, which are thermal neutron absorbers placed in selected fuel elements. In addition, the concentration of a soluble boron compound in the coolant water is varied to compensate for fuel depletion and fission product buildup. Banks of control rods are used for varying power level and for rapid shutdown. The control rods, which enter from the top of the core, occupy clusters of lattice vacancies in selected fuel assemblies.

To achieve high core outlet temperatures without boiling, the coolant must be maintained at high pressures. This pressure, typically about 2200 psia, in turn mandates that the reactor be placed in a thick-walled steel vessel. The indirect cycle system that transports heat away from the reactor consists of two to four parallel coolant loops, each consisting of the loop piping, a pump, and a steam generator. The working fluid is boiled on the secondary sides of the steam generator and transported through a conventional turbine-condenser system. A pressurizer controls the primary system pressure through electrical heater and spray cooling of the water inventory of a chamber. The pressurizer is connected to a coolant loop but is maintained at the saturation temperature that corresponds to the system pressure.

The boiling-water reactor, or BWR, shown in Fig. 16e, differs from the PWR in that bulk boiling of the coolant takes place as it passes through the

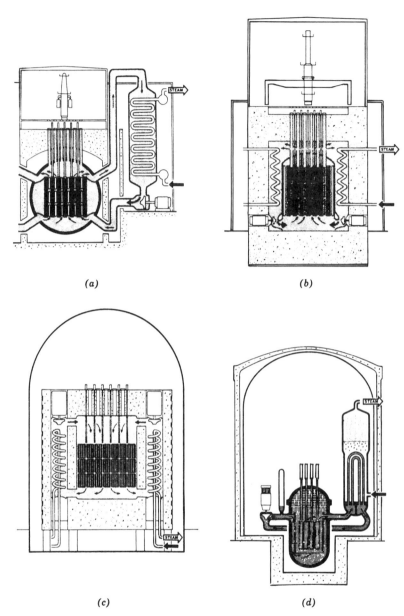

FIGURE 1-16 Primary systems of nuclear power plants. (*a*) Magnox reactor. (*b*) Advanced gas-cooled reactor. (*c*) High-temperature gas-cooled reactor. (*d*) Pressurized-water reactor. (*e*) Boiling-water reactor. (*f*) CANDU reactor. (*g*) Steam-generating heavy-water reactor. (*h*) Fast breeder reactor. From "Know Your Reactors", *Nucl. Eng. Int.*, **18**, 1973. Used by permission.

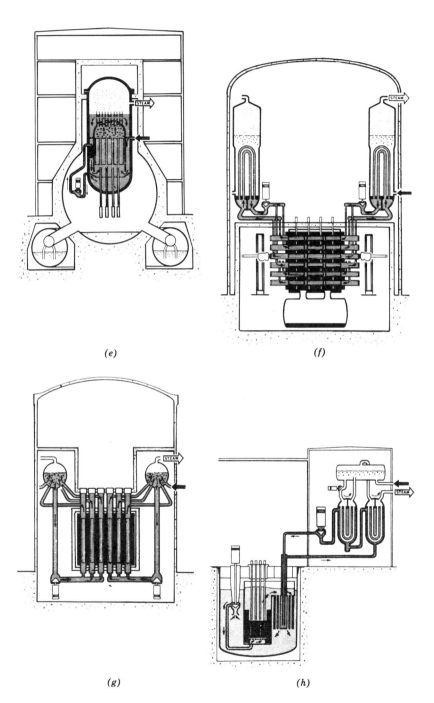

(e)

(f)

(g)

(h)

reactor core. The generation of the steam within the reactor vessel permits the use of a direct cycle system in which the need for separate steam generators is eliminated. The saturated steam is extracted from the two-phase core outlet flow by a series of separators in the upper reaches of the reactor vessel. It then flows through parallel steam lines to the turbine-condenser. The water inventory and pressure are controlled by the feed-water flow, which enters the vessel at about the elevation of the core outlet, and, if necessary, by the vessel's relief valves. The liquid phase of the core outlet flow mixes with the feedwater and is recirculated by forced convection back through the reactor core. In later designs than that shown in Fig. 1-16d, the size of the recirculation lines have been greatly reduced through the use of jet pumps, or completely eliminated by placing the recirculation pump impellers within the reactor vessel.

The pressure of a BWR is less than one-half that of a PWR, for although the two systems have comparable core outlet temperatures of about 600°F, the BWR is at the coolant saturation pressure. As a result, the BWR reactor vessel is substantially thinner than that of a PWR. On the other hand, the placement of the steam separation equipment results in the BWR vessel having a substantially larger volume.

The fuel elements of BWR's and PWR's are similar; both utilize uranium dioxide pellets enriched to 2 to 4%, clad in zircoloy tubes of about the same diameter. The BWR lattice also has a fuel-to-water volume ratio of about one-to-one, but in several other respects it tends to differ from a PWR. Unlike the PWR, the fuel assemblies are enclosed in channel boxes or shrouding to prevent the cross flow of the boiling liquid between assemblies. The shrouding also permits the control of the power-to-flow ratio within individual assemblies by orificing the channel box inlets. Curiform control rods, as indicated in Fig. 1-14a, are placed between the assemblies. These enter from the bottom of the core for power shaping purposes: The substantial void fraction tends to depress the neutron flux above the core midplane. Therefore placing the neutron-absorbing control rods in the lower reaches of the core creates a compensating flux depression that results in an overall flattening of the power distribution.

Like PWR's, BWR's are depressurized and the reactor vessel head removed every 12 to 18 months for partial refueling. In addition to using burnable poisons for reactivity control, the fuel depletion and fission product buildup is compensated for by gradually withdrawing the control rods. Power level adjustants are made by varying the recirculation flow, which in turn controls reactivity by varying the void fraction in the core.

The BWR has a neutronically loosely coupled core. Its power density is somewhat lower than that of a PWR, as indicated in Table 1-5, and it has a slightly longer neutron migration length due in part to the presence of the low-density steam.

Careful attention must be paid to water chemistry in BWR's, since the same water that serves as a working fluid passes through the core. As a result, oxygen and other impurities that are picked up as a result of condenser leaks or other steam cycle problems can cause corrosion problems throughout the primary system. This is in contrast to the indirect cycle system, where working fluid impurities attack the primary system envelope only from the secondary side of the steam generator tubes. Likewise, greater care must be taken in the removal of fission and activation products from the reactor coolant, since in a direct cycle system these may pass through and deposit on all of the steam cycle components, thereby causing unacceptable levels of radioactive contamination.

Heavy-Water Reactors

Although expensive, heavy water, or D_2O, is an attractive moderator because its extremely small thermal-neutron capture cross section permits reactors to be built utilizing natural uranium fuel. The smaller value of the slowing down power of D_2O, about 0.18 cm^{-1} as opposed to 1.28 cm^{-1} for H_2O, necessitates a small fuel-to-moderator ratio in the reactor lattice. This in turn leads to low power densities and large-volume cores.

Reactor vessels large enough to contain these large cores at high pressures have not been built. Rather, heavy-water reactors utilized pressure tube systems, which have been most highly developed in the Canadian CANDU reactor design shown schematically in Fig. 1-16f. The fuel assemblies are contained in parallel zirconium pressure tubes that pass through a large tank—known as a calandria—filled with heavy water. The tubes are thermally insulated from the calandria so that the large mass of heavy water remains at nominal temperatures. Therefore the heavy water is not required to be under high pressure, and the calandria is a thin-walled vessel.

Heavy water is also used as the coolant in this indirect cycle system. As in the PWR, it is maintained under high pressure by a pressurizer to prevent boiling. The coolant passes from the pressure tubes through U tube steam generators and is pumped back through the core. Each pressure tube contains several fuel assemblies placed in series. As pictured in Fig. 1-12e, each assembly consists of a cluster of uranium oxide pins with zirconium cladding.

Because heavy-water reactors utilize natural uranium, not enough excess reactivity can be built into the core to compensate for the long-term effects of fuel depletion and fission product buildup. As a result, more or less continual fueling is carried out during operation, through the use of remotely operated machines that connect to both ends of the pressure tubes to insert and withdraw fuel assemblies. The control rods are inserted in the moderator through penetrations in the calandria, and the level of the moderator also can be varied to make fine power adjustments. In addition to the use of the control rods, pressure tube reactors can be rapidly shut down by dumping the heavy water from the calandria tank into a dump tank located beneath the core.

Several variations on the foregoing system have been proposed and/or built as prototype plants in which the coolant is light water, in order to reduce the expense of heavy-water leakage from the high-pressure system, or organic liquids to reduce the required pressure in the primary system. Slightly enriched fuels have also seemed attractive since with them enough excess reactivity can be built into the fuel element so that the fueling can be carried out at scheduled shutdowns, as is the case in PWR's and BWR's.

One of the more promising heavy-water reactors is the SGHWR, or steam-generating heavy-water reactor, shown schematically in Fig. 1-16g. The intent of this concept is to combine the advantages of the CANDU system with those of the direct cycle boiling-water reactor. The core layout is similar to the CANDU system with the exception that the pressure tubes are vertical to accommodate the boiling of the light-water coolant. The uranium dioxide fuel is slightly enriched to compensate for the capture of thermal neutrons in the light water coolant and to provide for the option of off-line fueling.

After passing through the SGHWR pressure tubes, the two-phase mixture of steam and water flows to the steam drums above the core. There steam separation takes place; the steam passes to the turbine while the liquid mixes with incoming feed water and is pumped back to the core inlet. The problems of water chemistry control and contamination of the working fluid by fission and activation products are similar to those in the BWR, since in both cases the light water serves as both coolant and working fluid.

Gas-Cooled Reactors

Natural uranium can also be used as a reactor fuel in graphite-moderated systems. However, relatively small slowing-down power ($\xi \sum_s = 0.065$ cm^{-1}) that stems from the larger atomic weight of carbon results in the need for very large volume ratios of moderator to fuel. This leads, in turn, to low power densities and large volume cores. Moreover, because of these large volume ratios and the fact that neutrons have a long migration length in such cores, they are neutronically more tightly coupled than those of light-water-moderated reactions. At the same time, the significant thermal neutron capture takes place in the large mass of graphite. Therefore if natural uranium is to be used, neutron capture must be minimized in other constituents. Thus from a neutronic standpoint, the low atom densities of gases make them ideal coolants for graphite-moderated systems.

Gases have poorer thermal-hydraulic properties than liquids. As a result, a significant fraction of the reactor power output must be utilized to drive circulators that are powerful enough to achieve an adequate mass flow rate of the coolant through the core. Although direct cycle gas turbine systems are possible, to date, all gas-cooled reactors utilize indirect systems with water as the working fluid. Typically, each circulator is driven by a small turbine that extracts steam directly from the secondary side of steam generators.

Typical of the early natural uranium-fueled reactors that were developed in Great Britain and France was the carbon dioxide-cooled Magnox reactor, named after its magnesium-based cladding. As shown in Fig. 1-16a, these reactors are contained in primary systems consisting of large steel vessels and ductings. To minimize parasitic thermal neutron capture, the fuel rods consist of uranium metal, and the aforementioned magnesium alloy is used as a cladding. As indicated in Fig. 1-12b, the fuel elements are placed within coolant channels in the graphite moderator blocks that constitute the core. Often fins are used on the cladding to enhance the transfer of heat to the gas. As in other natural uranium power reactors, refueling is done on-line by remotely operated machinery. Control rods enter the core through separate holes in the graphite moderator.

The efficiency of Magnox reactors cannot be made as high as desirable for two reasons: First, the large size of the core results in the requirement for steel reactor vessels that are much larger than those used for water-cooled reactors. Consequently, the permissible pressure of the primary coolant is limited by the ability to fabricate large thick-walled vessels. In more recent gas-cooled systems these vessels have been superseded by prestressed concrete vessels that permit larger systems to be enclosed even as the coolant pressure is increased. Second, the metallurgy at the uranium metal and Magnox cladding limits severely the coolant temperatures at which this system can be operated. In later gas-cooled graphite-moderated reactors slightly enriched uranium is used. This has two effects. Ceramic fuels and cladding materials that can withstand higher temperatures can be used, since more thermal neutron capture can be tolerated. The slightly enriched fuel also allows the volume ratio of graphite to fuel to be reduced, leading to higher power densities and smaller core volumes.

Higher coolant temperatures, and therefore improved steam conditions, are achieved in the advanced gas-cooled reactor, or AGR, pictured in Fig. 1-16b. This carbon dioxide-cooled system utilizes both a prestressed concrete vessel and slightly enriched fuel. The fuel elements consist of uranium dioxide clad in stainless steel. Thirty-six of the elements are loaded into a graphite sleeve that constitutes a fuel assembly, and the assemblies are loaded into the channels in the graphite moderator blocks. On-line fueling is again adopted. However, the grouping of fuel pins into assemblies causes the operation to be simpler than the reloading of the individual pins in the Magnox reactors.

A further evolution of gas-cooled reactors is typified by the high-temperature gas-cooled reactor, or HTGR, shown schematically in Fig. 1-16c. In the most recently proposed of these systems the primary system is contained as a series of chambers and ducts that are entirely enclosed within the massive prestressed concrete vessel. The combination of slightly enriched fuel and the inert helium coolant allows the use of an all ceramic core, and this in turn permits high coolant outlet temperatures to be achieved. In these systems off-line refueling can be used. Even with the increased power densities, however, the volume of these cores is still much larger than for light water-moderated reactors, as indicated in Table 1-5.

The fuel consists of millions of spheres of highly enriched uranium carbide and of thorium carbide. These are coated with graphite and silicone carbide to form coated particles a few hundred microns in diameter. The coated particles are packed into axial cavities in the graphite blocks that form the reactor core. As shown in the lattice cells in Figs. 1-12c and 1-12d, the helium coolant flow is through other axial channels through these blocks. As in other gas-cooled reactors, control rods enter separate channels through the graphite from the top of the core.

Fast Reactors

The combination of uranium-235 and thorium-232 in the foregoing HTGR has the potential for converting large amounts of the thorium to the fissile element uranium-233. The nearly universal choice as a breeder reactor (i.e., one that converts more fertile material to fissile material than it consumes fissile material), however, is a fast neutron spectrum system fueled with plutonium and/or uranium-235, with uranium-238 as the fertile material. The requirements of high neutron flux levels and the need to minimize the amount of moderating material in such fast reactor systems result in cores consisting of tightly packed hexagonal lattices. The fuel elements are uranium–plutonium dioxide pellets that are enriched to 15 to 20% in stainless steel cladding of about a quarter of an inch in diameter. Around the core is a blanket of depleted uranium oxide pins that serve to enhance the breeding properties.

Both sodium and helium have been proposed as fast reactor coolants. In both cases the core designs have power densities that are much higher than those in thermal reactors, and therefore the cores are much smaller, as indicated in Table 1-5. Fast reactors are refueled off-line at relatively infrequent intervals. They do not, however, require large excess reactivities to operate for sustained periods. There are two reasons for this. The capture cross sections óf the fission products that build up in the core are very small for the fast neutron spectrum, and the depletion of fissile material is partially or completely countered by the buildup of plutonium-239. As a consequence, and because of the fact that fast reactors are not neutronically loosely coupled, relatively few control rods suffice to control the reactivity. The control rod arrangement for a proposed gas-cooled fast reactor is shown in Fig. 1-14b.

Proposed helium-cooled fast reactors utilize prestressed concrete vessels and indirect cycles similar to that shown for the HTGR in Fig. 1-16c. The core and blanket, however, are much smaller than for the thermal system. The excellent heat transfer properties of sodium, however, make it the more universal choice as fast reactor coolant, even though the sodium causes the neutron spectrum in these systems to be somewhat degraded from what can be obtained in a gas-cooled system. The high boiling point of sodium (1620°F at atmospheric pressure) allows sodium systems to operate at only a few atmospheres pressure and minimizes the danger of rupture of a pressurized system. Moreover, the reactor vessel is relatively thin walled, and a secondary guard vessel is normally placed closely around the reactor vessel so that even in the event of a vessel rupture the core would remain covered by sodium.

Because sodium reacts violently with either oxygen or water, the primary systems of liquid-metal-cooled reactors must be designed to prevent the coolant from coming into contact with either of these materials. Sodium-oxygen reactions are prevented by using argon as a cover gas over the sodium in the reactor vessel and by placing the entire primary system within vaults with inert atmospheres. To prevent the sodium–water reactions that would occur in the event of steam generator tube failures from directly affecting the reactor core, intermediate loops are used as shown schematically in Fig. 1-15c. The intermediate loops serve to isolate the steam generators from the primary system sodium. Thus should a generator accident take place, only the sodium in the intermediate loop is affected, and the damage should not propagate to the primary system.

Figure 1-16h shows an intermediate cycle fast reactor system. The primary system is a pot configuration in which the intermediate heat exchanger and the primary coolant pumps are located within the reactor vessel. This configuration minimizes the chance of a loss-of-coolant accident, since primary system coolant loops never pass outside the double-walled reactor vessel. Sodium-cooled fast reactors are also built with loop configurations in which the intermediate heat exchangers, coolant pumps, and connecting piping are located outside the reactor vessel. In these systems double-walled piping is sometimes used, for the possible rupture of a coolant pipe must be considered in the safety analysis of the plant.

1-6 SAFETY CHARACTERISTICS

The discussion of the preceding sections indicates that a variety of nuclear power plant concepts have been designed and constructed for the production of electrical power. For economical operation such plants must incorporate highly optimized energy transport paths for removing the heat produced in the nuclear fuel, transferring it away from the core and converting it to the high-temperature steam needed to drive the turbine-generator. At the same time, the safe operation of a plant requires that the large inventories of the radioactive by-products of the fission reaction that accumulate in the reactor fuel must be prevented from entering the environment. The isolation of the radioactive materials from the environment is accomplished by incorporating a series of fission product barriers into the energy transport system. For large power reactors there are normally at least three of these: the reactor fuel elements, the primary system envelope, and one or more containment structures.

Much of reactor safety is concerned with ensuring that the integrity of these fission product barriers is maintained under a wide spectrum of situations, ranging from normal operating conditions to severe accidents. An integral part of safety analysis therefore must be to categorize the types and magnitude of accidents that may befall a nuclear power plant and to assess the potential damage to these barriers. Invariably, situations are perceived in which active intervention would be required following the initiation of severe accidents in order to protect the integrity of the fission product barriers. For this reason a number of highly reliable safety systems must be incorporated into the design of the nuclear power plant.

Discussions of the complex interactions between accident transients, safety system operations, and fission product barrier integrity form a major part of the following chapters. As a prologue, however, we present here a brief discussion of the nature of the various fission product barriers, the classes of accidents that may threaten their integrity, and the functions of safety systems that are normally incorporated into nuclear power plants.

Fission Product Barriers

As previously stated, the barriers to the release of radioactive materials to the environment may normally be classified as the fuel elements, the primary system envelope, and containment structures. In most designs the fuel elements consist of ceramic pins in metal cladding. These provide a two-stage barrier to the release of fission products. So long as the cermic fuel does not melt, the fission products must first diffuse out of the fuel region to the fuel-cladding interface. They must then penetrate the cladding before reaching the coolant stream. In the coated-particle fuel configurations used in high-temperature gas-cooled reactors, the fission products must both penetrate the particle coatings, which serve as a cladding, and then diffuse through the graphite moderator blocks before reaching the coolant stream.

A fuel element is said to fail when the cladding is breached, allowing those fission products that have diffused out of the fuel region to escape through the cladding. Provided the fuel remains solid, it is only the noble gases and a few of the other more volatile fission products that will be available in the fuel-cladding gap in significant quantities. Because the fission product inventory in a large reactor is divided between thousands of independent fuel elements, the failure of a few fuel elements can be tolerated in most reactor systems without necessitating reactor shutdown. The number of such failures that is tolerable is dependent on the capacity of the radioactive waste treatment facilities to prevent the contamination in the cooling stream from becoming unacceptably large.

Of far more concern than random fuel element failures is the specter of gross overheating of large numbers of elements leading to fuel melting. For not only does the release rate of fission products from the fuel region increase sharply with temperature and make an abrupt jump when the fuel becomes molten, but cladding failures increase rapidly in elements containing molten fuel. Moreover, extensive melting or gross distortion of the cladding geometry resulting from overheating may cause both a loss of the large heat transfer surface-to-mass ratio of the fuel elements and the plugging of the coolant channels. In this situation the heat transport capability of the reactor core would be gravely impaired. Even though the reactor is shut down, the decay heat from the fission products may then be sufficient to cause a meltdown of the reactor core, and the destruction of the fuel elements as a fission product barrier.

The second fission product barrier is the primary system envelope, which encloses the inventory of reactor coolant. As illustrated for the variety of reactor systems shown in Fig. 1-16, the form of the envelope is strongly dependent on the properties of the reactor coolant as well as other functional design considerations. The most common configuration, shown in Figs. 1-16a and 1-16d, is a steel system consisting of a reactor vessel and external coolant loops, including the heat exchanger tubes, the pumps, and connecting piping. In pressure tube systems, such as the one shown in Fig. 1-16f, the reactor vessel is replaced by many independent pressure tubes connected in parallel and passing through the reactor core. In direct cycle systems, whether of the vessel or pressure tube design, as shown are in Figs. 1-16e and g, respectively, the reactor coolant also serves as the working fluid, passing through the turbine-condenser cycle. In these systems fission product escape is prevented by isolating the reactor from the turbine-condenser system by automatic valves placed in the feedwater and steam lines. The primary system thus is defined to include only that part of the envelope on the reactor side of these valves.

It is sometimes attractive to have the coolant pumps and heat exchangers enclosed within a large vessel that acts as the primary system envelope, thus eliminating reliance on large diameter external piping, pump casings, and other such components as a part of the fission product barrier. For liquid-metal-cooled systems, a pot configuration, shown in Fig. 1-16h, has the reactor core, pumps, and intermediate heat exchangers located inside a large steel vessel that serves as the fission product barrier. More recent gas-cooled reactors are designed such that the core, circulators, and heat exchangers are located entirely within the chambers of a single massive vessel of steel-lined prestressed concrete. Such vessels, as shown, for example, in Figs. 1-16b and c, serve as the fission product barrier.

To increase the integrity of a primary system envelope, a system of relief and safety valves must be provided to ensure that accidental overpressure will not cause a failure of the primary system envelope, but only a controlled release of coolant until the overpressure is relieved and the system can be resealed. A great deal of attention must be given in the design, construction, and operation to ensure that the probability of a major primary system envelope breach is reduced to an absolute minimum. As discussed in Chapter 8, reactor vessels are required to be designed and built to such high standards that the consequences of their massive failure are not assessed in detail for most reactor designs. On the other hand, the loss-of-coolant accident ensuing from ruptures of coolant piping, vessel penetrations, heat exchanger tubing, or other lesser components of the primary system envelope often is a major consideration in the overall evaluation of reactor safety.

The third fission product barrier—the reactor containment—is qualitatively different from the first two in that it may serve no functional requirement during normal plant operation. Unlike the fuel elements and primary system envelope, which are vital components of the reactor heat transport systems, the sole functions of the containment are to prevent fission products that escape from the primary system in the event of an accident from entering the environment and to protect the reactor from external forces arising from adverse weather, aircraft impact, or other site-related phenomena. In some reactor systems built outside of the United States the likelihood and consequences of a radioactive release from the primary system are considered to be sufficiently small that the large expense of an additional containment barrier for retention of fission products is not necessary. Such is the case, for example, for the Magnox and AGR reactors of Figs. 1-16a and b, which have been built in Great Britain.

Where a containment is required it must be nearly leaktight to fission products, particularly under accident conditions. A leakage of less than 0.1% per day of the containment atmosphere under the design pressure is a typical requirement. At the same time the energy produced by the reactor must be passed through the containment in steam and feedwater lines, and additional penetrations are needed for refueling, service water and electrical connections, maintenance access, and so on. A containment structure not only must be designed to reduce leakage to a minimum but must usually be equipped with an isolation system to close penetrations, which are needed during normal operations but which would lead to an unacceptable leak rate in the event of an accident.

To meet the stringent leakage requirements under accident conditions, containment structures must be designed to withstand the pressure and temperature generated by the accident postulated for reactors of a particular type; they must also be protected from blasts, missiles, or other forces arising internally. Thus containment concepts and layouts vary with reactor concepts. For example, in BWR's a vapor suppression system such as that shown in Fig. 1-16e is used to limit the containment pressure from the flashing of the water coolant in the event of a primary system rupture. In LMFBR's blasts from hypothesized nuclear excursions and pressures and temperatures generated from sodium fires become the primary considerations. In some systems a single containment vessel is used, whereas others have double—or primary and secondary—containments. A variety of containment concepts are discussed further in Chapter 9.

Accident Characterization

The thorough safety assessment of a nuclear power plant necessarily involves consideration of the potentiality and consequences of a large number of reactor accidents, ranging from minor mishaps to truly catastrophic events. The specifics of the accident sequences that must receive attention depend strongly on the reactor concept and on the details of the design and operation of heat transport systems, fission product barriers, and safety systems. Therefore a comprehensive list of accidents is not appropriate here. As a preliminary to the more thorough discussion of nuclear power plant behavior under accident conditions that constitute much of the remainder of this text, however, it is useful to divide reactor accidents into some broad categories. Although not mutually exclusive, the following four headings encompass most conceivable nuclear power plant accidents:

1. Reactivity accidents.
2. Cooling-failure accidents.
3. Fuel-handling accidents.
4. Site-induced accidents.

The first two categories have at their origin imbalances in the energy transport system of the nuclear power plant. In reactivity accidents the reactor becomes uncontrollably supercritical and produces energy at a rate in excess of the ability of the heat transport system to remove it from the fuel, even though the heat transport system may be in good working order. In cooling-failure accidents the ability to transport heat away from the reactor core is impaired to the point where overheating of the core results even though the rated reactor power is not exceeded.

Reactivity accidents are often subclassified as overpower transients or nuclear excursions, depending on whether they involve a relatively slow rise in the reactor power above its rated value or a very rapid power excursion. In either event the damage originates in the melting or even vaporization of fuel resulting from the inability to remove heat to the coolant fast enough. Severe nuclear transients may be expected to cause high pressure in the primary system once the heat reaches the coolant. In the extreme cases that are sometimes postulated, a substantial part of the fuel may be lost as a fission product barrier, and the primary system may ultimately be threatened by a number of mechanisms. Fuel meltdown may result in molten core materials melting through the primary system envelope. Alternately, some of the thermal energy that is rapidly deposited in the fuel may be effectively converted to destructive mechanical work. Both the expansion of vaporized fuel and vapor explosions resulting from rapid mixing of molten fuel with a higher vapor pressure coolant have been postulated as possible conversion mechanisms. In either case, blast may threaten the reactor vessel integrity.

Cooling-failure accidents may originate from a number of sources. Most of these may be seen by examining the flow diagrams in Fig. 1-15. In the primary system, adequate heat transport requires that coolant flow, temperature, and inventory be maintained. Loss-of-flow accidents can result either from obstruction in the coolant loops or from pumping failures. Flow obstructions are most likely to lead to local flow starvation in the core, whereas pumping failures are most likely to affect a major part or all of the core. Locking of a pump rotor may result in a sharp drop in flow rate. However, with parallel coolant loop design this is not as serious as a simultaneous loss of power to the coolant pumps in all the loops. The temperature and pressure of the whole primary system may rise to unacceptably high levels if the heat produced by the reactor cannot be removed from the primary system envelope. Accidents that obstruct the heat transport path through the steam generator—or steam lines in a direct cycle system—may arise from cooling failures in the secondary system: loss of flow (e.g., from a feedwater pump failure), loss of working fluid inventory (e.g., from a steam line or feedwater line break), or loss of heat sink (if the heat dump to the condenser becomes inoperative). Finally, the loss of coolant from the primary system can often be considered the ultimate difficulty, in the event of a large rupture in a coolant loop.

The preceding cooling-failure accidents are characterized by problems that are likely to be most severe when the reactor is at rated power. This should not, however, be taken to imply that a nuclear power plant is a priori safe when it is at low power, or even when the reactor is subcritical. As discussed in subsequent chapters, there are a number of potential dangers associated with nuclear excursions when the reactor is at very low power, or when fuel is being loaded during start-up procedures. In addition, fuel-handling accidents, whether refueling is done continuously at power or while the reactor is shut down, present some unique problems and are underscored as a third accident category.

Usually one, or at most a few, fuel assemblies may be expected to be involved in a fuel-handling accident. Therefore the total amount of radioactive material that may be released is only a fraction of the total core inventory. Nevertheless, such accidents may be significant because refueling necessarily requires removal of the fuel to a point outside of the primary system envelope. Under these circumstances failure of the fuel elements would release radioactive fission products directly to the containment atmosphere rather than within the primary system, as would be the case with most of the foregoing accidents. At the same time, the decay heat production is large enough to cause the spent fuel to melt should cooling be lost during the removal or storage procedures, and the fuel would also be subject to failure from mechanical impact should it be dropped during the refueling. Finally, the potential for an accidentally supercritical configuration arising either in the reactor vessel or in fuel storage pools must be examined in conjunction with fuel-handling accidents.

The fourth category of accidents includes those arising from external or site-induced phenomena, either natural or man-made: wind, flooding, tornadoes, landslides, earthquakes, aircraft impact, and so on. As discussed in Chapter 9, most of these have their primary effect on the containment structures that must protect the other fission product barriers from damage. Earthquakes are normally considered to present the threat requiring the most careful analysis, since the intense ground shaking may directly affect not just the containment, but every system within the nuclear power plant. With earthquakes, as with other site-induced accidents, the geographical location of the nuclear power station largely determines the severity of the forces to which it may be subjected.

Safety Systems

The probability of the aforementioned accidents can be reduced to extremely low levels through proper precautions in design, construction, and operational procedures. Nevertheless, a plant protection system is necessary to warn of impending trouble and to take automatic action to interfere with the sequence of an accident in order to prevent the accident from propagating, and to ensure the integrity of the fission product barriers. The protection system must detect adverse conditions through automatic sensing devices and actuate active safety systems. These systems may be divided into two classes: the reactor shutdown system and the so-called engineered safety features.

Perhaps the most important action to be taken in the majority of accident situations is the prompt shutdown of the nuclear chain reaction. This action rapidly reduces the power to the level of decay heat and therefore greatly alleviates the energy dissipation problem under adverse conditions. It also eliminates the later possibility of a nuclear excursion, should the changes in temperature, pressure, and chemical environment caused by the accident result in increased multiplication of the reactor core. In this regard it is extremely important that the shutdown system (alternately referred to as the trip or scram system) be fast enough that such a situation would not have time to develop before subcriticality is achieved, and that enough poison be available to maintain the system in a subcritical state, regardless of the conditions in the core following the initiation of the accident.

Once reactor shutdown is achieved, the primary problem is to assure an orderly path for removal of decay heat. Otherwise the energy imbalance would eventually lend to the meltdown of the core. Thus a major engineered safety feature of any nuclear power plant is the emergency core cooling system. The layout of such systems depends completely on whether the reactor is cooled by gas or high or low vapor pressure liquid. Many of the considerations are discussed in later chapters. Suffice it to say that such systems must be capable of responding rapidly to a wide spectrum of accidents that tend to be peculiar to the particular coolant and plant design. Equally important, these systems must also provide a continuous cooling capability over long periods of time in the event of a major accident, since the rate of decay heat production several months after shutdown may still be capable of melting fuel elements.

If an accident is of sufficient severity to be felt outside of the primary system envelope, then there is a need for a number of engineered safety features to guarantee that fission products do not escape to the atmosphere. Three of the most common are devices for containment isolation, containment depressurization, and fission product removal from the containment atmosphere. Because of steam lines, equipment transfer interlocks, and other containment penetrations, the containment leak rate under operating conditions may be considered too high to be acceptable under accident conditions, particularly if it is enhanced by the containment being pressurized by the accident. Therefore a containment isolation system may be provided to close many of these penetrations if an impending accident is sensed. The closure must ensure a low leak rate while at the same time permitting vital instruments to remain intact and, in particular, providing for the decay heat transport out of the system and for electrical power for the emergency cooling system.

Assuming containment isolation is achieved, it then is necessary to ensure that the design pressure and temperature limits of the containment structures are not exceeded, thereby preventing the possibility of the failure of the barrier. It is also desirable to reduce the internal pressure of the containment to a low level as rapidly as possible, since the leakage rate of the containment atmosphere to the environment is a rapidly increasing function of the internal pressure. Depending on the reactor type, containment pressurization may result from the rupture of high-pressure lines in the primary or secondary system, such as in water or gas-cooled systems, or from fires, such as in a sodium-cooled system. The depressurization methods are also diverse and may be either active or passive. If the pressurization is by water vapor, forcing passage of the vapor through a pool of cool water or an ice bed, or spraying of cool water droplets into the containment atmosphere can be used to condense the vapor. Heat exchangers in the containment atmosphere have also been used, along with the presence of large cool surfaces to absorb heat carried from the primary system. Inerting of the containment atmosphere or chemical means may be used to control the heat production by fire.

Just as fission product transport to the environment may be prevented by reducing the leak rate of containment atmosphere, it may also be reduced by eliminating the fission products from the containment atmosphere. For this purpose a number of engineered safety features grouped under the heading of fission product removal systems may be used. These range from placing chemicals in sprays or pools, which scrubs the undesirable elements from the atmosphere, to providing filter trains through which the atmosphere is circulated, to providing paints in the containment surfaces that adsorb selected fission products.

A reliable source of power is essential to operating most active plant safety systems. Therefore a carefully designed system must be incorporated to provide multiple independent supplies of power. These include, as a final backup, redundant on-site diesel generators that start automatically if the protection system senses an accident and banks of batteries to supply critical sensors, relays, and other devices.

In addition to the preceding engineered safety features, a great deal of care must be given to both the plant layout and the incorporation of passive systems to prevent the propagation of accidents. Structures and equipment must be designed to remain functional under temperatures and pressure loadings expected under accident as well as operating conditions. Missile shields must be provided to ensure that the containment structures are protected from fragments, pipe whip, or other mechanical impact arising from the primary system. And, in particular, the emergency core-cooling system and other safety features must be protected from mechanical or thermal damage during an accident. Finally, consideration is sometimes given to passive or active core-catching devices to prevent the melt-through of the primary system envelope and/or containment foundation in the event of a core meltdown.

PROBLEMS

1-1. Verify Eqs. (1-16) and (1-20).

1-2. A thermal reactor operates at a power of 3000 MW(t). Estimate the number of curies of I-131 and of Sr-90 in the reactor after 1 hr, 1 day, 1 week, 1 month, and 1 year.

1-3. Verify that Eq. (1-66) follows from Eqs. (1-62) through (1-65).

1-4. Verify Eq. (1-87).

1-5. Show that Eq. (1-114) follows from (1-113) provided that $f_z(z) = f_z(-z)$.

1-6. Consider a cylindrical reflected reactor with radius R and height H. With control rods partially inserted the power density distribution is approximated by

(The (r, z) origin is at the center of the reactor.)

a. Find A in terms of the reactor power P.

b. Plot $f_r(r)$ and $f_z(z)$.

c. Determine F_r, F_z, and F_q (assuming $F_l = 1.1$).

1-7. A uniform cylindrical reactor core has a height-to-diameter ratio of one. The reactor is reflected both radially and axially with the radial and axial reflector savings each being equal to M, the migration length of the core composition.

a. Show that the power peaking factor given in terms of $\mathcal{P}$, the core-diameter-to-migration-length ratio, is

$$F_q = \frac{1.889\left(\dfrac{\mathcal{P}}{\mathcal{P}+2}\right)^2}{J_1[2.405\mathcal{P}(\mathcal{P}+2)^{-1}]\sin\left[\dfrac{\pi}{2}\mathcal{P}(\mathcal{P}+2)^{-1}\right]}$$

b. Plot F_q against $\mathcal{P}$ between $\mathcal{P}=2$ and 100 as well as the results for the same reactor in the absence of a reflector.

c. Suppose the reflected system is to be a sodium-cooled fast reactor with $M = 5.0$ cm and a power of 2000 MW(t). If the thermal design limits the maximum allowable power density to 550 W/cm³, what is the minimum value that the core diameter can have, the corresponding value of the core volume, and the required value of k_∞ to maintain criticality.

d. Suppose that to increase the thermal safety margins, it is decided to reduce the maximum permissible power density by 10%. What are the percentage changes for the reflected reactor in diameter, volume, and value of k_∞, assuming that M remains constant.

e. Repeat parts (c) and (d) in the absence of the reflector.

1-8. You are to design a 3000 MW(t) pressurized-water reactor. The reactor is a uniform, bare cylinder with a height-to-diameter ratio of one. The fuel element to coolant volume ratio is one to one, in a square lattice. The volume occupied by the control and structural materials can be neglected. The core inlet coolant conditions are uniform at 550°F and 2500 psia.

The reactor must operate under three thermal constraints:

1. maximum power density $= 250$ W/cm³.
2. maximum cladding surface heat flux $= 125$ W/cm².
3. maximum core outlet temperature $= 625$°F.

Determine:

a. the reactor dimensions and volume.

b. the fuel element diameter and lattice pitch.

c. the approximate number of fuel elements.

d. the mass flow rate and average coolant velocity.

SAFETY ASSESSMENT

Chapter 2

2-1 INTRODUCTION

A major public health hazard would result if a significant fraction of the inventory of radioactive materials contained in a large power reactor were to be released to the environment in a populated area. As a result, it is nearly universal practice to legislate licensing and regulatory procedures through which governmental bodies place stringent requirements on the design, construction, operation, and location of power reactors in order to protect the public from large accidental releases of radioactive material. Efforts are made to prevent accidents through sound, conservative design that incorporates only the highest-quality components and allows wide safety margins for off-normal conditions. Highly reliable protection and safety systems are required to ensure that the transients anticipated from equipment failures or operator errors are safely accommodated by reactor shutdown and orderly decay heat removal. Containment and safety systems are required to prevent the atmospheric release of radioactive materials even in the event that one of a series of highly improbable postulated accidents should occur. Finally, demographic and meteorological analysis are made in evaluating reactor sites to limit the exposed population in the event of a large radioactive release.

Having taken all these precautions, if one then asks, "Is the reactor absolutely safe?" the answer still must be no. For the potential hazard is still present, residing in the large inventory of radioactive materials. Moreover, although some margin of safety may be provided by the remoteness of the reactor site, the primary protection is through a system engineered by man. To claim absolute safety, therefore, would imply both that the reactor designers were infallible and that the upper limits of the effects on the system of all natural disasters can be both predicted and protected against with complete certainty.

Because absolute safety is precluded, it is necessary to have some relative measure of safety in order to determine whether a reactor is safe enough. The most acceptable indicator of safety is usually in the form of risk to the public as measured, for example, in the probable number of deaths per year, disabilities per year, or property damage per year. If we assume that a quantitative measure of risk can be settled on, two problems remain: (1) The actual risk resulting from the presence of a reactor must be determined and (2) a level of risk that is acceptable to the public must be specified. With these two risks established, the reactor can then be judged to be safe enough if the actual risk does not exceed the acceptable risk. The problems of determining actual and acceptable risks are quite different. The actual risk created by a nuclear plant is estimated by scientific and engineering analysis. The second problem—that of determining acceptable risk—is one not of strictly technical analysis, but of public policy.

It is impossible to place a precise number on the risk from nuclear accidents that would be found acceptable by a society. Nevertheless, some idea can be gained as to the approximate level of acceptable risk by examining the risks resulting from other activities. Such risks can be determined from the statistics on deaths, injuries, or property damage from various categories of diseases and accidents. In what follows we utilize as a criterion the risk of death to the average individual. This is obtained simply by dividing the number of fatalities per year due to a particular cause by the total population, and it yields the risk to the average individual as a probability of death per person per year. In detailed studies other consequences, such as injury, disability, or property damage, may also be included. In the United States the average individual risk due to all accidents is 6×10^{-4} deaths per person per year (U.S. Atomic Energy Commission, 1974). The accident risk may be broken down by cause, as shown in Table 2-1.

Otway and Erdmann (1970) have studied data such as that in Table 2-1 to ascertain public attitudes toward risk and have made the following observations. It is difficult to find accidents with an individual death risk to the general public with a magnitude of 10^{-3} per person per year or more. This level of risk evidently is unacceptable, and actions are quickly taken to reduce hazards of this magnitude.

People are less inclined to take concerted action to further reduce risks at the level of 10^{-4} deaths per person per year. There is, however, a willingness to spend money to reduce the risks of this level, as, for example, in traffic control systems, fire departments, and fences around hazardous areas. At this risk level safety slogans containing an element of fear are often invoked, as, for example, "The life you save may be your own" (applied to motor vehicle operation).

Active recognition is given to accidental risks at the level of 10^{-5} deaths per person per year. Parents warn children of the dangers of drowning, firearms, poisoning, and so on, and a certain amount of inconvenience is

TABLE 2-1   Individual Risk of Acute Fatality by Various Causes
(U.S. Population Average 1969)

Accident Type	Total Number for 1969	Approximate Individual Risk: Acute Fatality Probability per Year^{-1}
Motor vehicle	55,791	3×10^{-4}
Falls	17,827	9×10^{-5}
Fires and hot substance	7,451	4×10^{-5}
Drowning	6,181	3×10^{-5}
Poison	4,516	2×10^{-5}
Firearms	2,309	1×10^{-5}
Machinery (1968)	2,054	1×10^{-5}
Water transport	1,743	9×10^{-6}
Air travel	1,778	9×10^{-6}
Falling objects	1,271	6×10^{-6}
Electrocution	1,148	6×10^{-6}
Railway	884	4×10^{-6}
Lightning	160	5×10^{-7}
Tornadoes	91[1]	4×10^{-7}
Hurricanes	93[2]	4×10^{-7}
All others	8,695	4×10^{-5}
All accidents		6×10^{-4}
Nuclear accidents (100 reactors)	0	3×10^{-9}

U.S. Atomic Energy Commission, 1974b.
[1] Based on total U.S. population, except as noted.
[2] (1953–1971 average).
[3] (1901–1972 average).

accepted by the public to avoid this risk level. At this risk level a precautionary ring dominates the safety slogans: "Keep Out of the Reach of Children"; "Never Swim Alone."

The average person is apparently not greatly concerned with accidents with a risk of 10^{-6} deaths per person per year. Although he is aware of them, he does not feel they will happen to him. Such accidents may even be thought to be due to stupidity; "Everyone knows you shouldn't stand under a tree during a lightning storm." Fatalism also frequently is associated with such accidents, commonly referred to as "acts of God."

The preceding observations indicate that a risk of 10^{-6} deaths per person per year is considered negligible whereas risks of a level of 10^{-3} deaths per person per year or more are unacceptable. Risk levels between 10^{-3} and 10^{-6} are acceptable to varying degrees. Otway and Erdmann (1970) conclude from such qualitative arguments that nuclear power plants should be acceptable if the risk to an individual living at the worst location on the site boundary is at a level of 10^{-7} deaths per person per year or less. The risk to the general population living in the vicinity of the reactor would then be much smaller.

Although in substantial agreement with these simple observations, more comprehensive studies (Starr 1972; U.S. Atomic Energy Commission, 1974b) have pointed out at least two additional factors regarding public attitude toward risk. The first is that the level of risk to which individuals subject themselves varies substantially. In sporting activities, people often voluntarily submit themselves to risk levels substantially in excess of those to which the general public is exposed. Similarly, limited groups expose themselves to occupational hazards that exceed the accidental death rate of the general populace. Even after discounting such strictly voluntary activities, however, it may be argued that by accepting inconvenience, an individual can further reduce his risk. Many of the activities associated with the fatalities listed in Table 2-1 may be avoided. For example, only about one-tenth of the victims of aircraft crashes are people on the ground killed by the falling aircraft. Thus by avoiding air travel the risk for such accidents is reduced to about 9×10^{-7} per person per year. More significantly, only about 1% of automobile fatalities are incurred by people who are neither in a vehicle or on a roadway. Therefore a person who never rides in a motor vehicle or enters a roadway would reduce his risk from this cause to about 3×10^{-6}. Taking other such reductions into account, about 6×10^{-6} deaths per person per year represents a lower limit of risk to the safety-conscious individual who is willing to eliminate all but the most involuntary of activities for the purpose of risk reduction. In an industrial society, however, very few people have a lifestyle that would allow them to make such a large reduction.

A second factor not yet accounted for is the magnitude of individual accidents. It has been argued (U.S. Atomic Energy Commission, 1974b) that accident causes resulting in similar numbers of fatalities per year may be viewed quite differently, depending on the magnitude of the individual events. Even though the resulting number of deaths may be identical, a single large accident appears to be less tolerable to society than many smaller accidents. This view is sustained, for example, by the news coverage given to single dam failures or large aircraft crashes as compared to the larger number of deaths from automobile accidents. This dimension of the problem is missed if the individual risk from power reactors is simply compared to death probabilities such as those contained in Table 2-1. The aversion of the public to catastrophic, albeit improbable, events makes it necessary also to consider reactor accidents in relationship to catastrophic but improbable accidents due to other causes. Data from such accidents are more difficult to obtain, since their frequency is small, and historical records are not adequate. Therefore theoretical predictions often must be used to extrapolate the rate of occurrence of highly improbable events from data for more frequent but less serious accidents. In Figs. 2-1 and 2-2 such data are plotted for both natural catastrophes and accidents arising from the failure of man-made systems. These figures indicate the probability of an accident causing more than N deaths during one year.

A qualitative argument can be made that if the risk due to large but improbable reactor accidents is substantially smaller than that due to either natural or other man-made catastrophes, then the reactors are probably safe enough. For then the health and safety of the public could be more

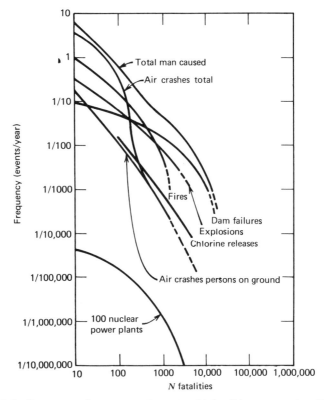

FIGURE 2-1 Frequency of man-caused events with fatalities greater than N. From U.S. Nuclear Regulatory Commission, "Reactor Safety Study," WASH-1400 (NUREG-75/104), Appendix VI, 1975.

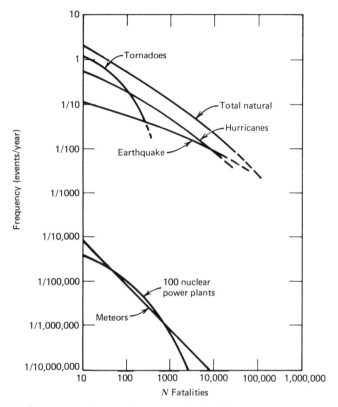

FIGURE 2-2 Frequency of natural events with fatalities greater than *N*. From U.S. Nuclear Regulatory Commission, "Reactor Safety Study," WASH-1400 (NUREG-75/104), Appendix VI, 1975.

effectively improved by taking additional measures to reduce the risk from the other events instead of spending additional money to decrease further the risk resulting from reactors. An estimate of the risk resulting from 100 water-cooled reactors of the type presently being built in the United States is included in Figs. 2-1 and 2-2.* With the assumption that the study (U.S. Nuclear Regulatory Commission, 1975) that led to this curve is valid, the foregoing premises lead to the conclusion that these reactors are safe enough.

---

* A catastrophic nuclear accident would also lead to an increased incidence of cancers for about 30 years following the event. See Section 10-5 for further discussion.

The foregoing discussion relates the risk from nuclear plants to those resulting from both natural disasters and other types of accidents. Nuclear plants may also be considered within a more specific framework. The risks incurred from nuclear plants can be compared to those resulting from the production of electricity by alternative methods. Fossil-fueled plants emit noxious gases, such as SO_2 and NO_x to the atmosphere that are known to cause respiratory problems and that may lead to increased incidence of lung or other cancers. Although precise comparisons are made difficult by a lack of adequate data on emission rates, dispersion properties, and health effects of combustion products of fossil-fired plants, several studies have been made in an attempt to relate the risks from nuclear plants to those of coal- and oil-fired units (Starr and Greenfield, 1973; Hull, 1971; Lave and Freeburg, 1973).

The primary health hazards associated with fossil-fired plants appear to be those due to routine operation during conditions of air stagnation. The resulting air pollution can be correlated to increased numbers of deaths from respiratory problems. The use of fossil fuels also may lead to large, albeit improbable, accidents that result in significant numbers of nonoccupational fatalities. An explosion and fire in the large storage capacity of an oil-fired plant, for example, could cause as many fatalities as the postulated nuclear accidents if it occurred under circumstances that combined a stagnant air mass, the presence of a variety of other airborne pollutants and chronic irritants (such as asthmagenic allergens), and the occurrence of a respiratory epidemic (such as influenza) (Starr and Greenfield, 1973).

Starr and Greenfield (1972, 1973) have compared oil-fired and nuclear plants and concluded that under routine operating conditions, nuclear plants present a public health risk that is a factor from 10 to 100 smaller than that for oil-fired plants, whereas the risks from accidents in oil and nuclear units are of the same order of magnitude and about a factor of 100,000 smaller than those for routine operation. Although no comparable study is available on accidents in coal-fired plants, the health effects of the routine emissions from these plants is known to far exceed those from nuclear or oil-fired plants (Lave and Freeburg, 1973).

Probabilistic methodologies for the quantitative assessment of reactor risk are in a rapid state of development. The point has not been reached, however, where regulatory decisions regarding the licensing and operation of reactors can be approached in terms of straightforward risk criteria. In the United States, as elsewhere, the decision-making process by which reactors are licensed has evolved to incorporate a complex array of legal and technical procedures and criteria. Although probabilistic risk arguments are utilized to some extent, reliance is also placed on codes of good practice, intermediate design criteria, analysis of standard hypothesized accidents, and a host of other requirements that enable the regulatory body to make a qualitative judgment of whether there is a sufficient level of confidence that the health and safety of the public are protected from undue risk.

In the following section we review current techniques for arriving at quantitative risk assessments. In Section 2-3 we then review the decision-making process by which reactors are licensed in the United States. Finally, as a prologue to the remainder of the text, we summarize some of the general design principles that are employed to ensure a high degree of reliability in the safety systems incorporated into nuclear power reactors to prevent and/or mitigate the consequences of serious accidents.

2-2 QUANTITATIVE RISK ASSESSMENT

Quantitative methods for the evaluation of reactor risk are becoming an increasingly important part of safety assessment. The methods found their first widespread use in the United Kingdom as a result of the work of Farmer (1967) and his colleagues. More recently such methods have gained visibility in the United States, particularly as a result of the comprehensive risk analysis of light-water reactors carried on under the direction of Rasmussen (U.S. Atomic Energy Commission, 1974b). Parallel advances in quantitative assessment are also being made in several other countries (International Atomic Energy Agency, 1973). An exposition of the statistical methodology required to carry out a detailed risk assessment is beyond the scope of this text. However, an outline of the general features of the more prominent quantitative techniques follows, since it serves to place in perspective much of the physical description of reactors under accident conditions that is contained in later chapters.

Before considering the total risk caused by the presence of a power reactor, suppose we consider the risk incurred because of a particular type of reactor accident, say type l. The risk $\mathcal{R}_l$ may be defined either as the risk to an individual living on the site boundary or as the risk to the total population of the surrounding area. Individual risk can be measured, for example, in the probability of death per year for the individual, whereas the corresponding risk for the entire population would be in the probable number of deaths per year. For brevity we denote either of these in units of deaths per year.

The risk from the accident may be divided into three factors

$$\mathcal{R}_l = \mathcal{P}_l C_l \mathcal{D}_l, \tag{2-1}$$

where

$\mathcal{P}_l$ = probability per year that the accident will occur (in accidents per year),

C_l = amount of radioactive material released to the atmosphere as a result of the accident (in curies per accident),

$\mathcal{D}_l$ = the risk resulting from the atmospheric release of one unit of radioactive material (deaths per curie).

Although the three factors on the right of Eq. (2-1) are not totally independent, they tend to depend on quite different considerations. To determine $\mathcal{P}_l$, the probability that the initiating event will occur must be estimated. The initiating event may be a break in a coolant pipe, the ejection of a control rod, loss of power to the coolant pumps, or any number of other component failures or operator errors.

With the accident established the radioactive release C_l must be determined. This normally requires a time-dependent analysis to determine the behavior of the nuclear power plant during the course of the accident transient. Elaborate neutronic and thermal-hydraulic modeling may be required to determine the extent, if any, to which fuel failure or melting takes place and to determine whether the integrity of the primary system envelope or the containment is violated. From this information the transport of fission products from the fuel to the atmosphere can be estimated and hence a value for C_l, the radioactive release from the accident, can be estimated.

In general, the value of $\mathcal{D}_l$, the risk per curie of radioactive material released, is a function both of the nature of the accident and of the reactor site characteristics. It is dependent on the nature of the accident because the radioactive material released to the atmosphere can be a mixture of many different fission products and actinides, and each of these may appear in elemental form, in a chemical compound, in aerosol particles or possibly in other forms. The release from different accidents may contain different proportions of the various radioactive materials, and it may be necessary to account for this in the consequence model used to predict the risk per curie released. Often, however, it has been found that one nuclide dominates the biological effects of the release; that is iodine-131. It has thus become a frequent practice, particularly in the United Kingdom (Farmer, 1967), to express the release in curies of iodine-131. If this is done, the release of other fission products is equated to some equivalent number of curies of iodine-131. The consequence model then becomes independent of the accident type and $\mathcal{D}_l \to \mathcal{D}$, since all accidents are characterized simply by C_l in curies of iodine-131 released per accident. Other, more detailed classifications also find frequent use. For example, in the United States it is common practice to group the fission products into noble gases, iodine and other halogens, and nonvolatile fission products, and calculate the release and consequences of each group separately.

Once the composition of the release is specified, the risk per curie is dependent primarily on site conditions. An atmospheric dispersion model dependent on local meteorological and geological data is used to calculate the dose at any point outside the plant. Health physics data are used to determine the probability of death from the dose. If individual risk is being used, the probability of death per curies released is calculated for the person at the worst location outside the plant boundary. If risk to the total population is to be calculated, demographic and evacuation data must be used to determine the total number of deaths expected per curie released taken over the entire population of the surrounding area.

The total risk $\mathcal{R}$ incurred from reactor accidents is obtained by summing Eq. (2-1) over all possible accident types. Thus

$$\mathcal{R} = \sum_l \mathcal{P}_l C_l \mathcal{D}_l. \tag{2-2}$$

If, for purposes of analysis, we assume that radioactivity releases to the atmosphere are adequately characterized as a single quantity—for example, curies of iodine-131—the consequences model for determining the risk per curie released is independent of the accident type. Therefore we may write

$$\mathcal{R} = \left(\sum_l \mathcal{P}_l C_l\right) \mathcal{D}. \tag{2-3}$$

This is a useful approximation because it permits the separation of siting effects, required to calculate $\mathcal{D}$, from the analysis of the nuclear power plant. Alternate reactor designs, the efficacy of safety systems, and other safety-related questions can be assessed in terms of the probable number of curies per year relased from the reactor:

$$C = \sum_l \mathcal{P}_l C_l. \tag{2-4}$$

This quantity is determined once the power plant design and operational procedures are fixed. The public risk incurred by placing the reactor at a specified location is then determined by combining C, the probable curie release, with the consequence model $\mathcal{D}$:

$$\mathcal{R} = C\mathcal{D}. \tag{2-5}$$

Deterministic approaches to reactor safety stress the evaluation of C_l for a series of specific hypothesized accidents. These are often called design basis accidents because they are analyzed to set conservative limits on pressures, temperatures, and other quantities that may be expected to cause problems under accident conditions. The limiting values of these quantities are then used to set specifications that must be met by plant components, fission product barriers, and safety systems. If a quantitative analysis of the risk resulting from a nuclear power plant is to be made, however, emphasis must be placed on the calculation of accident probabilities and in particular on the assurance that all accidents that may give rise to a significant release of radioactive material are included in the summation.

Event Trees

The consequences of an accident of a particular type depend not only on the initiating event but also on the operation of the reactor safety systems during the ensuing transient. Although the accident may have minimal effects if the safety systems function properly, it is necessary to examine the sequence of events for which one or more of the safety systems fails. For this purpose the event tree is quite useful.

To illustrate the use of an event tree we consider the case of a loss of coolant accident in a pressurized-water reactor, which has been treated in detail elsewhere (U.S. Atomic Energy Commission, 1974). Figure 2-3 is a simplified event tree for this accident. On the left is the initiating event, in this case a rupture of the primary cooling system. It is assigned a probability of occurrence, p_1. The next significant event is the availability of electric power. The failure of the electric power to operate the emergency safety systems is assigned a probability p_2. On the event tree this appears as a two-branched fork, with an upper branch for successful operation and a lower branch for failure. Since electric power either is or is not available, $1 - p_2$ must be the probability that power is available. If the power fails, none of the safety systems included in Fig. 2-3 will function. In this event it may be shown that the reactor core will melt and that there is the potential for a very large release of radioactivity to the atmosphere. The probability of this sequence of events is $p_1 \times p_2$. If electric power is available the next significant event is the operation of the emergency core-cooling system (ECCS). The failure and success probabilities for this system are taken as p_3 and $1 - p_3$, respectively. As indicated by the diagram, this procedure is continued through the fission product removal and containment integrity until all safety provisions have been considered.

In a detailed analysis many more events would be included in the tree, as, for example, the shutdown of the reactor. In addition, some of the events included in Fig. 2-3 would be subdivided. For example, the operating of the ECCS might be divided into initial ejection of water into the reactor core

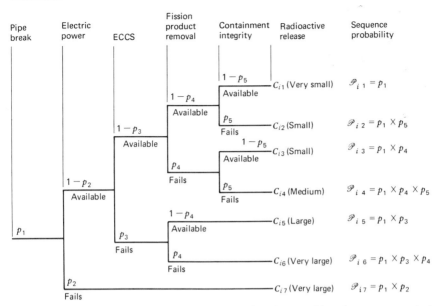

| Pipe break | Electric power | ECCS | Fission product removal | Containment integrity | Radioactive release | Sequence probability |

FIGURE 2-3 Simplified event tree for a loss-of-coolant accident in a water-cooled reactor.

and long-term cooling following the loss-of-coolant transient. Whereas more branches are added by increasing the detail of the analysis, branches may be eliminated by accounting for dependences between the probabilities. For example, in Fig. 2-3, failure of containment integrity is assumed to follow from the failure of the ECCS, since analysis would indicate that ECCS failure would lead to fuel meltdown, and this would cause the containment to be overpressurized. Hence two branches that would otherwise be present are eliminated.

The overall probability of each sequence of events can be calculated, provided a method is available for evaluating the failure probabilities of each system on the event tree. In calculating these sequence probabilities, the failure probabilities p_i are taken to be small, so that the nonfailure probabilities $1 - p_i$ are replaced by 1. The analysis of the accident transient associated with each sequence then allows the radioactive release from each sequence to be estimated. The results may then be incorporated into the risk model as follows: If the l again refers to the accident type (a loss of coolant in our example) and the sequences in the event tree are enumerated with a subscript i, then for each sequence we have a radioactive release C_{li} and a probability $\mathscr{P}_{li}$ as shown on the right of Fig. 2-3. The probable release from an accident of type l is then determined by summing over the possible sequences of events that may follow the initiation of the accident:

$$\mathcal{P}_l C_l = \sum_i \mathcal{P}_{li} C_{li}. \tag{2-6}$$

From Eq. (2-3) the total risk for the reactor resulting from all accident types may be seen to be

$$\mathcal{R} = \left(\sum_l \sum_i \mathcal{P}_{li} C_{li} \right) \mathcal{D}. \tag{2-7}$$

Limit Lines

The detailed analysis of every mishap that could befall a reactor would be a prodigious if not impossible task. For this reason the evaluation of reactor risk normally involves an iterative approach. In a first analysis, many of the initiating events may be eliminated on the basis of qualitative considerations or rough calculations that show that the accidents do not have the potential for significant radioactive releases. Emphasis can then be placed on more detailed analysis of a relatively few accidents that have the greatest risk potential and therefore dominate the summation of Eq. (2-3). It then may be argued that the uncertainties in estimating the risks in the most dominant accident types are sufficiently large as to mask the effects of other lesser accidents that are not considered.

Once the dominant accident types have been isolated, it may be considered prudent to put limits directly on the probable radioactive release per year for each accident type. These limits are set sufficiently low that when the limits are summed over several most dominant accident types, the total risk is still acceptable. The probability–consequence diagram and limit line concepts developed by Farmer (1967) are extremely useful not only in evaluating the risk for a given accident, but also in analyzing the effects of changes in design and operating procedure and in assessing the required accuracy of safety calculations. A diagram of probability versus consequence is plotted in Fig. 2-4. The probability is measured in accidents per year and the consequences are expressed in terms of the release of radioactive material in curies of iodine-131.

Suppose that the event tree, such as Fig. 2-3, is constructed for the accident sequences arising from a particular initiating event. Each branch on such a diagram gives rise to a radioactive release and a probability and therefore may be plotted as a point on the probability–consequence diagram. In examining such points, plotted on Fig. 2-4, it will be noted that accident sequences resulting in points near the origin are of low risk to the public, for they result in very small releases of radioactivity, and they occur

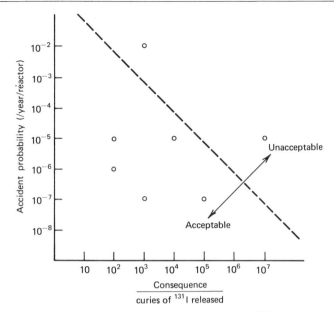

FIGURE 2-4   Accident consequences as a release of curies of ^{131}I at ground level versus accident probability occurrences. Adapted from F. R. Farmer, "Reactor Safety and Siting: A Proposed Risk Criterion," *Nucl. Saf.*, **8**, 539–548 (1967).

very infrequently. In contrast, points that approach the upper right-hand corner of the graph are very serious, for they both have large fission product releases, and they occur relatively frequently. If one selects as a measure of risk for a given accident sequence the number of curies released to the atmosphere per year, then sequences of equal risk will be connected by a line with slope of -1. As shown in Fig. 2-4, a possible safety criterion is that no accident sequence shall result in a risk above some fixed value. Then all accident sequences should lie underneath the locus of sequences having this risk. This line of equal risk is then referred to as the *limit line*. The direction of increasing risk is then taken to be normal to the equal risk limit line as shown.

Using this type of analysis, it is clear that to improve the safety of the reactor, measures should be taken to move the accident points that are far from the origin, in the high-risk areas, toward the origin. This may be accomplished in two ways. First, adding additional fission product barriers or safety features that reduce the fission product release moves the points to the left. Second, improving the reliability of the present configuration of the reactor to reduce the probabilities either that the accident will happen in the first place or that the fission product barriers or safeguard systems will fail under accident conditions moves the points downward. Note that as far as determining the accident risk from these criteria is concerned, there is no need to obtain greater precision in the estimate of the fission product release than the precision that can be attributed to the probability of that release.

If an additional containment barrier or engineered safety feature is used to reduce the magnitude of the fission product release from an accident by many orders of magnitude, the reliability of that system should be of at least the same order of magnitude. This is schematically illustrated in Fig. 2-5. Suppose we consider an accident whose result at point A is unacceptable. We add an engineered safety system (e.g., possibly an emergency core-cooling system for a loss-of-coolant accident) that when properly functioning results in a reduction of the fission product release by three orders of magnitude as shown at point B. With the addition of the safety system, however, we have created an additional branch in the accident event tree; we must not only consider the end point with the safeguard working properly, but also that when it fails. If there is 1 chance in 10 that the safety system would fail completely under accident conditions, an additional accident termination point must be added at C. This branch, with the safety system failure, now becomes the primary concern, because it is in a location of greater risk than is point B. On the other hand, if the safeguard is made more reliable so that the failure probability under accident conditions is

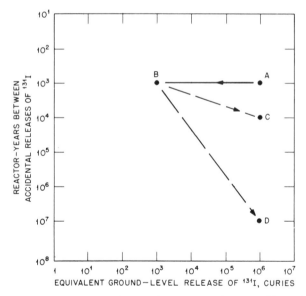

FIGURE 2-5 Reliability of protective equipment used to modify accident consequences. Adapted from F. R. Farmer, "Reactor Safety and Siting: A Proposed Risk Criterion," *Nucl. Saf.*, **8**, 539–548 (1967).

10^{-4}, then the terminal point of the event tree branch corresponding to a safety failure is at point D. In this situation the risk due to the safety system failure is smaller in terms of curies per year to which the public is exposed than the accident in which the safeguard functions properly. Thus the safeguard is sufficiently reliable.

It is not necessary that the limit line coincide with a line of equal risk, as is the case in Fig. 2-4. Regulatory bodies may take into account considerations other than the risk as defined earlier. For example, because of adverse public reaction to very large accidents, it may be thought prudent to place the greatest emphasis on the prevention of high-release albeit low-probability accidents by increasing the negative slope of the limit line. Such a curve, with a slope of -1.5, has been proposed by Farmer (1967) and is shown as Fig. 2-6. In this curve a transition in the slope between 10 and 1000 Ci is also incorporated to minimize the frequency of small releases. This is done to decrease the economic loss that would result from plant downtime and damage caused by accidents that are more frequent but have negligible public health consequences.

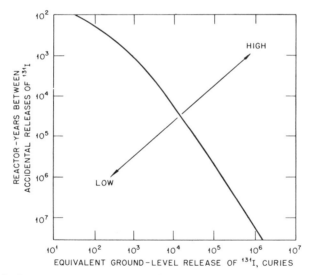

FIGURE 2-6 Proposed release criterion. Adapted from F. R. Farmer "Reactor Safety and Siting: A Proposed Risk Criterion," *Nucl. Saf.*, **8**, 539–548 (1967).

Fault Trees

Central to any attempt to estimate risk from reactor accidents is the ability to evaluate the probability of accident-initiating faults and the failure probabilities for the reactor safety systems. Direct determination of these probabilities from reactor-operating experience faces a twofold difficulty. First, the nuclear industry is young; therefore relatively little data have been accumulated, and much of these data are on systems that are in a rapid state of evolution. Second, reactor systems must be designed so that the probabilities of major initiating faults and of safety system failures are very small. Therefore it is unlikely that operating experience in the foreseeable future will provide sufficient direct data to evaluate these small failure probabilities with sufficient accuracy for risk analysis. Hence other methods must be used to predict the necessary probabilities.

The prediction methods used depend on the nature and probability of the failure. Perhaps the most difficult involve the prediction of initiating faults involving a catastrophic but highly improbable failure of a single large mechanical component, as, for example, the fracture of a large coolant pipe or even of the reactor vessel itself. Failure rate data are not likely to be available for such occurrences. Nevertheless, study groups have been able to use historical experience in the nuclear as well as other industries and a thorough understanding of the design and operating procedures in both the nuclear and conventional industries to set bounds on what the probability of failure will be. For example, the Advisory Committee on Reactor Safeguards (U.S. Atomic Energy Commission, 1974) has made an extensive review of the integrity of reactor vessels for light-water power reactors. From this study they were able to estimate the probability of disruption failure of such vessels at less than 10^{-6} per vessel-year of operation.

Many of the faults with the potential for causing a major reactor accident differ from the aforementioned vessel rupture in that they involve the failure of a complex system consisting of many active and passive components. Reactor accidents may be caused, for example, by the failure of a coolant pump, the station power supply, or the reactivity control system for the reactor core. Safety system failures following the initiation of an accident similarly arise from the malfunction of a complex set of components. Such systems are constructed primarily of components that are widely used in several industries. Therefore extensive failure rate data from both operating experience and laboratory testing may be available for the basic components: the valves, switches, relays, and so on. The failure probability of the system may then be synthesized from the failure rate of the components by a number of methods, most of which involve the use of fault trees (cf. U.S. Atomic Energy Commission, 1974; Green and Bourne, 1972; Pasternack, 1975).

Fault-tree analysis is a graphic method for tracing back from an undesirable event to any of the many possible causes. Most often the fault tree is drawn in the form of a tree trunk for the event to be avoided and tree roots for the contributing events or actions. The fault-tree logic is nearly the reverse of that of the event tree, because the fault tree starts with the undesired event and attempts to find its causes or roots, whereas the event tree starts with the undesired event and attempts to follow all its possible consequences or branches.

Fault trees are useful both as a qualitative and a quantitative analytical technique. Qualitatively, they enable one to examine closely the logical structure of a system and therefore ensure that it has the functional capability required, provided all the components work. Often critical components requiring high reliability also can be located merely by drawing the fault tree. Quantitatively, the fault tree enables one to relate the failure probability of the system back to the failure rate data of subsystems or individual components.

The use of fault trees is best illustrated by a simple example. In Fig. 2-7 is shown a rudimentary fault tree for the failure of an electrical power supply system. The roots of the tree are related through the two types of logic gates shown in the diagram. The OR gate is activated if any of the input events are

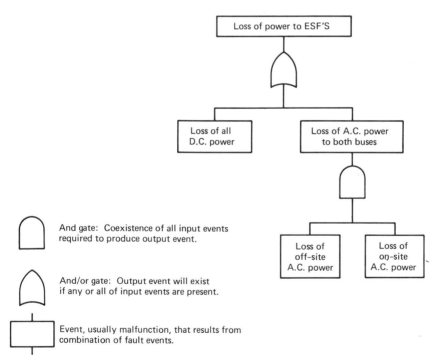

FIGURE 2-7 Simplified fault tree for loss of power to engineered safety features. Adapted from U.S. Atomic Energy Commission, "Reactor Safety Study," WASH-1400 (Draft), 1974.

present, and activation of the AND gate requires that all the inputs be present. The OR gate in Fig. 2-7 indicates that failure of either A.C. or D.C. power will cause the failure of the engineered safety features, because the A.C. power provides the energy needed for their operation whereas the systems are energized by the D.C. power. On-site and off-site power from the utility network and on-site power from diesel generators must be lost before the A.C. power supply fails.

The system failure probability can be estimated by working upward through the tree, provided the failure probabilities of the fault-tree roots are small compared to one. This is accomplished by adding the failure probabilities from the inputs at the OR gates and multiplying the failure probabilities at the AND gates. In Fig. 2-7, for example, we take the probabilities of failure of the D.C., off-site A.C., and on-site A.C. power to be, respectively, p_1, p_2, and p_3. The system failure probability is then $p_1 + p_2 p_3$.

The use of a fault tree to evaluate the system failure probability requires that the tree roots be developed to finer and finer detail. The tip of each root must be an equipment failure or human error for which it is possible for reliability engineers familiar with the components to assign component failure rates and experts on human factors to assign probabilities of human errors. The detail required in a fault tree may differ greatly between systems and even between the roots of a single system. For example, failure information may be available on the overall performance of diesel generators that make up the on-site A.C. power block in Fig. 2-7. Only one additional level to the fault-tree root is then required. On the other hand, it may be necessary to go to the level of much more elementary components before failure data are available. In many systems it is often necessary to construct fault trees down to such elementary components that many pages are required to draw the tree.

If the quantitative results of fault-tree analysis are to be valid, extreme care must be taken to assure that the failure rates are not underestimated. There are several causes of such underestimates, which must be guarded against (cf. U.S. Atomic Energy Commission, 1974). First, component failure rates based on laboratory data rather than data taken in a more poorly controlled operating environment may result in failure rates that are an order of magnitude or more too low. Second, human errors, which cause failure rates from $10^{-1} - 10^{-2}$, must be taken properly into account in the operation, testing, and maintenance of equipment. A third problem involves the spread in available failure data, which usually is of an order of magnitude or more. For example, the failure probability of a typical active component may be between 10^{-2} and 10^{-3}. Using mean values for such data in complex fault trees may result in gross errors in the system failure probability estimate. A more reliable method is to use a range of values for each failure probability in a computerized simulation model to obtain a spread of system failure probabilities. These can then be characterized by a mean value and a confidence interval.

Finally, great care must be taken in the application of fault trees to ensure that dependencies between failure rates are taken into account. Assuming that component failure rates are independent when in fact they are even slightly related may cause common-mode failures to be ignored that may dominate the overall failure rate of the system. To illustrate, suppose that the on-site A.C. power supply consists of four diesel generators of identical design placed in parallel such that any one of them will supply the requisite power. On the fault tree, on-site A.C. power then appears at the output of an AND gate, with the diesel failure probability at the input. If the failure probability for one diesel generator is 10^{-1}, the failure rate of the on-site power is 10^{-4}! This value assumes that all failures are independent. In reality there may be a common-mode failure or failure produced by causes that affect all four diesels simultaneously. Such failures may be caused, for example, by common adverse environmental exposure or incompetent maintenance by the same crew on all four diesels. Although it may be possible to make such common modes a small part of the failure rate of the individual diesels, they are likely to dominate the failure rates of a highly redundant system. In our example, a common-mode failure probability of 10^{-3} would be insignificant for a single diesel, but would dominate the behavior of the on-site A.C. power system.

2-3 REACTOR LICENSING

It is nearly universal practice to require either government ownership or strict licensing of privately owned nuclear facilities. The possibility for diverting materials for weapons purposes as well as the potential for releases of large amounts of radioactive materials to the environment necessitates close regulation to protect the health and safety of the public. In what follows, we discuss government regulation as it pertains to power reactor safety in the United States. Discussion of the regulation of nuclear facilities in other countries may be found elsewhere (cf. International Atomic Energy Agency, 1974; Richardson, 1976).

In the United States the licensing and regulation of power reactors issues from the charge "to regulate the peaceful use of atomic energy to protect the health and safety of the public" that was given by law to the Atomic Energy Commission through the Atomic Energy Act of 1954. The licensing requirements for reactors are not specified in any detail in the Atomic Energy Act. Rather, the Atomic Energy Commission (AEC) has carried out its statutory responsibility by promulgating a body of rules, procedures, and standards that are applied by a staff of the AEC's Division of Reactor Licensing in determining the suitability of nuclear facilities for licensing. Since the splitting of the AEC in 1974 into the Energy Research and Development Adminstration (ERDA) and the Nuclear Regulatory Commission (NRC), these duties have been taken over by NRC. The AEC rules and regulations, which have the force of law, are published in the Code of Federal Regulations, Title 10 (referred to as 10 CFR) and its amendments. Several parts of 10 CFR relate to reactor safety. The most important are parts 20, 50, and 100: 10 CFR 20, Standards for Protection against Radiation; 10 CFR 50, Licensing of Production and Utilization Facilities; and 10 CFR 100, Reactor Site Criteria. Several other parts of Title 10 are also relevant to reactor safety, such as, for example, 10 CFR 55, Operator Licenses.

Licensing Procedures

Most of the NRC's regulations concerning the licensing of nuclear power reactors and other nuclear facilities are contained in 10 CFR 50, Licensing of Production and Utilization Facilities. Procedures for preparation, filing, and processing applications are specified and the technical information that must be provided concerning the plant and its operation is stipulated. Appendixes to Part 50 provide further guidance on several important aspects of the license application. Appendix A, General Design Criteria, establishes a set of minimum requirements for the design criteria. Appendix B of Part 50 contains quality assurance criteria. Other appendixes deal with such items as environmental impact, reporting deficiencies, and so on. At present, several of these appendixes are strictly applicable only to water-cooled nuclear power plants. In addition to statutory law contained in Title 10 of the Code of Federal Regulations, the regulatory staff of the NRC publishes regulatory guides. These guides do not carry the force of law. They are, rather, established solutions to safety issues in licensing that are acceptable to the licensing authorities. Solutions other than those contained in the guide may also be found to be acceptable.

The complex decision-making procedure by which NRC licences are granted for the construction and operation of nuclear power plants involves the designers of the power plant, the utility that will own and operate the plant, the Nuclear Regulatory Commission, and in some cases the public. The procedure is illustrated by the flow sheet for obtaining a construction permit shown in Fig. 2-8. With this construction permit the utility is allowed to build the power plant with the understanding that safety problems coming to light during the review procedure must be resolved before the plant may go into operation. A similar procedure is followed to obtain an operating license.

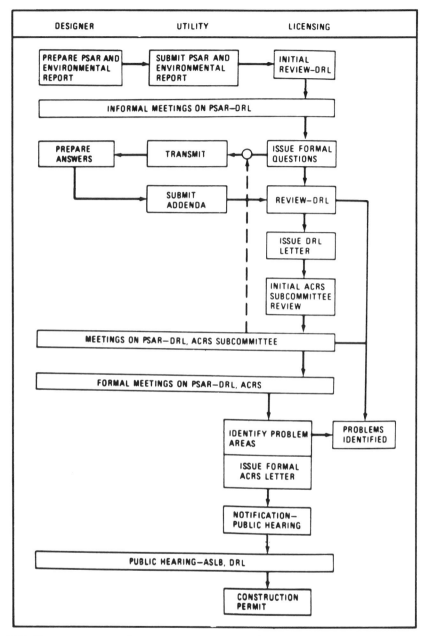

FIGURE 2-8 Nuclear safety licensing process. PSAR, preliminary safety analysis report. DRL, Division of Reactor Licensing. ACRS, Advisory Committee on Reactor Safeguards. ASLB, Atomic Safety and Licensing Board. From G. G. Lawson, "Emergency Core-Cooling Systems for Light-Water-Cooled Power Reactors," Oak Ridge National Laboratory Report, ORNL-NSIC-24, October 1968.

The utility that decides to build a nuclear power plant first submits a preliminary safety analysis (PSAR) to the NRC. This report, along with a required environmental impact statement, is prepared with the help of the plant designers: the reactor vendor and the architectural engineers. Concurrently with the preparation of the PSAR, informal meetings may take place with the NRC Division of Reactor Licensing (DRL) concerning siting plans.

The licensing staff of the NRC performs a comprehensive review of the report and submits questions to the utility concerning unresolved issues. The answers to these questions become part of the report as an addendum. The preliminary safety analysis report and its addendum become part of the public record, and the review continues until DRL is satisfied that the plant can be built without undue risk to the public.

The application along with the DRL analysis is submitted to the Advisory Committee on Reactor Safeguards (ACRS). The ACRS is a committee established by law to advise the NRC on safety aspects of reactors. It is composed of scientists and engineers who are considered to be eminently qualified in the various fields of reactor technology. The committee's purpose is to provide disinterested and expert third-party opinion in matters of reactor safety.

After the ACRS has studied the applicant's PSAR, along with the DRL analysis, it releases a report to the public and to the NRC. A public hearing is then held, usually near the proposed reactor site, by a three-person NRC-appointed Atomic Safety and Licensing Board. Public "intervenors" may raise questions about the proposed plant at this point. After reviewing the testimony and the DRL and ACRS analysis, the board decides for or against the granting of a construction permit for the reactor.

The construction permit is provisional; complete technical data will be required later to complete the safety analysis. Moreover, as construction proceeds, periodic inspections take place by the NRC's compliance division to assure that the requirements of the construction permit are maintained.

When the final design of the plant is completed and operational plans have been completed, the applicant submits a final safety analysis report (FSAR). Once again evaluations are made by DRL and independently by the ACRS. Based on these analyses the NRC decides whether an operating permit should be granted. During both construction and operation of the plant inspections are made by the DRL to assure that the reactor is built, maintained, and operated in compliance with the license.

Radiological Criteria

The criteria by which the DRL evaluates the consequences of accidents are contained in 10 CFR 20 and 100. In 10 CFR 20 the limits of the release of radioactive materials that must be maintained during planned operation are specified. Reactor Siting Criteria, largely determined by the release of radioactive material under severe accident conditions, are contained in 10 CFR 100.

The radiological criteria contained in 10 CFR 20, pertaining to normal plant operation, place limits on the amount of radioactivity released in three categories:

1. Radioiodine and particulate nuclides with half-lives greater than 8 days.
2. Noble gases and elements of similar behavior.
3. Tritium and fission and corrosion products released with liquid effluents.

The consequences of less severe accidents can be measured with respect to the amounts of these effluents that are permissible under nonaccident conditions. The criteria in 10 CFR 100 are more relevant to the consequences of severe accidents.

The Reactor Siting Criteria, 10 CFR 100, express radiological limits in terms of an "exclusion area," a "low population zone," and a "population center distance." These are described as follows:

(a) "Exclusion area" means that area surrounding the reactor, in which the reactor licensee has the authority to determine all activities including exclusion or removal of personnel and property from the area. This area may be traversed by a highway, railroad, or waterway, provided these are not so close to the facility as to interfere with normal operations of the facility and provided appropriate and effective arrangements are made to control traffic on the highway, railroad, or waterway, in case of emergency, to protect the public health and safety. Residence within the exclusion area shall normally be prohibited. In any event, residents shall be subject to ready removal in case of necessity. Activities unrelated to operation of the reactor may be permitted in an exclusion area under appropriate limitations, provided that no significant hazards to the public health and safety will result.

(b) "Low population zone" means the area immediately surrounding the exclusion area which contains residents, the total number and density of which are such that there is a reasonable probability that appropriate protective measures could be taken in their behalf in the event of a serious accident. These guides do not specify a permissible population density or total population within this zone because the situation may vary from case to case. Whether a specific number of people can, for example, be evacuated from a specific area, or instructed to take shelter, on a timely basis will depend on many factors such as location, number and size of highways, scope and extent of advance planning, and actual distribution of residents within the area.

(c) "Population center distance" means the distance from the reactor to the nearest boundary of a densely populated center containing more than about 25,000 residents.

To meet the criteria of 10 CFR 100 the license applicant must convince the DRL that for the worst accidents considered (i.e., the design basis accidents) the following criteria are met:

(1) An exclusion area of such size that an individual located at any point on its boundary for two hours immediately following onset of the postulated fission product release would not receive a total radiation dose to the whole body in excess of 25 rem or a total radiation dose in excess of 300 rem to the thyroid from iodine exposure.

(2) A low population zone of such size that an individual located at any point on its outer boundary who is exposed to the radioactive cloud resulting from the postulated fission product release (during the entire period of its passage) would not receive a total radiation dose to the whole body in excess of 25 rem or a total radiation dose in excess of 300 rem to the thyroid from iodine exposure.

(3) A population center distance of at least one and one-third times the distance from the reactor to the outer boundary of the low population zone. In applying this guide due consideration should be given to the population distribution within the population center.

Note that the criteria are given in terms of individual doses at the exclusion area and low population zone boundaries. The assumption, which is normally quite conservative, is that if these individual dose criteria are met, the cumulative dose of the whole population will be very small, for normal population distribution. Low cumulative doses are further assured by the population center distance requirement. A greater distance than the one-and-one-third times the low population zone distance may be required by the DRL where large cities are involved. Moreover, in more densely settled regions population distribution data may be incorporated into atmospheric dispersion calculations to estimate both the cumulative dose within the low population zone and the dose to the total population.

Safety Evaluation Criteria

The philosophy by which reactor safety is assessed is closely tied to the contents required by the NRC to appear in the PSAR and FSAR and by the yardstick used by the DRL staff in evaluating these reports. This evaluation in the United States is carried out on the basis of the defense-in-depth concepts (U.S. Atomic Energy Commission, 1973). The regulatory staff assures that the plant design is within the regulations and provides a multiplicity of lines of defense against accidents. The first line of defense is that of accident prevention. The plant must be designed, constructed, and operated so that the probability of an accident is very small. Second, a plant protection system is required to ensure that safe conditions are not exceeded during normal operation or as a result of operational transients or other anticipated occurrences. Finally, improbable accidents are hypothesized to occur, and it must be demonstrated that consequences can be mitigated through engineered safety features incorporated into the plant design to the point that the public would not be subjected to undue risk.

The implementation of the defense-in-depth policy may be better understood by examining the required contents of the safety analysis reports submitted in application for construction permits and operating licenses. The format for these reports has been standardized; the required chapter headings are shown in Table 2-2. These headings are applicable to both

TABLE 2-2 Content of a Safety Analysis Report

1. Introduction and general plant description	10. Steam and power conversion system
2. Site characteristics	11. Radioactive waste management
3. Design criteria	12. Radiation protection
4. Reactor	13. Conduct of operations
5. Reactor coolant system	14. Initial tests and operation
6. Engineered safety features	15. Accident analysis
7. Instrumentation and controls	16. Technical specifications
8. Electric power	17. Quality assurance
9. Auxiliary systems	

preliminary and final safety analysis reports; the difference in the contents of PSAR's and FSAR's is primarily that the PSAR may not contain some supporting data that is obtained during the period between the submittal of the PSAR and the FSAR. (Both types of reports are referred to as SAR's.)

Design Criteria. Emphasis in the SAR is placed on demonstrating that the series of generalized design criteria contained in Appendix A of 10 CFR 50 are met, thereby making the probability of an accident very low and providing for the mitigation of consequences should an accident occur. The

64 design criteria contained in Appendix A for water-cooled reactors are divided into six groups:

   I. Overall Requirements.
  II. Protection of Multiple Fission Product Barriers.
 III. Protection by Reactivity Control Systems.
 IV. Fluid Systems.
  V. Reactor Containment.
 VI. Fuel and Radioactivity.

The criteria are for the most part stated qualitatively. They set functional and other requirements that serve to enumerate the safety-related features that must be met to avoid an overlooked weakness in the plant. An example best illustrates the nature of these criteria:

> *Criterion 34*—Residual heat removal. A system to remove residual heat shall be provided. The system safety function shall be to transfer fission product decay heat and other residual heat from the reactor core at a rate such that specified acceptable fuel design limits and the design conditions of the reactor coolant pressure boundary are not exceeded.
>
> Suitable redundancy in components and features, and suitable interconnections, leak detection, and isolation capabilities shall be provided to assure that for onsite electric power system operation (assuming offsite power is not available) and for offsite electric power system operation (assuming onsite power is not available) the system safety function can be accomplished, assuming a single failure.

In addition to the 64 NRC-specified design criteria, other design criteria are contained in Chapter 3 of a SAR to represent a broad frame of reference within which more detailed design effort is to proceed and against which design objectives will be judged. Substantial attention is given to formulating criteria to protect against site-induced accidents: earthquake, wind and tornado, flood, missiles. The basic data used in the consideration of such accidents is contained in Chapter 2 of the PSAR, Site Characteristics, where the geography, demography, meteorology, hydrology, geology, and seismology of the site are described.

    Chapters 4 to 12 provide a detailed description of the nuclear power plant with emphasis on demonstrating that the design meets the appropriate criteria under all planned operating conditions and that safety systems can be expected to function properly under accident conditions. Chapters 13 and 14 concern operational procedures both initially, to verify that the plant has been constructed to meet the design criteria, and later, to assure that the plant is operated by a technically competent and safety-oriented staff.

Safety Analysis. The accident analysis constituting Chapter 15 of the SAR is the focal point of the safety analysis of the reactor system. In this analysis all possible events that could lead to an accident sequence are enumerated. The accidents are then analyzed, drawing on the data given in the other chapters, to show that the reactor meets the design criteria and that the consequences of the accidents are within acceptable limits. Chapter 16 provides quantitative specifications on the operation of the plant to assure that no operation occurs that has not been thoroughly analyzed in the safety analysis. Chapter 17 ensures that the plant will be constructed to meet the criteria built into the design.

The safety analysis report for a typical power reactor is massive, typically containing from 10 to 20 volumes of material, Even then it does not contain all of the safety-related material. Extensive use is made of standards and codes developed by the ASME, IEEE, ASTI, and other professional engineering organizations to ensure adequate design of pressure vessels, piping, instrumentation, and other components. In addition, generic reports are sometimes submitted to the NRC by reactor vendors detailing designs and methods of analysis. These reports make unnecessary the reproduction of information that may be common to more than one nuclear power plant in each new SAR. This trend is being accelerated by the use of Reference Safety Analysis reports in which most of the chapters are applicable to all plants with major components of a standard design.

At present, probabilistic risk analysis is not used explicitly in a quantitative evaluation of the safety analysis reports. Considerations frequently found in quantitative risk analysis often appear in the safety reports and their evaluation, however, particularly the concern that the more severe the consequences of an accident are, the smaller its probability of occurrence be made.

Because it is not possible to consider all possible accidents in detail, a spectrum of accidents or potential accidents is chosen for analysis. The selected accidents serve as a bounding envelope in likelihood and consequence to indicate in a qualitative way the risk incurred from the power plant. These accidents may be classed into three categories (U.S. Atomic Energy Commission, 1973):

1. Events of moderate frequency (anticipated operational occurrences) leading to no abnormal radioactive releases from the facility.
2. Events of small probability with the potential for small radioactive release from the facility.
3. Potentially severe accidents of extremely low probability, postulated to establish performance requirements of engineered safety features and used in evaluating the acceptability of the facility site.

The events in category 1 may reasonably be expected to occur during the life of any given reactor. They include such things as turbine trips, loss of off-site power, loss of a single reactor coolant pump or of a feedwater pump, and so on. From an economic as well as a safety viewpoint these events of moderate frequency should result in a plant shutdown with no radioactive release from the reactor fuel. The plant should be capable of returning to power after the necessary corrective actions are taken. There should be no radioactive release to the environment beyond the limits established for normal operation.

The second category of accidents consists of less likely but more serious failures, such as complete loss of normal forced-coolant flow, small breaches of the primary system, and major leaks from radioactive decay storage tanks. These failures would normally not be expected to occur during the lifetime of any given plant, but might be expected to occur somewhere in the industry sometime over a 30- to 40-year time span. Evaluation of this category of accidents must demonstrate that the safety system and containment barriers will function effectively under accident conditions to eliminate—or at least reduce to an insignificant level—the potential for releasing radioactive material to the environment. These accidents may be sufficiently serious to prevent the resumption of plant operation for an extended time because of the potential for failure of some of the fuel elements and the consequent requirement for replacement and clean-up. Although such accidents may cause a radioactive release somewhat in excess of normal operating levels permitted by 10 CFR 20, the safety and containment features of the plant must provide for minimal radiological consequences compared to the standard set for the 10 CFR 100.

The third category of accidents consists of highly unlikely accidents such as major ruptures of reactor coolant pipes or of secondary system steam lines, severe nuclear transients, and so on. These accidents are postulated to occur even though steps are taken to prevent them, and although they are extremely unlikely to occur, protection against them provides additional defense in depth. These hypothetical events are analyzed using pessimistic assumptions often with inconsistencies introduced to further assure that the analysis will pessimistically bound any more realistic analysis. These accidents provide the most adverse conditions under which the plant safety and containment features must operate. They normally provide the limits for which these systems must be designed, and they are therefore referred to as design basis accidents. Extensive fuel failure and other plant damage may result from such hypothesized design basis accidents. It must be shown, however, that the radiological consequences of the accident are within the guidelines set in the siting criteria 10 CFR 100.

Further conservatism is built into the analysis of design basis accidents by assuming degraded performance of safety systems and sometimes by deliberately introducing physical inconsistencies into the calculations. The object is to provide a pessimistic envelope on the entire spectrum of accidents for which the safety and containment systems must effectively function.

The manner in which the pessimism is introduced into accident analysis is best illustrated by an example. In water-cooled reactors the severest of the design basis accidents normally results from a hypothesized guillotine rupture of a reactor coolant loop. In analysis of this loss-of-coolant accident it must be demonstrated that the emergency core-cooling system can remove the decay heat following the accident. Extensive cladding failure may take place, but it must be shown that the emergency cooling system is designed to prevent the loss of cooling geometry that could lead to a subsequent melting of the reactor core. This must be done, moreover, using pessimistic models for heat transport, materials failure, and so on, and under the assumption that the cooling system performance is degraded by the loss of off-site power and by the failure of the single most important component of the emergency core-cooling system.

An inconsistency is introduced into the calculations in evaluating the containment structure and associated fission product removal system. Even though the emergency core-cooling system is designed to prevent core meltdown, fission product releases to the containment are assumed at a level that could result only from melting of the core—100% of the noble gases, 25% of the halogens, and 1% of all other fission products. Pessimistic assumptions are again made concerning the containment leak rate and the performance of the fission product removal systems, including the failure of off-site power and the failure of the single most important safety system component. The fission product release rate to the atmosphere is determined, and the most adverse weather conditions are then assumed in a meteorological model for atmospheric dispersion of the fission products. Under these conditions the thyroid and whole-body dose to persons living on the exclusion area and low population zone boundaries must be shown to be within the limits specified by 10 CFR 100.

Probabilistic Considerations.   From the preceding discussion it can be seen that the various transients and accidents analyzed in safety assessment procedures in some sense can be placed on a probability–consequence diagram and a limit line drawn for the acceptable radiological consequences for accidents of various frequencies of occurrence. With increasing category number the acceptable accident probability decreases, and the consequences increase, the radiological limits ranging between the guidelines of

10 CFR 20 and 10 CFR 100, and the probability from everyday occurrences to the very remote possibility of a design basis accident. Thus far no mention has been made of accidents that lie beyond the design basis and therefore could result in radiological consequences beyond the limits stipulated by 10 CFR 100. Although the likelihood of such an accident is extremely remote, it cannot be made impossible because several combinations of physically plausible circumstances could lead to a core meltdown and subsequent breach of the primary and containment systems. In water-cooled reactors two such mechanisms would be the fast fracture of the reactor pressure vessel and a large break of a reactor coolant loop followed by a failure of the emergency core-cooling system. The present safety

TABLE 2-3   Reactor Facility Classification of Postulated Accidents and Occurrences (U.S. Atomic Energy Commission, 1973.)

Class Number	Description	Example(s)
1	Trivial incidents	Small spills Small leaks inside containment
2	Miscellaneous small releases outside containment	Spills Leaks and pipe breaks
3	Radwaste system failures	Equipment failure Serious malfunction or human error
4	Events that release radioactivity into the primary system	Fuel defects during normal operation Transients outside expected range of variables
5	Events that release radioactivity into the secondary system	Class 4 and heat exchanger leak
6	Refueling accidents inside containment	Drop fuel element Drop heavy object onto fuel Mechanical malfunction or loss of cooling in transfer tube
7	Accidents to spent fuel outside containment	Drop fuel element Drop heavy object onto fuel Drop shielding cask—loss of cooling to cask, transportation incident on site
8	Accident initiation events considered in design basis evaluation in the safety analysis report	Reactivity transient Rupture of primary piping Flow decrease—steamline break
9	Hypothetical sequences of failures more severe than Class 8	Successive failures of multiple barriers normally provided and maintained

objective of the Nuclear Regulatory Commission is to assure that all accidents not included in the design basis envelope should have an average recurrence interval of not less than 1000 years for all nuclear plants combined (U.S. Atomic Energy Commission, 1973a). Assuming that there will be on the order of 1000 plants in the United States by the year 2000, this criterion would require that the probability of an accident exceeding the 10 CFR 100 guidelines be less than one chance in a million per year of reactor operation.

In the United States these accidents that lie beyond the design basis have come to be referred to as Class 9 accidents. This stems from the classification of accidents shown in Table 2-3. In this classification, presently used in the preparation of environmental impact statements, accident consequences are categorized in terms of the penetration of radioactive materials through the fission product barriers.

2-4 RELIABILITY CONSIDERATIONS

A major part of reactor safety assessment is concerned with the reliability of the structures, systems, and components that make up nuclear power plants. A high degree of reliability must be achieved both in the fission product barriers and in heat transport, reactivity control, and other systems required for plant operations. Only then can the incidence of faults with the potential for initiating accidents be kept at an acceptably low level. Similarly, high reliability is required of the safety systems incorporated into the plant to mitigate the consequences of accidents if they should occur, because the probability of an accident whose consequences are aggravated by the failure of one or more safety systems must be reduced to an extremely small value. In this section we focus attention primarily on the reliability of safety systems. Plant design reliability for the purpose of reducing the frequency of accident-initiating failures is not addressed explicitly in what follows, because such considerations appear in subsequent chapters in conjunction with specific accident analyses.

The nomenclature regarding safety systems varies substantially (O'Brien, and Walker, 1974). For our purposes it is convenient to distinguish between three types of safety systems. The protection system encompasses the sensors, instrumentation channels, logic circuits and all other circuitry, electrical and mechanical devices involved in producing the signals that actuate the shutdown (or trip) system, and the engineered safety systems (alternately referred to as engineered safeguards or engineered safety features). The shutdown system consists of the electrical and mechanical devices actuated by the protection system signal for emergency reactivity reduction. Control rods are the most commonly used method of reactivity reduction. The engineered safety systems include all safety systems, other than the shutdown system, that are actuated by the protection system. Systems for emergency core-cooling, containment isolation, emergency power, containment depressurization, and fission product removal from the containment atmosphere are among the more common engineered safety systems. Passive structures such as the containment vessel and core-catching devices are also safety systems in a broad sense. The following discussion, however, is limited to the protection, shutdown, and active engineered safety systems.

For a safety system to be reliable, it must possess three characteristics: functional capability, availability, and serviceability. Functional capability implies that if a system functions correctly (i.e., according to design), it will be capable of accomplishing its assigned task. Availability is the probability that the system will operate correctly when called on to do so. Serviceability requires that the system will not function spuriously when there has been no requirement for its operation. Functional capability is associated with system performance requirements, availability with the low probability of unsafe system failures, and serviceability with the absence of safe failures.

The three concepts are simply illustrated by using a control rod shutdown system as an example. A performance requirement for the control rod system is that it shut down completely the chain reaction when the rods are fully inserted into the core. If the control rods contain insufficient neutron absorber to accomplish this task, the system is not functionally capable. If at a given time the failure of one or more components of the shutdown system makes it impossible for the rods to be driven into the core on demand, the unsafe failure has caused the system to be unavailable to shut down the reactor. The shutdown system is considered serviceable if it does not interfere with the normal operation of the plant. To achieve serviceability the probability of fail-safe faults that cause the rods to be spuriously inserted into the core when there is no need for reactor shutdown must be made small. We first take up functional capability and then discuss the reliability concepts that are closely related to the achievement of availability and serviceability: redundance, coincidence, maintenance, and diversity.

Functional Capabilities

The performance requirements (or functional capabilities) of safety systems most often are set by specifying the mitigating effects that they must have on the wide variety of accidents that may befall the reactor. These accidents are hypothesized by first enumerating the various classes of events—reactivity insertions, flow failures, primary system breaks, and so on. Within each class a spectrum of accidents is chosen for consideration, with the severest accident determined by one of two criteria. Either a more severe accident may be physically impossible, or it may be argued that although a more severe accident is possible, it is so improbable as not to affect significantly the public risk even though the safety systems may be incapable of dealing with it. The performance requirements for the safety systems are typically stated by specifying the limits on the levels of damage acceptable to the fission product barriers, the limits of permissible releases of radioactive materials through the fission product barriers, and other related criteria. Because the performance criteria are most often set by the severest accident in each class, these often are referred to as design basis accidents.

To illustrate, the performance requirements on emergency core-cooling systems for water-cooled reactors are largely determined from the analysis of the loss-of-coolant accidents that result from ruptures of the primary system coolant loops. For these accidents the emergency core-cooling system must prevent the reactor core from reaching such high temperatures that the cladding will fail structurally. Although some cracking in the cladding, causing some of the fission products to leak out, is considered acceptable for the more severe accidents, structural failures cannot be tolerated because they could cause the highly optimized heat transport configuration of the core to be lost; decay heat then could not be removed, and a core meltdown would result. The performance criterion that the coolable geometry be maintained is applied to a spectrum of coolant-loss accidents, including the rupture of any pipes in the primary system ranging from those of smallest diameter up to and including an instantaneous circumferential fracture of the largest-diameter coolant pipes. The emergency core-cooling system is not, however, required to protect against the most severe accident, the fracture of the reactor vessel itself. This event is argued to have been made sufficiently improbable, by reliable vessel design and operation, that mitigation of the consequences of the accident through emergency core cooling is not required.

Accident Detection. The performance requirements for the protection system are that all accidents within the design basis be unequivocally detected and that the signals be delivered to the actuators of the shutdown system, and appropriate engineered safety systems. These signals must be delivered in sufficient time to assure the successful operation of the systems. At the same time the protection system must not interfere with normal plant operations by causing shutdown or safety system actuation when no accident has occurred. The unequivocal detection of accidents requires that measureable plant parameters that will indicate each of the possible initiating events be isolated. Safety limits that must not be exceeded if unacceptable consequences are not to result are then fixed on these parameters, and these limits become the trip points to actuate the shutdown system and appropriate engineered safety systems.

The choice of parameters and trip points is difficult, because many of the temperatures, power densities or other qualities that give the most direct indications of trouble are not directly measurable. Thus, these quantities must be inferred from neutron flux, core inlet and outlet temperature, primary system pressure, and other measurements. Precision and response time of the parameters in detecting the accident often become particularly important considerations in the detection process. For example, the power level of a reactor may be most accurately determined from the average coolant temperature rise through the core. And this indeed may give the best indication of an overpower transient that only slightly exceeds the normal reactor power. If the transient develops rapidly, however, the heat transport time from the fuel to the core outlet causes this measurement to lag the reactor power by too long a time span for effectively terminating the accident by fast shutdown. Alternately, neutron flux can be used to detect the over power transient without time delay, because the flux is proportional to the instantaneous fission rate. Flux measurements, however, are localized within the core and thus are susceptible to variation in space caused by changes in the power distribution either during normal operation or during the transient. As a result, the flux measurements are not as precise, and thus trip levels must be set further from the normal full power value to prevent spurious shutdown of the system. Finally, if the overpower is very localized in the form of an excessive power peak caused by flow blockage, fuel loading error, or some other such phenomenon, neither the flux nor average core outlet temperature may provide clear warning of the situation. In such instances, local outlet temperature measurements may be necessary, or alternately, detectors of excessive fission product activity or delayed neutrons in the coolant stream may be required to detect the onset of local fuel failures resulting from the power peak.

Similar difficulties arise in determining methods for detecting flow failures, loss-of-heat sink, and other classes of accidents in time to take effective protective action. As a result, anticipatory variables may be used. For example, the voltage and/or speed of coolant pumps may be sensed, and the reactor shut down if these variables indicate that the pumps are not functioning properly. Likewise, loss of the load to the electrical generator in most plants is used on a trip signal for reactor shutdown. A further complication arises from the fact that it is often necessary to vary the trip settings with reactor power or other operating conditions, and a particular difficulty—discussed in Chapters 3 and 5—appears when the reactor is at zero-power level where only the neutron flux measurements can be used to monitor the state of the reactor.

Protection Logic. Once sensors and corresponding instrument channels, which will produce appropriate signals for all accidents, have been specified, it must be decided how to arrange the protection system logic so that the correct systems are actuated. To illustrate some of the considerations, suppose we consider a hypothetical reactor in which the following systems are actuated by the protection system:

1. Reactor shutdown.
2. Emergency power supplies.
3. Emergency core cooling.
4. Containment isolation.
5. Containment spray (for fission product removal or depressurization).

In Fig. 2-9 a simplified logic diagram is shown for sensing plant parameters and actuating appropriate safety systems.

Any one or more of the plant parameters entering an unacceptable range of values will trip the reactor and cause an orderly shutdown of the nuclear power plant. For many situations, however, actuation of all the other safety systems may be neither necessary nor desirable. Some cooling failure accidents may require the use of the emergency core-cooling system. The loss-of-coolant accident, for example, would be detected by primary system

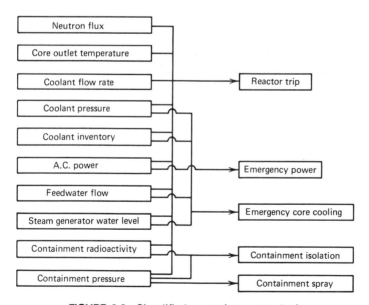

FIGURE 2-9 Simplified protection system logic.

depressurization or loss of the primary system coolant inventory. A reactor trip may be adequate for other cooling failure transients. For example, if the forced coolant circulation is lost, decreasing the reactor coolant flow, natural convection may be adequate to remove the decay heat following the reactor shutdown. The actuation of an engineered safety system may actually have adverse safety effects in some situations. For example, if an overpower transient causes the reactor flux and core outlet temperature to rise, the reactor must be shut down, but it may be undesirable to actuate the emergency core-cooling system, because the ejection of cool water into the primary system might then cause undesirable thermal shock from the ensuing pressure and temperature transients.

Provided a transient does not result in a breach of the primary system envelope, it may be inadvisable to isolate the containment if such action causes a disruption of the normal heat transport paths from steam generators to turbine-condenser, thereby requiring the use of emergency cooling. Similarly, the use of the containment spray may require an extended plant shutdown following the accident for clean-up purposes. Thus in the diagram the spray is actuated only if a loss-of-coolant accident, steam-line break, or other such accident causes containment pressurization.

The reactor shutdown and engineering safety systems may also be actuated manually, permitting operator judgement to be used in dealing with slower and less severe situations. If the radioactive level, but not the pressure of the containment were to rise because of a spent fuel handling accident outside the primary system, for example, reactor shutdown and containment isolation might be sufficient, for the normal filtration system may be capable of removing the activity from the containment atmosphere. On the other hand, the operator has the option of using the containment spray to remove the activity rapidly, realizing that it will result in a longer plant shutdown, for purposes of cleanup.

The logical structure of a real protection system is likely to be substantially more complex than is indicated in Fig. 2-9. As examples, there may be a backup as well as a primary reactor shutdown system. Different trip signals may be required, depending on whether zero-power testing or intermediate- or full-power operation is taking place. The emergency core-cooling system may be designed to take different actions, depending on whether a large or small primary system break or a failure in the secondary system heat transport path is indicated. There may also be a number of standby actions tied to sensor information. In the event of an accident, emergency diesel generators may be started up but not connected to the A.C. power grid, so that they are immediately available should off-site power be disrupted. Similarly, pumps in emergency cooling or containment spray systems may be started, requiring only the later opening of valves should the systems be needed to deal with the transient. For accidents confined to the primary system the containment may be partially isolated, plugging all penetrations (such as those used for maintenance and refueling) not required for immediate use, but leaving the steam and feedwater lines open for heat transport by the normal path. Finally, abnormal conditions that are not deemed to require complete reactor shutdown may be programmed to require a power cutback or at least a block so that reactivity cannot be added until the situation is remedied.

Regulation Versus Protection. From these considerations it is apparent that the logic of a protection system must be able to deal with many diverse events in order to assure that radioactivity will not be released in the event of an accident. At the same time the protection should not interfere with normal operation to the extent that plant economics suffer. In this regard the relationship of the protection system to the regulating systems that control the normal operation of the plant is one of the primary considerations in setting performance criteria for both systems (cf. Hagan, 1973; Pearson and Lennoy, 1964).

The regulating system provides operational control of the reactor behavior. Electrical demand on the generator is translated into a demand for a steam flow rate within narrow limits in pressure and temperature to the turbine. The regulating system must control reactor power level, coolant and feedwater flow, steam generator pressure, and other process variables to accomplish the orderly flow of energy through the primary and secondary systems to the turbine and condenser. These variables must be regulated to accommodate load following power demands, as well as constant power operations, without causing excess temperatures, pressures, heat fluxes, and so on, to be incurred at any point within the system. In addition, the regulating system is expected to provide for the orderly start-up and shutdown of the system, and to compensate for changes in the system such as those caused by fuel depletion, xenon buildup, programmed pump start-up, loop isolation, and so on. Although the slower operations such as programmed loop isolation may be carried out manually, with safety constraints imposed by mechanical or electrical interlocks, a substantial part of the regulating system is usually automated.

The details of the regulating system must be taken carefully into account in the design of the protection system, and, conversely, safety considerations place bounds on the accuracy and speed of the control system. Even though the two systems may share sensors and possibly other components, the protection system must be independent of the regulating system so that protective action will always override the regulating system to shut down the plant safely. The ability of the protection system to override regulatory actions is particularly important where accident conditions fool the regulating system sensors or logic into calling for increased power or decreased cooling levels or where the accident originates in a malfunction of the regulating system.

In general, the higher the performance requirements of the regulating system, the more difficult the design of an adequate protection system becomes. A high-performance regulating system that is capable of precisely maneuvering the power plant to follow rapid or large variations in electrical load tends to have features that are detrimental from a safety viewpoint. The rapid compensation for changes in core conditions to maintain the correct power level may mask deterioration of core components that is just cause for shutting the plant down. The extreme accuracy and time response in such a regulating system lead to requirements for increasingly complex sensors, logic circuitry, and so on. With this increase in complexity the regulating system tends to become less reliable and more susceptible to instabilities, thus increasing the likelihood of an accident induced by a regulating system failure. Moreover, in the event of a malfunction the resulting transient is likely to be of increased severity, because the high performance of the system typically requires the ability to make rapid increases in reactivity, sharp changes in feedwater flow rates, and so on. These, in turn, require early detection and swift action by the protection system if the transient is to be terminated before plant damage results.

Performance Requirements. The performance requirements on the shutdown system may be stated relatively simply. The shutdown system must be able to bring the reactor to a subcritical state within some specified period of time and to maintain it in that state. Failure to bring the reactor subcritical may result from underestimating the maximum amount of excess reactivity that is present in the core under any set of conditions or from not correctly calculating the reactivity worth of the control rods or other neutron poisons. Failure to bring the reactor subcritical fast enough may result from problems in estimating incremental worth because of the power-shaping considerations discussed in Section 3-4. Failure to insert rods fast enough, or at all in the event of an accident, may result from the inability to predict the conditions that suddenly develop in the core. Unanticipated distortion of the control rod sleeves, the generation of high pressure in the core, or other possible effects of rapidly developing transients could make the shutdown system functionally incapable if the possibility of such conditions is not accounted for in the design.

The performance criteria for the engineered safety features tend to be substantially more difficult to set than for the shutdown system. The safety features involve active components that often require substantial amounts of A.C. power. They may be required to come on line very rapidly even though off-site power has been lost, therefore causing their operation to be delayed until emergency power sources can be brought to full strength. At the same time most safety features are required to operate for long periods of time if not indefinitely, because they may be required for the removal of decay heat. The conditions in the power plant during the course of an extended transient may be difficult to calculate with any precision, and therefore wide design margins may be necessary to ensure that functional capability is met even under the severest of transients in pressure, temperature, and other variables.

Some of these difficulties in setting performance criteria are illustrated by using as an example the water reactor emergency core-cooling system discussed earlier in conjunction with setting a design basis. The pressure and temperature transients in the reactor core during loss-of-coolant accidents are difficult to predict, making the performance criteria on pressure head, flow rates, and other parameters subject to uncertainty. Cooling may be required before pumps can be started, requiring the use of gas accumulator or other stored-energy devices to force the first few seconds of cooling. Even then, the power requirements for the cooling pumps at later times must be considered carefully in terms of the ability of the emergency power system to shed nonessential loads and pick up the required engineered safety features without overloading. Once in operation the cooling must continue indefinitely to prevent core meltdown. This requires an assured reservoir of cooling water, and often requires operator judgement in shifting from one reservoir to another, or in shifting to a recirculation mode in which water lost from the primary system is collected, recirculated through a heat exchanger, and pumped back into the emergency cooling system.

Some of the difficulties in assuring the functional capabilities of the safety system are apparent from the foregoing discussion. A thorough dynamic analysis is in order for the entire spectrum of design basis accidents, because the lack of understanding of how the nuclear plant will behave during these accidents can lead to any one of a number of design errors. These errors can render the safety system incapable of performing its functions even though the system operates as designed. The designer may fail totally to consider a particular accident for which protection is required. An accident may proceed to a point where the shutdown or engineered safety systems are inadequate because the sensors provided are incapable of detecting the accident at an early enough stage. The designer may fail to recognize all the safety actions necessary to maintain the damage level within the design requirements. Or the unanticipated severity of the reactor conditions during the course of the accident may render the safety system incapable of carrying out its functions.

Reliability

To discuss the quantitative estimates of the likelihood of failure of either safety or operating systems because of component or subsystem failures, it is necessary to introduce some elementary statistical concepts. In doing so we first examine the reliability of a system as a whole, and then examine its behavior in terms of subsystems or components in order to understand the effects of redundancy, coincidence, and diversity.

Failure Probability.   Suppose we let

$\mathscr{P}(t)$ = probability that a system that is operational at time 0 will have failed by time t

This is a cumulative probability distribution that has the properties $\mathscr{P}(0) = 0$, and $\mathscr{P}(\infty) = 1$. Most often reliability data is available not in terms of $\mathscr{P}(t)$ but of $\theta(t)$, the failure rate, defined as follows:

$\theta(t) \, dt$ = conditional probability that a system that is operational at t will fail between t and $t + dt$.

The failure rate may be used to determine the change in $\mathscr{P}(t)$ between t and $t + dt$. The probability of a system failure by $t + dt$ is just equal to the probability of a failure by t plus the probability of failure between t and $t + dt$. The probability of failure between t and $t + dt$ is just equal to $1 - \mathscr{P}(t)$, the probability that the system has not failed by time t, times $\theta(t) \, dt$, the probability that an unfailed system will fail between t and $t + dt$. Thus we have

$$\mathscr{P}(t + dt) = \mathscr{P}(t) + [1 - \mathscr{P}(t)]\theta(t) \, dt. \qquad (2\text{-}8)$$

Using the definition of the derivative, we obtain the differential equation

$$\frac{d\mathscr{P}(t)}{dt} = [1 - \mathscr{P}(t)]\theta(t). \qquad (2\text{-}9)$$

Solving for $\mathscr{P}(t)$, with $P(0) = 0$, yields

$$\mathscr{P}(t) = 1 - \exp\left(-\int_0^t \theta(t') \, dt'\right), \qquad (2\text{-}10)$$

which is the required relationship.

A third quantity of interest is the probability distribution function

$f(t) \, dt$ = probability that a system that is operational at $t = 0$ will fail between t and $t + dt$.

Obviously,

$$f(t) \, dt = \mathcal{P}(t + dt) - \mathcal{P}(t), \qquad (2\text{-}11)$$

or

$$f(t) = \frac{d}{dt} \mathcal{P}(t). \qquad (2\text{-}12)$$

Therefore, using Eq. (2-10), we may also write

$$f(t) = \theta(t) \exp \left(- \int_0^t \theta(t') \, dt' \right). \qquad (2\text{-}13)$$

From the preceding, it is clear that $\mathcal{P}(t)$ and $f(t)$ can be estimated once sufficient failure rate data is available to determine $\theta(t)$. Typical engineering devices tend to have failure rates with the general characteristics shown in Fig. 2-10. Three phases appear. In the first phase a high failure rate is expected because of production, test, or assembly faults. In the last phase the failure rate rises sharply because the device is wearing out. The intermediate phase, characterized by a low failure rate, is often termed the *useful life* of the device. Most equipment may be designed and operated to eliminate phases 1 and 3. Phase 1 is eliminated by preoperational testing and phase 3 by early replacement of components. Thus it is often possible to approximate $\theta(t)$ by a constant θ of relatively small magnitude, as shown, for

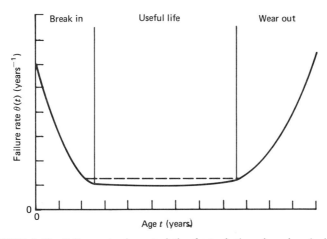

FIGURE 2-10   Failure rate characteristic of a typical engineering device.

example, by the dotted line in Fig. 2-10. With a constant value of θ we have

$$\mathscr{P}(t) = 1 - e^{-\theta t} \qquad (2\text{-}14)$$

and

$$f(t) = \theta\, e^{-\theta t}. \qquad (2\text{-}15)$$

Finally, when equipment is made sufficiently reliable that the probability of failure over the period of time that it is used in a nuclear plant is small, we have

$$\mathscr{P}(t) \simeq \theta t \qquad (2\text{-}16)$$

and

$$f(t) \simeq \theta. \qquad (2\text{-}17)$$

The preceding expressions for $\mathscr{P}(t)$ are most useful in predicting the probability that an operating system will fail over some period of reactor operation. For example, if θ is the failure rate for loss of off-site power in failures per year, then Eqs. (2-10), (2-14), or (2-16) yield the probability that a loss-of-off-site power accident will occur within t years of reactor operation.

Availability, Testing, and Repair.   For safety systems a different reliability parameter is of more importance than $\mathscr{P}(t)$. It is the unavailability or fractional dead time:

Ξ = fraction of reactor operating time during which a system is in a failed state.

The importance of this parameter is attached to the fact that most safety systems are not required during normal operation, but only in the event of an accident. If a safety system responds correctly when its action is demanded, it is said to be in an *operational* state; if it does not respond correctly, it is said to be in a *failed* state. If accidents are infrequent and randomly distributed over the operating cycle of a reactor, the probability of a safety system failure per demand is just equal to the fraction of operating time that the system is in a failed state: The probability of a failure per demand is just equal to Ξ.

To find Ξ in terms of θ, we first specify a time of operation τ beginning at $t = 0$. If a failure takes place at t, the system will be in a failed state for a fraction $(1 - t/\tau)$ of the total operating time. Since $f(t)\, dt$ is the probability that the failure will take place between t and $t + dt$, failures during this time interval will make a contribution of

$$\left(1 - \frac{t}{\tau}\right) f(t)\, dt \tag{2-18}$$

to the unavailability. Integrating over failures at all times, we have for the unavailability

$$\Xi = \int_0^\tau \left(1 - \frac{t}{\tau}\right) f(t)\, dt. \tag{2-19}$$

Using Eq. (2-12), and integrating by parts, we may simplify this expression to

$$\Xi = \frac{1}{\tau} \int_0^\tau \mathscr{P}(t)\, dt. \tag{2-20}$$

If the failure rate is sufficiently small that $\theta\tau \ll 1$, Eq. (2-16) can be used for $\mathscr{P}(t)$, and we have

$$\Xi \simeq \frac{\theta\tau}{2}. \tag{2-21}$$

The unavailability of the safety systems is strongly affected by the procedures used for testing, monitoring, maintaining, and repairing the systems over the life of the plant. To illustrate these effects we consider two types of idealized systems. The first is one in which faults are not detected by continuous monitoring and therefore go unrevealed until a deliberate test of the system is performed. The unavailability of such a system may be expressed by Eq. (2-21) with the following modifications. If the reactor is shut down and the system is tested and repaired at a time interval τ, it may be assumed that it is renewed to a like-new state. Thus the unavailability is given by Eq. (2-20), or by Eq. (2-21) if $\theta\tau \ll 1$, where τ is now the test and repair interval instead of the total lifetime of the system.

In the preceding example it is assumed that the reactor is shut down during testing and repair. Therefore only the periods of time over which the system is in a state of unrevealed failure and not the test and repair times contribute to the unavailability. The second idealized type of system has the contrasting property that it is continuously monitored while in operation, and therefore neither unrevealed failure time not test time contribute to the unavailability. If the system is vital to reactor safety, the reactor would be shut down and the system repaired whenever a failure is detected. The availability would then be one. Often, however, safety systems are highly redundant, and it is permissible to continue operation even though one or more of the subsystems may remain in a failed state for some period of time before it is repaired. To estimate the effect of such a subsystem on the reliability of the overall safety system it is necessary to determine the availability of the subsystem.

The availability of a continuously monitored subsystem, for which faults are repaired while the reactor is on-line, may be estimated as follows. Suppose T is the time of reactor operation, and τ_r is the mean time required to repair a failure. Then if there are N failures during the time T, we have for the unavailability

$$\Xi = \frac{N\tau_r}{T} \qquad (2\text{-}22)$$

If the subsystem reliability is adequately described by a small constant failure rate θ, the number of failures is equal to θ multiplied by the time during which the system is in an unfailed state,

$$N = \theta(T - N\tau_r). \qquad (2\text{-}23)$$

Thus

$$N = \frac{\theta T}{1 + \theta\tau_r} \qquad (2\text{-}24)$$

provided $N\tau_r < T$. Combining Eqs. (2-22) and (2-24), we have for the unavailability

$$\Xi = \frac{\theta\tau_r}{1 + \theta\tau_r}. \qquad (2\text{-}25)$$

In the situation where the failure rate and repair times are small ($\theta\tau_r \ll 1$) we may simplify the estimate of the availability by reducing Eq. (2-25) to

$$\Xi \simeq \theta\tau_r. \qquad (2\text{-}26)$$

The two preceding examples are highly idealized in their assumptions. Frequently, subsystems have a more complex behavior. Failures may be unrevealed, requiring testing and repair. At the same time it may be undesirable to shut down the reactor for testing or repair, provided that the safety systems are sufficiently redundant. Therefore the time intervals during which unrevealed failures exist, during which testing takes place, and during which repairs are being made all contribute to the unavailability of the subsystem. Exact expressions for such situations are difficult to derive. If each of the contributors to the unavailability is small, however, the total unavailability may be estimated simply by adding the unavailabilities due to the separate contributions. The unavailabilities due to unrevealed faults and on-line repair are given by Eqs. (2-21) and (2-26), respectively. If the test time τ_t is small compared to the test interval τ, it contributes τ_t/τ to the unavailability. Therefore we have

$$\Xi \approx \frac{\theta\tau}{2} + \frac{\tau_t}{\tau} + \theta\tau_r. \tag{2-27}$$

Obviously the unavailability of such a subsystem is reduced by minimizing the failure rate, test time, and repair time. However, the smallest test interval τ is not the optimum, because if τ is chosen too small, the unavailability will be dominated by the testing time over which the subsystem is out of service. The optimum τ is determined by finding the minimum of Ξ with respect to changes in τ. We have

$$\tau = \left(\frac{2\tau_t}{\theta}\right)^{1/2}. \tag{2-28}$$

We then obtain for the minimum unavailability

$$\Xi = (2\theta\tau_t)^{1/2} + \theta\tau_r. \tag{2-29}$$

Other considerations also are pertinent to determining the testing interval. One such consideration is that too frequent testing of active components may lead to excessive wear, causing the failure rate to increase as indicated by Fig. 2-10. A second consideration is that the imperfect testing procedure may have a finite probability of causing system failure and that if such a procedure is applied too frequently, test-induced failures may dominate the unavailability.

Redundancy

Both essential operating systems and safety systems often must consist of redundant subsystems designed to meet performance requirements even though one or more of the subsystems may have failed. This is deemed necessary because the extremely small failure probabilities required to maintain both the incidence of accidents and the unavailability of safety systems at an acceptably low level may not be realistically achieved if a single component or subsystem failure can cause a failure of the total system. Indeed, safety system design criteria typically state that the failure of any single component or subsystem within a system shall not cause the total system to fail. In addition to satisfying these single-failure criteria, redundancy permits increased reliability by permitting testing and repair of redundant components while the reactor is on-line.

To discuss the system reliability in terms of the reliability of the redundant subsystems, we need to develop relationships between failure probabilities of the system and its subsystems (Green and Bourne, 1972). Suppose we have n independent but identical subsystems, and the probability that a single subsystem will fail is p. The probability that some particular combination of m of these subsystems will fail then is just

$$p^m(1-p)^{n-m}. \tag{2-30}$$

There are, however,

$$\binom{n}{m} \equiv \frac{n!}{m!(n-m)!} \tag{2-31}$$

different possible combinations of m subsystem failures (i.e., if $m = 2$ and $n = 3$ subsystems, 1 and 2, 2 and 3, or 1 and 3 may fail). Because each of these combinations is equally probable, the probability that any of the combinations of the m subsystems will fail is

$$\mathscr{P}_m = \binom{n}{m} p^m(1-p)^{n-m} \tag{2-32}$$

The probability that m or more subsystem failures will take place is then

$$\mathscr{P}_{\geq m} = \sum_{j=m}^{n} \binom{n}{j} p^j(1-p)^{n-j}. \tag{2-33}$$

Finally, if the subsystem failure probability is small, $\mathscr{P} \ll 1$, we may approximate this equation by

$$\mathscr{P}_{\geq m} \approx \binom{n}{m} p^m, \tag{2-34}$$

which is to say that the contribution from more than m failures can be neglected.

l/n Systems. It is common to denote redundant systems as (l/n) if the function of l or more of the subsystems meets the system performance criteria. The system failure probability is determined by noting that the system failure requires that at least

$$m = n - l + 1 \tag{2-35}$$

subsystems fail. Thus, for example, if p is small, the (l/n) system failure probability is given by

$$\mathscr{P}_{\geq n-l+1} = \binom{n}{n-l+1} p^{n-l+1}. \tag{2-36}$$

The choice of the (l/n) redundance used in an operating or safety system may depend on many factors. Probably $(\frac{1}{2})$ systems are most popular, because they meet the single failure criteria with a minimum number of subsystems. If it is desirable to have the capability to remove a subsystem from service for testing and repair during reactor operation, then at least a $(\frac{1}{3})$ system is required in order also to comply with the single failure criteria. Often other design considerations lead to a specification of more than two or three subsystems. Then $(n-1)/n$ or $(n-2)/n$ are likely to be used. For example, if operational specifications lead to the requirement for four independent control rod banks in a reactor, it may be required that any three banks be capable of shutting the reactor down [i.e., the shutdown system is $(\frac{3}{4})$]. In this case a $(\frac{2}{4})$ system would not be specified, because on-line testing of the control rods or their drives is normally not possible. A contrasting example is that of auxiliary feedwater pumps that constitute part of the emergency cooling system. If steam generator sizing considerations lead to a primary system with four coolant loops, it may be expedient just to have one auxiliary feedwater pump to feed each steam generator, or $n = 4$. In this situation, however, a $(\frac{2}{4})$ system is likely to be chosen, because this would permit the single failure criteria to be met, while at the same time allowing one of the pumps to be removed from service for maintenance or repair. The failure probability of an (l/n) system can be further decreased by making n much larger than l. The achievement of extremely small failure probabilities by this method, however, eventually runs into two limitations that are discussed later: serviceability limitations caused by fail-safe events and common-mode failures.

Estimates of the reliability parameters for redundant (l/n) systems can be made using the failure rate relationship derived in the preceding subsection. Suppose, by analogy to Eq. (2-16) we let

$$p(t) \simeq \theta t \qquad (2\text{-}37)$$

be the probability that a subsystem that is operational at $t = 0$ will have failed by time t. The probability that an (l/n) system that is in perfect order at $t = 0$ will have failed by time t is then

$$\mathcal{P}_{\geq n-l+1} = \binom{n}{n-l+1}(\theta t)^{n-l+1}. \qquad (2\text{-}38)$$

If a system is tested at time intervals τ for unrevealed failures, the unavailability or fractional dead time of an (l/n) system caused by unrevealed faults is found by substituting this equation into Eq. (2-20) to obtain

$$\Xi\left(\frac{l}{n}\right) = \frac{1}{\tau} \int_0^\tau \mathcal{P}_{\geq n-l+1}(t) \, dt, \qquad (2\text{-}39)$$

or

$$\Xi\left(\frac{l}{n}\right)=\left(\frac{n}{n-l+1}\right)\frac{(\theta\tau)^{n-l+1}}{n-l+2}. \qquad (2\text{-}40)$$

Simultaneous testing of the subsystems of a redundant safety system can be carried out only if the reactor is shut down during tests. Equation (2-40) is also applicable, however, if the tests on the subsystem are carried out successively one after another, as long as the total testing time for all the subsystems is extremely small compared to the test interval τ.

Further decreases in the unavailability may be achieved by using not simultaneous testing but some other sequence such as random or perfectly staggered testing of the redundant system. Results for some common (l/n) systems are given in Table 2-4. Note that for any testing pattern $\theta\tau$ is raised to the $n-l+1$ power; however, the coefficient of the unavailability expressions vary significantly.

TABLE 2-4 Unavailability as a Function of Logic Configuration and Testing Schedule

	Unannounced Unavailability		
Logic	Simultaneous Testing	Random Testing	Perfectly Staggered Testing
$\frac{1}{2}$	$\frac{1}{3}\theta^2\tau^2$	$\frac{1}{4}\theta^2\tau^2$	$\frac{5}{24}\theta^2\tau^2$
$\frac{2}{2}$	$\theta\tau$	$\theta\tau$	$\theta\tau$
$\frac{1}{3}$	$\frac{1}{4}\theta^3\tau^3$	$\frac{1}{8}\theta^3\tau^3$	$\frac{1}{12}\theta^3\tau^3$
$\frac{2}{3}$	$\theta^2\tau^2$	$\frac{3}{4}\theta^2\tau^2$	$\frac{2}{3}\theta^2\tau^2$
$\frac{3}{3}$	$\frac{3}{2}\theta\tau$	$\frac{3}{2}\theta\tau$	$\frac{3}{2}\theta\tau$
$\frac{1}{4}$	$\frac{1}{5}\theta^4\tau^4$	$\frac{1}{16}\theta^4\tau^4$	$\frac{251}{7680}\theta^4\tau^4$
$\frac{2}{4}$	$\theta^3\tau^3$	$\frac{1}{2}\theta^3\tau^3$	$\frac{3}{8}\theta^3\tau^3$
$\frac{3}{4}$	$2\theta^2\tau^2$	$\frac{3}{2}\theta^2\tau^2$	$\frac{11}{8}\theta^2\tau^2$

Marquis and Jacobs, 1973.

The situation is more complex when the testing and repair times as well as the times during which unrevealed faults exist contribute significantly to the unavailability of a redundant system. An indication of the form of the unavailability can be seen, however, by considering the case in which the testing is carried out on each subsystem with a mean interval τ, but in which the test on the various subsystems is randomly staggered. In these circumstances the unavailability, or probability that a particular subsystem is inoperable, is given approximately by Eq. (2-27). For present purposes we let this subsystem unavailability be denoted by

$$p = \frac{\theta\tau}{2} + \frac{\tau_t}{\tau} + \theta\tau_r. \tag{2-41}$$

Now assuming that the failures and the testing of the various subsystems are uncorrelated to one another, the probability that more than m subsystems are in a failed state is related to the probability that any one of the systems is in a failed state by Eq. (2-33). If we take $m = n - l + 1$, however, $\mathscr{P}_{\geq m}$ is then just the system unavailability, or, using Eq. (2-41),

$$\Xi\left(\frac{l}{n}\right) \simeq \left(\begin{array}{c} n \\ n-l+1 \end{array}\right)\left\{\frac{\theta\tau}{2} + \frac{\tau_t}{\tau} + \theta\tau_r\right\}^{n-l+1}, \tag{2-42}$$

where we have assumed that the subsystem unavailability, given by Eq. (2-41), is small compared to one.

Fail-Safe and Fail-to-Danger. Spurious operation of the system not only adversely affects plant economics by increasing the reactor down time, but may also have adverse safety implications since unwanted operation of some safety features may cause undue thermal shock and stress transients, interference with normal energy transport paths, and so on. Since the effect of such spurious failures most often is to shut down the plant safely, they are referred to as fail-safe modes.

Spurious operation of safety systems can originate from faults within either the protection system or the shutdown or engineered safety systems. For example, short circuits in motor-operated control rod drives or isolation valves could cause spurious operation. By and large, however, the most common reason for spurious operation is the generation of spurious signals within the protection system that reach the actuators of shutdown or engineered safety systems. Therefore it is in the design of the protection system that the most care must be taken to prevent failures-to-safety that are so frequent as to impede seriously the economic viability of the power plant. At the same time, reduction of fail-to-safety modes must not increase the likelihood of fail-to-danger modes to the point where the protection system reaches an unacceptable level of unavailability.

The trade-off between fail-safe and fail-to-danger behavior is made more difficult by the fact that many of the design alterations that can be made to reduce the fail-to-danger probabilies increase the fail-safety probability. The classic example of this relates to intrumentation power supplies. Power supply failures often contribute significantly to the total failure rate of instruments, logic circuits, and so on. Therefore such systems are often designed on the principle of deenergization. That is to say, a power supply failure does not cause the device to become incapable of generating or transmitting a signal, but rather causes it to transmit a signal. Thus power supply failures are eliminated as a contributor to unavailability, but added as contributions to fail-safe spurious events.

To maximize safety system availability while limiting the chance of spurious operation $(2/n)$ systems are popular in protection system logic. The trade-off between fail-to-danger and fail-safe events for various redundant configurations can be seen by examining (l/n) logic. Suppose p and p_s represent the probabilities of fail-to-danger and fail-safe events. From Eq. (2-36) we already have shown that the fail-to-danger probabilities of the system are

$$\mathcal{P}_{\geq n-l+1} \approx \binom{n}{n-l+1} p^{n-l+1}. \tag{2-43}$$

For the (l/n) system to fail-safe there must be l or more safe failures of subsystems. Thus the fail-safe probability is

$$\mathcal{P}_{\geq l} = \binom{n}{l} p_s^l. \tag{2-44}$$

The problem with the use of simple $(1/n)$ redundancy is readily seen. As n is increased the fail-to-danger probability decreases,

$$\mathcal{P}_{\geq n} = p^n, \tag{2-45}$$

but the fail-safe probability increases linearly with n,

$$\mathcal{P}_{\geq 1} = np_s \tag{2-46}$$

Thus unless p_s is very small the system may be dominated by spurious operation. On the other hand, a $(2/n)$ system yields a larger fail-to-danger probability,

$$\mathcal{P}_{\geq n-1} = np^{n-1}, \tag{2-47}$$

but the fail-safe probability is on the order of p_s^2:

$$\mathcal{P}_{\geq 2} = \frac{n(n-1)}{2} p_s^2.$$

When both p_s and p are small compared to one, $(2/n)$ logic often represents a satisfactory compromise.

In the foregoing discussion the safety systems are assumed to consist of redundant systems that are completely separated from one another. In practice, this often is not compatible with the complex logic required in the protection system. Coincidence requires, for example, that the independent instrument channels meet in common logic circuits. Likewise, if the protection system actuates redundant shutdown or engineered safety systems, it is desirable that the signal from any one of the redundant protection systems channels be capable of actuating all subsystems.

Even with these complications it is possible to build highly redundant systems that meet, for example, the single failure criteria. An example is shown in Fig. 2-11, where a protection system is designed so that either temperature or neutron flux sensors can actuate a redundant $(\frac{4}{5})$ shutdown system. The parallel structure of the system ensures that the failure of no single instrument, logic circuit, or other device can prevent a trip signal from reaching the control rod actuators. The probabilistic analysis of such complex systems to estimate the availability of the system in terms of the component failure data is beyond the scope of this text. However, fault trees that utilize truth tables, Boolean algebra, and particularly Monte Carlo simulation techniques may be applied to estimate the necessary reliability parameters (U.S. Atomic Energy Commission, 1974b).

Common-Mode Failures

The availability estimates derived in the preceding subsection are predicated on the assumption that the subsystems are truly independent of one another. A common-mode failure is said to occur when this assumption is violated by a single failure event that prevents the subsystems from performing independently in accordance with design. Thus to each of the expressions for the unavailability or other reliability parameters must be added a contribution caused by common-mode failures. The assumption in using (l/n) redundant subsystems is that this common-mode contribution can be made vanishingly small compared to the unavailability caused by independent subsystem failures. However, if highly redundant systems are used to make unavailability small, it becomes increasingly difficult to guarantee that there does not exist an unanticipated common-mode failure mechanism that, although quite improbable, will be the dominant contributor to the failure probabilities.

The possibility of common-mode failure can be factored into fault trees or other reliability evaluation methods through the use of appropriate conditional probability techniques (U.S. Atomic Energy Commission, 1974b). The more difficult problem is to assure that all possible common-mode mechanisms have been recognized and to place meaningful bounds on their frequency of occurrence. Prerequisite for such recognition are a study of past accident and failure experience and a thorough knowledge of the safety system under consideration and of the environment in which it must function.

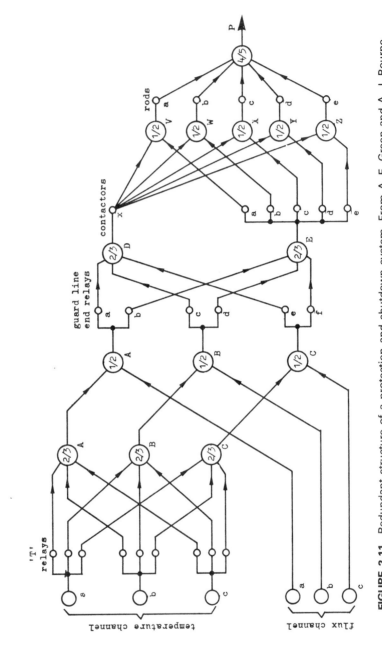

FIGURE 2-11 Redundant structure of a protection and shutdown system. From A. E. Green and A. J. Bourne, *Reliability Technology*, Wiley, New York, 1972. Used by permission.

More detailed discussions of possible common-mode failure mechanisms are given elsewhere (O'Brien and Walker, 1974; U.S. Atomic Energy Commission, 1974b). Only some of the better understood causes are outlined here. Probably the most likely cause of a common-mode failure is an unrecognized dependence of the redundant subsystems on a common component: power supplies, electrical grounds, ventilating equipment, support structures, and so on. Redundant systems may also be linked through testing and maintenance procedures; an improperly trained technician or poorly calibrated test equipment may cause failures in all the subsystems. Failures originating in a single component may propagate to redundant components in the protection system if instrument channels do not provide sufficient isolation. For example, it is conceivable that a design oversight in the system shown in Fig. 2-11 could allow an electrical failure in one of the flux sensors to result in an overload or other failure of both of the $(\frac{2}{3})$ logic circuits. Likewise, if instrument cables are not physically separated, fires, corrosive chemical spills, or other mechanisms may lead to common-mode failures. Interaction between the operating system and the protection system channels is another potential source of common-mode failure.

Common-mode failures may also arise from unanticipated conditions, particularly during accident transients, or from external catastrophes. Extreme transients may cause sensor saturation, amplifier overload, or any number of other protection system problems that are common to all redundant instrument channels. High temperature and pressure may cause problems with instrumentation cables, electric motors, and so on, and, in particular, steam jets and missiles can result in common failures of systems that are not physically isolated. Finally, forces from external causes, such as earthquake, fire, or flooding, may cause unanticipatedly harsh environmental conditions that are common to the redundant systems.

A number of design and operational procedures can reduce the risk of common-mode failures. Isolation of redundant systems, both physically and electrically, is one such step. Equally important is inclusion of diversity in the safety systems. This may take the form of diversity in equipment, in function, and/or in test and maintenance procedures (O'Brien and Walker, 1974).

Equipment diversity can be provided by using different sensor types to measure the same variable, different power supplies, diverse pump or valve designs to carry out the same functions, and so on. The diverse equipment may be based on different operating principles or simply may be manufactured by different suppliers. The diversity provides some defense against unrecognized design deficiencies that may cause failure in the operating or accident environment of the reactor.

Functional diversity may be provided both in the methods by which accidents are detected and in the principles used by the shutdown and engineered safety systems to carry out their assigned tasks. For example, it is frequently required that at least two diverse plant parameters be sensed by the protection system, each of which is capable of indicating the initiating event of a given accident. For example, in the diagram shown in Fig. 2-11 both neutron flux and temperature are used to actuate the shutdown system. Similarly, a backup shutdown method is often required in addition to control rod insertion; the injection of liquid neutron poison into the coolant stream is one example of such a system. Likewise, both air filtration and liquid spray systems may be incorporated into the engineered safety system for fission product removal from the containment atmosphere. In all such systems the diversity of equipment is automatically implied. Functional diversity provides defense against maintenance errors, equipment design deficiencies, external forces, and, in addition, is one of the best means of avoiding performance failure resulting from lack of insight into plant environment under accident conditions.

Diversity in test and maintenance procedures also may be used to prevent the type of common-mode failures that may result from a single defective test instrument or incompetent technician servicing all the redundant subsystems. This may be accomplished by assigning separate technicians and test equipment to each redundant subsystem or by having a second person independently check the work of the first. Alternatively, it may involve staggering the testing and maintenance of redundant systems so that not all systems are serviced on a single day during which instruments may be out of calibration or the technician may be not fully alert because of illness, lack of sleep, or other factors.

PROBLEMS

2-1. The following system is designed to deliver emergency cooling to a primary system:

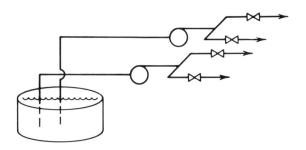

In the event of an accident the protection system delivers an actuation signal to the two identical pumps and the four identical valves. The pumps then start up, the valves open, and liquid coolant is delivered to the primary system. The following failure probabilities are found to be significant:

$p_{ps} = 10^{-5}$ probability that protection system will not deliver a signal to the pump and valve actuators.

$p_p = 2 \times 10^{-2}$ probability that a pump will fail to start up when the actuation signal is received.

$p_v = 10^{-1}$ probability that a valve will fail to open when the actuation signal is received.

$p_r = 0.5 \times 10^{-5}$ probability that the reservoir will be empty at the time of the accident.

a. Draw a fault tree for the failure of the system to deliver any coolant to the primary system in the event of an accident.

b. Evaluate the probability that such a failure will take place in the event of an accident. (You may assume that each failure probability is small compared to one.)

2-2. Verify Eq. (2-20).

2-3. The design criterion for the A.C. power system for a reactor is that it must have a failure probability of less than 2×10^{-5} year^{-1}. Off-site power failures may be expected to occur about once in five years. If the on-site A.C. power system consists of two independent diesel generators, each of which is capable of meeting the A.C. power requirements, what is the maximum failure probability per year that each diesel generator can have if the design criterion is to be met? If three independent diesel generators are used in parallel, what is the value of the maximum failure probability?

2-4. A reactor pressure vessel is equipped with eight relief valves. Pressure transients in the vessel can be controlled successfully by using any three of these. If the probability that any one of these valves will fail to operate on demand is .01, what is the probability on demand that the relief valve system will fail to control a pressure transient?

2-5. The failure rate for the system that supplies power to the coolant pumps of a reactor is 0.005 year^{-1}. The failure rate for the protection system that must detect the accident and deliver scram signals to the shutdown system actuators is 10^{-4} year^{-1}. The shutdown system consists of four independent subsystems, each consisting of a control rod bank and its associated drives and actuators. Insertion of any three of the banks will bring the reactor subcritical. The failure rate for an individual shutdown subsystem is 0.5×10^{-2} year^{-1}.

Determine the probability per year that there will be a loss-of-flow accident in which the reactor fails to scram under the following conditions:

a. The reactor is shut down once per year and the protection and shutdown system are thoroughly tested and repaired and maintained.

b. Same as (a), except the testing, repair, and maintenance take place every six months.

2-6. Suppose that in problem 2-3 one-fourth of the diesel generator failures are caused by common-mode effects and therefore incapacitate all the parallel systems. Under these conditions what is the maximum failure probability (i.e., random and common-mode) that is allowable if two diesel generators are used? If three diesel generators are used?

2-7. The failure rate for a reactor component is given empirically by

$$\theta(t) = 0.001(1 + 2e^{-2t} + e^{t/40}) \, (years^{-1})$$

where t is in years. If the system is installed at $t = 0$, calculate the probability that it will have failed by time t'. Plot your results for 40 years.

2-8. After studying the safety analysis report for a PWR or a BWR, discuss in some detail the measures taken to ensure:

a. A reliable supply of power for the safety systems.

b. A reliable supply of water for emergency and containment cooling.

REACTOR KINETICS

Chapter 3

3-1 INTRODUCTION

The power produced during the course of a reactor accident is one of the most important factors determining the degree of damage that may result to the plant. The rate at which power is produced is closely linked to the critical state of the reactor and therefore to the multiplication k, as defined in Chapter 1. For small deviations about the critical state of $k = 1$, the power tends to behave exponentially as given in Eq. (1-27), provided the multiplication remains constant long enough for the asymptotic behavior to emerge. In reality, the situation is more complex for transients, since the deviations of the multiplication about one may be neither small nor time independent.

To determine the time-dependent behavior of a neutron chain reaction, a distinction must be made between the behavior of prompt and delayed neutrons. Over 99% of fission neutrons are prompt in that they are emitted instantaneously at the time of fission. The emission of the remaining fission neutrons is delayed by times ranging from less than a second to a minute or more. These delayed neutrons result when fission product beta decay leads to a highly excited daughter nucleus that then emits a neutron. Although the fraction β of total fission neutrons produced by such fission product decay is small, the delayed neutrons have dramatic effects on the behavior of chain reactions.

The time-dependent power production is related to the multiplication and delayed neutron properties through the reactor kinetics equations. Many different forms of these equations may be formulated, depending on the amount of rigor that is included in treating the transport of neutrons in space and energy. Our primary purpose is to underscore the nature of the physical effects that occur during transients rather than to be able to calculate the time-dependent power distribution with great accuracy. Therefore we use a rather simple reactor model in order to limit the amount of mathematical apparatus that must be developed. More rigorous treatments may be found elsewhere (Henry, 1975; Stacey, 1969); the following discussion is confined to the simple one-energy-group neutron diffusion model developed in Chapter 1. Although this model permits examination of effects in space and

129

in energy—through spectral effects in the one-group cross sections—it neglects the coupling between the two variables. Consistent with this level of approximation, we make other assumptions that would not be justifiable if the kinetics equations were to be developed using the more sophisticated neutronics model. The intent is to express the results in the simplest physical terms: the local mutiplication, migration length, neutron lifetime, and so on.

In this chapter we first derive the one-group reactor kinetics equations and then make approximations that allow the spatial dependence to be eliminated. The resulting set of coupled ordinary differential equations, referred to as the *point kinetics approximation*, relates the time dependence of the reactor power to a relatively few parameters: the reactivity, the prompt neutron lifetime, and the fractions and decay constants of the delayed neutrons. In this form the equations are applicable to all reactors, provided the fuel remains stationary, although the parameters vary significantly, depending on the reactor type and the fuel being considered. Fast reactors, for example, have much smaller prompt neutron lifetimes than thermal reactors, and plutonium-fueled reactors have smaller delayed neutron fractions than uranium-fueled reactors.

In Section 3-3 the properties of the solutions of the point kinetics equations are examined for known time dependences of the reactivity, and the chapter is concluded with discussions of start-up accidents, which originate from very low powers, and of the rates at which reactivity may be withdrawn to shut a chain reaction down rapidly. With the exceptions of these two topics, it is impossible to describe the course of a nuclear transient without first examining the reactivity feedback effects caused by the changes of temperatures and densities that take place within a reactor core whenever significant power levels are reached. Therefore these reactivity feedback effects are discussed in detail in Chapter 4, before a general examination of reactivity accidents is presented in Chapter 5.

3-2 THE ONE-GROUP KINETICS EQUATIONS

In this section the one-group model for steady-state behavior developed in Chapter 1 is generalized to treat the time-dependent behavior of nuclear reactors. Space–time separability is then assumed to obtain the familiar point kinetics approximation.

Derivation

The time-dependent diffusion equation simply states that the rate of change of the number of neutrons per unit volume is just equal to the production rate of neutrons in the volume minus the loss rate. In general, production is due to neutrons born in fission and from external sources, whereas losses are due to absorption and to the net diffusion of neutrons out of the volume. Thus if $\tilde{N}(r, t)$ is the neutron density, or number of neutrons per unit volume, we have

$$\frac{\partial}{\partial t}\tilde{N}(\mathbf{r}, t) = S_f(\mathbf{r}, t) + S_e(\mathbf{r}, t) - \Sigma_a(\mathbf{r}, t)\phi(\mathbf{r}, t) + \nabla D(\mathbf{r}, t)\nabla\phi(\mathbf{r}, t),$$

$$(3\text{-}1)$$

where S_f and S_e are the fission and external source strengths. The remaining notation is the same as that used in Chapter 1, with the exception that the energy-averaged neutron flux and cross sections are now time dependent. The density is eliminated by defining a mean neutron speed as

$$\tilde{v} \equiv \frac{\phi(\mathbf{r}, t)}{\tilde{N}(\mathbf{r}, t)}.$$

$$(3\text{-}2)$$

Strictly speaking, $\tilde{v}$ is a function of space and time. However, in power reactors it tends to vary very little in these variables. Hereafter we assume it to be a constant.

The fission source has two contributions; one is due to prompt neutrons, produced instantaneously at the time of fission, and the other is due to delayed neutrons, produced by the decay of the neutron-emitting fission products. To determine these fission sources quantitatively, we first note from Eq. (1-53) that the fission rate per unit volume becomes $\Sigma_f(\mathbf{r}, t)\phi(\mathbf{r}, t)$ if the cross sections and neutron flux are allowed to be time dependent. The rate of fission neutrons production per unit volume is therefore $\nu(\mathbf{r})\Sigma_f(\mathbf{r}, t)\phi(\mathbf{r}, t)$, where $\nu(\mathbf{r})$ is the number of neutrons per fission. The fraction of fission neutrons produced promptly is $1 - \beta(\mathbf{r})$, and therefore the prompt fission source is $[1 - \beta(\mathbf{r})]\nu(\mathbf{r})\Sigma_f(\mathbf{r}, t)\phi(\mathbf{r}, t)$. The number of neutrons per fission and the delayed neutron fractions have different values for each fissionable nuclide, as is apparent from Table 3-1. Therefore $\nu(\mathbf{r})$ and $\beta(\mathbf{r})$ are assumed to be space dependent to include situations where they are averages over more than one nuclide and the ratios of the fissionable nuclide concentrations are not constant over the entire reactor core.

Contributions to the delayed neutron source arise from as many as 20 fission products that may be precursors of delayed neutron emission. Sufficient accuracy can usually be obtained, however, by dividing these fission products into six groups (Keepin, 1965). In each group the precursors decay exponentially with a characteristic half-life that determines the rate of delayed neutron emission. The decay constants and abundances of each of the groups are given in Table 3-1. Let the precursor decay constants for the

TABLE 3-1 Delayed Neutron Properties

Approximate Half-life (sec)	Delayed Neutron Fraction		
	U^{233}	U^{235}	Pu^{239}
56	0.00023	0.00021	0.00007
23	0.00078	0.00142	0.00063
6.2	0.00064	0.00128	0.00044
2.3	0.00074	0.00257	0.00069
0.61	0.00014	0.00075	0.00018
0.23	0.00008	0.00027	0.00009
Total delayed fraction	0.00261	0.00650	0.00210
Total neutrons/ fission	2.50	2.43	2.90

ith group be denoted by λ_i and the corresponding precursor concentrations, in nuclei per unit volume, by $C_i(\mathbf{r}, t)$. The rate at which delayed neutrons are produced per unit volume is then $\sum_i \lambda_i C_i(\mathbf{r}, t)$. Substituting the expressions for the prompt and delayed neutron sources, and Eq. (3-2), into Eq. (3-1), we have

$$\frac{1}{\tilde{v}}\frac{\partial}{\partial t}\phi(\mathbf{r}, t) = \nabla D(\mathbf{r}, t)\nabla\phi(\mathbf{r}, t) + [1 - \beta(\mathbf{r})]\nu(\mathbf{r})\sum_f(\mathbf{r}, t)\phi(\mathbf{r}, t)$$

$$- \sum_a(\mathbf{r}, t)\phi(\mathbf{r}, t) + \sum_{i=1}^{6}\lambda_i C_i(\mathbf{r}, t) + S_e(\mathbf{r}, t). \qquad (3\text{-}3)$$

The precursor equations are nearly identical to those given for fission products in Chapter 1. The precursor production rate per unit volume is $\beta_i(r)\nu(r)\sum_f(r, t)\phi(r, t)$ where $\beta_i(r)$ is the delayed neutron fraction for the ith precursor. The decay rate is $\lambda_i C_i(r, t)$. Therefore

$$\frac{\partial}{\partial t}C_i(\mathbf{r}, t) = \beta_i(\mathbf{r})\nu(\mathbf{r})\sum_f(\mathbf{r}, t)\phi(\mathbf{r}, t) - \lambda_i C_i(\mathbf{r}, t), \qquad i = 1, 2, \ldots, 6.$$

$$(3\text{-}4)$$

In this formulation we have assumed that the delayed neutron precursors remain stationary during their lifetime, an assumption that would require modification for the treatment of circulating fuel reactors. For stationary fuel reactors Eqs. (3-3) and (3-4) are the conventional space–time kinetics equations. For a reactor of volume V and surface $\mathcal{S}$ the flux and precursor concentration must satisfy the conditions

$$\phi(\mathbf{r}, t) > 0, \quad C_i(\mathbf{r}, t) > 0; \quad \mathbf{r} \in V, \tag{3-5}$$

$$\phi(\mathbf{r}, t) = 0, \quad C_i(\mathbf{r}, t) = 0; \quad \mathbf{r} \in \mathcal{S}. \tag{3-6}$$

For purposes of physical interpretation it is helpful to express the kinetics equations in terms of local multiplication, migration length, and neutron lifetime. We therefore generalize the definitions given in Chapter 1 to include time dependence:

$$k_\infty(\mathbf{r}, t) = \frac{\nu(\mathbf{r}) \Sigma_f(\mathbf{r}, t)}{\Sigma_a(\mathbf{r}, t)}, \tag{3-7}$$

$$M^2(\mathbf{r}, t) = \frac{D(\mathbf{r}, t)}{\Sigma_a(\mathbf{r}, t)}, \tag{3-8}$$

$$l_\infty(\mathbf{r}, t) = \frac{1}{\tilde{v} \Sigma_a(\mathbf{r}, t)}. \tag{3-9}$$

Combining these definitions with Eqs. (3-3) and (3-4) we have, after some algebra,

$$\frac{\partial}{\partial t} \phi(\mathbf{r}, t) = l_\infty^{-1}(\mathbf{r}, t)\{\nabla M^2(\mathbf{r}, t)\nabla \phi(\mathbf{r}, t) - M^2(\mathbf{r}, t)[\nabla \ln l_\infty(\mathbf{r}, t)]\nabla \phi(\mathbf{r}, t)$$

$$+ [1 - \beta(\mathbf{r})]k_\infty(\mathbf{r}, t)\phi(\mathbf{r}, t) - \phi(\mathbf{r}, t)\} + \sum_{i=1}^{6} \lambda_i \tilde{v} C_i(\mathbf{r}, t) + \tilde{v} S_e(\mathbf{r}, t)$$

$$\tag{3-10}$$

and

$$\frac{\partial}{\partial t} \tilde{v} C_i(\mathbf{r}, t) = l_\infty^{-1}(\mathbf{r}, t)\beta_i(\mathbf{r})k_\infty(\mathbf{r}, t)\phi(\mathbf{r}, t) - \lambda_i \tilde{v} C_i(\mathbf{r}, t), \quad i = 1, 2, \ldots, 6.$$

$$\tag{3-11}$$

Note that in the time-independent case where the reactor is just critical, these equations reduce to Eq. (1-69).

Space-Independent Kinetics Equations

Even for the simple one-group diffusion model incorporated into Eqs. (3-10) and (3-11), solution for the detailed space–time dependence of the flux and precursor concentrations requires extensive numerical calculations, often to the point of being prohibitive. Consequently, it is often desirable to solve only for the reactor power as a function of time and treat the time dependence of the power distribution either by a less rigorous approximation or not at all. Although not as accurate, such simplification provides a great deal of insight into the nature of nuclear transients, and moreover, through more approximate treatments of the spatial distribution the main characteristics of the power distribution often may be obtained to an accuracy that is consistent with the extent to which the physical state of the reactor under accident conditions can be estimated.

The spatial dependence in Eqs. (3-10) and (3-11) may be removed by assuming that the energy-averaged neutron flux and the precursor concentrations are approximately separable in space and time:

$$\phi(\mathbf{r}, t) \simeq \psi(\mathbf{r}) P(t), \tag{3-12}$$

$$\tilde{v} C_i(\mathbf{r}, t) \simeq \psi(\mathbf{r}) \tilde{C}_i(t). \tag{3-13}$$

The spatial dependence is then formally eliminated by substituting these equations into Eqs. (3-10) and (3-11). The equations are then multiplied by $l_\infty(\mathbf{r}, t)$ times a weight function $\psi_0(\mathbf{r})$, which has the properties $\psi_0(\mathbf{r}) > 0; \mathbf{r} \in V$ and $\psi_0(\mathbf{r}) = 0; r \in \mathcal{S}$. Finally, the integral $\langle \cdot \rangle$ is performed over the reactor volume. After applying Green's theorem to the leakage term, we obtain

$$\langle \psi_0 l_\infty(t) \psi \rangle \frac{dP(t)}{dt} = \langle \psi_0 [1 - \beta] k_\infty(t) \psi \rangle P(t) - \langle \psi_0 \psi \rangle P(t)$$

$$- \langle M^2(t)(\nabla \psi_0)(\nabla \psi) \rangle P(t) - \langle \psi_0 M^2(t)[\nabla \ln l_\infty(t)] \nabla \psi \rangle P(t) \tag{3-14}$$

$$+ \langle \psi_0 l_\infty(t) \psi \rangle \sum_{i=1}^{6} \lambda_i \tilde{C}_i(t) + \tilde{v} \langle \psi_0 l_\infty(t) S_e(t) \rangle$$

and

$$\langle \psi_0 l_\infty(t) \psi \rangle \frac{d\tilde{C}_i(t)}{dt} = \langle \psi_0 \beta_i k_\infty(t) \psi \rangle P(t) - \lambda_i \langle \psi_0 l_\infty(t) \psi \rangle \tilde{C}_i(t), \qquad i = 1, 2, \ldots, 6, \tag{3-15}$$

where for brevity the $\mathbf{r}$ indicating the spatial dependence of quantities under the volume integrals is deleted. Equations (3-14) and (3-15) may be written in the more compact form

$$\frac{dP(t)}{dt} = \frac{\{[1 - \bar{\beta}(t)]k(t) - 1\}}{l(t)} P(t) + \sum_{i=1}^{6} \lambda_i \tilde{C}_i(t) + \tilde{S}_e(t) \tag{3-16}$$

$$\frac{d\tilde{C}_i(t)}{dt} = \frac{\bar{\beta}_i(t)k(t)}{l(t)} P(t) - \lambda_i \tilde{C}_i(t), \qquad i = 1, 2, \ldots, 6 \tag{3-17}$$

where

$$k(t) = \bar{k}_\infty(t)\bar{P}_N(t) \tag{3-18}$$

$$l(t) = \bar{l}_\infty(t)\bar{P}_N(t) \tag{3-19}$$

$$\bar{k}_\infty(t) = \frac{\langle \psi_0 k_\infty(t)\psi \rangle}{\langle \psi_0 \psi \rangle} \tag{3-20}$$

$$\bar{l}_\infty(t) = \frac{\langle \psi_0 l_\infty(t)\psi \rangle}{\langle \psi_0 \psi \rangle} \tag{3-21}$$

$$\bar{P}_N(t) = \left\{ 1 + \frac{\langle M^2(t)(\nabla \psi_0)\,\nabla \psi \rangle + \langle \psi_0 M^2(t)[\nabla \ln l_\infty(t)]\nabla \psi \rangle}{\langle \psi_0 \psi \rangle} \right\}^{-1} \tag{3-22}$$

$$\bar{\beta}_i(t) = \frac{\langle \psi_0 \beta_i k_\infty(t)\psi \rangle}{\langle \psi_0 k_\infty(t)\psi \rangle} \tag{3-23}$$

$$\bar{\beta}(t) = \sum_{i=1}^{6} \bar{\beta}_i(t) \tag{3-24}$$

$$\tilde{S}_e(t) = \frac{\tilde{v}\langle \psi_0 l_\infty(t)S_e(t)\rangle}{\langle \psi_0 l_\infty(t)\psi \rangle}. \tag{3-25}$$

The foregoing procedure results in kinetics equations that are functions only of time. It leaves unanswered, however, the question of how the coefficients appearing in Eqs. (3-16) and (3-17) are to be determined, since the evaluation of the volume integrals in Eqs. (3-18) through (3-25) requires the spatial shape functions $\psi_0(\mathbf{r})$ and $\psi(\mathbf{r})$ to be known.

The weight function ψ_0 is taken to be the flux distribution before the initiation of the transient. Thus for the one-group model it must satisfy Eq. (1-69):

$$\nabla M^2(\mathbf{r})\,\nabla \psi_0(\mathbf{r}) - M^2(\mathbf{r})[\nabla \ln l_\infty(\mathbf{r})]\,\nabla \psi_0(\mathbf{r}) + [k_\infty(\mathbf{r}) - 1]\psi_0(\mathbf{r}) = 0, \tag{3-26}$$

where $k_\infty(\mathbf{r})$, $M^2(\mathbf{r})$, and $l_\infty(\mathbf{r})$ are the initial values. This definition of ψ_0 has the following property. If a spatially uniform core, such as the one discussed in Chapter 1, is considered, Eq. (3-26) reduces to Eq. (1-70), and the separability approximation becomes exact, provided we take

$$\psi(\mathbf{r}) = \psi_0(\mathbf{r}). \tag{3-27}$$

Furthermore, the spatial independence of k_∞, M^2, and l_∞ causes P_N to reduce to the nonleakage probability as defined by Eq. (1-80), and the multiplication to revert to Eq. (1-83). Thus one can interpret the state of the reactor in terms of the infinite medium multiplication, migration length, and geometric buckling.

In accident analysis, spatial nonuniformities must be taken into account, for even if the reactor were uniform initially, changes in control rod position, fuel temperature, and other variables lead to spatially dependent cross sections. Nevertheless, in a large number of situations the spatial dependence of the local core parameters k_∞, M^2, l_∞ does not change enough during the course of the transient to perturb the flux greatly from the initial distribution ψ_0. Hence the assumption that $\psi = \psi_0$ can be retained at all times without introducing unacceptable errors in evaluating the volume integrals appearing in the coefficients of the kinetics equations. If Eq. (3-27) is used in evaluating the volume integrals, then Eqs. (3-16) and (3-17) are called the point kinetics approximation. As a rule, this approximation is used most successfully in reactor cores that are not neutronically loosely coupled, for when the ratio of the migration length to the core dimensions becomes small, nonseparable spatial effects tend to become more pronounced (Yasinsky and Henry, 1965).

Use of the point kinetics equations permits the examination of the effects of localized perturbations on the core parameters in terms of their effect on the multiplication. In many situations neglecting the last term in the denominator of Eq. (3-22) results in no significant loss of accuracy because it is appreciable only when there are large inhomogeneities in $l_\infty(t)$. Moreover, without this term, the multiplication can be expreseed in the same form as Eq. (1-83), where the parameters are given as core-averaged values. From Eqs. (3-18) and (3-22) we have

$$k(t) = \frac{\bar{k}_\infty(t)}{1 + \bar{M}^2(t)\bar{B}_g^2},$$
(3-28)

where the core-averaged parameters are given by

$$\bar{k}_\infty(t) = \frac{\langle k_\infty(\mathbf{r}, t)\psi_0^2(\mathbf{r})\rangle}{\langle \psi_0^2(\mathbf{r})\rangle}$$
(3-29)

$$\bar{M}^2(t) = \frac{\langle M^2(\mathbf{r}, t)[\nabla\psi_0(\mathbf{r})]^2\rangle}{\langle[\nabla\psi_0(\mathbf{r})]^2\rangle},$$
(3-30)

$$\bar{B}_g^2 = \frac{\langle[\nabla\psi_0(\mathbf{r})]^2\rangle}{\langle \psi_0^2(\mathbf{r})\rangle}.$$
(3-31)

From these expressions it is apparent that increases in the multiplication result from increases in $\bar{k}_\infty(t)$ and decreases in $\bar{M}^2(t)$ or $\bar{B}_g^2$. Generally speaking, $\psi_0(\mathbf{r})$ is large near the center of the reactor, whereas $|\nabla\psi_0(\mathbf{r})|$ becomes small, vanishing completely at the geometrical center of the core. Conversely, as the outer surface of the reactor is approached, $|\nabla\psi_0(\mathbf{r})|$

reaches its maximum value whereas $\psi_0(\mathbf{r})$ becomes vanishingly small. Thus we can see that increases in $k_\infty(\mathbf{r}, t)$ have their largest effect in increasing k near the center of the core. Increases in $M^2(\mathbf{r})$ decrease k by the greatest amount in the high flux gradient regions near the edge of the core. Changes in the geometric buckling can occur only if the flux shape or the dimensions of the reactor change, with increases in reactor volume causing decreases in B_g^2 and therefore increases in k.

Improvements over point kinetics are likely to be required in the event that control rod motion, coolant boiling, or any number of other localized physical effects cause substantial changes in the spatial distribution of the flux during the course of a transient. Some improvement may be obtained by calculating $\psi(\mathbf{r})$ at a time when the flux distribution is more representative of an averaged value over the course of a transient. At a particular time t, the multiplication and shape factor can be calculated by solving the equation

$$\nabla M^2(\mathbf{r}, t)\, \nabla \psi(\mathbf{r}) - M^2(\mathbf{r}, t)[\nabla \ln l_\infty(\mathbf{r}, t)]\nabla \psi(\mathbf{r}) + \left\{ \frac{k_\infty(\mathbf{r}, t)}{k} - 1 \right\} \psi(\mathbf{r}) = 0$$

(3-32)

with

$$\psi(\mathbf{r}) > 0, \qquad \mathbf{r} \in V \tag{3-33}$$

and

$$\psi(\mathbf{r}) = 0, \qquad \mathbf{r} \in \mathscr{S}. \tag{3-34}$$

and where the core parameters k_∞, M^2, and l_∞ are evaluated at t. The preceding equation is just an extension of the one-group diffusion equation to noncritical states in such a manner that it is consistent with Eqs. (3-18), (3-19), and (3-22) in defining the multiplication k in terms of the local core parameters.

Further detail may be gained by recalculating $\psi(\mathbf{r})$ with Eq. (3-32) at several different times during the course of a transient in order to update the coefficients in the kinetics equations. If this is done continuously, an approximation results that is closely related to the so-called adiabatic model (Henry, 1958).

In using the adiabatic or other more elaborate methods for incorporating spatial effects into the kinetics equations, a more rigorous study of the errors introduced in arriving at the space-independent kinetics equations is necessary to assure the validity of the results. The efficacy of. any treatment of spatial effects within the kinetics equations is to a great extent dependent on the clever modeling of the reactor for the particular class of transients under consideration; an insight into such modeling is best gained through consultation of the applications literature (Stacey, 1969).

Some further manipulations can be carried out on Eqs. (3-16) and (3-17) to put them in a more convenient form. They are often rewritten in terms of the reactivity and neutron generation time, defined by

$$\rho(t) = \frac{[k(t)-1]}{k(t)} \tag{3-35}$$

and

$$\Lambda(t) = \frac{l(t)}{k(t)}, \tag{3-36}$$

respectively. Combining these equations with Eqs. (3-16) and (3-17), we obtain

$$\frac{dP(t)}{dt} = \frac{[\rho(t)-\bar{\beta}(t)]}{\Lambda(t)} P(t) + \sum_{i=1}^{6} \lambda_i \tilde{C}_i(t) + \tilde{S}_e(t), \tag{3-37}$$

$$\frac{d\tilde{C}_i(t)}{dt} = \frac{\bar{\beta}_i(t)}{\Lambda(t)} P(t) - \lambda_i \tilde{C}_i(t), \qquad i = 1, 2, \dots, 6. \tag{3-38}$$

Note that both the generation time and the delayed neutron fraction are time dependent. In the great majority of situations the spatially averaged values $\Lambda(t)$ and $\bar{\beta}_i(t)$ vary no more than a few percent over the course of a transient. Consistent with the level of the other approximations, it is usually justifiable to ignore the slight time dependence in these quantities and assume that they are constants, hereafter denoted as Λ and $\bar{\beta}_i$. A similar argument cannot be made for $\rho(t)$, the reactivity. A reactor is typically taken to be just critical at the initiation of a transient; thus $\rho(0) = 0$, $k(0) = 1$. Even though $k(t)$ varies only slightly, the time dependence of $\rho(t)$ must be taken carefully into account, since, as shown in subsequent sections, the time dependence of the reactivity about zero profoundly affects the behavior of the kinetics equation.

In making the separability approximation of Eqs. (3-12) and (3-13), it is desirable to normalize $\psi(\mathbf{r})$ such that $P(t)$ is the reactor power. Recalling that the reactor power is just

$$P(t) = \gamma \langle \Sigma_f(\mathbf{r}, t)\phi(\mathbf{r}, t)\rangle, \tag{3-39}$$

where γ is the energy produced per fission, we then obtain, from Eq. (3-12), the normalization

$$\gamma \langle \Sigma_f(\mathbf{r}, t)\psi(\mathbf{r})\rangle = 1. \tag{3-40}$$

Strictly speaking, $P(t)$ can be forced to be equal to the power only at one point in time, if $\Sigma_f(\mathbf{r}, t)$ is time dependent. Consistent with the approximations that Λ and $\bar{\beta}$ are time independent, however, we may ignore the few percent change in $\sum_f (\mathbf{r}, t)$ over the course of a transient and assume that if Eq. (3-40) is satisfied at $t = 0$, $P(t)$ can be equated with the reactor power at later times without a significant loss of accuracy.

3-3 KINETICS EQUATIONS SOLUTIONS WITH KNOWN REACTIVITY

During the course of a nuclear transient the reactivity $\rho(t)$ is a complicated function of the entire physical state of the reactor. Reactivity contributions arise not only from the cause of the accident, whatever that may be, but also from the reactor control and shutdown systems and from inherent reactivity feedback as the materials in the reactor core change temperature and density with the production of power. The reactivity normally can be precalculated only if the reactor is at such a low power level that the inherent temperature and density feedback mechanisms are insignificant, and even then only if the time dependence of the reactivity insertion and control system can be modeled.

It is, nevertheless, very helpful to examine the behavior of the point kinetics equations, (3-37) and (3-38), for known reactivity behavior. For such analysis not only reveals a great deal of the behavior of the kinetics equations, but also provides a simplified model by which the important parameters for several classes of accidents can be isolated. In the remainder of this chapter we assume that the reactivity is known and that the prompt neutron generation time and delayed neutron fractions are constants. Equations (3-37) and (3-38) may then be rewritten as

$$\frac{dP(t)}{dt} = \frac{\rho(t) - \bar{\beta}}{\Lambda} P(t) + \sum_{i=1}^{6} \lambda_i \tilde{C}_i(t) + \tilde{S}_e(t), \qquad (3\text{-}41)$$

$$\frac{d\tilde{C}_i(t)}{dt} = \frac{\bar{\beta}_i}{\Lambda} P(t) - \lambda_i \tilde{C}_i(t), \qquad i = 1, 2, 3, 4, 5, 6. \qquad (3\text{-}42)$$

Step Reactivity Changes

We first consider the situation where there is a step change in reactivity applied to a reactor that is initially just critical. This simple example serves to illustrate the importance of delayed neutrons in the kinetic behavior and to emphasize the concept of prompt criticality.

Assume that a source-free reactor is operating at a steady-state power $P(0)$ when $t < 0$, and therefore $\rho = 0$ for $t \leq 0$. The initial conditions are found by setting the derivative in Eq. (3-42) equal to zero:

$$\tilde{C}_i(0) = \frac{\bar{\beta}_i}{\lambda_i \Lambda} P(0), \qquad i = 1, 2, 3, 4, 5, 6. \tag{3-43}$$

If a step change in reactivity ρ_0 is made at $t = 0$, then the equations become time dependent. Assume that the solution of the point kinetics equations has the form

$$P(t) = P_\omega e^{\omega t}, \tag{3-44}$$

$$\tilde{C}_i(t) = C_{i\omega} e^{\omega t}, \qquad i = 1, 2, 3, 4, 5, 6, \tag{3-45}$$

where ω is a parameter, referred to as an eigenvalue, to be determined. Inserting these equations into Eqs. (3-41) and (3-42), we obtain

$$\omega P_\omega = \frac{\rho_0 - \bar{\beta}}{\Lambda} P_\omega + \sum_{i=1}^{6} \lambda_i C_{i\omega}, \tag{3-46}$$

$$\omega C_{i\omega} = \frac{\bar{\beta}_i}{\Lambda} P_\omega - \lambda_i C_{i\omega}, \qquad i = 1, 2, 3, 4, 5, 6. \tag{3-47}$$

Writing Eq. (3-47) as

$$C_{i\omega} = \frac{\bar{\beta}_i P_\omega}{\Lambda(\omega + \lambda_i)}, \tag{3-48}$$

and eliminating the $C_{i\omega}$ between Eqs. (3-46) and (3-48) yields a transcendental equation for the ω:

$$\rho_0 = \omega \Lambda + \sum_{i=1}^{6} \frac{\omega \bar{\beta}_i}{\omega + \lambda_i}. \tag{3-49}$$

It may be shown that this equation always has seven real distinct solutions or eigenvalues, say, $\omega_1 > \omega_2 \ldots > \omega_7$. The solution may therefore be written in the form of the superposition

$$P(t) = \sum_{l=1}^{7} P_l e^{\omega_l t}, \tag{3-50}$$

$$\tilde{C}_i(t) = \frac{\bar{\beta}_i}{\Lambda} \sum_{l=1}^{7} \frac{P_l}{(\omega_l + \lambda_i)} e^{\omega_l t}, \tag{3-51}$$

where Eq. (3-48) is utilized to eliminate the $C_{i\omega}$, and $P_{\omega l}$ is written as P_l for brevity. Once the ω_l are determined, the coefficients $P_l, l = 1, 2, \ldots, 7$ can be determined in terms of $P(0)$ by equating Eqs. (3-50) and (3-51) to Eq. (3-43) at $t = 0$.

The behavior of the reactor power is better understood by examining the values of the ω_l. It may be shown that the largest eigenvalue will always have the same sign as the reactivity ρ_0, whereas the remaining will always be negative. Thus after sufficient time has elapsed, the first term in Eq. (3-50) will dominate the solution, yielding a simple exponential behavior. This may be seen by plotting $P(t)$ on a semilog scale for positive and negative values of ρ_0, as shown in Figs. 3-1 and 3-2. The asymptotic behavior is characterized by the reactor period $\mathcal{T}$, defined by

$$\mathcal{T} = \omega_1^{-1}$$ (3-52)

or

$$P(t) \sim e^{t/\mathcal{T}}.$$ (3-53)

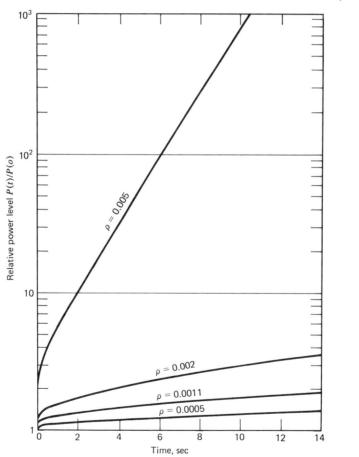

FIGURE 3-1 Relative power level of U^{235} reactor versus time for positive step function reactivity changes, $\Lambda = 10^{-4}$ sec. Adapted from M. A. Schultz, *Control of Nuclear Reactors and Power Plants*, McGraw-Hill, New York, 1961. Used by permission.

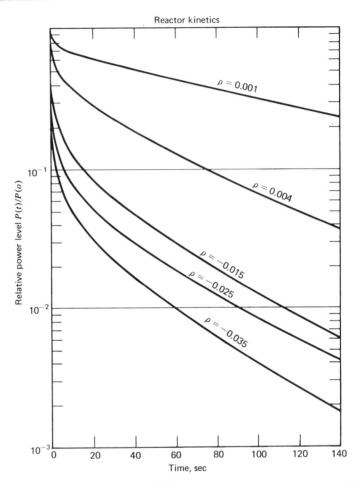

Reactor kinetics

FIGURE 3-2 Relative power level of U^{235} reactor versus time for negative step function reactivity changes, $\Lambda = 10^{-4}$ sec. Adapted from M. A. Schultz, *Control of Nuclear Reactors and Power Plants*, McGraw-Hill, New York, 1961. Used by permission.

The period is the length of time that is required for the power to increase by a factor of e; in the case of negative reactivity, the period is negative, and its magnitude is then the length of time required for the power to decrease by a factor of e.

From the standpoint of reactor safety, very small positive periods must be avoided, because with them the rate at which the power level increases may become so large that shutdown systems, such as the control rods, may not be able to terminate the transient fast enough to prevent damage to the reactor core. The decrease in reactor period with increasing reactivity is illustrated in Fig. 3-3 for U^{235} delayed neutron data and a number of different prompt neutron generation times. Note that for small positive reactivity additions the period is long and nearly independent of the neutron generation time. There is then a precipitous drop in the period length about the line marked prompt critical, with the short periods associated with reactivities above the prompt critical line being approximately proportional to Λ.

The behavior of Fig. 3-3, and in particular the significance of prompt criticality, may be examined by considering the solution of Eq. (3-49) when ω is replaced by the largest eigenvalue $1/\mathscr{T}$:

$$\rho_0 = \frac{\Lambda}{\mathscr{T}} + \sum_{i=1}^{6} \frac{\bar{\beta}_i}{1 + \lambda_i \mathscr{T}}. \tag{3-54}$$

For small reactivity insertions $\mathscr{T}$ is large, and therefore the first term in the denominator can be neglected compared to $\lambda_i \mathscr{T}$. Therefore

$$\mathscr{T} \simeq \frac{1}{\rho_0}\left(\Lambda + \sum_{i=1}^{6} \frac{\bar{\beta}_i}{\lambda_i}\right). \tag{3-55}$$

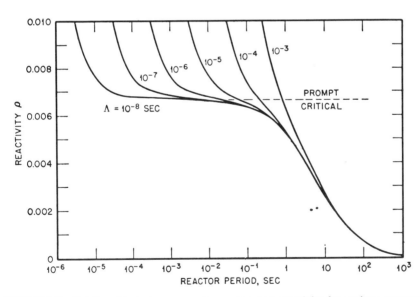

FIGURE 3-3 Relationship between reactor period and reactivity for various neutron generation times. Adapted from *Physics of Nuclear Kinetics* by G. R. Keepin with the permission of publishers, Addison-Wesley/W. A. Benjamin Inc., Advanced Book Program, Reading, Mass., U.S.A.

For any power reactor the sum in the parenthesis dominates the prompt generation time Λ by well over an order of magnitude. Thus the period is determined by the delayed neutron parameters $\bar{\beta}_i$ and λ_i, and not by the prompt generation time Λ. The quantity in parenthesis is sometimes referred to as the effective generation time; however, to avoid confusion we reserve the term *generation time* for the prompt generation time Λ unless otherwise specified.

If we consider large reactivities, where $\mathcal{T}$ is small, the λ_i in the denominator of Eq. (3-49) can be neglected,

$$\mathcal{T} \simeq \frac{\Lambda}{\rho_0 - \bar{\beta}}. \tag{3-56}$$

The behavior becomes independent of the delayed neutron parameters and strongly dependent on the prompt generation time. The reason for the sharp change in behavior may be explained as follows. If the reactivity becomes greater than $\bar{\beta}$, the chain reaction can be sustained even in the absence of delayed neutrons, as witnessed by the fact that the first term on the right of Eq. (3-41) becomes positive when $\rho_0 > \bar{\beta}$. Thus for situations in which $\rho_0 > \bar{\beta}$ the neutron kinetics equations are governed predominantly by the behavior of the prompt neutrons, and therefore by the prompt generation time Λ, which ranges from 10^{-3} sec to 10^{-7} sec for various power reactor systems. Because it is governed by the prompt generation time, the power rise becomes so rapid that the relatively sluggish behavior of the delayed neutron precursors lags the transient and becomes insignificant by the time the asymptotic solution emerges. Thus the reactor kinetics equations can be approximated adequately by eliminating the effects of delayed neutrons entirely:

$$\frac{dP(t)}{dt} \simeq \frac{(\rho - \bar{\beta})}{\Lambda} P(t). \tag{3-57}$$

As stated previously, the condition $\rho = \bar{\beta}$ is known as prompt critical, and Eq. (3-57) is the prompt critical approximation. Probably no concept is more important to the prevention of destructive nuclear excursions than that of prompt criticality, for if a reactor becomes superprompt critical (i.e., $\rho > \bar{\beta}$), the period becomes so short that electromechanical shutdown systems become much too slow to terminate the resulting transient. Only prompt inherent feedback mechanisms, discussed in the following chapter, can be relied on to terminate prompt critical excursions if damage to the core is to be avoided. Reactivity is often referred to in terms of prompt criticality because of its importance: $\rho/\bar{\beta}$ is defined as the reactivity in dollars or $100\,\rho/\bar{\beta}$, the reactivity in cents, with one dollar or 100 cents being equal to prompt critical.

In contrast to the foregoing situation, when the reactivity is only a small fraction of a dollar, the kinetics equations are only very weakly dependent on Λ. This behavior can be illustrated most easily by the following simplification of the kinetics equations. Suppose that instead of considering all six groups of delayed neutrons, we approximate the behavior of the kinetics equations by lumping all the delayed neutrons into a single group. The decay constant for this group is found by weighting the half-lives $t_{1/2i} = 0.693/\lambda_i$ of each group by the corresponding delayed fraction:

$$\lambda^{-1} = \bar{\beta}^{-1} \sum_{i=1}^{6} \frac{\bar{\beta}_i}{\lambda_i}. \tag{3-58}$$

The kinetics equations with one delayed group are then

$$\frac{dP(t)}{dt} = \frac{(\rho - \bar{\beta})}{\Lambda} P(t) + \lambda \tilde{C}(t) \tag{3-59}$$

and

$$\frac{d\tilde{C}(t)}{dt} = \frac{\bar{\beta}}{\Lambda} P(t) - \lambda \tilde{C}(t). \tag{3-60}$$

By differentiation, $\tilde{C}(t)$ can be eliminated between these equations, and then in the limit $\Lambda \to 0$ we obtain

$$\frac{dP(t)}{dt} = \frac{1}{\Lambda^*}\left(\rho + \frac{1}{\lambda} \frac{d\rho}{dt}\right) P(t), \tag{3-61}$$

where

$$\Lambda^* = \frac{(\bar{\beta} - \rho)}{\lambda}. \tag{3-62}$$

Moreover, if $|\rho| \ll \bar{\beta}/\lambda$, we may make the additional approximation

$$\Lambda^* = \frac{\bar{\beta}}{\lambda}. \tag{3-63}$$

where Λ^* is known as the effective generation time. Equation (3-61), known as the zero lifetime or prompt jump approximation, is a useful simplification in analyzing mild transients such as some of those encountered under normal operating procedures.

The behavior of the reactor power following negative reactivity insertions is also important to reactor safety considerations, since if there is a sudden failure of the ability of the coolant system to remove the heat produced by the reactor, the power must be decreased rapidly in order to avoid damage of the core. Equation (3-58) is valid for small negative as well as positive reactivities. To decrease reactor power rapidly, however, large negative reactivities are required in the form of banks of control rods or other poisoning devices. For large negative reactivity insertions the delayed neutron precursor with the largest half-life (i.e., smallest λ_i) places an upper limit on how fast the power can be decreased. It may be shown that for situations in which $\rho_0 \ll -\bar{\beta}$

$$\mathcal{T} \simeq -\frac{1}{\lambda_1}\left[1 + \frac{\bar{\beta}}{|\rho_0|}\right] \tag{3-64}$$

where λ_1 is the time constant for the precursor with the largest half-life. From Table 3-1 we thus see that the shortest negative period possible is ~ -80 sec. For negative reactivity insertions, equally as important as the asymptotic period is the negative jump due to the prompt neutrons. This is illustrated in Fig. 3-2, where the importance of large negative reactivities in decreasing power is shown. Of course, after the reactor power has fallen below a few percent of its initial value, the predominant heat source will be from decay heat, as discussed is Chapter 1.

In the foregoing situations, where only time-independent reactivities are considered, the reactor kinetics equations always give rise to an asymptotic solution in the form of an exponential. The reactor is then said to be on a stable period. The concept of the reactor period has also been found to be useful in situations where not enough time has elapsed for an exponential asymptotic solution to emerge or where the reactivity is changing in time. Thus the reactor period is defined more generally as

$$\mathcal{T}(t) = \frac{P(t)}{\dfrac{dP(t)}{dt}} = \left[\frac{d}{dt}\ln P(t)\right]^{-1}, \tag{3-65}$$

which can be measured instantaneously. This definition reduces to the constant values obtained above only if the asymptotic behavior for step reactivity changes is considered.

Ramp Reactivity Insertions

The step reactivity changes discussed in the preceding subsection are a mathematical idealization, valuable for the insight they provide into the behavior of the kinetics equations, but difficult to achieve in reality. For a reactivity change to be instantaneous it would be required to take place over a time span that is short compared to the prompt neutron generation time. This is clearly impossible using reactor control or shutdown systems, and in practice it is not even approached by any of the physical reactivity insertion mechanisms that have been postulated to cause severe accidents. Nevertheless, when only small changes in reactivity are involved, the step reactivity insertion sometimes may be a reasonable approximation. For in such situations the reactor periods are very long, and relatively small power change may result during the time span over which the reactivity is inserted. The stable period calculated from the step insertion approximation may be quite accurate in describing the behavior after the reactivity insertion is completed.

The situation is quite different for the large reactivity insertions postulated in many safety analyses. In these cases the reactor period may become so short that the reactor power may rise to orders of magnitude before the maximum value of $\rho(t)$ is reached. Particularly in situations where the total excess reactivity available is sufficient to bring the reactor above prompt critical, it is the rate at which reactivity is inserted that becomes of paramount importance rather than the total reactivity that is available. The reasons for this are twofold. First, the rate of reactivity addition determines how much time is available to detect the accident and shut down the reactor before prompt critical is reached. Second, if the reactor reaches prompt critical, the power rise will be so rapid that electromechanical shutdown systems cannot act sufficiently fast to terminate the excursion; rather, either inherent feedback mechanisms or core destruction will terminate the prompt-critical excursion before all the available reactivity is inserted. As discussed in Chapter 5, the amount of core destruction caused by an excursion is closely correlated to the ramp reactivity insertion rate and not the total magnitude of available reactivity.

For the present we do not consider the aforementioned feedback reactivity effects but examine the solution of the kinetics equations when there is only a simple ramp increase of reactivity,

$$\rho(t) = \dot{\rho}t, \tag{3-66}$$

where $\dot{\rho}$ is a constant, referred to as the ramp rate. From the standpoint of safety, the ramp rate may be considered to be due to any number of causes, depending on the hypothesized accident: the uncontrollable withdrawal of a simple control rod or an entire bank of control rods, the voiding of the liquid metal coolant from a fast reactor, or the compaction of a highly enriched core.

Solutions to the kinetics equations with ramp reactivity insertions may be obtained by substituting Eq. (3-66) into Eqs. (3-41) and (3-42) with $S_e(t) = 0$. Numerical results are shown for several ramp rates and neutron lifetimes in Fig. 3-4. As would be anticipated, the curves are concave upward on the logarithmic ordinate because of the ever-decreasing value of the reactor

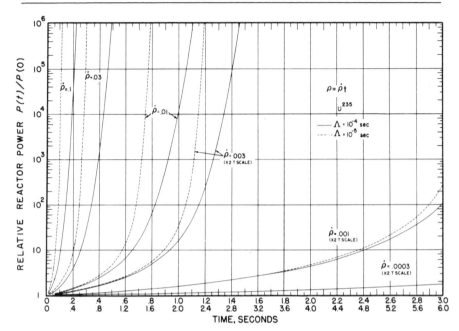

FIGURE 3-4 Power response to linear time variation of reactivity in U^{235} systems characterized by prompt neutron generation times in the range 10^{-4} sec to 10^{-5} sec. Adapted from *Physics of Nuclear Kinetics* by G. R. Keepin with the permission of publishers, Addison-Wesley/W. A. Benjamin Inc., Advanced Book Program, Reading, Mass., U.S.A.

period. The rate of power rise increases both with increased ramp rate and with decreased prompt generation time. If one calculates the time at which the reactor becomes prompt critical (i.e., one dollar),

$$t_p = \frac{\bar{\beta}}{\dot{\rho}}, \tag{3-67}$$

two additional characteristics can be seen. First, before the time, and therefore the reactivity, reaches about two-thirds of the prompt-critical value, the curves are independent of the neutron generation time, but beyond this point in time the period is nearly proportional to Λ; this is consistent with the behavior for step insertions shown in Fig. 3-3. Second, the ratio of the power at which prompt critical is reached to the initial power, $P(t_p)/P(0)$, decreases as the ramp rate increases.

The decrease in $P(t_p)/P(0)$ with increased ramp rate has important safety repercussions. If a potentially destructive prompt critical excursion is to be prevented, either the accident must be detected at a time sufficiently long before t_p for the reactor shutdown system to become effective before prompt critical is reached or inherent negative reactivity feedback mechanisms must override the insertion mechanism before prompt criticality is reached. For large ramp rates it can be exceedingly difficult to terminate the excursion before prompt critical is reached by either of these mechanisms if the initial power of the reactor is small compared to the full rated power of the reactor. Inherent thermal feedback effects tend to be negligible if the reactor power is not at least a few percent of the full power value. Thus if $P(0)$ is several orders of magnitude below the full power value, as is normally the case under start-up or zero-power testing conditions, the reactor would likely reach prompt critical at a power well below that at which any of the negative temperature feedback mechanisms would become significant. Moreover, none of the instrumentation measuring the conventional thermal parameters, such as core outlet temperature or primary system pressure, would detect the accident before prompt critical in this situation.

From the preceding it can be surmised that if a reactor is at a very low power level, the only way to prevent a prompt-critical excursion in the event of a ramp reactivity insertion of unlimited duration is to detect the accident by neutron flux measurements while the reactor is still near its initial power. This in itself can present difficult problems. For although neutron flux is invariably monitored by redundant instruments over the power range of the reactor, these instruments may not be adequate at very low power levels. What is necessary is a flux trip that can be set just above the initial flux level. This may require neutron counting as opposed to continuous current devices at zero power criticality. Moreover, the setting on the device or devices must be changed as the power is increased over many orders of magnitude. These frequent changes in the flux level for reactor trip may in turn lead to errors in which the trip level is set too high, resulting in a partial loss of protection, or a trip level that is too low, resulting in a spurious trip. An alternative is a flux period meter that measures $\mathcal{T}(t)$ as defined by Eq. (3-65), which will detect the accident at an early stage, regardless of the initial power level. These meters, however, are quite susceptible to noise and therefore may lead to an unacceptable number of spurious trips.

The Start-up Accident

In the absence of proper precautions, the start-up of a new reactor can be a potentially dangerous situation. The core design may incorporate features whose characteristics have been studied primarily through computational techniques, with complete experimental verification not being available until the reactor is brought critical. As fuel is loaded or control rods withdrawn, ramp reactivity insertions must be made into the core, therefore creating the potential of an overshoot, causing the core to become supercritical on too short a period. Moreover, as the reactor is brought critical it will be at an extremely low power—many orders of magnitude below the rated full power value—thus presenting the undesirable kinetics characteristics described in the preceding paragraphs. Under these circumstances it is only through careful design of maximum control rod reactivity worths and withdrawal rates, highly reliable neutron monitoring instrumentation, and a well-rehearsed crew that unquestioned safety is achieved.

The start-up procedure most often consists of alternately making incremental loadings of fuel into the reactor with the control rods inserted and withdrawing control rods to check on the reactivity of the fuel configuration. During this procedure a neutron flux is maintained in the reactor through the presence of an external source. To determine the equilibrium flux, or power, we set the derivatives in Eqs. (3-41) and (3-42) equal to zero. With a negative reactivity $\rho = -|\rho_0|$, we have

$$0 = -\frac{|\rho_0|+\bar{\beta}}{\Lambda}P(0) + \sum_{i=1}^{6} \lambda_i \tilde{C}_i(0) + \tilde{S}_e, \qquad (3\text{-}68)$$

$$0 = \frac{\bar{\beta}_i}{\Lambda}P(0) - \lambda_i \tilde{C}_i(0), \qquad i = 1, 2, \ldots, 6, \qquad (3\text{-}69)$$

or solving for the power,

$$P(0) = \frac{\Lambda \tilde{S}_e}{|\rho_0|} = \frac{l \tilde{S}_e}{1-k}, \qquad (3\text{-}70)$$

where $l = k\Lambda$ is the prompt neutron lifetime. At the very low power levels that exist during start-up, the power is detected only as being proportional to the neutron count rate. However, for fixed source strength the inverse count rate gives an indication of how rapidly one is approaching criticality relative, say, to the weight of fuel that has been loaded. This is shown in Fig. 3-5. All neutron counters will not represent identical curves because there will be some shifting of the neutron distribution in the core as fuel is added.

In general, as more fuel is added and the control rods are withdrawn, the power level rises initially and then comes to an equilibrium value. The closer k is to one, the longer it will take the equilibrium to be established. If the reactor is just critical then the source will cause P to increase linearly with time until some additional action is taken. If the source is withdrawn from a just critical system, the reactor will stabilize at a fixed power level. In practice it may be difficult to bring the reactor just critical. If the reactor is brought slightly supercritical the flux will rise more rapidly, with a stable period being established either as the power level rises to a point where the contribution from the source neutrons is insignificant or when the source is removed.

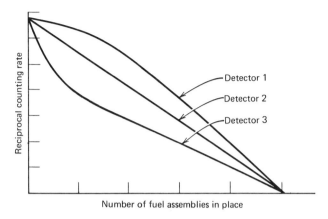

FIGURE 3-5 Reciprocal counting rates during an approach to a critical configuration.

The accident most often hypothesized in conjunction with start-up procedures consists of an insertion of reactivity at some fixed rate bringing the reactor from some subcritical initial state through criticality and on through prompt critical if nothing is done to terminate the ramp. Such an accident might by hypothesized as being due to the uncontrollable withdrawal of a control rod or a bank of control rods resulting from equipment failure or operator error. The analysis of such accidents can do much toward setting limits on control rod worths and withdrawal rates as well as specifying instrument reliability and operator procedures. Such an accident can be analyzed using the reactor kinetics equations, with Eq. (3-70) providing the initial condition on the power level at $t = 0$, and using a reactivity insertion given by $\rho(t) = -|\rho_0| + \dot{\rho}t$, where $-|\rho_0|$ is the reactivity state of the subcritical system before the insertion begins. The power trace for a typical set of data is shown in Fig. 3-6. The appearance of the curve is very similar to the ramp results from an initially critical reactor. Once again it is found that as the ramp rate is increased, the power at which the reactor reaches prompt criticality becomes smaller.

The start-up accident must be terminated by a reactor trip. The trip signal is most typically generated when the reactor passes above some unacceptable power level P_t. A critical parameter is the period of the reactor as the power passes through P_t, since if the period is too short, the power may rise to destructive levels before the trip system can become effective. For a reactor in a given subcritical initial state, the solution of the kinetics equations results in a curve such as that shown in Fig. 3-7. From this we see that it is desirable to have a small rate of reactivity insertion—which may be accomplished either by having small control rod worths, small withdrawal speeds, or both—and the smallest value of $P_t/P(0)$ possible. The latter is

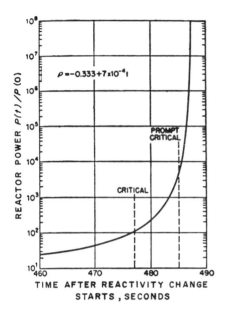

$\rho = -0.333 + 7 \times 10^{-4} t$

PROMPT CRITICAL

CRITICAL

REACTOR POWER $P(t)/P(0)$

TIME AFTER REACTIVITY CHANGE STARTS, SECONDS

FIGURE 3-6 Relative power level versus time for ramp function input. Adapted from M. A. Schultz, *Control of Nuclear Reactors and Power Plants*, McGraw-Hill, New York, 1961. Used by permission.

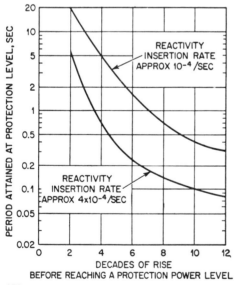

REACTIVITY INSERTION RATE APPROX 10^{-4}/SEC

REACTIVITY INSERTION RATE APPROX 4×10^{-4}/SEC

PERIOD ATTAINED AT PROTECTION LEVEL, SEC

DECADES OF RISE
BEFORE REACHING A PROTECTION POWER LEVEL

FIGURE 3-7 Period attained by a reactor during a start-up accident at a given fixed protection level versus the number of decades below the protection level reactivity insertion started. Startup from −13% in reactivity. Adapted from M. A. Schultz, *Control of Nuclear Reactors and Power Plants*, McGraw-Hill, New York, 1961. Used by permission.

152

accomplished in two ways. First, since $P(0)$ is proportional to the external source strength, as shown in Eq. (3-70), one would like as large an external source as possible. In this regard, once the reactor has been operated at high power, the external source may be substantially enhanced by photo neutrons produced from the gamma rays of fission product decay. Second, as stated earlier, it may be desirable to have a variable trip level to keep P_t as small as possible consistent with the constraints placed by the occurrence of spurious trips. It should be emphasized that this discussion has not included the effects of temperature coefficients on other inherent feedback phenomena. These substantially alter the effects of the transients if the power level rises to a significant fraction of the reactor's rated power.

Reactor Trip Reactivity

In view of the foregoing discussions, it is apparent that early accident detection and rapid reactor shutdown are essential features in protection against reactor damage. The effectiveness of reactor trip systems in poisoning the chain reaction depends both on the speed with which the rods can be driven into the core and on their incremental worth, defined as the decrease in reactivity per unit length of rod insertion. In reactor trip systems some fraction of a second is required to transmit the electrical signals from the flux detectors to the control rod actuators. A second time interval, often several seconds, is required to drive the rods far enough into the core to bring the reactor subcritical.

The effectiveness of a control rod versus time can be better understood by considering the situation illustrated in Fig. 3-8. If the rod is dropped into the reactor under the acceleration of gravity with no significant frictional forces present, the rod position versus time has the parabolic appearance shown in Fig. 3-8a. Assuming the rod does not greatly perturb the flux distribution within the reactor, incremental rod worth is greatest when the rod tip is passing through the high flux region near the core midplane. Thus the fraction of the total rod worth versus the distance inserted increases most rapidly near the core midplane, as shown in Fig. 3-8b. The fraction of the total rod worth available to shut the reactor down versus time is found by combining the results of Figs. 3-8a and b to obtain Fig. 3-8c. The total negative reactivity contribution from the control rods is then obtained by multiplying the curve in Fig. 3-8c by the total rod worth. It is only at the time when this negative contribution causes the multiplication to decrease below 1 that the reactor falls subcritical.

There are a number of complications in predicting the rate at which reactivity can be removed from the core by trip systems. For example, suppose the power distribution in a reactor is skewed toward the bottom of

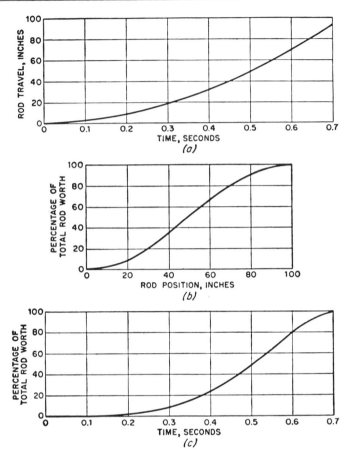

FIGURE 3-8   Rod effectiveness in a gravity drop trip situation. (a) Rod position versus time. (b) Rod worth versus position. (c) Rod worth versus time. Adapted from M. A. Schultz, *Control of Nuclear Reactors and Power Plants*, McGraw-Hill, New York, 1961. Used by permission.

the core. Then for control rods entering from the top, the curve in Fig. 3-8b is translated toward the right. The effect of this, in turn, is to move the curve in Fig. 3-8c to the right; it thus has the effect of further delaying the bulk of the negative reactivity insertion. A second important mechanism occurs in neutronically loosely coupled cores when banks of control rods are employed. As the rods enter the core they perturb the power distribution, tending to push it toward the bottom of the core. This again has the effect of moving the curves in Figs. 3-8b and c to the right, thus slowing the effectiveness of the reactor shutdown.

PROBLEMS

3-1. Show that Eqs. (3-10) and (3-11) follow from the preceding equations.

3-2. Verify the zero lifetime approximation, Eq. (3-61).

3-3. Consider a critical reactor that is initially operating at a power level $P(0)$. At $t = 0$ a step reactivity change ρ is made. Using the point kinetics equations with one group of delayed neutrons and the assumption that

$$\frac{\bar{\beta}}{\lambda \Lambda}|\rho - 1| \gg 1,$$

a. Show that

$$P(t) = P(0)\left[\frac{\rho}{\rho - \bar{\beta}}e^{[(\rho - \bar{\beta})t/\Lambda]} + \frac{\bar{\beta}}{\bar{\beta} - \rho}e^{[\lambda\rho/(\bar{\beta} - \rho)]t}\right]$$

b. Show that for long times the solution is independent of λ when $\rho > 1$ and independent of Λ when $0 < \rho < 1$, and when $\rho < 0$.

c. Taking $\bar{\beta} = 0.007$, $\Lambda = 5 \times 10^{-5}$ sec, and $\lambda = 0.08$ sec^{-1}, make a graph of the reactivity (in dollars) versus the reactor period for reactivities between -5 dollars and $+5$ dollars. Indicate the region on the graph where you expect the results of part (a) to be poor.

3-4. A reactor has no neutrons present. At $t = 0$ an external source $\tilde{S}_{ex}$ is inserted in the reactor. Use the point kinetics equations, with one delayed neutron group, to determine the reactor power as a function of time for $\rho < 0$, $\rho = 0$, $\rho > 0$. Qualitatively sketch your results.

3-5. Repeat problem 4, but assume that instead of a source being inserted, the reactor is pulsed with a source of external neutrons [i.e., $\tilde{S}_{ex} = S_{ex}\delta(0^+)$].

3-6. A reactor operating at power level P_0 displays a power oscillation. The cause is a reactivity oscillation of the form $\rho(t) = \rho_0 \exp(i\omega t)$, where $\rho_0 \ll \bar{\beta}$. Use the zero lifetime approximation to determine the normalized amplitude

$$\frac{P(t)}{P_0}\bigg|_{max} - \frac{P(t)}{P_0}\bigg|_{min}$$

of the power oscillation. Plot this normalized amplitude versus ω when $\rho_0 = 0.0005$, $\lambda = 0.08$ sec^{-1}, $\bar{\beta} = 0.0065$. Give a physical explanation of the ω dependence of your graph.

3-7. After studying the safety analysis report for a power reactor of your choice, discuss in some detail the precautions in design, instrumentation, and procedure that are taken to preclude the possibility of a start-up accident.

REACTIVITY FEEDBACK EFFECTS

Chapter 4

4-1 INTRODUCTION

In the preceding chapters the neutronic description of a nuclear reactor is treated independently of the thermal characteristics of the reactor core. The microscopic cross sections and atomic densities of each material are considered to be known quantities so that once the size and composition of the reactor are specified, its critical state can be determined. The equations from which the criticality condition is derived specify the distribution of the neutron flux within the core, but the flux normalization may be chosen arbitrarily. As indicated in Eq. (1-85), the power may be specified as a flux normalization. Hence the foregoing procedure allows the power to be chosen independently of the criticality condition. As discussed in Section 1-4, the temperatures in the reactor core are then determined from the power as well as the core inlet temperature and the mass flow rate of the coolant.

So long as the power level in a reactor is very low, the independence of the criticality determination from the power level implied by this analysis remains valid. However, as soon as power is produced at a level sufficient to cause significant temperature rises in the fuel elements, coolant channels, and other core constituents, a reactivity feedback loop is established, for the atom densities and microscopic cross sections that were assumed to be fixed when neutronics is treated independently are invariably temperature dependent.

In this chapter the link of the neutron multiplication to the thermal state of a reactor core is examined. The problem of modeling reactivity feedback is first discussed, and a simple lumped parameter model applicable to all current reactor types is formulated. In Sections 4-3 and 4-4 we examine in some detail the mechanisms that cause the more pronounced feedback effects in thermal and in fast reactors. Having gained a better understanding of reactivity feedback for the various reactor types, we return to our generalized model in Sections 4-5 and 4-6 to examine the composite feedback effects that determine a reactor's reactivity characteristics under steady-state operating conditions. The role of reactivity feedback in accident transients is taken up in the following chapter after a trasient heat transport model is formulated to link thermal-hydraulic to neutronic core behavior under transient conditions.

4-2 REACTIVITY MODELING

It is apparent from the discussion of reactor kinetics in the preceding chapter that the power produced by a reactor core is determined by the time-dependent multiplication $k(t)$, or correspondingly, the reactivity

$$\rho(t) = \frac{k(t) - 1}{k(t)}. \tag{4-1}$$

In a few situations it is possible to determine a priori the reactivity versus time. Such is often the case, for example, when a control rod of known properties is moved in a prescribed manner in a reactor operating at very low neutron flux levels. If the reactor produces enough power to raise the temperatures in the core above their ambient values, however, the densities of the materials and some of the microscopic cross sections in the reactor core become affected. Since the multiplication of the core depends on these densities and cross sections, a reactivity feedback loop is established in which $\rho(t)$ is dependent on temperatures and densities that are determined in turn by the reactor power history. The understanding of the nature of the mechanisms leading to reactivity feedback is essential to the analysis of power reactor behavior under both anticipated and accident conditions.

In relating incremental reactivity changes to the reactor multiplication, the following approximation is nearly universal:

$$d\rho = \frac{dk}{k^2} \simeq \frac{dk}{k}. \tag{4-2}$$

Since even in the severest of conditions k differs from one by no more than a few percent, little error is introduced into ρ. At the same time k is often expressed as a product of terms, and the logarithmic differential

$$\frac{dk}{k} = d(\ln k) \tag{4-3}$$

permits these terms to appear as additive contributions to ρ.

The examination of reactivity effects is often facilitated by expressing the multiplication in terms of infinite medium and leakage effects. Thus, following Eq. (1-83) or (3-28), we have

$$k = \frac{\bar{k}_\infty}{1 + \bar{M}^2 \bar{B}_g^2},\qquad(4\text{-}4)$$

where $\bar{k}_\infty$, $\bar{M}^2$, and $\bar{B}_g^2$ are the core-averaged values of the infinite medium multiplication, the migration area, and the geometrical buckling. Taking the differential of k, and utilizing Eq. (4-2), we have

$$d\rho = \frac{d\bar{k}_\infty}{\bar{k}_\infty} - \frac{\bar{M}^2 \bar{B}_g^2}{1 + \bar{M}^2 \bar{B}_g^2}\left(\frac{d\bar{M}^2}{\bar{M}^2} + \frac{d\bar{B}_g^2}{\bar{B}_g^2}\right).\qquad(4\text{-}5)$$

The sign of each contribution to the reactivity change is intuitively obvious. A change in composition that leads to a larger infinite medium multiplication also increases the reactivity of a finite system. An increase in $\bar{M}$, which is proportional to the distance over which neutrons migrate between birth and death, results in an increased fraction of neutrons leaking from a reactor of fixed size, thereby decreasing the reactivity. Likewise, a decrease in the reactor dimensions that gives rise to increased buckling increases the leakage and subtracts reactivity.

The leakage effects represented by the fractional changes in $\bar{M}^2$ and $\bar{B}_g^2$ are weighted by a factor

$$P_L = \frac{\bar{M}^2 \bar{B}_g^2}{1 + \bar{M}^2 \bar{B}_g^2}.\qquad(4\text{-}6)$$

From the definition of the nonleakage probability given in Eq. (1-80), this weight is seen to be just $1 - P_N$, or the probability that a neutron born in fission will leak from the reactor. For large, neutronically loosely coupled systems, P_L is quite small. From Table 1-5 it may be inferred that the leakage effects tend to be smallest for thermal reactors, particularly those with water as the coolant.

Effects of Temperature on Reactivity

When power is produced, the constituents of a reactor core will increase in temperature, causing them to undergo thermal expansion, and the volume of the reactor core will tend to increase. Suppose we first ask what the reactivity effect is if this expansion is uniform for all the core components.

That is to say, what if the total atom density

$$\bar{N} = \sum_i \bar{N}_i \tag{4-7}$$

decreases, but the ratios $\bar{N}_i/\bar{N}$, $i = 1, 2, \ldots$, of each constituent density to the total density remains constant? We first note that Eqs. (1-36) through (1-38) can be used to write the infinite medium multiplication for a uniform reactor as

$$k_\infty = \frac{\displaystyle\sum_i \bar{N}_i \int_0^\infty \nu\sigma_f^i(E)\varphi(E)\,dE}{\displaystyle\sum_i \bar{N}_i \int_0^\infty \sigma_a^i(E)\varphi(E)\,dE}. \tag{4-8}$$

If numerator and denominator are divided by $\bar{N}$, it is obvious that k_∞ is a function only of the $\bar{N}_i/\bar{N}$ ratios and not of $\bar{N}$. Therefore a uniform change in core density has no effect on k_∞, and the core expansion affects reactivity only through the leakage term.

In contrast to k_∞, the migration area M^2 is strongly dependent on the atomic density $\bar{N}$. Combining Eqs. (1-36), (1-37), (1-54), (1-61), and (1-64), we have

$$M^2 = \frac{1}{\bar{N}^2}\left[\int_0^\infty \frac{\varphi(E)\,dE}{3\sum_i (\bar{N}_i/\bar{N})\sigma^i(E)}\right]\left[\sum_i \frac{\bar{N}_i}{\bar{N}} \int_0^\infty \sigma_a^i(E)\varphi(E)\,dE\right]^{-1}. \tag{4-9}$$

Thus while the bracketed terms remain constant, $M^2 \propto \bar{N}^{-2}$ or

$$\frac{dM^2}{M^2} = -2\frac{d\bar{N}}{\bar{N}}. \tag{4-10}$$

It is convenient to relate dM^2 directly to the change in volume that the reactor undergoes as a result of the expansion. The total mass and therefore the total number of atoms in the reactor—$\bar{N}V$, where V is the volume—remains constant. Therefore

$$\frac{d\bar{N}}{\bar{N}} = -\frac{dV}{V}, \tag{4-11}$$

or

$$\frac{dM^2}{M^2} = 2\frac{dV}{V}. \tag{4-12}$$

The geometrical buckling also changes with the increase in reactor volume. From the definitions in Chapter 1 it is easily shown that once the shape of a uniform reactor is fixed, $B_g^2 \propto R^{-2}$, where R is a characteristic dimension. However, volume is related to the characteristic dimension by $V \propto R^3$. Therefore $B_g^2 \propto V^{-2/3}$, and we may write

$$\frac{dB_g^2}{B_g^2} = -\frac{2}{3}\frac{dV}{V}. \tag{4-13}$$

If we now substitute Eqs. (4-12) and (4-13) into Eq. (4-5), we have

$$d\rho = -\frac{4}{3}P_L\frac{dV}{V}. \tag{4-14}$$

Thus for a uniform expansion, the reactivity effect is negative. Conversely, if the reactor core should be suddenly compacted because of external pressure, a reactivity increase would result. In either case the magnitude of the result decreases with increased core volume because of the decreasing magnitude of P_L, the leakage probability.

In some early experimental reactors, particularly those consisting of a solid mass of fissionable material, the reactivity loss from uniform expansion played a dominant role in causing the reactor to shut down before excessive heating could take place. In large power reactors, however, the small value of the leakage tends to make this expansion effect quite small relative to other reactivity feedback effects that are present. These reactivity feedback mechanisms are discussed in some detail for thermal and for fast reactors in the following two sections. A few additional introductory remarks, however, are in order at this point.

In a power reactor the fuel, coolant, moderator, and other core constituents are not a homogeneous mixture but are physically separated from one another as illustrated, for example, in the lattice cell configurations of Fig. (1-12). Since the core constituents also vary greatly in compressibility, uniform heating results in a variety of expansion rates. Only in the event that the components are rigidly bonded, causing them to move in unison, would the atom ratios $\bar{N}_i/\bar{N}$ remain constant, causing the differing thermal expansivities to generate internal thermal stresses in the constituents. The only reactivity effect from expansion would then be from the change in the total atom density discussed previously.

In power reactors the coolant is free to move relative to the fuel elements, and often the fuel, cladding, solid moderator, or other constituents are not rigidly joined. The atom ratios $\bar{N}_i/\bar{N}$ may therefore change, and this effect is often more important than the uniform density change, since k_∞ is sensitive to these ratios, as seen from Eq. (4-8). For example, if the temperature of the reactor is raised uniformly but the pressure of the coolant is kept approximately constant, the atom ratios of coolant to the solid core constituents decreases for the following reason. The liquid or gaseous coolant has a substantially larger coefficient of thermal expansion than the solid fuel element constituents. Therefore, although the changes in fuel element diimensions are quite small, the coolant expands axially out the ends of the core, leaving fewer atoms in the coolant channels. Should boiling of a liquid coolant take place, the decrease in the coolant atom density would become even more pronounced.

Further changes in the core atom density rates are caused by the fact that rarely do the fuel element and coolant temperatures change at the same rate. Under both operating and accident conditions, ceramic fuel may undergo temperature swings of thousands of degrees, whereas coolant temperature rarely varies by more than a few hundred degrees. In these circumstances fuel and cladding expansion effects may become important, even though the thermal expansivity is much smaller than that of the coolant. As an example, in early fast reactor designs the axial expansion of the metal fuel during fast power transients was an important shutdown mechanism, because it tended to decrease the ratio of fuel to coolant in the core (Hummel and Okrent, 1970).

Several more subtle effects may come into play in determining the reactivity change from changes in temperature and density. It will be noted from Eqs. (4-8) and (4-9) that k_∞ and M^2 depend on the neutron energy spectrum $\varphi(E)$ and on the microscopic cross sections $\sigma_x^i(E)$ as well as the atom densities. Changes in the atom density ratios also affect the neutron spectrum. As discussed in Chapter 1, decreasing the amount of lighter atomic weight coolant or moderator material relative to the fuel tends to shift $\varphi(E)$ upward in energy. In a fast reactor this tends to increase both k_∞ and M^2, with the dominant effect depending on the size and composition of the core. In a thermal reactor M^2 again tends to increase from the upward energy shift of $\varphi(E)$, but the effect on k_∞ may be positive or negative, depending on the initial moderator to fuel ratio.

In addition to the reactivity changes resulting from density-induced shifts in the neutron spectrum, there are situations where the energy dependence of the microscopic cross sections $\sigma_x^i(E)$ changes with temperature. Probably the most important example of such a change is the Doppler broadening in the resonances of fissionable material cross sections. Doppler broadening is discussed in more detail in the following sections. Suffice it to say that increase in fuel temperature has the effect of causing $\sigma_a(E)$ of fertile materials to increase in the eV to keV energy range. The result is that k_∞ decreases and reactivity is lost.

A Simplified Core-Averaged Feedback Model

The accurate determination of reactivity changes caused by temperature or density feedback from the fuel, coolant, or other core constituents is sometimes quite difficult, since the feedback effects depend not just on the core-averaged values of the feedback variables but on their distribution through the reactor core. And, moreover, if the feedback effect is strong and localized, it may lead to substantial changes in the spatial distribution of the neutron flux. For purposes of analysis it is useful to make two simple assumptions:

1. The feedback variables may be written as separable functions of space and time.
2. The reactivity feedback causes no significant changes in the spatial distribution of the neutron flux.

With these assumptions, simplified models may be used that provide a good deal of physical insight into the nature of reactivity feedback while often giving sufficient accuracy provided they are applied with care.

For purposes of illustration, suppose we consider the simple thermal hydraulic reactor model developed in Chapter 1. We assume that although the core-averaged values of the fuel element and coolant temperatures may change with time, the spatial distributions of these temperatures about their core-averaged values remain fixed. Thus, combining Eqs. (1-107), (1-109), (1-112), and (1-114), we find that the temperature distributions are

$$\tilde{T}_{fe}(r, z, t) = [\bar{T}_{fe}(t) - \bar{T}_c(t)] f_r(r) f_z(z)$$

$$+ [\bar{T}_c(t) - \bar{T}_i(t)] f_r(r) \frac{2}{\tilde{H}} \int_{-\tilde{H}/2}^{z} f(z') \, dz' + \bar{T}_i(t) \tag{4-15}$$

and

$$T_c(r, z, t) = [\bar{T}_c(t) - \bar{T}_i(t)] f_r(r) \frac{2}{\tilde{H}} \int_{-\tilde{H}/2}^{z} f(z') \, dz' + \bar{T}_i(t). \tag{4-16}$$

As in Chapter 1, $\bar{T}_{fe}$, $\bar{T}_c$, and $\bar{T}_i$ are the core-averaged fuel element, coolant, and inlet temperatures, and $f_r(r)$ and $f_z(z)$ are the normalized radial and axial power shapes as defined by Eqs. (1-89), (1-91), and (1-92). Thus if $\bar{T}_{fe}$, $\bar{T}_c$, and $\bar{T}_i$ are known as a function of time, the entire temperature state of the reactor is known. We assume that the reactivity feedback depends only on these temperature distributions. Therefore reactivity feedback changes may be represented in terms of the three core-averaged temperatures as

$$d\rho_{fb} = \frac{1}{k}\frac{\partial k}{\partial \bar{T}_{fe}}\, d\bar{T}_{fe} + \frac{1}{k}\frac{\partial k}{\partial \bar{T}_c}\, d\bar{T}_c + \frac{1}{k}\frac{\partial k}{\partial \bar{T}_i}\, d\bar{T}_i. \tag{4-17}$$

The temperature coefficients $(1/k)(\partial k/\partial \bar{T}_x)$ (where $x = fe, c$, or i) must be evaluated if reactivity changes are to be predicted. In some situations these may be estimated experimentally from the measurable composite coefficients described in Section 4-5. We here utilize the simple reactor physics model developed in Chapter 3 to investigate further the temperature coefficients. Using Eq. (4-4) to express k in terms of the core-averaged parameters, we have

$$\frac{1}{k}\frac{\partial k}{\partial \bar{T}_x} = \frac{1}{\bar{k}_\infty}\frac{\partial \bar{k}_\infty}{\partial \bar{T}_x} - \frac{\bar{M}^2 \bar{B}_g^2}{1+\bar{M}^2 \bar{B}_g^2}\left(\frac{1}{\bar{M}^2}\frac{\partial \bar{M}^2}{\partial \bar{T}_x} + \frac{1}{\bar{B}_g^2}\frac{\partial \bar{B}_g^2}{\partial \bar{T}_x}\right). \tag{4-18}$$

The terms on the right consist of partial derivatives of the core-averaged values $\bar{k}_\infty$ and $\bar{M}^2$ and $\bar{B}_g^2$, with respect to the core-averaged temperatures $\bar{T}_{fe}$ and $\bar{T}_c$ and $\bar{T}_i$. The derivations of the following sections, 4-3 and 4-4, however, make it clear that physically meaningful expressions for reactivity feedback effects are invariably obtained in the form of partial derivatives of the local, space-dependent values of k_∞ and M^2 with respect to the local temperatures $\tilde{T}_{fe}$ and T_c. It is therefore necessary to express the derivatives on the right in terms of the partial derivatives of

$$k_\infty = k_\infty(\mathbf{r}, \tilde{T}_{fe}, T_c) \tag{4-19}$$

and

$$M^2 = M^2(\mathbf{r}, \tilde{T}_{fe}, T_c) \tag{4-20}$$

with respect to $\tilde{T}_{fe}$ and T_c.

The local and core-averaged values of $\bar{k}_\infty$, $\bar{M}^2$, and $\bar{B}_g^2$ are related by Eqs. (3-29), (3-30), and (3-31). From these equations we obtain

$$\frac{1}{\bar{k}_\infty}\frac{\partial \bar{k}_\infty}{\partial \bar{T}_x} = \frac{1}{\langle \psi_0^2 k_\infty \rangle}\left\langle \psi_0^2 \frac{\partial k_\infty}{\partial \bar{T}_x}\right\rangle \tag{4-21}$$

$$\frac{1}{\bar{M}^2}\frac{\partial \bar{M}^2}{\partial \bar{T}_x} = \frac{1}{\langle (\nabla\psi_0)^2 M^2 \rangle}\left\langle (\nabla\psi_0)^2 \frac{\partial M^2}{\partial \bar{T}_x}\right\rangle \tag{4-22}$$

$$\frac{1}{\bar{B}_g^2}\frac{\partial \bar{B}_g^2}{\partial \bar{T}_x} = 0, \tag{4-23}$$

where we have assumed that the reactor volume and therefore the value of $\bar{B}_g^2$ does not change. For our three-temperature parameter model the required derivatives of Eqs. (4-21) and (4-22) can be represented in terms of the fuel element and coolant local temperatures as

$$\frac{\partial k_\infty}{\partial \bar{T}_x} = \frac{\partial k_\infty}{\partial \tilde{T}_{fe}} \frac{\partial \tilde{T}_{fe}}{\partial \bar{T}_x} + \frac{\partial k_\infty}{\partial T_c} \frac{\partial T_c}{\partial \bar{T}_x} \tag{4-24}$$

and

$$\frac{\partial M^2}{\partial \bar{T}_x} = \frac{\partial M^2}{\partial \tilde{T}_{fe}} \frac{\partial \tilde{T}_{fe}}{\partial \bar{T}_x} + \frac{\partial M^2}{\partial T_c} \frac{\partial T_c}{\partial \bar{T}_x}, \tag{4-25}$$

where $x = fe$, c, or i.

Now, substituting Eqs. (4-15) and (4-16) into these expressions, and then substituting the result into Eqs. (4-21) and (4-22) yields core-averaged temperature coefficients:

$$\frac{1}{\bar{k}_\infty} \frac{\partial \bar{k}_\infty}{\partial \bar{T}_{fe}} = \frac{1}{\langle \psi_0^2 k_\infty \rangle} \left\langle \psi_0^2 \tilde{\alpha} \frac{\partial k_\infty}{\partial \tilde{T}_{fe}} \right\rangle \tag{4-26}$$

$$\frac{1}{\bar{k}_\infty} \frac{\partial \bar{k}_\infty}{\partial \bar{T}_c} = \frac{1}{\langle \psi_0^2 k_\infty \rangle} \left\langle \psi_0^2 \left[(\tilde{\beta} - \tilde{\alpha}) \frac{\partial k_\infty}{\partial \tilde{T}_{fe}} + \tilde{\beta} \frac{\partial k_\infty}{\partial T_c} \right] \right\rangle \tag{4-27}$$

$$\frac{1}{\bar{k}_\infty} \frac{\partial \bar{k}_\infty}{\partial \bar{T}_i} = \frac{1}{\langle \psi_0^2 k_\infty \rangle} \left\langle \psi_0^2 (1 - \tilde{\beta}) \left(\frac{\partial k_\infty}{\partial \tilde{T}_{fe}} + \frac{\partial k_\infty}{\partial T_c} \right) \right\rangle \tag{4-28}$$

$$\frac{1}{\bar{M}^2} \frac{\partial \bar{M}^2}{\partial \bar{T}_{fe}} = \frac{1}{\langle (\nabla \psi_0)^2 M^2 \rangle} \left\langle (\nabla \psi_0)^2 \tilde{\alpha} \frac{\partial M^2}{\partial \tilde{T}_{fe}} \right\rangle \tag{4-29}$$

$$\frac{1}{\bar{M}^2} \frac{\partial \bar{M}^2}{\partial \bar{T}_c} = \frac{1}{\langle (\nabla \psi_0)^2 M^2 \rangle} \left\langle (\nabla \psi_0)^2 \left[(\tilde{\beta} - \tilde{\alpha}) \frac{\partial M^2}{\partial \tilde{T}_{fe}} + \tilde{\beta} \frac{\partial M^2}{\partial T_c} \right] \right\rangle \tag{4-30}$$

$$\frac{1}{\bar{M}^2} \frac{\partial \bar{M}^2}{\partial \bar{T}_i} = \frac{1}{\langle (\nabla \psi_0)^2 M^2 \rangle} \left\langle (\nabla \psi_0)^2 (1 - \tilde{\beta}) \left(\frac{\partial M^2}{\partial \tilde{T}_{fe}} + \frac{\partial M^2}{\partial T_c} \right) \right\rangle, \tag{4-31}$$

where the spatial weight functions are

$$\tilde{\alpha} = f_r(r) f_z(z) \tag{4-32}$$

and

$$\tilde{\beta} = f_r(r) \frac{2}{\tilde{H}} \int_{-\tilde{H}/2}^{z} f_z(z') \, dz'. \tag{4-33}$$

At this point the integrals of Eqs. (4-26) through (4-31) may be evaluated. If the steady-state power distribution is used in evaluating the neutron flux, then ψ_0 and $\nabla\psi_0$ are expressed in terms of f_r and f_z as given by Eqs. (1-84) and (1-89). The results may then be substituted into Eqs. (4-17) and (4-18) to obtain $d\rho_{fb}$ in terms of the changes in $\bar{T}_{fe}$, $\bar{T}_c$, and $\bar{T}_i$.[*]

The foregoing simple lumped parameter feedback model includes under the heading of fuel element feedback both the effects of the fuel and the cladding regions. For typical oxide-fueled metal-clad fuel elements the feedback effects of the fuel temperature tend to dominate those caused by the cladding. Thus to a first approximation it is often acceptable to ignore the cladding effects and equate the fuel element feedback with that of the fuel region of the element; this is done in the following sections.

In fast reactors, where there is no moderator, representing the feedback in terms of the fuel element and coolant temperatures is appropriate. Likewise, in those thermal reactors for which the coolant also serves as the moderator (such as in the core in many water-cooled systems) the foregoing representation of the feedback effects in terms of $\tilde{T}_{fe}(r, z)$ and $T_c(r, z)$ is appropriate. In thermal reactors for which the moderator is separate from the coolant a separate term for the moderate temperature $T_m(r, z)$ must be included. In gas-cooled thermal systems the coolant feedback effects are often small, and therefore $T_m(r, z)$ may simply replace $T_c(r, z)$ as a local feedback variable, provided appropriate relationships are available to replace Eqs. (4-15) and (4-16) in relating $\tilde{T}_{fe}(r, z)$ and $T_m(r, z)$ to the core-averaged temperature $\bar{T}_{fe}$, $\bar{T}_c$, and $\bar{T}_i$ and the core-averaged moderate temperature $\bar{T}_m$. Then Eq. (4-17) can be replaced by

$$d\rho_{fb} = \frac{1}{k}\frac{\partial k}{\partial \bar{T}_{fe}} d\bar{T}_{fe} + \frac{1}{k}\frac{\partial k}{\partial \bar{T}_m} d\bar{T}_m + \frac{1}{k}\frac{\partial k}{\partial \bar{T}_c} d\bar{T}_c + \frac{1}{k}\frac{\partial k}{\partial \bar{T}_i} d\bar{T}_i. \qquad (4\text{-}34)$$

In reactors such as the light-water-cooled heavy-water-moderated pressure tube system, where the liquid coolant is separate from the moderator, both of the local temperature distributions $T_m(r, z)$ and $T_c(r, z)$, as well as $\tilde{T}_{fe}(r, z)$, must be included in the evaluation of the coefficients of Eq. (4-34). Finally, if boiling water reactors are being treated, it is more appropriate to work directly in terms of the coolant density distribution $N_c(r, z)$; the average coolant temperature $\bar{T}_c$ is then most often replaced by the core-averaged void fraction α, which is defined as the ratio of the volume of coolant vapor to the total coolant volume in the core.

---

[*] In many applications where the axial flux and power distributions are symmetric about the core midplane and where $f_r(r) \approx 1$, the right-hand sides of Eqs. (4-28) and (4-31) become small (see problem 4-4). Therefore $(1/k)/(\partial k/\partial \bar{T}_i)$ can often be set equal to zero in Eq. (4-17) and subsequent expressions without introducing significant errors.

Localized Reactivity Feedback

In the foregoing discussion reactivity feedback is expressed in terms of the relatively few core-averaged parameters. Such a representation is valid, however, only if the deviations of the power distribution from a separable function in space and time can be ignored and the temperature distributions can be expressed in terms of a few spatially averaged parameters. Even then a fair amount of algebra is required to get from the local reactivity feedback data to a core-averaged feedback model. In the event that either of the assumptions at the beginning of the preceding subsection breaks down, adequate modeling of the feedback becomes considerably more difficult. This is particularly true for transients where localized changes in temperature or density are important. Such is the case, for example, in localized flow blockage accidents where overheating is initially confined to the core region downstream from the blockage. In such situations core-averaged parameters are poor indicators of the feedback, and distributed parameter models must be used. One approach is to take fuel and coolant temperature averages over subregions of the core. If the spatial distribution of the power changes radically during the course of the transients, as is prone to happen in large loosely coupled cores when strong, local reactivity effects take place, the definitions of $\bar{k}_\infty$, $\bar{M}^2$, and $\bar{B}_g^2$ given in Eqs. (3-29) through (3-31) are no longer appropriate. One must then resort to more sophisticated space–time dynamics methods (Stacey, 1969) to solve the reactor kinetics equations.

At the opposite extreme from the core-averaged model presented in the preceding subsection, it is often necessary in safety work to estimate the reactivity effect resulting from extremely localized perturbations. It is often desirable to know, for example, the relative effect of local boiling of the coolant at any point in a core. Suppose, for example, that we are required to determine the effect of voiding some small fraction of the coolant from the core. Assuming that the small void does not appreciably affect the neutron flux distribution, we may use Eq. (4-5) to express the reactivity change as

$$\delta\rho_{fb} = \frac{\langle\psi_0^2\delta k_\infty\rangle}{\langle\psi_0^2 k_\infty\rangle} - \frac{\bar{M}^2\bar{B}_g^2}{1+\bar{M}^2\bar{B}_g^2}\frac{\langle(\nabla\psi_0)^2\delta M^2\rangle}{\langle(\nabla\psi_0)^2 M^2\rangle}. \tag{4-35}$$

Now suppose that the changes in the local values of k_∞ and M^2 resulting from the replacement of the liquid by the vapor phase of the coolant are Δk_∞ and ΔM^2, respectively. If a unit volume of the reactor core localized about position $\mathbf{r}_0$ is voided, we have

$$\delta k_\infty = \Delta k_\infty \delta(\mathbf{r}-\mathbf{r}_0); \qquad \delta M^2 = \Delta M^2 \delta(\mathbf{r}-\mathbf{r}_0). \tag{4-36}$$

Evaluating the Dirac delta function $\delta(\mathbf{r} - \mathbf{r}_0)$, we have

$$\delta\rho_{fb} = \frac{\psi_0(\mathbf{r}_0)^2 \Delta k_\infty}{\langle \psi_0^2 k_\infty \rangle} - \frac{\bar{M}^2 \bar{B}_g^2}{1 + \bar{M}^2 \bar{B}_g^2} \frac{[\nabla \psi_0(\mathbf{r}_0)]^2 \Delta M^2}{\langle (\nabla \psi_0)^2 M^2 \rangle}. \tag{4-37}$$

It is clear from this expression that the sign and magnitude of a localized reactivity effect may be strongly position dependent. If the change in k_∞ predominates, the largest effects will be seen when the perturbation takes place near the center of the core, where the flux is large but the flux gradient tends to be small. Conversely, if the change in M^2 predominates, the effect is larger for perturbations in the outer regions of the core where ψ_0 is smaller but $|\nabla \psi_0|$ becomes larger. In the case where the two terms are of comparable magnitude but opposite in sign, the net reactivity effect may change signs as the perturbation is moved from the center to the periphery of the core.

Probably the most widely studied local perturbation has been the sodium void coefficient, because of its extreme importance to liquid-metal-cooled

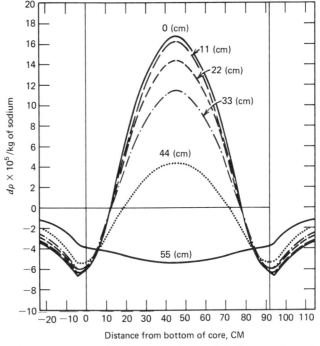

FIGURE 4-1 Sodium void worth curves at various radial distances from the Fast Flux Test Facility centerline. Adapted from *Lecture Series in Fast Reactor Safety Technology and Practices*, Vol. 2, P. L. Hoffman (Ed.), Batelle Northwest Laboratory Report, BNWL-SA-3093.

fast reactor design. As indicated in Section 4-4, removing sodium from a fast reactor causes both k_∞ and M^2 to increase. Thus the sign of the first term of Eq. (4-37) is positive, whereas that of the second is negative. Thus if sodium boiling takes place near the core periphery a reactivity decrease results, whereas if the boiling occurs near the center of the core, reactivity increases. This effect is illustrated in Fig. 4-1, where plots are shown of the reactivity increase per unit mass of sodium removed from the Fast Flux Test Facility. Note that the reactivity changes sign, becoming negative as the distance from the core midplane increases, and it remains negative in the reflecting blankets above and below the core. Likewise, as one proceeds radially outward through increasing values of r_i, the effect of voiding changes sign and remains negative within the radial blanket, which begins at $r = 60$ cm.

In addition to determining the effect of local sodium boiling, it is desirable to know the effect of voiding the entire core. The sign of the net reactivity effect of whole-core boiling depends on the core dimensions. For small cores the leakage effect is dominant, and the voiding results in a reactivity loss. However, as the size of the core is increased, the effect of the increase becomes more pronounced, and for large fast reactor cores, a reactivity increase results from voiding the entire core.

4-3 THERMAL REACTORS

The models developed in the preceding section express the reactivity feedback for a reactor core in terms of partial derivatives of the local values of k_∞ and M^2 with respect to the local temperatures of the fuel, coolant, and other core constituents. In this and the following section we examine the local reactivity coefficients for thermal and for fast reactors, respectively. In Section 1-3 we define the local quantities k_∞ and M^2 at a point $\mathbf{r}$ in a reactor of finite size to be the multiplication and migration area that would occur in a uniform infinite reactor whose composition is the same as that occurring in the immediate vicinity of point $\mathbf{r}$. Specifically, $k_\infty(\mathbf{r})$ and $M^2(\mathbf{r})$ at any point in the reactor are calculated from an infinite lattice of cells, each consisting of an arrangement of fuel pin, coolant channel, and other constituents that is identical to the cell containing point $\mathbf{r}$. We therefore confine our attention to infinite lattices for which the cells are of identical material composition, geometrical arrangement, and component temperatures and densities, and we seek to determine the partial derivatives of k_∞ and M^2 with respect to the component temperature and/or densities.

In large thermal power reactors, leakage effects tend to be small, as shown, for example, by Table 1-5. Thus, as indicated by Eq. (4-18), changes in the migration area and buckling make a relatively minor contribution to reactivity changes, since they are multiplied by the leakage probability. For this reason it is much more important to determine changes in k_∞ accurately than it is to determine those in M^2. In what follows we confine our attention to the changes in k_∞; we here simply reiterate that M^2 is expected to increase with reactor temperature, since it is inversely proportional to the square of the atom density.

For brevity, we confine our treatment to a simple, two-region cell consisting of fuel and moderator regions, thus avoiding the more complex notation required to take into account cladding effects or to treat three region cells containing separate coolant and moderator regions. We begin by deducing the widely used four-factor formula, (Lamarsh, 1966), and then view the fuel and moderator temperature derivatives of k_∞ in terms of the contributions from each of the four factors.

The Four-Factor Formula

In Chapter 1 the infinite medium multiplication at any point in the core is expressed as

$$k_\infty = \frac{\int_0^\infty \nu\Sigma_f(E)\varphi(E)\,dE}{\int_0^\infty \Sigma_a(E)\varphi(E)\,dE}. \tag{4-38}$$

Since this definition states that k_∞ is just equal to the total rate of neutron production (by fission) divided by the total rate of neutron destruction (by absorption), it is referred to as the neutron balance definition of k_∞ (Weinberg and Wigner, 1958). For natural uranium or slightly enriched thermal reactors of the type used for electrical power production, several approximations can be made to Eq. (4-38) that permit k_∞ to be viewed more intuitively in terms of a neutron life cycle (Weinberg and Wigner, 1958). These approximations lead to the traditional four-factor formula.

In a natural uranium or slightly enriched system, the neutron spectrum may be roughly divided into three energy regions, each having distinct patterns of fission and capture: a thermal energy region from 0 to about 1 eV, an intermediate region between about 1 eV and 1 MeV, and the fast region above 1 MeV. We denote reaction rates in terms of their thermal (T), intermediate (I), and fast (F) contributions as

$$\int_0^\infty \Sigma_x(E)\varphi(E)\,dE = \int_T \Sigma_x(E)\varphi(E)\,dE + \int_I \Sigma_x(E)\varphi(E)\,dE$$

$$+ \int_F \Sigma_x(E)\varphi(E)\,dE. \tag{4-39}$$

The properties of the three neutron energy regions may be understood in terms of the relative magnitudes of the capture and fission cross sections of the fissile, fertile, and moderator nuclides. The scattering cross sections enter only indirectly since they do not result in the production or destruction of neutrons, but only affect the neutron energy spectrum $\varphi(E)$. Microscopic cross sections of a fissile and a fertile nuclide are given in Figs. 1-7 and 1-8. In a slightly enriched system the atom density ratio of fissile to fertile nuclides is of the order of a few percent or less. A typical plot of $\varphi(E)$ for a thermal reactor is given in Fig. 1-5.

In the thermal energy spectrum the macroscopic capture cross sections of fissile, fertile, and moderator materials are comparable to one another and to the fission cross section; therefore all must be taken into account. The capture cross section of the moderator in the intermediate and fast regions is negligible. Likewise, because of its small atomic density, both capture and fission in the fissile nuclide may be neglected in the intermediate and fast regions. In the intermediate region capture in the large resonance cross sections of the fertile nuclides is the only nonscattering process that need be considered. In the fast region the fission cross section of the fertile material becomes nonzero above a threshold energy. This fast fission in the fertile material along with associated absorption reactions must be treated, although the treatment may be quite approximate since fast fission accounts for only a few percent of the neutron production and fast absorption for even less of the neutron destruction.

In devising an approximation for k_∞, the net effect of the fast fission and absorption is lumped into a multiplication correction ε on the thermal fission neutron production rate. That is to say, if $\int_T \nu \sum_f (E)\varphi(E)\,dE$ is the rate of neutron production by thermal fission, then the total rate at which neutrons are produced and made available for slowing down through the intermediate energy region is $\varepsilon \int_T \nu \sum_f (E)\varphi(E)\,dE$, where ε is called the fast fission factor. The fast fission factor may be estimated from approximate techniques given in standard reactor theory texts (Lamarsh, 1966). It typically has a value between 1.00 and 1.05 and has very little effect on our further discussions of reactivity feedback.

In an infinite medium, all the fission neutrons made available for slowing down through the intermediate energy region by elastic collision with the moderator are destroyed in one of two ways: by absorption in the thermal energy region or by absorption in the fertile material capture resonances near the bottom of the intermediate energy region. Thus, using the fast fission factor to approximate the behavior in the fast region k_∞—the ratio of neutron production to destruction—becomes

$$k_\infty = \frac{\varepsilon \int_T \nu \sum_f (E)\varphi(E)\,dE}{\int_T \sum_a (E)\varphi(E)\,dE + \int_I \sum_a (E)\varphi(E)\,dE}. \qquad (4\text{-}40)$$

The ratio of thermal to thermal plus intermediate absorption

$$p = \frac{\int_T \Sigma_a(E)\varphi(E)\,dE}{\int_T \Sigma_a(E)\varphi(E)\,dE + \int_I \Sigma_a(E)\varphi(E)\,dE},$$ (4-41)

may be interpreted as the probability that fission neutrons, slowing down into the intermediate region, will escape capture in the resonances of the fertile material and reach the thermal energy region; thus p is called the resonance escape probability. From Eqs. (4-40) and (4-41) we may write

$$k_\infty = \varepsilon p \frac{\int_T \nu \Sigma_f(E)\varphi(E)\,dE}{\int_T \Sigma_a(E)\varphi(E)\,dE}.$$ (4-42)

To complete the factorization into the four factors and to express k_∞ in a form more amenable to analysis for the feedback effects, account must be taken of the fact that the infinite medium multiplication is defined for a reactor lattice and not for a homogeneous mixture of nuclides. The lattice may have, for example, one of the configurations shown in Fig. 1-12. The fuel regions consist of the mixture of heavy atomic weight fertile and fissile materials, most often in the form of an oxide, carbide, or other ceramic. The moderator region consists of the light atomic weight nuclei for neutron slowing down. In water-cooled reactors the moderator also serves as coolant. If a solid moderator, such as graphite, is used, then there must be a third distinct region occupied by the coolant. In our analysis we confine attention to a two-region lattice consisting of fuel and moderator regions. This is strictly applicable only to reactors where the moderator and coolant are the same. However, it is often a very good approximation for gas-cooled reactors, because the atom density of the gas coolant is so small that neutronically the coolant channels can be treated as voids, or at least treated through correction on the basic two-region formulas. The effects of structural materials and the oxygen or carbon in the ceramic fuel area may be treated as correction terms.

As stated in Section 1-4, the cross sections and neutron spectra for a reactor lattice are the spatial averages over a lattice cell. Thus the cell averaged reaction rates must consist of the sum of contributions from the fuel and from the moderator regions. If f and m refer to fuel and moderator regions, the reaction rates averaged over the volume of the cell appear as

$$\int_T \Sigma_x(E)\varphi(E)\,dE = \frac{1}{V_{\text{cell}}}\left[V_f \int_T \Sigma_x^f(E)\varphi_f(E)\,dE \right.$$

$$\left. + V_m \int_T \Sigma_x^m(E)\varphi_m(E)\,dE \right],$$ (4-43)

where $\varphi_f(E)$ and $\varphi_m(E)$ are the neutron fluxes that are space-averaged over the fuel and moderator volumes, respectively; V_f, V_m are the corresponding volumes, and to a first approximation the thermal cross sections $\Sigma_x^f(E)$ and $\Sigma_x^m(E)$ for the reaction x are assumed to be space-independent within the fuel and moderator regions, respectively.

For the two-region lattice under consideration there is no fission in the moderator region, whereas there is absorption in both fuel and moderator regions. The integrals of Eq. (4-43) thus reduce to

$$\int_T \nu \Sigma_f(E)\varphi(E)\,dE = \frac{V_f}{V_{\text{cell}}} \nu \bar{\Sigma}_f^f \bar{\varphi}_f, \tag{4-44}$$

and

$$\int_T \Sigma_a(E)\varphi(E)\,dE = \frac{V_f}{V_{\text{cell}}} \bar{\Sigma}_a^f \bar{\varphi}_f + \frac{V_m}{V_{\text{cell}}} \bar{\Sigma}_a^m \bar{\varphi}_m, \tag{4-45}$$

where the energy-averaged thermal cross sections for reaction x in fuel or moderator regions are

$$\bar{\Sigma}_x^f = \bar{\varphi}_f^{-1} \int_T \Sigma_x^f(E)\varphi_f(E)\,dE, \tag{4-46}$$

$$\bar{\Sigma}_x^m = \bar{\varphi}_m^{-1} \int_T \Sigma_x^m(E)\varphi_m(E)\,dE, \tag{4-47}$$

and the energy-averaged thermal fluxes in fuel and moderator regions are

$$\bar{\varphi}_f = \int_T \varphi_f(E)\,dE, \tag{4-48}$$

$$\bar{\varphi}_m = \int_T \varphi_m(E)\,dE. \tag{4-49}$$

The preceding expressions now may be used to obtain the conventional form of k_∞. Multiply and divide Eq. (4-42) by the absorption rate in the fuel,

$$\int_T \Sigma_a^f(E)\varphi(E)\,dE = \frac{V_f}{V_{\text{cell}}} \bar{\Sigma}_a^f \bar{\varphi}_f. \tag{4-50}$$

The result may be expressed in terms of the four factors:

$$k_\infty = \epsilon p \eta f, \tag{4-51}$$

if we define

$$\eta = \frac{\nu \bar{\Sigma}_f^f}{\bar{\Sigma}_a^f} \tag{4-52}$$

and

$$f \equiv \frac{V_f \overline{\Sigma}_a^f \bar{\varphi}_f}{V_f \overline{\Sigma}_a^f \bar{\varphi}_f + V_m \overline{\Sigma}_a^m \bar{\varphi}_m}. \tag{4-53}$$

The factoring of the thermal reaction rates into η and f aids physical insight. The thermal regeneration rate η is just the number of fission neutrons produced per thermal neutron absorbed in the fuel region; it is predominantly dependent on the thermal cross sections of the fissile and fertile materials. The thermal utilization f is the ratio of absorption in the fuel to total absorption. It therefore may be viewed as the probability that a neutron that has been slowed down to thermal energies will be absorbed in the fuel. The thermal utilization is often expressed as

$$f = \frac{\overline{\Sigma}_a^f}{\overline{\Sigma}_a^f + (V_m / V_f)\zeta \, \overline{\Sigma}_a^m}, \tag{4-54}$$

where the ratio of neutron flux in the moderator to that in the fuel

$$\zeta = \frac{\bar{\varphi}_m}{\bar{\varphi}_f} \tag{4-55}$$

is referred to as the thermal disadvantage factor.

The resonance escape probability can also be written in a more useful form for a thermal reactor lattice by explicitly representing its dependence on fuel and moderator region parameters. First, Eq. (4-41) may be rewritten as

$$p = 1 - \frac{\int_I \Sigma_a (E)\varphi(E) \, dE}{\int_T \Sigma_a (E)\varphi(E) \, dE + \int_I \Sigma_a (E)\varphi(E) \, dE}. \tag{4-56}$$

Because the nonthermal absorption is due predominantly to the resonance capture cross sections in the fuel element fertile material, we may write

$$\int_I \Sigma_a (E)\varphi(E) \, dE \simeq \frac{V_f}{V_{\text{cell}}} N_{fe} \int_I \sigma_c^{fe}(E)\varphi_f(E) \, dE, \tag{4-57}$$

where N_{fe} and σ_c^{fe} are the atom density and microscopic capture cross section of the fertile material. The total absorption rate in the denominator of Eq. (4-56) must just be equal to the total number of neutrons slowed down below the fast fission threshold of the fertile nuclides. To a good approximation the number of neutrons slowing down can be given by (Lamarsh, 1966)

$$\int_T \Sigma_a (E)\varphi(E) \, dE + \int_I \Sigma_a(E)\varphi(E) \, dE \simeq \frac{V_m}{V_{\text{cell}}} \xi \, \Sigma_s^m \, [E\varphi_m(E)], \tag{4-58}$$

where ξ is the moderator slowing-down decrement as defined in Chapter 1 and Σ_s^m is the moderator scattering cross section. The quantity $E\varphi_m(E)$ is relatively independent of energy above about 50 KeV. Moreover, because it can be shown to be directly proportional to the number of neutrons slowed down from fission energies, $E\varphi_m(E)$ acts as a normalized factor on the $\varphi_f(E)$ calculation; we take $E\varphi_m(E) = 1$ in the intermediate energy range. With this normalization condition understood, Eqs. (4-56), (4-57), and (4-58) can be combined to yield

$$p = 1 - \frac{V_f N_{fe}}{V_m \xi \Sigma_s^m} I,$$ (4-59)

where I is the resonance integral defined by

$$I = \int_I \sigma_c^{fe}(E) \varphi_f(E) \, dE.$$ (4-60)

In many systems p is sufficiently close to one that Eq. (4-59) is a reasonable two-term expansion for the exponential function

$$p = \exp\left(-\frac{V_f N_{fe}}{V_m \xi \Sigma_s^m} I\right).$$ (4-61)

Reactivity Coefficients of k_∞

In thermal reactors both fuel and moderator conditions play important roles in determining reactivity feedback effects. Since the fuel is a solid that can be considered incompressible, the temperature coefficient $(1/k_\infty)/(\partial k_\infty/\partial T_f)$ is sufficient to describe fuel reactivity feedback. In the case of a solid moderator or for a liquid that is nearly incompressible the moderator temperature coefficient $(1/k_\infty)(\partial k_\infty/\partial T_m)$ adequately describes the reactivity feedback. The fuel and moderator temperature coefficients are considered in more detail later. However, it is first necessary to emphasize that there are important categories of events for which moderator reactivity feedback is not adequately described by T_m, since for most feedback effects it is the moderator atom density that is of primary importance.

If a liquid moderator boils, then the density may change drastically at essentially constant temperature. It is then necessary to include a second reactivity feedback parameter to describe the situation. The most usual is the void coefficient $(1/k_\infty)(\partial k_\infty/\partial \alpha)$, where the void fraction α is just the fraction of the cell moderator volume occupied by the vapor phase. At constant pressure or temperature, specification of the void fraction also specifies the atom density. However, if the reactor is undergoing a severe pressure or temperature transient in which boiling is involved, a second thermodynamic variable in addition to the void fraction is necessary to specify the moderator density. For this purpose, pressure coefficients are often used in boiling systems.

Fuel and moderator coefficients are conveniently expressed in terms of the behavior of the components of the four-factor formula:

$$\frac{1}{k_\infty}\frac{\partial k_\infty}{\partial T_f} = \frac{1}{\varepsilon}\frac{\partial \varepsilon}{\partial T_f} + \frac{1}{p}\frac{\partial p}{\partial T_f} + \frac{1}{\eta}\frac{\partial \eta}{\partial T_f} + \frac{1}{f}\frac{\partial f}{\partial T_f}, \tag{4-62}$$

$$\frac{1}{k_\infty}\frac{\partial k_\infty}{\partial T_m} = \frac{1}{\varepsilon}\frac{\partial \varepsilon}{\partial T_m} + \frac{1}{p}\frac{\partial p}{\partial T_m} + \frac{1}{\eta}\frac{\partial \eta}{\partial T_m} + \frac{1}{f}\frac{\partial f}{\partial T_m}. \tag{4-63}$$

There are three effects that may make substantial contributions to these reactivity coefficients with increased temperature. These are the Doppler broadening of the resonance cross sections in the fertile materials, the thermal expansion of the coolant, and the hardening of the thermal spectrum. We discuss these with respect to the fuel and moderator temperature coefficients.

Fuel Temperature Coefficient

The overwhelming contribution to the fuel temperature coefficient in relatively low enrichment thermal power reactors is due to the Doppler broadening of the resonance capture cross sections of the fertile material. This effect appears as a decrease in the resonance escape probability with increased temperature. There is no Doppler effect on ε, its contribution coming from energies well above where fuel resonance cross sections occur. There may be relatively minor changes in η and f in the case where a fissile element, such as plutonium-239, is present, since it has a resonance in the thermal neutron range. The latter effects, however, tend to be small compared to the change in p and will be neglected.

The Doppler effect arises from the fact that neutron cross sections are a function of the relative speed between neutron and nuclei. The resonance cross sections are sharply peaked in energy as shown, for example, in Fig. (1-8). For a given neutron speed the cross section must be averaged over the range of relative speeds that result from the Maxwell–Boltzmann distribution of fuel atom velocities. This averaging has the net effect of slightly smearing the resonances in energy, making them appear wider and less peaked. The smearing becomes more pronounced as the fuel temperature rises, as shown in the exaggerated cross section curves in Fig. 4-2.

If we were to consider systems in which there were only a very dilute mixture of resonance absorber atoms in a moderator, there would be no

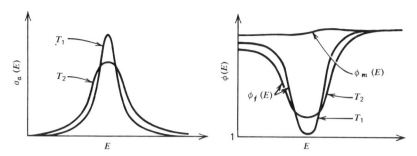

FIGURE 4-2 Energy-dependent resonance cross section and neutron flux distributions versus energy at two temperatures, $T_2 > T_1$.

reactivity effects from this Doppler broadening of the resonance peaks: the area under the resonance cross section, shown in Fig. 4-2, is independent of temperature. In addition, for a very dilute mixture the flux distribution $\varphi_f(E)$ is unperturbed by the presence of the resonance, and can be considered essentially energy independent over the width of importance for any single resonance. Consequently, each individual resonance integral could be written in the temperature-independent form

$$I \simeq \varphi_f(E_i) \int_I \sigma_c(E)\, dE, \qquad (4\text{-}64)$$

where E_i is the neutron energy at which the resonance occurs.

In nuclear reactors the concentration of resonance absorbers in the fuel is large. The effect is then to depress the neutron flux in the fuel both spatially, making $\varphi_f(E)/\varphi_m(E) < 1$, and in energy, as shown in Fig. 4-2. These flux depressions tend to be most pronounced where the resonance cross section is largest, and therefore the effect is referred to as resonance self-shielding. It may be shown by more rigorous analysis (Bell and Glasstone, 1970) that as the fuel temperature increases, the flux depression or self-shielding becomes less pronounced, as sketched in Fig. 4-2. The net effect is to increase the absorption rate and also the resonance integral. The increased resonance integral may be seen from Eq. (4-59) to lead to a decrease in p and subsequently to a decrease in reactivity.

To be more quantitative, we assume that the resonance escape probability p is the only term in the four-factor formula to be significantly affected by Doppler broadening. With this assumption we may write

$$\frac{1}{k_\infty} \frac{\partial k_\infty}{\partial T_f} \simeq \frac{1}{p} \frac{\partial p}{\partial T_f}, \qquad (4\text{-}65)$$

then utilizing Eq. (4-61) we have

$$\frac{1}{k_\infty}\frac{\partial k_\infty}{\partial T_f} = -\frac{V_f N_{fe}}{V_m \xi \sum_s^m}\frac{\partial I}{\partial T_f} = -\ln\left(\frac{1}{p}\right)\frac{1}{I}\frac{\partial I}{\partial T_f}. \qquad (4\text{-}66)$$

The temperature dependence of the resonance integral has been extensively studied both experimentally and analytically. For thermal reactors it is most often expressed with respect to a reference absolute temperature T_0 as

$$I(T_f) = I(T_0)[1 + \gamma(\sqrt{T_f} - \sqrt{T_0})], \qquad (4\text{-}67)$$

where γ is a function of the surface to mass ratio (S/M) of the fuel element (Lamarsh, 1966):

$$\gamma = C_1 + C_2\left(\frac{S}{M}\right). \qquad (4\text{-}68)$$

Typical values for C_1 and C_2 are given in Table 4-1, where T_0 is taken at $300°K$. The differentiation of Eq. (4-67) yields

$$\frac{1}{I}\frac{\partial I}{\partial T_f} = \frac{\gamma}{2\sqrt{T_f}}\frac{I(T_0)}{I(T_f)}. \qquad (4\text{-}69)$$

This result may then be combined with Eqs. (4-61) and (4-66) to yield the fuel temperature coefficient

$$\frac{1}{k_\infty}\frac{\partial k_\infty}{\partial T_f} = -\frac{\gamma}{2\sqrt{T_f}}\ln\left[\frac{1}{p(T_0)}\right]. \qquad (4\text{-}70)$$

Note that as the fuel temperature becomes higher, the magnitude of the coefficient becomes smaller.

TABLE 4-1   Resonance Integral Constants for Eq. (4-68)

Fuel	$C_1 \times 10^4$	$C_2 \times 10^4$
U^{238}	48	64
$U^{238}O_2$	61	47
Th	85	134
ThO_2	97	120

From J. R. Lamarsh, *Introduction To Nuclear Reactor Theory*, Addison-Wesley, Reading, Mass., 1966. Used by permission.

Other effects may make small contributions to the fuel temperature coefficient. Fuel expansion, for example, causes perturbations on ε, p, and f

as well as causing changes in the core leakage. Because of the small magnitude of the change in k_∞ and the small leakages in thermal reactors, however, fuel expansion effects tend to be minor.

Moderator Temperature Coefficient

The predominant contributions to the moderator temperature coefficient are changes in the moderator density and changes in the thermal neutron energy spectrum. The moderator atom density changes can be related to the temperature coefficient through the use of the volumetric coefficient of thermal expansion at constant pressure

$$\beta_m \equiv -\frac{1}{N_m}\frac{\partial N_m}{\partial T_m}, \tag{4-71}$$

which may then be used explicitly wherever the moderator atom density N_m appears. The causes of the hardening, or moving upward in energy, of the thermal neutron spectrum may be more subtle. The neutrons never establish complete thermal equilibrium with the moderator atoms because they do not make enough collisions in the thermal energy range before they are absorbed in the fuel or moderator. Nevertheless, their distribution in energy resembles the classic Maxwell–Boltzmann distribution, and to a first approximation the mean neutron energy is proportional to the absolute moderator temperature. The behavior of the thermal spectrum is complicated by the tendency for spectral hardening or movement away from thermal equilibrium as the ratio of moderator to fuel atoms decreases, this effect being present even in boiling situations, where there may be a change in density but not in temperature. Typical thermal neutron spectrum hardening resulting from temperature increases are shown in Fig. 4-3. The spectral hardening causes changes in the energy-averaged cross sections and fluxes, as defined by Eqs. (4-46) through (4-49), because of changes in $\varphi_f(E)$ and $\varphi_m(E)$. As a rule, $\varphi_f(E)$ and $\varphi_m(E)$ in the thermal range are strong functions of the moderator conditions but tend to be only weakly dependent on the fuel temperature. Hence, in the following elementary treatment, the thermal spectrum in the fuel as well as in the moderator is assumed to be independent of fuel temperature.

Consider now the dependence of the four factors on the moderator temperature. The fast fission factor increases somewhat with decreased moderator density because the neutrons are not slowed down below the energy range where fast fission may take place as effectively. The effect is small, however, compared to that on the resonance escape probability and thermal utilization. The resonance escape probability is not affected by the thermal neutron spectral changes, but a decrease in moderator density does decrease the effectiveness by which neutrons are slowed down through the

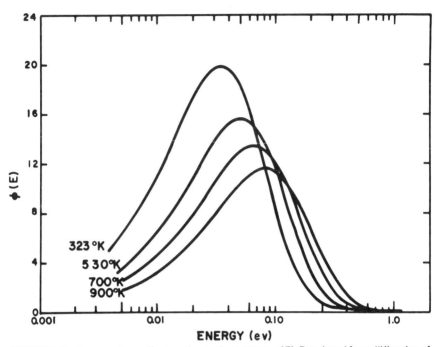

FIGURE 4-3   Temperature effect on thermal spectrum $\phi(E)$. Reprinted from "Kinetics of Solid Moderator Reactors" by H. B. Stewart and M. M. Merril in *The Technology of Nuclear Reactor Safety*, Vol. 1, T. J. Thompson and J. G. Beckerly (Eds.), 1973, by permission of the M.I.T. Press, Cambridge, Mass.

resonance region. Hence the resonance absorption increases, and the resonance escape probability decreases. This may be seen by taking

$$\frac{\partial p}{\partial T_m} = \frac{\partial N_m}{\partial T_m}\frac{\partial p}{\partial N_m}. \tag{4-72}$$

Substituting Eq. (4-61) for p, performing the density derivative, and utilizing Eqs. (4-61) and (4-71) to simplify the result, we obtain

$$\frac{1}{p}\frac{\partial p}{\partial T_m} = -\beta_m \ln\left(\frac{1}{p}\right). \tag{4-73}$$

Proceeding to the thermal utilization, we first write Eq. (4-53) in terms of the microscopic cross sections:

$$f = \frac{\tilde{e}\bar{\sigma}_a^{fi} + (1-\tilde{e})\bar{\sigma}_a^{fe}}{\tilde{e}\bar{\sigma}_a^{fi} + (1-\tilde{e})\bar{\sigma}_a^{fe} + (N_m/N_f)(V_m/V_f)\zeta\bar{\sigma}_a^{m}}, \tag{4-74}$$

where $\tilde{e}$ is the enrichment, here defined as the ratio of the fissile atom density to the total fuel (fissile plus fertile) atom density, and $\bar{\sigma}_x^{fi}$ and $\bar{\sigma}_x^{fe}$ denote the fissile and fertile microscopic cross sections for reaction x. We assume that the total number of fuel atoms $N_f V_f$ is independent of the moderator temperature. The moderator absorption cross section, in addition to being small, most often has the $1/v$ dependence on the neutron speed v that is insensitive to changes in the thermal neutron energy spectrum. Therefore we ignore the temperature dependence of moderator absorption cross section $\bar{\sigma}_a^m$.

With the preceding assumptions Eq. (4-74) may be differentiated with respect to the moderator temperature to obtain, after some algebra,

$$\frac{1}{f}\frac{\partial f}{\partial T_m} = (1-f)\left[\beta_m - \frac{1}{\zeta}\frac{\partial \zeta}{\partial T_m} + \frac{1}{\bar{\sigma}_a^f}\frac{\partial \bar{\sigma}_a^f}{\partial T_m}\right], \tag{4-75}$$

where we use the definition

$$\bar{\sigma}_x^f \equiv \tilde{e}\bar{\sigma}_x^{fi} + (1-\tilde{e})\bar{\sigma}_x^{fe} \tag{4-76}$$

for the fuel absorption cross section.

The first of the three terms in Eq. (4-75) corresponds to the decreased absorption in the moderator resulting from the smaller density of atoms. This term may be expected to be much larger for liquid than for solid moderators. The second term is due to the change in the disadvantage factor or, correspondingly, to the flux depression in the fuel. Normally, as the spectrum hardens, the disadvantage factor tends to decrease, leading to a positive contribution to the temperature coefficient. The last term is more difficult to predict. It tends to be negligibly small if only uranium-235 is involved, but may have a positive contribution if a significant amount of plutonium-239 or uranium-233 is present.

To determine the temperature dependence of η we consider, for simplicity, a fuel consisting of a single fertile and a single fissile isotope. Equation (4-52) may then be rewritten as

$$\eta = \eta^{fi}\Bigg/\left[1 + \frac{(1-\tilde{e})\bar{\sigma}_a^{fe}}{\tilde{e}\bar{\sigma}_a^{fi}}\right], \tag{4-77}$$

where

$$\eta^{fi} \equiv \frac{\nu\bar{\sigma}_f^{fi}}{\bar{\sigma}_a^{fi}}. \tag{4-78}$$

The temperature coefficient arises from the fact that the fuel cross sections for fission and absorption do not have the same energy dependence. Thus changes in the neutron spectrum can lead to changes in η^{fi} and in $\sigma_a^{fe}/\sigma_a^{fi}$.

The temperature coefficient of η^{fi} is negative for uranium-235 and plutonium-239 and positive for uranium-233. For a standard thermal spectrum at 100°C the temperature coefficients are about $-3 \times 10^{-5}/°C$, $-5 \times 10^{-5}/°C$, and $+4 \times 10^{-5}/°C$, respectively (Lamarsh, 1966).

As indicated in Eq. (4-63) the temperature coefficient of the moderator consists of the sum of four terms, even when leakage effects are neglected. We have seen that these terms are not all expected to have the same sign. Moreover, the term that predominates may well depend on the choice of moderator and of fuel configuration. Therefore it is difficult to draw conclusions concerning the sign or magnitude of the moderator coefficients without referring to a particular class of reactor designs. It is illustrative to examine two reactors, one with a liquid and one with a solid moderator. The discussion is simplified somewhat by choosing a pressurized-water reactor (for which the coolant and moderator are identical) and a gas-cooled graphite moderator (for which the coolant effects may be neglected).

Examples of Temperature Coefficients

Pressurized-Water Reactor. In a liquid moderated reactor of fixed fuel pin design, the net effect of the moderator temperature on reactivity is primarily a function of the atom ratio of moderator to fuel $(V_m N_m)/(V_f N_f)$. Thus a more global view often can be obtained if the multiplication is plotted as a function of this ratio, as is sketched, for example, in Fig. 4-4. The ratio changes either by varying the moderator temperature, and therefore N_m, or by varying the fuel pin spacing, and therefore V_m/V_f. As shown, there tends to be an optimal value of $(V_m N_m)/(V_f N_f)$ for which k_∞ is a maximum. For values of $(V_m N_m)/(V_f N_f)$ smaller than this optimum, the core is said to be undermoderated, and conversely, for larger values, overmoderated. For a fixed lattice one moves to the left with increased temperatures. Therefore the moderator temperature coefficient is negative only if the lattice is undermoderated. Note that the temperature coefficient may well change signs as the moderator is heated from room to operating temperature. This

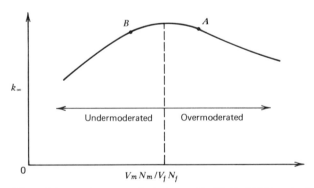

FIGURE 4-4   Effect of the moderator-to-fuel atom ratio $(V_m N_m/V_f N_f)$ on k_∞ in a thermal reactor lattice.

would occur if at room temperature the core were overmoderated, say at point A, but as the moderator temperature is raised to operating temperature, the decrease in N_m is sufficient to bring the lattice to an undermoderated condition, say point B.

In general, a thorough evaluation of the temperature coefficients must include the interactions of the coefficients with the neutron absorbers used for controlling the reactor and with fuel depletion and fission product buildup effects. The situation is particularly instructive for present-day pressurized-water reactors because they incorporate two control mechanisms: (1) control rods, used primarily to regulate the reactor power as well as the power distribution, and (2) soluble boron in the coolant, primarily to compensate for fuel burnup effects. The effect of water temperature changes on the effectiveness of these two forms of control poison is opposite in sign. When reactivity is controlled by adding boron poison to the water, the effect of raising the water temperature is to decrease the atomic density of both the boron poison and the moderator in the core. This expulsion of poison with coolant expansion causes the reactivity worth of this control mechanism to decrease with increasing temperature when the relative boron concentration remains the same. Thus the presence of soluble boron causes the moderator temperature coefficient to be more positive, as illustrated in Fig. 4-5. To a first approximation the reactivity worth of a control rod is proportional to the migration length of the surrounding lattice (Dietrich, 1964). As the

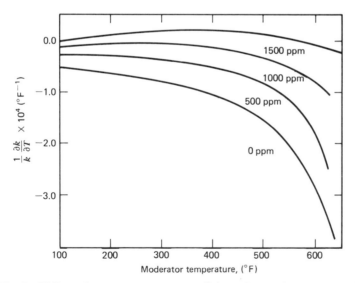

FIGURE 4-5 PWR moderator temperature coefficients for varying concentrations of soluble boron poison. Adapted from Commonwealth Edison Company, "Final Safety Analysis Report—Zion Station," USAEC Docket No. 50-295, 1970.

moderator density decreases, the migration length increases, as can be seen from the considerations of the preceding section. Thus the effect of a control rod tends to become larger with increasing moderator temperature, and the presence of control rods makes a negative contribution to the temperature coefficient. This is illustrated by the curves for rodded and unrodded cores in Fig. 4-6.

From this example we can see that the change in temperature coefficient with burnup may tend to become more or less negative, depending on whether burnup compensation is provided by decreasing the concentration of soluble boron with time or by withdrawing control rods. These factors by themselves, however, do not provide a complete picture of the change in temperature coefficient with burnup. For, in addition, the composition of the fuel changes with burnup to the point that by the end of the core life a

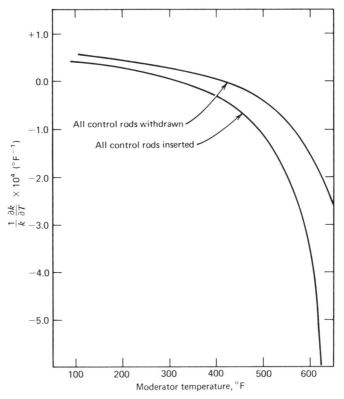

FIGURE 4-6   Control effect on PWR moderator temperature coefficient. Adapted from Commonwealth Edison Company, "Final Safety Analysis Report—Zion Station," USAEC Docket No. 50-295, 1970.

substantial fraction of the fissioning may be in the plutonium that has been produced by neutron capture in the uranium-238. This may have substantial effects on the thermal neutron spectrum and on the effective values of $\nu\sigma_f^f$ and σ_a^f. Finally, fission products—particularly xenon and samarium—are built up, and these also have effects on the temperature coefficients.

High-Temperature Gas-Cooled Reactor. For solid moderated reactors the magnitude of the moderator temperature coefficient tends to be much smaller than for liquid moderated systems because the density changes of the solid moderator with temperature are very small. The prominent changes in p and f resulting from β_m, as indicated by Eqs. (4-73) and (4-75), nearly disappear. The remaining contributions to the moderator coefficient are due to the effects of spectral hardening on f and η. In Fig. 4-7 is shown the moderator temperature coefficient for a graphite-moderated high-temperature gas-cooled reactor. Note that although the moderator temperature coefficient is relatively small in magnitude, it is positive in sign. It is positive because the fuel loading contains substantial amounts of uranium-

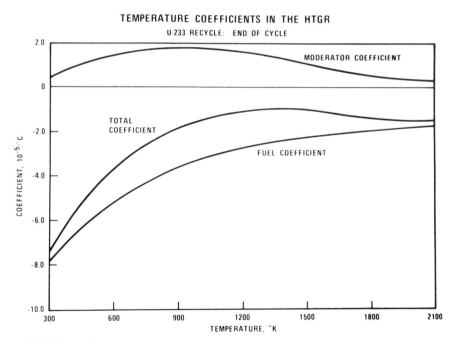

FIGURE 4-7 Typical temperature coefficients in the HTGR. From R. C. Dahlberg, "Physics of Gas-Cooled Reactors," *Proc. Nat. Top. Meeting in New Developments in Reactor Physics and Shielding*, Kiamesha Lake, N.Y., CONF-270901, September 1972.

233, a nuclide for which the temperature coefficient of η is positive. The fuel temperature coefficient is also plotted to emphasize an important point. A small positive moderator temperature coefficient in itself may not be detrimental to the plant safety. If the fuel temperature coefficient is sufficiently negative to result in the net temperature coefficient being negative, then raising the temperature of the whole core will still result in a reactivity loss.

4-4 FAST REACTORS

Fast reactors differ from thermal reactors in the several respects alluded to in Chapter 1. The enrichment of the fuel is higher than that in a thermal system, with typical enrichments ranging from 10 to 20%. There is also an absence of low atomic weight moderator material, resulting in a fast neutron spectrum such as that shown in Fig. 1-5. As a rule, cross sections tend to decrease with increasing neutron energy, causing the neutron spectrum-averaged cross sections in a fast reactor to be smaller than those in a thermal system.

The smaller cross sections have at least two effects that are important to the consideration of reactivity feedback. First, the lattice cell dimensions are small compared to a mean free path. From neutron transport considerations it may be shown that under these circumstances the flux spectrum in fuel and coolant regions are nearly identical (Hummel and Okrent, 1970). We thus have

$$\varphi(E) = \varphi_f(E) = \varphi_c(E), \qquad (4-79)$$

where f and c signify the space-averaged values in the fuel and coolant regions. Reaction rates become simple volume averages:

$$\int_0^\infty \Sigma_x(E)\varphi(E)\,dE = \frac{V_f}{V_{\text{cell}}}\bar{\Sigma}_x^f + \frac{V_c}{V_{\text{cell}}}\bar{\Sigma}_x^c, \qquad (4-80)$$

where

$$\bar{\Sigma}_x^y = N^y\bar{\sigma}_x^y, \qquad (4-81)$$

$$\bar{\sigma}_x^y = \int_0^\infty \sigma_x^y(E)\varphi(E)\,dE, \qquad (4-82)$$

$$\int_0^\infty \varphi(E)\,dE = 1, \qquad (4-83)$$

and N^y signifies the atom density in region y.

The second consequence of the smaller cross sections is that the migration lengths in fast reactors tend to be larger than those for thermal systems, as is readily observable from Eq. (4-9). At the same time, fast reactors are characterized by higher power densities. Therefore for a fixed power rating they have smaller dimensions and larger values of the geometric buckling than thermal systems. Thus $M^2 B_g^2$, and therefore the leakage effects, is more substantial in fast systems, particularly compared to the contributions from the infinite medium reactivity components, which often tend to be quite small.

Fuel Temperature Coefficient

Both the Doppler effect and thermal expansion contribute to the fuel temperature coefficient. As fast reactor designs have evolved toward lower enrichment oxide fuels and softer neutron spectra, however, the Doppler coefficient has come to be the dominant contribution. The Doppler effect has practically no effect on leakage properties, but affects the fuel temperature coefficient primarily by causing changes in k_∞. Unlike thermal systems, the multiplication cannot be divided into separate physical processes through the use of the four-factor formula. Rather, we must directly utilize the neutron balance form given in Eq. (1-38):

$$k_\infty = \frac{\int_0^\infty \nu \Sigma_f(E)\varphi(E)\, dE}{\int_0^\infty \Sigma_a(E)\varphi(E)\, dE}. \tag{4-84}$$

For brevity we ignore the effects of cladding, the oxygen in the ceramic fuel, and other core constituents, and consider each of the reactor lattice cells to consist of a fuel region, with a fissile and a fertile material, and a coolant region. Generalizing Eqs. (4-80) through (4-83) to account for fissile and fertile materials, we have

$$k_\infty = \frac{\tilde{e}\nu^{fi}\bar{\sigma}_f^{fi} + (1 - \tilde{e})\nu^{fe}\bar{\sigma}_f^{fe}}{\tilde{e}\bar{\sigma}_a^{fi} + (1 - \tilde{e})\bar{\sigma}_a^{fe} + (V_c N_c / V_f N_f)\bar{\sigma}_a^c}, \tag{4-85}$$

where $\tilde{e}$ is again the enrichment and fi and fe signify fissile and fertile material properties.

In determining the temperature dependence of k_∞ resulting from the Doppler effect, several simplifications may be made. The threshold in the fertile material above which fission takes place is well above the energy range where resonances are present. Therefore the temperature dependence of $\bar{\sigma}_f^{fe}$ can be ignored. Unlike thermal systems, the concentration of fissile material is large enough that it could potentially make significant contributions to the Doppler effect. However, the contributions from the fission and capture resonances tend to cancel one another, leaving the capture resonances in the fertile material as the dominant contributor to the Doppler coefficient. We therefore assume that only $\bar{\sigma}_a^{fe}$ contributes to the Doppler coefficient and obtain

$$\frac{1}{k_\infty}\frac{\partial k_\infty}{\partial T_f} \simeq -\frac{(1-\tilde{e})}{\tilde{\sigma}_a}\frac{\partial \bar{\sigma}_a^{fe}}{\partial T_f}, \tag{4-86}$$

where

$$\tilde{\sigma}_a \equiv \tilde{e}\bar{\sigma}_a^{fi} + (1-\tilde{e})\bar{\sigma}_a^{fe} + \left(\frac{V_c N_c}{V_f N_f}\right)\bar{\sigma}_a^c. \tag{4-87}$$

Even with these simplifications the determination of the Doppler coefficient is complicated by the fact that in a fast reactor neutron spectrum most of the contribution comes from the upper energy part of the resonance region where measurements cannot be made with sufficient resolution to resolve individual resonances. Therefore statistical distributions of parameters must be extrapolated from data measured at lower energies where the resonance parameters can be resolved.

The fertile isotope cross section $\bar{\sigma}_a^{fe}$ increases with fuel temperature because of Doppler broadening of the resonance peaks. The explanation is the same as that for Doppler broadening in thermal reactors given in the preceding section. However, the contribution to decreased reactivity with increased fuel temperature is substantially smaller than that for thermal systems, and the magnitude of the resulting temperature coefficient may be quite sensitive to the details of the core composition.

These two properties may be understood qualitatively by examining the energy spectrum $\varphi(E)$ of a typical fast reactor. Two such spectra are shown in Fig. (4-8). The first is a typical spectrum for a sodium-cooled, uranium-oxide-fueled fast reactor. The second is for the same reactor with the sodium removed. The Doppler effect becomes significant only at energies below about 25 keV, where the resonances appear, the contribution to the Doppler effect becoming increasingly important with decreasing neutron energy. Only the low-energy tail of the neutron spectra occur in this energy range. Moreover, more detailed studies of spectral effects indicate that the neutrons in this resonance range are rather unimportant to the chain reaction. This is due, in part, to the fact that the number of fission neutrons produced per neutron absorbed in the fuel is relatively small at the low-energy end of the fast reactor spectrum. It is difficult to give a single expression for the Doppler coefficient in a fast reactor because of the complex dependence on the core composition. As in thermal systems, the magnitude of the coefficient decreases with increased absolute fuel temperature. The parametric

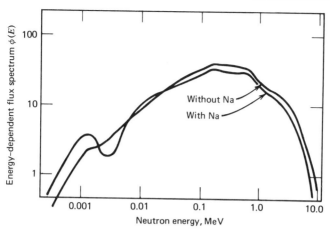

FIGURE 4-8 Neutron spectra in sodium-cooled fast reactor with and without sodium present.

expression most often used for fast reactors is

$$\frac{1}{k_\infty}\frac{\partial k_\infty}{\partial T_f} = -\frac{\gamma}{T},\tag{4-88}$$

where γ is a positive constant that depends on the core composition.

The Doppler effect in fast systems also would be expected to be smaller than that in the thermal reactor because the enrichment is larger and therefore the atom density of fertile material in the fuel is smaller. This appears in Eq. (4-86) as a smaller weighting factor $(1-\tilde{e})$ on the temperature derivative of $\bar{\sigma}_a^{fe}$. If the enrichment is increased, the atom density of fertile material becomes yet smaller, and there is a decrease in the importance of the Doppler coefficient. This decreased importance with increased enrichment is more rapid than can be attributed strictly to the atom density change, however. A more important phenomenon is that with increased enrichment, the neutron spectrum shifts upward in energy, causing a smaller fraction of the neutrons to be in the energy range where the resonance cross sections occur.

The sensitivity of the Doppler coefficient to the changes in core composition other than variation in enrichment also may be understood in terms of changes in the neutron spectrum. Any increase in the amount of light or intermediate atomic weight material to the core causes an increase in neutron slowing down due to elastic scattering collisions, and this in turn results in a softening or shifting downward of the neutron energy spectrum. With such spectral softening, more neutrons appear in the energy range below 25 keV, where resonance absorption becomes important, and therefore the Doppler coefficient becomes larger. Coolant expansion or the loss of the coolant in a fast reactor results in a decrease in the relatively light coolant nuclei present in the core. Therefore there is a spectral hardening, a consequent shifting of the neutrons away from the resonance region, and a resultant decrease in the negative Doppler coefficient. These coolant density effects are more significant for liquid-metal-cooled systems than for gas-cooled systems because of the greater atom density of the liquid coolants. The typical change in the neutron spectrum resulting from the complete loss of a sodium coolant is shown in Fig. 4-8.

Coolant Density Coefficient

Transients that overheat fast reactor coolants lead to decreased coolant density. This density decrease is through one of two mechanisms. If the coolant remains as a single phase and the temperature rises at approximately constant pressure, the temperature coefficient may be related to the density change by

$$\frac{1}{k}\frac{\partial k}{\partial T_c} = -N_c\beta_c\frac{1}{k}\frac{\partial k}{\partial N_c}, \tag{4-89}$$

where

$$\beta_c = -\frac{1}{N_c}\frac{\partial N_c}{\partial T_c} \tag{4-90}$$

is the coolant volumetric coefficient of expansivity at constant pressure. The second mechanism is through the boiling of a liquid coolant such as sodium. In this case the change in density is most conveniently measured through the void fraction α, which is defined as the fraction of the coolant volume occupied by the vapor phase. Thus we may write

$$N_c = (1-\alpha)N_c^l + \alpha N_c^v, \tag{4-91}$$

where N_c^l and N_c^v are the atom densities of the liquid and vapor phases, respectively. Changes in void fraction and in coolant density may thus be related by first writing

$$\frac{1}{k}\frac{\partial k}{\partial \alpha} = \frac{\partial N_c}{\partial \alpha}\frac{1}{k}\frac{\partial k}{\partial N_c}, \tag{4-92}$$

and then inserting Eq. (4-91) to obtain

$$\frac{1}{k}\frac{\partial k}{\partial \alpha} = -(N_c^l - N_c^v)\frac{1}{k}\frac{\partial k}{\partial N_c}. \tag{4-93}$$

Density decreases may appear in either liquid-metal-cooled or gas-cooled reactors, but the seriousness of the effects are of different orders of magnitude. Because the gas coolants have such low atomic densities to begin with, even a complete depressurization would have only a minimal effect on reactivity. In contrast, the larger changes in coolant densities resulting from rising coolant temperatures, and in particular to coolant boiling, in liquid-metal-cooled fast reactors gives rise to much larger reactivity effects. In either system the calculation of the net effect of decreased coolant density is made more difficult by the presence of two effects with opposite signs: (1) an increase in k_∞ caused by hardening of the neutron spectrum and (2) an increase in the neutron leakage caused by an increase in the migration length. As may be surmised from Eq. (4-5) the sign of the net effect may depend on the size as well as the composition of the reactor core, with negative leakage effects dominating in small cores and positive, infinite medium effects dominating in large cores. In addition, the density changes may be nonuniform, with the result that the reactivity feedback may change sign, depending on the distribution of coolant temperature in the core. In discussing these infinite medium and leakage effects in more detail, we use the void coefficient from a sodium-cooled reactor as an example, since the liquid metal boiling from which voids would result presents one of the largest feedback effects in present-day fast reactor designs. To begin with, we consider the effect on k_∞ of the introduction of a void fraction α into a coolant. A convenient starting point is Eq. (4-85), which, if differentiated with respect to the void fraction, yields, after some algebra,

$$\frac{1}{k_\infty}\frac{\partial k_\infty}{\partial \alpha}=[\tilde{e}\nu^{fi}\bar{\sigma}_f^{fi}+(1-\tilde{e})\nu^{fe}\bar{\sigma}_f^{fe}]^{-1}\times\left\{e\left[\frac{\partial}{\partial\alpha}(\nu^{fi}\bar{\sigma}_f^{fi})-k_\infty\frac{\partial\bar{\sigma}_a^{fi}}{\partial\alpha}\right]\right.$$

$$\left.+(1-\tilde{e})\left[\frac{\partial}{\partial\alpha}(\nu^{fe}\bar{\sigma}_f^{fe})-k_\infty\frac{\partial\bar{\sigma}_a^{fe}}{\partial\alpha}\right]-\frac{V_cN_c}{V_fN_f}k_\infty\left[\frac{\partial\bar{\sigma}_a^c}{\partial\alpha}-(N_c^l-N_c^v)\frac{\bar{\sigma}_a^c}{Nc}\right]\right\}.$$

$$(4\text{-}94)$$

The very last term on the right is due to decreased coolant absorption with increased void fractions. It tends to be quite minor because of the small coolant absorption cross section. The derivatives of the microscopic cross sections with respect to void fraction arise from changes in the neutron energy spectrum. Thus, for example,

$$\frac{\partial\bar{\sigma}_a^c}{\partial\alpha}=\int_0^\infty \sigma_a^c(E)\frac{\partial\varphi(E)}{\partial\alpha}\,dE. \tag{4-95}$$

As noted in the treatment of Doppler coefficients, a decrease in coolant density tends to shift the neutron spectrum upward in energy because of the decreased density of lighter-weight atoms that slow down neutrons through elastic collisions.

Neglecting the absorption effects in the coolant, Eq. (4-94) may be written to exhibit the effects of the spectral hardening with increased void fraction:

$$\frac{1}{k_\infty}\frac{\partial k_\infty}{\partial \alpha} = [\tilde{e}\nu^{fi}\bar{\sigma}_f^{fi} + (1 - \tilde{e})\nu^{fe}\bar{\sigma}_f^{fe}]^{-1}$$ (4-96)

$$\times \left\{ \tilde{e}\int_0^\infty [\nu^{fi}\sigma_f^{fi}(E) - k_\infty\sigma_a^{fi}(E)]\frac{\partial \varphi(E)}{\partial \alpha} dE \right.$$

$$\left. + (1 - \tilde{e})\int_0^\infty [\nu^{fe}\sigma_f^{fe}(E) - k_\infty\sigma_a^{fe}(E)]\frac{\partial \varphi(E)}{\partial \alpha} dE \right\}.$$

If the normalization condition of Eq. (4-83) is to hold, we also must have

$$\int_0^\infty \frac{\partial}{\partial \alpha}\varphi(E) dE = 0.$$ (4-97)

The spectral hardening, exhibited in Fig. 4-8 with all coolant removed, is equivalent to stating that the derivative $(\partial/\partial\alpha)\varphi(E)$ is positive for energies above about 10 keV and negative below. We can understand the behavior of the void coefficient by taking $k_\infty = 1$ and examining the parameter $\nu\sigma_f(E) - \sigma_a(E)$, which is shown in Fig. 4-9 for fertile and fissile isotopes.

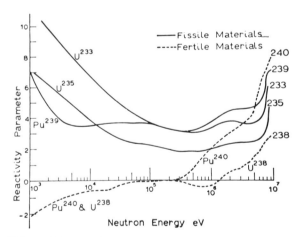

FIGURE 4-9   Values of the reactivity parameter $[\nu\sigma_f(E) - \sigma_a(E)]$ as a function of energy for fissile and fertile materials. From D. Jakeman, *Physics of Nuclear Reactors*, American Elsevier, New York, 1966. Used by permission.

The most obvious effect in Fig. 4-9 is that the hardening of the energy spectrum increases the neutron flux above the fission threshold of the fertile elements, and in general shifts the spectrum to larger values of $\nu\sigma_f^{fe} - \sigma_a^{fe}$. The behavior is more subtle for fissile materials where a valley $\nu\sigma_f^{fi} - \sigma_a^{fi}$ exists between about 0.1 and 1 MeV; the result is that there are compensatory effects from removing flux from lower energies and placing it at higher energies. As a result, for example, because $\nu\sigma_f - \sigma_a$ does not rise as sharply with decreasing energy for plutonium-239 as for uranium-235, the void feedback would be expected to be more positive for cores fueled with plutonium. Finally, because $\nu\sigma_f - \sigma_a$ tends to rise more sharply with energy for uranium-238 than for the fissile nuclei, one would expect that decreases in enrichment would lead to more positive void coefficients. The net result of these effects is to increase k_∞ with the introduction of coolant voids over the range of enrichments found in current fast power reactor designs.

The effect of coolant voids on the value of the migration length is also positive. This may be illustrated as follows. For a uniform lattice the definition for M^2 given by Eq. (1-61) is space independent. Combining Eq. (1-61) with Eqs. (1-54) and (1-64), we then have

$$M^2 = \int_0^\infty \frac{\varphi(E)\, dE}{3 \sum(E)} \times \left(\int_0^\infty \Sigma_a(E) \varphi(E)\, dE \right)^{-1}. \qquad (4\text{-}98)$$

Now we use Eqs. (4-80) through (4-83) to separate $(3 \sum)^{-1}$ and Σ_a into fuel and coolant contributions. Hence

$$M^2 = \tfrac{1}{3} V_{\text{cell}}^2 \int_0^\infty \frac{\varphi(E)\, dE}{N_f V_f \sigma^f(E) + N_c V_c \sigma^c(E)} \times \left[\int_0^\infty [N_f V_f \sigma_a^f(E) \right.$$
$$\left. + N_c V_c \sigma_a^c(E)]\, \varphi(E)\, dE \right]^{-1} \qquad (4\text{-}99)$$

Again the effects of coolant voiding are twofold: N_c is decreased and $\varphi(E)$ is shifted to higher energies. Since N_c appears in the denominator of both terms, decreases in N_c result in a larger value of M^2. Likewise, the spectral hardening tends to cause the energy-averaged values of the microscopic cross sections to become smaller. Since these cross sections also appear in the denominator of the two terms, the result also is an increase in M^2.

It is apparent from Eq. (4-5) that the increases in k_∞ and M^2 with coolant density decreases have opposite effects on reactivity, since both the infinite medium multiplication and the core leakage are increased. The core-averaged values of the coolant temperature and void coefficients become more positive as the size of the core increases, since the negative leakage effect becomes smaller with increased dimensions. In considering the effects of local coolant voiding, the location within the core, as well as the core size, becomes important in determining the sign as well as the magnitude of the reactivity feedback. As illustrated for sodium at the end of Section 4-2, voiding near the core periphery may well cause a reactivity decrease, whereas voiding near the center of the core has the opposite effect.

The evolution of fast reactor core design has resulted in substantial shifts in the relative importance of reactivity feedback mechanisms. Early fast reactor cores were relatively small in size and had metal fuels with rather high enrichments. As a result, the neutron spectra were quite hard. These characteristics lead to extremely small Doppler coefficients, with the primary contribution to the small fuel temperature coefficient coming from thermal expansion and bowing effects. At the same time the small core dimensions caused coolant void effects to be negative.

As liquid-metal-cooled fast reactor designs have tended toward larger cores with lower enrichment oxide fuels, the neutron spectra have become softer. As a result, the negative Doppler contribution to the fuel temperature coefficient has become much larger and dominates over expansion effects. With the larger core size, however, the effects of coolant voiding have become more positive in their contribution to reactivity changes.

4-5 COMPOSITE COEFFICIENTS

In Section 4-2 reactivity changes are expressed in terms of core-averaged temperatures in the form

$$d\rho_{fb} = \frac{1}{k} \frac{\partial k}{\partial \bar{T}_{fe}} d\bar{T}_{fe} + \frac{1}{k} \frac{\partial k}{\partial \bar{T}_c} d\bar{T}_c + \frac{1}{k} \frac{\partial k}{\partial \bar{T}_i} d\bar{T}_i, \qquad (4\text{-}100)$$

and the core-averaged reactivity coefficients are modeled in terms of the derivatives of k_∞ and M^2 with respect to local component temperatures. The physical phenomena leading to these local reactivity coefficients are examined in the two preceding sections. Before Eq. (4-100), or similar models, can be applied to the study of power reactor dynamics, however, a means must be devised for determining the core-averaged temperatures. Frequently this requires modeling the thermal transient through the use of coupled sets of differential equations in time and obtaining solutions for the temperatures simultaneously with the solution of the reactor kinetics equations. One simplified model for describing such thermal transients is presented in the following chapter.

There are a number of important situations for which a priori relationships can be derived for the core-averaged temperatures and a single independent variable used to deduce a composite reactivity coefficient. The prompt and isothermal temperature coefficients and the power coefficient are three of the more important composite reactivity coefficients. The prompt coefficient is used in investigating superprompt-critical nuclear excursions. The isothermal and power coefficients are necessary for an understanding of reactivity control and are also useful in analyzing slower power transients.

Prompt Reactivity Coefficient

In the event of a very large reactivity insertion the reactor power may change significantly over periods of time that are short compared to that required to transfer heat from the fuel to the coolant. Over such time periods the coolant temperature does not change appreciably, and the only feedback reactivity comes directly from the heating of the fuel. Hence we take $d\bar{T}_c = d\bar{T}_i \approx 0$, and obtain from Eq. (4-100),

$$d\rho_{fb} = \frac{1}{k}\frac{\partial k}{\partial \bar{T}_{fe}}\,d\bar{T}_{fe}. \qquad (4\text{-}101)$$

For times so short that heat cannot be transferred to the coolant, the reactor fuel elements may be considered to behave adiabatically. Thus if M_{fe} is the total mass of fuel elements in the reactor core and c_{fe} is the fuel element specific heat per unit mass, we have

$$M_{fe}c_{fe}\,d\bar{T}_{fe} = P(t)\,dt, \qquad (4\text{-}102)$$

where $P(t)$ is the reactor power. Combining the two preceding equations, we obtain

$$d\rho_{fb} = \mu P(t)\,dt, \qquad (4\text{-}103)$$

where

$$\mu \equiv \frac{1}{M_{fe}c_{fe}}\frac{1}{k}\frac{\partial k}{\partial \bar{T}_{fe}}, \qquad (4\text{-}104)$$

is often referred to as the prompt reactivity coefficient. If μ is temperature independent, then Eq. (4-103) may be integrated, with $\rho_{fb}(0) = 0$, to yield

$$\rho_{fb}(t) = \mu E(t), \qquad (4\text{-}105)$$

where

$$E(t) \equiv \int_0^t P(t)\,dt \qquad (4\text{-}106)$$

is the energy release since the initiation of the transient. Finally, we note that μ may also be expressed as

$$\mu = \frac{d\rho_{fb}}{dE}. \qquad (4\text{-}107)$$

Reactor cores must be designed so that μ has a negative value, for a positive prompt coefficient would lead to an inherently unstable system that could undergo autocatalytic excursions. As may be surmised from the preceding sections, the Doppler effect is most often the primary contributor to μ, although fuel expansion, bowing, or other effects may also be significant. In the event that all feedback effects emanating from the fuel heating are extremely small, even the direct heating of the coolant through neutron-scattering collisions may contribute significantly to the prompt coefficient.

In evaluating prompt coefficients care must be taken to ensure that the averaging techniques used in obtaining $(1/k)(\partial k/\partial \bar{T}_{fe})$ are valid. Where very localized reactivity insertions are made in neutronically loosely coupled cores, extreme power peaking is prone to occur in the vicinity of the insertion. In such a situation the quasi-steady-state fuel temperature distribution discussed in Section 4-2 must be replaced by a spatial temperature distribution that is applicable during the course of the transient.

Isothermal Temperature Coefficient

In most power reactors the temperature of the primary system is brought slowly from room temperature to the operating core inlet temperature before operation at power begins. This may be accomplished either by heating the system with an external heat source, such as friction from the coolant pumps, or by bringing the reactor critical and heating the primary system at a very low power level. Likewise, the cooldown of the primary system following operation is done slowly, since either too rapid heating or too rapid cooling may lead to unacceptable thermal stress on the reactor vessel or other components.

Substantial reactivity changes are associated with heat-up and cooldown of the primary system. Since the power level is very low—if not zero—during such operations, the reactivity feedback can be estimated by assuming that fuel element, coolant, and inlet temperatures are essentially equal to a single primary system averaged temperature $\bar{T}$. Therefore we have

$$\bar{T}_{fe} = \bar{T}_c = \bar{T}_i = \bar{T},$$

and Eq. (4-100) reduces to

$$\frac{d\rho_{fb}}{d\bar{T}} = \frac{1}{k}\frac{\partial k}{\partial \bar{T}_{fe}} + \frac{1}{k}\frac{\partial k}{\partial \bar{T}_c} + \frac{1}{k}\frac{\partial k}{\partial \bar{T}_i}, \tag{4-108}$$

which is referred to as the isothermal temperature coefficient.

The isothermal temperature coefficient has a particularly simple relationship to the local fuel and coolant reactivity coefficients. Combining Eq. (4-108) with Eqs. (4-18) and (4-26) through (4-31) yields

$$\frac{d\rho_{fb}}{d\bar{T}} = \frac{1}{\langle \psi_0^2 k_\infty \rangle} \left\langle \psi_0^2 \left(\frac{\partial k_\infty}{\partial \tilde{T}_{fe}} + \frac{\partial k_\infty}{\partial T_c} \right) \right\rangle$$

$$- \frac{\bar{M}^2 \bar{B}_g^2}{1 + \bar{M}^2 \bar{B}_g^2} \frac{1}{\langle (\nabla \psi_0)^2 M^2 \rangle} \left\langle (\nabla \psi_0)^2 \left(\frac{\partial M^2}{\partial \tilde{T}_{fe}} + \frac{\partial M^2}{\partial T_c} \right) \right\rangle. \quad (4\text{-}109)$$

Power Coefficient

When a reactor is operating at more than a few percent of its rated power, it is no longer reasonable to assume that the temperature throughout the core is uniform. The coolant temperature increases as it passes through the core, and the fuel element temperature necessarily must be substantially higher than that of the coolant to facilitate fuel-to-coolant heat transport. The incremental change of reactivity with power resulting from these temperature differences is called the power coefficient. It is obtained by dividing Eq. (4-100) by the differential power change dP:

$$\frac{d\rho_{fb}}{dP} = \frac{1}{k} \frac{\partial k}{\partial \bar{T}_{fe}} \frac{d\bar{T}_{fe}}{dP} + \frac{1}{k} \frac{\partial k}{\partial \bar{T}_c} \frac{d\bar{T}_c}{dP} + \frac{1}{k} \frac{\partial k}{\partial \bar{T}_i} \frac{d\bar{T}_i}{dP}. \quad (4\text{-}110)$$

In evaluating the derivatives with respect to power on the right, we assume that the changes in power are carried out quasi-statically so that the core-averaged component temperatures are governed by the steady-state relationships developed in Chapter 1. Assuming that the reactor is operated at constant inlet temperature, we have from Eqs. (1-112) and (1-114)

$$\frac{d\bar{T}_{fe}}{dP} = R + \frac{1}{2Wc_p} \quad (4\text{-}111)$$

$$\frac{d\bar{T}_c}{dP} = \frac{1}{2Wc_p} \quad (4\text{-}112)$$

$$\frac{d\bar{T}_i}{dP} = 0 \quad (4\text{-}113)$$

where R and W are the core thermal resistance and mass flow rate, and c_P is the coolant specific heat per unit mass. The power coefficient is

$$\frac{d\rho_{fb}}{dP} = \left(R + \frac{1}{2Wc_p} \right) \frac{1}{k} \frac{\partial k}{\partial \bar{T}_{fe}} + \frac{1}{2Wc_p} \frac{1}{k} \frac{\partial k}{\partial \bar{T}_c}. \quad (4\text{-}114)$$

4-6 REACTIVITY CONTROL

The signs and magnitudes of the composite reactivity coefficients discussed in the preceding section have a strong influence on the control and stability characteristics of a power reactor. Clearly, if either the isothermal temperature coefficient or the power coefficient is positive, the reactor will be unstable. Increases in temperature or power level would lead to increased reactivity, which in turn would cause the power to increase more rapidly, and so on. Even in the event that these composite coefficients are negative, however, it is still necessary to examine individually the contributions from the reactor fuel, coolant, moderator, and other constituents.

Prompt and Delayed Reactivity Feedback

For purposes of analysis it is helpful to divide reactivity coefficients into prompt and delayed contributions. The prompt reactivity feedback appears immediately with changes in reactor power and usually is related directly to changes in fuel temperature. The delayed contribution to reactivity feedback comes from the effects of changes in the temperatures of coolant, separate moderator region, or other core constituents outside the reactor fuel. For these constituents there is a delay between the time of power production and the time of component temperature rise that is due to the finite times required to transport heat from the fuel to the components.

Core designs in which both the net and prompt contributions to the isothermal and power coefficients are negative may be acceptable even though the delayed reactivity feedback may be positive. In such cores any tendency for the reactor to exceed the desired power production is immediately countered by the prompt negative effect; moreover, this negative effect is large enough to override the positive reactivity feedback that appears later as the heat is transported to the other core constituents. Some power reactors have been built in which a delayed positive reactivity effect is large enough to override the prompt negative effect and result in a net positive isothermal or power coefficient. The argument to justify such designs is that because the positive feedback is sufficiently delayed, the reactor power level can be controlled by a fast-acting control system. Such arguments, however, often are frowned on by licensing authorities. If such a situation occurs only in the isothermal temperature coefficient, it may be required to heat the reactor externally out of the temperature range where the positive coefficient occurs before the reactor is brought critical.

Any situation where the prompt contribution to a composite reactivity coefficient is positive is unacceptable even though the net value may be negative. If such cores were allowed, one could easily encounter situations in which a rate of power increase would develop that is large enough that the negative delayed effect would never be able to overtake the prompt positive contribution, and an autocatalytic power increase would result.

Temperature and Power Defects

The magnitudes as well as the signs of the temperature and power coefficients have safety implications. To understand these we introduce the concepts of temperature and power defects.

Suppose we are asked how much reactivity must be added by a control system to bring the primary system of a power reactor from room temperature to the operating inlet temperature. The control reactivity must just compensate for the feedback reactivity. The change in reactivity may be found by integrating the isothermal temperature coefficient, given in Eq. (4-108), between room temperature $\bar{T}_r$ and the zero-power operating temperature $\bar{T}_i$:

$$\rho(\bar{T}_i) - \rho(\bar{T}_r) = \int_{\bar{T}_r}^{\bar{T}_i} \left(\frac{1}{k} \frac{\partial k}{\partial \bar{T}_{fe}} + \frac{1}{k} \frac{\partial k}{\partial \bar{T}_c} + \frac{1}{k} \frac{\partial k}{\partial \bar{T}_i} \right) dT. \qquad (4\text{-}115)$$

If the isothermal temperature coefficient contained in the brackets on the right is negative, the reactivity at operating temperature is less than at room temperature. The net loss of reactivity in going from room to operating temperature is defined as the temperature defect D_T. For our simple lumped parameter model:

$$D_T \equiv - \int_{\bar{T}_r}^{\bar{T}_i} \left(\frac{1}{k} \frac{\partial k}{\partial \bar{T}_{fe}} + \frac{1}{k} \frac{\partial k}{\partial \bar{T}_c} + \frac{1}{k} \frac{\partial k}{\partial \bar{T}_i} \right) dT. \qquad (4\text{-}116)$$

It is equal to the amount of reactivity that must be added to the reactor through withdrawal of control rods or other means in order to bring the reactor from room to operating temperature. These two states normally are referred to as cold critical and hot zero-power critical, respectively.

As the reactor power is increased above the hot zero-power critical condition, to the rated reactor power, a negative power coefficient causes the reactivity to decrease further by an amount that is defined as the power defect

$$D_P \equiv - \int_0^P \left[\left(R + \frac{1}{2Wc_p} \right) \frac{1}{k} \frac{\partial k}{\partial \bar{T}_{fe}} + \frac{1}{2Wc_p} \frac{1}{k} \frac{\partial k}{\partial \bar{T}_c} \right] dP, \qquad (4\text{-}117)$$

which is the negative integral of the power coefficient given by Eq. (4-113). The temperature and power defects bear an important relationship to the magnitude of the nuclear excursions that may result from postulated control system failures. An amount of reactivity equal to $D_T + D_P$ must be added to the core in order to bring it to hot critical and then to power. This is accomplished by the withdrawal of control rods or other neutron absorbers. If some fraction χ of this required absorber were to be suddenly lost, say through a control rod ejection, while the reactor is near or in a cold critical state, the reactivity gain would be $\chi(D_T + D_P)$. With power and temperature defect increases, the fraction of the total control poison that can be ejected without the reactor becoming prompt critical becomes smaller. Put another way, increasing $D_T + D_P$ also increases the poison requirements. Therefore if the sudden loss of any one control rod is not to cause a prompt-critical excursion, many low-worth rods will be needed. By analogy, if soluble poisons are used to hold down reactivity in the cold reactor core, then a design that increases $D_T + D_P$ will require an increased poison concentrate to keep the cold core subcritical. Thus if a fixed mass of unpoisoned coolant is accidentally introduced into the core, the reactivity insertion will be larger because of the larger decrease in average poison concentration.

If a large fraction of the temperature and power coefficients is attributable to prompt negative effects, then the dangers associated with larger control poison requirements are greatly mitigated by the larger negative feedback that would be encountered if some of the poison were ejected from the core. If a large part of the temperature or power defect is due to delayed negative reactivity feedback, however, the accident potential is greater; the control poison requirement has not decreased. At the same time if some fraction of this poison is now ejected from the core, the resulting excursion will not immediately be countered by the delayed negative contribution to D_T and D_P, and therefore a more serious accident might result.

Excess Reactivity

For any well-defined state of a reactor core, the excess reactivity ρ_{ex} is defined as that value of ρ that would exist if all movable control poison was instantaneously eliminated from the core. Large excess reactivities are undesirable from the standpoint that extreme care must be taken to eliminate the possibility of accident events that could lead to the rapid ejection of even relatively small fractions of the total quantity of neutron poisons. From the preceding subsection it is apparent that increases in the magnitude of negative temperature or power coefficients lead to increased excess reactivity when the reactor is at room temperature or at hot zero power.

In addition to countering negative temperature coefficients, excess reactivity must be built into power reactor cores to compensate for a number of other reactivity losses that occur over the life of the core. These include the depletion of the fuel and the buildup of neutron-absorbing fission products.

These losses may be compensated for to a greater or lesser extent by the buildup of plutonium-239 and some of the other actinides that result from neutron capture in fertile materials. These interrelated effects cause the excess reactivity to be a function both of the temperature and power level state of the core and of the fuel exposure, or total energy produced by the core since the beginning of life.

The effects of the various reactivity loss mechanisms on the excess reactivity requirement are illustrated by solid lines of excess reactivity versus fuel exposure for the various operating states given in Fig. 4-10. Assuming the temperature and power defects are positive, the minimum excess reactivity for a given fuel exposure is at full power. The excess reactivity requirement at any exposure is set by the requirement that at the exposure corresponding to the end of core life the reactor is just critical at full power with all control poisons removed. The shape of the exposure-dependent curves is a function of fuel enrichment, neutron spectrum, and flux level, since these factors determine the concentration of fission products and the rate of buildup of the various actinides.

If a thermal reactor has just been brought to power, but the relatively short half-lived neutron-absorbing fission product xenon has not yet built up, a larger excess reactivity will be present as shown. At hot zero power the reactivity is further increased by the amount of the power defect. The maximum reactivity occurs when the reactor is at room temperature, the increase being equal to the temperature defect. The temperature and power

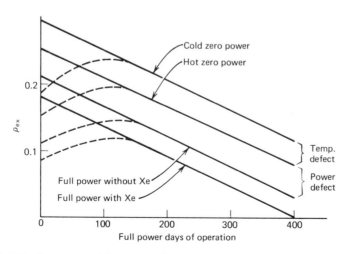

FIGURE 4-10   Excess reactivity versus time for a typical thermal reactor. ———) no burnable poison; – – – –) with burnable poison.

defects may vary somewhat during core life, since they exhibit some sensitivity to actinide buildup, neutron spectrum, and other factors that vary with exposure.

Shutdown Margin

A fundamental requirement in reactor operation is that there must always be sufficient control poison available to bring the reactor subcritical with some margin to spare. This negative reactivity following the trip of the reactor is referred to as the shutdown margin. Usually the even more stringent requirement is made that shutdown be obtainable even though one or more control rods are stuck outside of the core. The negative reactivity obtainable following trip is then referred to as the stuck rod shutdown margin.

Because of the economic penalties and safety restrictions often encountered in compensating for large reactivities solely with control rods, a number of additional mechanisms are in use for reactivity hold-down. In pressurized-water reactors, for example, the movable neutron poison is divided between control rods and soluble boron in the coolant. The reactivity worth of the control rods is required only to bring the reactor subcritical rapidly in the hot zero-power condition. Negative reactivity is then added more slowly by increasing the boron concentration in the coolant. Since thermal stress considerations place strict limitations on the maximum rate of primary system temperature decrease, the slower rate of boron addition is sufficient to maintain the reactor in a subcritical state during the cooldown to room temperature. Similarly, in pressure tube systems control rods can be used to bring the reactor subcritical at hot zero power, and then the D_2O moderator dumped slowly as the system cools down.

A second method used to mitigate the undesirable effects of a large excess reactivity is to place a fixed neutron absorber in the reactor fuel or other core constituent. These are called burnable poisons, since they provide negative reactivity at the beginning of the core life, but then deplete or "burn-up" by absorbing neutrons so that they are no longer present near the end of core life when the positive reactivity is needed to compensate for fuel depletion and fission product buildup. The dotted lines in Fig. 9-10 illustrate the effect that the presence of a burnable poison might have on the excess reactivity curves.

A major consideration in the design and operation of a reactor is the ability to predict the shutdown margin, and in particular to know in advance whether a fuel exposure state is being approached for which the shutdown margin may be lost. This is often not a simple task since it requires that one be able to predict in advance the cold and hot zero-power curves, even though operation has been proceeding at full power for some time. Determining these curves requires accurate calculations of how fast the burnable poisons deplete, since more rapid poison depletion results in larger peak reactivity. It also requires that changes in the temperature and power defects with fuel exposure be understood. Finally, the prediction of shutdown margin requires that exposure-dependent variations in the reactivity worth of the control rods be taken into account. Because of the difficulties inherent in making such predictions accurately, a large shutdown margin may be called for to compensate for the uncertainties.

PROBLEMS

4-1. Verify Eq. (4-5).

4-2. Verify Eqs. (4-26) through (4-28).

4-3. The flux and power distribution in a uniform cylindrical reactor of radius $\tilde{R}$ and height $\tilde{H}$ can be approximated as

$$\Phi(r, z) = A \cos\left(\frac{\pi z}{\tilde{H}}\right)\left[1 - \left(\frac{r}{\tilde{R}}\right)^2\right]$$

The fuel Doppler effect causes the local multiplication to vary with fuel temperature as

$$\frac{1}{k_\infty} \frac{\partial k_\infty}{\partial T_{fe}} = -\alpha_D$$

(where α_D is a positive constant) and has no effect on the core leakage properties. Estimate the Doppler coefficient $(k^{-1})(\partial k / \partial \bar{T}_{fe})$, where $\bar{T}_{fe}$ is the core-averaged fuel element temperature.

4-4. Consider a cylindrical power reactor in which the properties k_∞ and M^2 as well as the flux and power distribution are axially symmetric [e.g., $\psi(r, z) = \psi(r, -z)$ and $f_z(z) = f_z(-z)$]. The local temperature coefficients

$$\frac{1}{k_\infty} \frac{\partial k_\infty}{\partial \tilde{T}_{fe}}, \quad \frac{1}{k_\infty} \frac{\partial k_\infty}{\partial T_c}, \quad \frac{1}{M^2} \frac{\partial M^2}{\partial \tilde{T}_{fe}}, \quad \frac{1}{M^2} \frac{\partial M^2}{\partial T_c}$$

are constants.

a. Show that Eqs. (4-28) and (4-31) reduce to

$$\frac{1}{\bar{k}_\infty}\frac{\partial k_\infty}{\partial \bar{T}_i} = \frac{\langle k_\infty \psi_0^2(1-f_r)\rangle}{\langle k_\infty \psi_0^2\rangle}\left(\frac{1}{k_\infty}\frac{\partial k_\infty}{\partial \tilde{T}_{fe}} + \frac{1}{k_\infty}\frac{\partial k_\infty}{\partial T_c}\right),$$

$$\frac{1}{\bar{M}^2}\frac{\partial \bar{M}^2}{\partial \bar{T}_i} = \frac{\langle M^2(\nabla\psi_0)^2(1-f_r)\rangle}{\langle M^2(\nabla\psi_0)^2\rangle}\left(\frac{1}{M^2}\frac{\partial M^2}{\partial \tilde{T}_{fe}} + \frac{1}{k_\infty}\frac{\partial k_\infty}{\partial T_c}\right).$$

b. If the radial power peaking is made small, $f_r(r)\approx 1$, show that

$$\frac{1}{k}\frac{\partial k}{\partial \bar{T}_i}\approx 0,$$

and therefore no feedback reactivity change will take place even though $\bar{T}_i$ is changing, provided that $\bar{T}_{fe}$ and $\bar{T}_c$ remain constant.

4-5. A sodium-cooled fast reactor lattice is designed that has the following properties: migration length, 6 cm; maximum power density, 500 W/cm^3. Local sodium voiding results in the following reactivity effects:

$$\frac{\Delta k_\infty}{k_\infty} = +0.01\ \text{cm}^{-3}$$

$$\frac{\Delta M}{M} = +0.01\ \text{cm}^{-3}$$

Three bare cylindrical cores with height-to-diameter ratio of one are to be built, with power ratings of 300 MW(t), 1000 MW(t) and 3000 MW(t).

a. Find H, the core height, B_g^2, and k_∞ for each of these reactors.

b. For each of these cores plot the reactivity change caused by voiding 1 cm^3 of sodium from the core centerline versus z/H, where z is the distance from the core midplane.

4-6. A large boiling-water reactor lattice has slightly enriched UO$_2$ fuel pins with a diameter of $\frac{1}{2}$ in. The fuel-to-water-volume ratio is $1:1$. At hot zero power (i.e., fuel and moderator at 550°F), the resonance escape probability is .75. The system pressure is about 1000 psi.

a. Estimate the fuel temperature coefficient at hot zero power.

b. Estimate the fuel temperature coefficient when the fuel temperature is raised to its operating value of 1500°F, but with no boiling in the coolant channels.

c. Suppose that boiling is now allowed in the coolant channels, resulting in a 40% void fraction. Estimate the fuel temperature coefficient with the fuel again at 1500°F.

4-7. Verify Eqs. (4-94) and (4-96).

4-8. A 3000 MW(t) PWR has the following specifications: fuel mass, 240,000 lb; fuel specific heat, $0.075 \, \text{Btu}/(\text{lb})(°F)^{-1}$; coolant flow, $150 \times 10^6 \, \text{lb/hr}$; coolant specific heat, $1.3 \, \text{Btu}/(\text{lb})(°F)^{-1}$; core thermal resistance, $0.25°\text{F/MW(t)}$. The Doppler coefficient is given by

$$\frac{1}{k}\frac{\partial k}{\partial \bar{T}_f} = -\frac{4 \times 10^{-4}}{\sqrt{460 + T}}(°F)^{-1},$$

and the coolant temperature coefficient by

$$\frac{1}{k}\frac{\partial k}{\partial \bar{T}_m} = (66.7T - 0.167T^2) \times 10^{-8}(°F)^{-1}.$$

a. Over what temperature range is the core overmoderated?

b. What is the value of the temperature defect? (Assume a room temperature of 100°F and an operating coolant inlet temperature of 550°F.)

c. What is the value of the power defect?

d. If a superprompt-critical excursion is initiated while the reactor is at room temperature, how much reactivity will be lost because of prompt reactivity feedback before the fuel reaches its normal operating temperature?

4-9. Use the thermal model given in the text to derive the power coefficient for a reactor that is operated at constant outlet temperature in terms of

$$\frac{1}{k}\frac{\partial k}{\partial \bar{T}_{fe}}, \qquad \frac{1}{k}\frac{\partial k}{\partial \bar{T}_c}, \qquad \frac{1}{k}\frac{\partial k}{\partial \bar{T}_i}.$$

4-10. A LMFBR core is characterized by the following known parameters:

R = core thermal resistance [°F/MW(t)],

W_0 = initial mass flow rate [lb_m/sec],

c_p = coolant specific heat [$\text{MW(t)} \, \text{sec}/(°F)(\text{lb}_m)^{-1}$],

P_0 = initial power [MW(t)],

$$\frac{1}{k}\frac{\partial k}{\partial \bar{T}_f} = \text{average fuel temperature coefficient } [°F^{-1}],$$

$$\frac{1}{k}\frac{\partial k}{\partial \bar{T}_c} = \text{average coolant temperature coefficient } [°F^{-1}].$$

The fuel temperature coefficient is a negative constant; the coolant temperature coefficient is also constant, but may be either positive or negative.

The reactor undergoes a quasi-static flow coastdown to one-half of its initial mass flow rate. Neither the control nor the shutdown system function to reduce the reactor power.

a. Assuming the core inlet temperature remains constant, what will the reactor power be at the end of the flow transient?

b. What are the conditions on $(1/k)(\partial k/\partial \bar{T}_c)$ for which the power will (i) increase, (ii) decrease, (iii) remain the same?

REACTIVITY-INDUCED ACCIDENTS

Chapter 5

5-1 INTRODUCTION

With the background provided by the two preceding chapters, we are prepared to examine the dynamic behavior of power reactors during reactivity-induced transients. We recapitulate the general framework within which such dynamic analysis is performed before proceeding with the detailed discussion. Whether an accident is induced by equipment failure, instrumentation or control malfunction, or human error, the dynamic analysis ascertains whether the chain reaction can be terminated and a path established for the orderly transport and dissipation of decay heat. If these twin goals of reactor shutdown and orderly decay heat removal are not met, the degree of damage to the fuel elements and other fission product barriers must be determined and the consequent escape of radioactive materials assessed.

Accident categorization is facilitated by viewing reactor transients as mismatches between energy production and removal. Clearly, the net effect of any mismatch in which the production rate exceeds the removal capability is to store the excess energy in the reactor core, coolant, and/or other reactor components, causing them to overheat. If the overheating is severe enough, the fuel elements, primary system envelope, or other systems may fail by thermal or mechanical loading. The paths through which the energy transport mismatch leads to overheating and possible failure differ for each initiating event and accident sequence. It is useful, however, to categorize these as either reactivity or cooling-failure accidents.

In reactivity accidents the reactor becomes supercritical, causing the power to increase to levels beyond the steady-state removal capabilities of the heat transport system, even though that system may be in perfect operating order. Reactivity transients are further characterized by their speed and the extent of the power mismatch. Overpower transients usually involve situations in which the reactivity does not approach prompt critical, but where the power level nevertheless exceeds the core heat removal capability for some period of time that is at least comparable to that required for heat to be transported from the core. In milder overpower transients, the reactor core may appear to behave quasi-statically from both neutronic and thermal viewpoints. Nuclear excursions refer to more severe, often superprompt-critical, power transients in which the power level rises very rapidly compared to the times required to transport heat out of the fuel. In the severest situations an entire excursion may take place over a short enough time span that the core may be considered to behave adiabatically, with no significant heat transport taking place.

In a cooling-failure accident the core overheating is caused by inability to transport energy away from the core, even though power production may not be taking place at an excessive rate. In some cooling-failure accidents the reactor may be assumed to be tripped at the onset of the transient. The analysis then centers about the ability of the faulty heat transport system to dissipate the decay heat without causing unacceptable overheating or stress on the fuel elements or other fission product barriers. More often than not, however, the energy from the chain reaction is a significant contribution to the heat production during at least the early part of a cooling-failure transient—particularly if the initiating event does not result in a strong negative reactivity feedback, and the reactor trip is delayed or fails. Some cooling-failure events, moreover, may lead to positive reactivity insertions. In the absence of a quick reactor trip these cooling-failure accidents may then produce the initiating event for an overpower transient or nuclear excursion. This would occur, for example, in some liquid-metal-cooled fast reactors if a failure of the coolant pumps were coupled with a failure of the trip system.

In this chapter we first draw together in Section 5-2 the reactor kinetics and reactivity feedback models developed in Chapters 3 and 4 and a lumped parameter thermal model of the core to form an overall basis for viewing core transients. We then examine the potential initiating events for reactivity accidents in Section 5-3, and in Sections 5-4 and 5-5 we discuss reactivity accidents in the order of increasing rates of reactivity insertion, starting with quasi-static reactivity changes and ending with superprompt-critical excursions. To this point the general discussion is applicable to all stationary-fueled reactors, and examples are drawn from a number of particular power reactors. In the final section nuclear excursions that are so severe as to be terminated only by the explosive disassembly of the reactor core are examined. The treatment of disassembly accidents is exclusively for fast reactor cores, which are the only power reactors for which such accidents are thought to be possible.

5-2 REACTOR DYNAMICS

The complex interplay between the neutron chain reaction, thermal-hydraulic reactivity feedback, and reactor control and shutdown system behavior that determines the dynamic behavior of a reactor is central to the understanding of the reactivity accidents treated in this chapter and often plays an important—if not central—role in the analysis of the cooling-failure transients treated in subsequent chapters. The detailed modeling of the dynamic behavior of a reactor frequently requires one to work with the most sophisticated of space-time methods in neutronic and thermal-hydraulic modeling. Many of these methods are being developed rapidly, and they are well beyond the scope of this text. In what follows we concentrate on the use of simple lumped parameter models, which tend to be less accurate but often provide a clearer physical picture of the interplay of the various phenomena.

A Simplified Dynamic Model

The schematic diagram shown in Fig. 5-1 indicates the primary relationships between the physical processes that are most important in modeling the transient behavior of the reactor core. The description of the neutron chain reaction is contained in the reactor kinetics equations. If explicit treatment

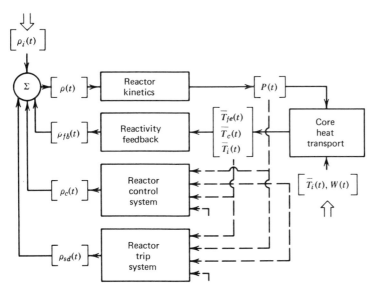

FIGURE 5-1 Simplified reactor dynamics model.

of spatial effects is not required, the point kinetics model may be used. Thus from Eqs. (3-41) and (3-42) we have

$$\frac{dP(t)}{dt} = \frac{\rho(t) - \bar{\beta}}{\Lambda} P(t) + \sum_{i=1}^{6} \lambda_i \tilde{C}_i(t), \tag{5-1}$$

$$\frac{d\tilde{C}_i(t)}{dt} = \frac{\bar{\beta}_i}{\Lambda} P(t) - \lambda_i \tilde{C}_i(t), \qquad i = 1, 2, 3, 4, 5, 6, \tag{5-2}$$

where external sources are neglected. These equations give the reactor power $P(t)$ in terms of the time-dependent reactivity $\rho(t)$ and the following core parameters, which usually may be considered constant: Λ, the prompt neutron generation time, $T_{1/2i} = 0.693/\lambda_i$, the delayed neutron half-lives, and $\bar{\beta} = \sum \bar{\beta}_i$, the delayed neutron fraction. As indicated in Chapter 3, extreme situations often justify further approximations. For very slow transients in which the reactivity does not approach the prompt-critical condition of $\rho = \bar{\beta}$, the zero prompt lifetime approximation of Eq. (3-61) may be adequate. For fast, superprompt-critical excursions, delayed neutron effects often may be neglected and the kinetics equations reduced to the prompt-critical approximation given by Eq. (3-57).

The reactivity $\rho(t)$ that drives the transients is the net effect of contributions arising from several mechanisms. In Fig. 5-1 these are represented as

$$\rho(t) = \rho_i(t) + \rho_{fb}(t) + \rho_c(t) + \rho_{sd}(t), \tag{5-3}$$

where $\rho_i(t)$ is the reactivity caused by the initiating event, $\rho_{fb}(t)$ the reactivity from thermal–hydraulic feedback, $\rho_c(t)$ the reactivity from the reactor power control system, and $\rho_{sd}(t)$ the shutdown or trip reactivity.

The reactivity insertion caused by the initiating event often may be represented as an explicit function of time. Such is the case, for example, when the transient is due to the uncontrolled withdrawal of a control rod bank or the ejection of a single rod. In other transients the initiating reactivity may not be amenable to explicit representation. The reactivity insertion from the coolant boiling that results from a flow failure, for example, often falls into this category. As a rule the reactivity insertion rate and the total reactivity available are the most important characteristics of $\rho_i(t)$. Consequently, if the details of the insertion for a postulated accident are not well defined, the reactivity is commonly represented as

$$\rho_i(t) = \dot{\rho}t, \qquad 0 \leq t \leq t_0, \tag{5-4}$$

where $\dot{\rho}$ is a constant ramp rate of insertion and $\dot{\rho}t_0$ is the total insertion.

The feedback reactivity $\rho_{fb}(t)$ is dependent on the fuel, coolant, and other temperature and/or density coefficients for the reactor core under consideration. If these reactivity feedback effects can be approximated in terms of a relatively few core-averaged variables, the reactivity can be represented compactly. For example, the feedback model developed in Section 4-2 expresses $\rho_{fb}(t)$ in terms of $\bar{T}_{fe}$, $\bar{T}_c$, and $\bar{T}_i$, the core-averaged fuel element, coolant, and coolant inlet temperatures. Integrating the incremental feedback equation obtained from this model, Eq. (4-17), we have

$$\rho_{fb}(t) = \int_{\bar{T}_{fe}(0)}^{\bar{T}_{fe}(t)} \frac{1}{k} \frac{\partial k}{\partial \bar{T}_{fe}} d\bar{T}_{fe} + \int_{\bar{T}_c(0)}^{\bar{T}_c(t)} \frac{1}{k} \frac{\partial k}{\partial \bar{T}_c} d\bar{T}_c + \int_{\bar{T}_i(0)}^{\bar{T}_i(t)} \frac{1}{k} \frac{\partial k}{\partial \bar{T}_i} d\bar{T}_i, \quad (5\text{-}5)$$

where we initialize by taking $\rho_{fb}(0) = 0$.

To determine the feedback reactivity effect, the core temperatures must be determined simultaneously with the power and reactivity during the course of the transient. In certain well-defined situations this task may be simplified by using power, isothermal temperature, or prompt reactivity coefficients, as discussed in Chapter 4. Most often, however, it is necessary to utilize a transient heat transport model in which the required temperatures appear in sets of coupled differential equations. These must be solved simultaneously with the reactor kinetics equations. One such model for the average fuel element and coolant temperature is developed in the following subsection. This, or other similar models, couples the core temperatures to the reactor power and inlet coolant conditions. The determination of these inlet conditions—most often $\bar{T}_i(t)$ and $W(t)$, the inlet coolant temperature and the coolant mass flow rate, must in turn be obtained either from a model of the primary system thermal hydraulics or from some simplifying assumptions concerning the behavior of that system.

In addition to the reactivity caused by the initiating event and that caused by the inherent feedback of thermal-hydraulic effects, the reactor control and shutdown systems may play important roles in some transients. These systems remove reactivity by inserting movable neutron poisons. They are actuated by instrumentation measuring power, temperature, or other variables, as indicated by the dotted lines in Fig. 5-1. In very fast transient, notably the superprompt-critical excursions, control and shutdown are likely to be too sluggish to have an appreciable effect, the primary determinants of the course of such transients being the interplay between initiating reactivity insertion and prompt reactivity feedback. In more slowly evolving transients the trip system is the primary means of bringing about an early termination of a reactivity accident. The control system function, in contrast, is to meet power demand rather than to shut the reactor down if an accident occurs. As a result its effects during accident transients are more varied. In some cases the control system may reduce the consequences of overpower transients by keeping the reactor at constant power. As we see in Section 5-4 however, the control system may tend to mask the early stages of an accident, preventing early shutdown by the trip system. Likewise, failures in control systems are one of the more frequent causes of accidental reactivity insertions.

A Core Heat Transport Model

Before proceeding with the analysis of reactor accidents, we must incorporate a transient heat transport module into the reactor dynamics model shown in Fig. 5-1.

The accurate thermal modeling of a reactor core under accident conditions frequently requires the numerical solution of large systems of partial differential equations for transient heat conduction, fluid flow, convection, and a host of other phenomena likely to occur during the course of a transient. Simple lumped parameter models, such as the one that follows, however, often provide sufficient accuracy when the primary objectives are to gain an understanding of the interaction of the various phenomena involved and to make rough estimates of the consequences of the accident.

We base our lumped parameter model on two energy balances, one for the energy contained in all the fuel elements of the reactor core, and a second for the energy stored in the coolant within the core volume. Suppose that M_{fe} is the total fuel element mass in the reactor core and c_{fe} is the specific heat per unit mass of the fuel elements. The energy balance for the fuel elements may be written as

$$M_{fe}c_{fe}\frac{d\bar{T}_{fe}}{dt} = P(t) - P_1(t),\qquad(5\text{-}6)$$

where $\bar{T}_{fe}$ is the core-averaged fuel element temperature, $P(t)$ the reactor power, and $P_1(t)$ the rate at which energy is transported across the cladding surfaces into the coolant. Similarly, suppose that M_c is the mass of coolant within the reactor core and c_p is the coolant specific heat per unit mass. The energy balance for the coolant within the core may be written as

$$M_c c_p \frac{d\bar{T}_c}{dt} = P_1(t) - P_2(t),\qquad(5\text{-}7)$$

where $\bar{T}_c$ is the average coolant temperature and $P_2(t)$ is the net rate of convective heat transport out of the coolant channels.

Thus far, Eqs. (5-6) and (5-7) are exact. We must now determine $P_1(t)$ and $P_2(t)$ to obtain simple differential equations that involve only the core-averaged temperatures. We approximate these quantities by requiring that the preceding energy balances reduce to the steady-state relationships of Chapter 1 when the time derivatives are set equal to zero. Equations (5-6) and (5-7) reduce to Eqs. (1-112) and (1-114) if we make the approximations

$$P_1(t) = \frac{1}{R}[\bar{T}_{fe}(t) - \bar{T}_c(t)] \qquad (5\text{-}8)$$

and

$$P_2(t) = 2W(t)c_p[\bar{T}_c(t) - \bar{T}_i(t)], \qquad (5\text{-}9)$$

where R is the core thermal resistance and W is the mass flow rate. Combining these expressions with Eqs. (5-6) and (5-7), we have

$$M_{fe}c_{fe}\frac{d\bar{T}_{fe}}{dt} = P(t) - \frac{1}{R}[\bar{T}_{fe}(t) - \bar{T}_c(t)] \qquad (5\text{-}10)$$

and

$$M_c c_p \frac{d\bar{T}_c}{dt} = \frac{1}{R}[\bar{T}_{fe}(t) - \bar{T}_c(t)] - 2W(t)c_p[\bar{T}_c(t) - \bar{T}_i(t)]. \qquad (5\text{-}11)$$

In Chapter 6 expressions for R and c_{fe} are developed in terms of the material properties and dimensions of cylindrical fuel elements. If in addition to this information, the time dependence of the reactor power, coolant flow rate, and inlet temperature is known, these equations can be solved for $\bar{T}_{fe}(t)$ and $\bar{T}_c(t)$. Some additional physical insight into the transient heat transport process may be gained by rewriting Eq. (5-10) as

$$\frac{d\bar{T}_{fe}}{dt} = \frac{P(t)}{M_{fe}c_{fe}} - \frac{1}{\tau}[\bar{T}_{fe}(t) - \bar{T}_c(t)]. \qquad (5\text{-}12)$$

The quantity

$$\tau = RM_{fe}c_{fe} \qquad (5\text{-}13)$$

must have units of time, and we refer to it as the core time constant. It is a measure of how much time is required to transport heat from fuel to coolant. For UO_2 metal-clad fuel elements τ is typically of the order of a few seconds. For HTGR lattices it is often much larger, since the heat must be·conducted through a large fraction of the graphite moderator before reaching the coolant channels. Some typical values of τ along with other heat transfer parameters are given in Chapter 6.

The significance of τ is illustrated by a simple example. Suppose that a reactor is initially operating at steady state. For $t > 0$ we set $P(t) = 0$, and ask how quickly the sensible heat stored in the fuel elements will be dumped into the coolant following the shutdown if the coolant temperature remains constant. We have from Eq. (5-12),

$$\bar{T}_{fe}(t) = \bar{T}_c + [\bar{T}_{fe}(0) - \bar{T}_c] e^{-t/\tau}. \qquad (5\text{-}14)$$

Hence there is an exponential decay of the fuel element temperature toward the value of the coolant temperature. Moreover, at the time 0.693τ, one-half of the sensible heat stored in the fuel element has been removed to the coolant.

In some circumstances it is reasonable to make further approximations to Eqs. (5-11) and (5-12) to clarify the dominant contributions to the transient effects and to simplify the heat transport model. In many power reactors the two following conditions reasonably approximate existing situations:

1. The heat capacity $M_{fe}c_{fe}$ of the solid fuel is much larger than the heat capacity $M_c c_p$ of the coolant within the core.
2. The average fuel temperature drop $\bar{T}_{fe} - \bar{T}_c$ is much larger than the coolant temperature rise $\bar{T}_c - \bar{T}_i$.

Under these assumptions it is reasonable to neglect the changes in heat stored in the coolant channels, since $(M_c c_p)(d\bar{T}_c/dt)$ is likely to be small compared to the other terms in the equations. We may therefore approximate Eq. (5-11) by the algebraic relationship

$$\bar{T}_{fe}(t) - \bar{T}_c(t) = 2W(t)c_p R[\bar{T}_c(t) - \bar{T}_i(t)]. \qquad (5-15)$$

If the temperature difference on the left is assumed to be much larger than that on the right we must have

$$2W(t)c_p R \gg 1. \qquad (5-16)$$

This condition tends to be met most closely in ceramic-fueled, liquid-cooled reactors for which R and W are both large.

Utilizing Eqs. (5-12) and (5-15) along with the condition (5-16) we can obtain approximate, uncoupled first-order equations for the fuel and coolant temperatures:

$$\frac{d\bar{T}_{fe}}{dt} = \frac{P(t)}{M_{fe}c_{fe}} - \frac{1}{\tau}[\bar{T}_{fe}(t) - \bar{T}_i(t)], \qquad (5-17)$$

and

$$\frac{d}{dt}[\bar{T}_c(t) - \bar{T}_i(t)] = \frac{P(t)}{2W(t)c_p\tau} - \left(\frac{1}{\tau} + \frac{1}{W}\frac{dW}{dt}\right)[\bar{T}_c(t) - \bar{T}_i(t)]. \qquad (5-18)$$

These are easily solved in terms of P, W, and $\bar{T}_i$. From these equations we can see that the core time constant τ dominates the transient heat transport behavior unless there is a flow transient in progress that causes $(1/W)(dW/dt)$ to be of a magnitude comparable to $1/\tau$. The time constant

τ in fact is a good indicator of the types of temperature distributions and of the nature of the reactivity feedback effects that are likely to exist during the course of a transient. If the reactor power, for example, varied only slowly compared to τ, steady-state temperature distributions yield reasonably accurate results, and the power coefficient may be used to predict reactivity feedback. On the other hand, if the power is increasing very rapidly compared to τ, the core may be considered to behave essentially adiabatically, and the use of the prompt reactivity coefficient is called for. Of course, a great many transients evolve over time scales comparable to τ, and a more detailed assessment of the thermal transient must be made. Likewise, the speed of inlet temperature and flow transients must be considered, and particular attention is called for when R—and therefore τ—changes because of a boiling crisis or other disruption of the heat transport path.

The foregoing equations may be incorporated into a dynamics model such as that shown in Fig. 5-1 to estimate time-dependent core-averaged temperatures along with reactor power transients. Following such transient analysis, a determination of the damage to the fuel elements, primary system envelope, and other reactor components is normally in order. Such estimates require that local values of the temperatures be known, since the local maxima often determine whether there will be fuel melting, cladding failure, or other related damage. Provided the spatial distributions of the fuel element and coolant temperatures about the core-averaged values do not depart markedly from the steady-state distributions, Eqs. (4-15) and (4-16) may be used to estimate local temperatures; in particular, the peaking factors F_r and F_z, defined in Chapter 1, may be useful in determining the maximum local temperatures reached during the transient.

5-3 REACTIVITY INSERTION MECHANISMS

Any mechanism by which the multiplication of a chain reaction may be increased is a potential source of a reactivity-induced accident. Only through the detailed examination of each new reactor core design can possible reactivity insertion mechanisms be understood with sufficient detail to prevent the occurrence of such accidents. Although an exhaustive exposition of the insertion mechanisms found in the variety of power reactor designs now in use is not possible, a discussion of some of the more frequently appearing mechanisms provides a base for the more detailed examination of the idosyncracies of each reactor core. We concentrate on those mechanisms that have the greatest potential for causing significant accidents: those with large total reactivity that may be inserted rapidly. In the following discussion such mechanisms are conveniently divided into four categories: rapid neutron poison removal, coolant density effects, changes in core component configurations, and introduction of extraneous materials.

Neutron Poisons

The neutron-absorbing material contained in a reactor core may take one or more of several forms: control rods of the regulating, shim, or safety variety, soluble poisons in the coolant stream, or burnable poisons incorporated into the fuel elements or other core constituents. The total reactivity worth of these materials required in the core depends on the fuel loading, cycle length requirements, reactivity feedback effects, and other factors discussed in this and the preceding chapters. Several dollars or more of poison reactivity worth are not uncommon in reactor cores. Hence it is necessary to ensure that even under accident conditions only small fractions—preferably less than one dollar—of such poisons can be lost from the core, and that they can be lost only at rates that are slow enough for effective action to be taken to detect the malfunction and shut down the reactor before undue damage results.

Control Rod Worths. The control rod is the most widely used form of neutron poison, and control rod malfunctions have resulted in a number of reactor accidents, albeit primarily in relatively low-powered experimental and test reactors (Thompson, 1964). Two important parameters govern the seriousness of control rod ejection accidents: the maximum worth of the individual control rod and the rate at which the rod is withdrawn or ejected. In some designs relatively few high-worth safety rods are used to shut down the reactor. If this is the case, interlocks are included to ensure that these rods are out of the core before the reactor is brought critical. Otherwise, their inadvertent withdrawal from an initially critical state could lead to an unacceptable reactivity addition. Reactivity adjustments while the reactor is operating are made by shim and regulating rods. The reactivity required for these adjustments should be divided among a sufficiently large number of rods so that the loss of any single rod will not result in an unacceptable power transient. Even with a sufficient number of rods, care must be taken that the local multiplication in the vicinity of a rod does not build up to a point where an excessive rod worth results. Several situations can be envisioned where this could take place: the burnup of a fixed burnable poison more rapidly than anticipated, the unexpectedly fast buildup of plutonium, fuel enrichment errors, or the presence of locally fresh fuel because of long-term neutron flux depression caused by the presence of the rod. Care must also be taken in properly accounting for the dependence of the control rod worths on the positioning of neighboring rods. Such effects, usually referred to as *shadowing*, are particularly important in neutronically loosely coupled cores, such as those found in present-day water-cooled reactors. As alluded to in Chapter 1, the placement and withdrawal sequences of control rods in large, neutronically loosely coupled cores give rise to unique considerations. As indicated in Fig. 1-14*a* for a boiling-water reactor, such cores are controlled by many identical control rods that are used for regulation, shim, and shutdown purposes. Many rods are required because the effects of any one rod are felt only within a few migration lengths of its location. Moreover, the reactivity worth of each rod is strongly affected by the positions of the neighboring rods. These shadowing effects make it essential that withdrawal sequences and control rod patterns be carefully chosen so that no single rod has an excessive worth. Normally this may be accomplished by withdrawal of the rods in checkerboard arrays so that the neutron absorber concentration over the *x*-*y* plane of the reactor is relatively uniform.

The potential for a situation with an extremely high-worth rod most often appears when the reactor is first being brought critical. If Fig. 1-14*a* represents a freshly loaded core, there may be sufficient excess reactivity that the reactor becomes critical during the withdrawal of only the second or third rod, if these rods are withdrawn from contiguous locations. Furthermore, the worth of these initially withdrawn rods tends to be large, since an extremely large power peak results in the thin cylindrical unpoisoned region. For this reason operators must avoid errors in withdrawing off-sequence rods that could create the situation in which the ejection of a single rod would then bring the reactor superprompt critical with a very large reactivity insertion. Because of the seriousness of this problem, electronic rod block interlocks may be incorporated into the control system. These allow rods to be withdrawn only in prescribed checkerboard sequences.

Control Rod Faults.   The faults leading to the ejection of control rods vary considerably with both the mechanical design of the control rods and the instrumentation and control that lead to the actuation of rod withdrawal. Some of the more prominent problems can be illustrated with a schematic diagram: As shown in Fig. 5-2, a typical control rod is linked to a drive mechanism by a connecting shaft sometimes referred to as the *control extension*. The drive is a mechanism for raising and lowering the rod in the core at a speed limited by the maximum desired rate of reactivity change. In addition to the normal movement of the rod, the drive mechanism must provide a means of rapidly driving it into the core if a reactor trip is called for. The driving force must be provided by stored energy, since prudence indicates that loss-of-power to the drive mechanism be required to cause the insertion of the rods into the core. The simple drop of the rods driven by

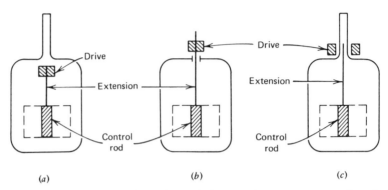

FIGURE 5-2   Control rod configurations relative to primary system envelope (not to scale).

gravity is the most widely used method for emergency insertion of control rods. Other methods must be required, however, if rods enter horizontally or from the bottom of the core. The use of stored gas in accumulators is the most prominent of these.

The layout of the control rods and their drives relative to the primary system envelope has particular safety implications for rod ejection. Except in certain pressure tube systems, the control rod must lie within the primary system envelope. Thus either the entire drive mechanism must be canned in such a way that it can be entirely enclosed in the primary system, or the control rod equipment must pass through the envelope. The latter situation calls for particular attention when the primary system is at high pressure. The connecting shaft may be designed to pass through a shaft seal in the reactor vessel, as shown in Fig. 5-2b. Alternately, the shaft may be enclosed in a sleeve that is part of the pressure envelope and operated by an external drive through a magnetic coupling system, as shown in Fig. 5-2c (Esselman et al., 1973).

A second consideration with safety implications involves the disposition of the control rod system during refueling. In some reactors—particularly those with on-line refueling—the control rods and their drives are intact throughout the fueling process. However, many systems are refueled by removing the pressure vessel head. Therefore if the control rods enter from the top of the core they must be decoupled from the drives. Even in systems in which refueling is done through smaller penetrations and the drives need not be removed, the connecting shafts are likely to obstruct the paths necessary for the movement of the fuel assemblies and refueling machinery.

With this background, some potential mechanism for control rod ejection becomes apparent. If the connecting shaft passes through the reactor pressure vessel, the end of the control rod is exposed to the reactor coolant pressure and the external end of the shaft is exposed to atmospheric pressure. If a failure of the rod hold-down mechanism were to occur, a large force would then be available to accelerate the rod out of the core. In some reactors drag devices are incorporated into the control rods to limit the ejection velocity in the event of such an accident.

If rods are uncoupled from their connecting shafts during refueling, design interlocks are required to ensure that they are securely seated in the core. Moreover, mechanical manipulation during the decoupling or coupling of the rod to the shaft and/or drive mechanism must not permit an inadvertent pulling of the rod from the core. It appears to have been just such a jerking of a control rod from the core, in the process of reconnecting it to its drive, that caused the SL-1 excursion that killed three maintenance people (Thompson, 1964).

A potentially dangerous situation also exists if a reactor is operated with rods in the core that are decoupled from the connecting shaft or drive mechanism. This might result from improper securing of the system follow-ing refueling or from breaking the linkage in an attempt to withdraw a stuck rod from the core. Such decoupled rods, even though they may initially be stuck in place, experience a variety of dynamic forces during the life of the core. Thermal expansion, bowing, or other mechanisms may free such a stuck rod, causing it to fall from the core or be lifted by the flow of coolant. In particular, the rod could be ejected as the result of pressure differentials created under accident situations. Because of the dangers of uncoupled rods it is desirable to have reliable indication of both rod position and drive setting, and also to incorporate mechanical design features that prevent the rod from falling freely from the core or being lifted by the force of the coolant flow.

In addition to avoiding the preceding mechanical problems, care must be taken in the drive design to eliminate potential faults that could cause the drive motor to malfunction and withdraw the rod without command. It is common practice to include governors or other overspeed protection to ensure that even if an unwanted withdrawal begins, an absolute limit is placed on the withdrawal speed and therefore the reactivity insertion rate.

Thus far we have considered only failure mechanisms associated with the control rods and their drive systems. The drives in many cores are actuated by the reactor control as well as the protection system, and therefore a control system failure may lead to accidental rod withdrawal. Faults in the control system may be particularly serious if the control rods are operated in banks, since the withdrawal is then not necessarily limited to one rod. A critical trade-off must be made between setting performance requirements for the control system and limiting the maximum rate with which reactivity can be inserted (Pearson and Lennox, 1964). If the control system is required to make rapid changes in power level, to meet swings in electrical demand, it must be able to place the reactor on short periods by rapidly adding or subtracting reactivity. On the other hand, safety constraints dictate the maximum number of control rods that can be included in a bank and maximum bank movement rates so that a failure in the control system instrumentation or logic will not cause an excessive rate of reactivity addition.

Soluble Poisons. In addition to control rods, neutron absorbers dissolved in the coolant stream are often used to control reactivity in pressurized-water reactor cores, with reactivity being inserted when the concentration of absorber in the coolant is decreased. The absorber concentration is decreased by replacing poisoned with unpoisoned water, with the neutron absorber being extracted from the water by a demineralizer, so that it can be reused. The maximum rate of reactivity increase in such systems is control-led by the size of the pipes, pumps, and other equipment in this purification system. Emergency cooling water must also be heavily poisoned to ensure that only negative reactivity effects result from its uses. Finally, if coolant loops or other large volumes of the primary system may be isolated from the coolant passing through the core, procedural or mechanical interlocks must be used to ensure that an isolated loop or reservoir containing unpoisoned coolant is not suddenly brought into hydraulic contact with the core, since a large quantity of pure coolant would then enter the core at a time when the reactivity was being held down by heavily poisoned coolant.

Coolant Density Effects

Regardless of whether a reactor has a positive or negative coolant density coefficient, potential faults in the heat transport system may lead to coolant-induced reactivity insertions. Such insertions tend to be very small in gas-cooled reactors, since the coolant densities are small to begin with. They are much larger when there are changes in the temperature of a liquid coolant, and the largest effects are likely to occur with the large density changes associated with transitions between liquid and vapor phase, whether this be in the form of coolant boiling or condensation. The thermal-hydraulic mechanisms leading to undesirable density changes in coolant density are discussed in some detail in subsequent chapters. It is nevertheless useful at this point to summarize briefly the more prominent effects.

In a reactor with a negative density coefficient (positive temperature and void coefficient), reactivity insertion results from increased coolant temperature. Such a temperature rise would result from a coolant flow failure that increased the temperature rise across the core, even though the inlet temperature remained constant. A coolant temperature rise also results from an increase in the inlet coolant temperature, such as would appear if the heat transport path out of the primary system were impaired. The most serious insertion of reactivity is likely to result if the overheated coolant reaches its boiling point, causing a precipitous drop in the coolant density over some region of the reactor core.

In a reactor with a positive density coefficient (negative temperature and void coefficient), a reactivity insertion results from decreased coolant temperature or from bubble collapse in the case of a boiling coolant. Decreased coolant temperature results from a decrease in inlet temperature or an increase in coolant flow rate. Both of these mechanisms are usually lumped under the heading of cold-water accidents. In this scenario a coolant loop that has been hydraulically isolated from the core, by either closed valves or pump shutdown, is assumed to contain coolant that is significantly cooler than that in the remainder of the primary system. If the valves are suddenly opened, or the pump started, this cool water enters the core causing a drop in the coolant temperature. Cold-water accidents may also result from a sudden increase in the heat removal rate from the primary system. Increased feedwater flow, loss of a feedwater heater, or other secondary system problem also may cause such an increase. In boiling-water reactor cores an even more severe reactivity insertion may result if a pressure surge in the primary system causes the sudden collapse of the coolant voids in the core. In direct-cycle systems such surges may result, for example, from the sudden closure of isolation valves that separate the reactor vessel from the turbine.

Rearrangement of Core Components

Insertion mechanisms also may be caused by changes in the physical arrangement of those solid components that remain stationary relative to one another during normal operation. Such rearrangements may result from wearing away of components or from structural instabilities or collapse. For example, reactivity may be gained from the corroding away of burnable poisons, cladding, or other neutron-absorbing materials. In the EBR-1 reactor, the thermally induced inward bowing of the fuel elements led to a reactor meltdown (Thompson, 1964). Structural failure of the loading machine during refueling may lead to the dropping of fuel assemblies into the core.

Probably the most severe examples of reactivity increase caused by the rearrangement of core materials come in the study of fast reactors when the reactor core is hypothesized to melt because of a severe accident. Fast reactors are not in their most reactive state during operation, and should forces arise that tend to compact a core or eliminate nonfissionable materials, a reactivity increase is likely. Limiting cases of core compaction have traditionally served to set upper bounds on the severity of fast reactor accidents. For the most part, these stem from the original Bethe–Tait (1956) model for fast reactor explosions. In this model the reactivity insertion was assumed to arise from a collapse of the upper half of the core into the lower half with the full acceleration of gravity.

An understanding of the reactivity consequences of core meltdown must take into account a variety of phenomena. The reactivity consequences of the cladding melting and being swept away must be accounted for, in addition to the effects of fuel movement. Fuel movement estimates must account for whether the fuel is swept away by the coolant stream, drops free from the core, or congeals in a more compact mass within the core volume. Moreover, vapor generation, which might explosively compact the fuel in some region of the core cannot be ruled out a priori as a cause of a reactivity increase. Finally, the disposition of the molten fuel mass following the collapse of the core structure must be considered under conditions by which secondary criticality might be reached as a result of a reconcentration of the core debris.

Extraneous Materials

In some circumstances the introduction of extraneous materials into a reactor core may lead to increased reactivity. The two most likely materials to be introduced in significant quantity are air, in the event of a breach of the primary system, and water, in the event that the heat transfer interface between primary system and working fluid is breached. Lubricants, corrosion products, and other impurities that may be present in the primary system also may have significant reactivity effects.

For some coolants, particularly liquid metals, the danger of combustion in the presence of oxygen or water is an overriding consideration to the point that it appears impossible that these elements could be introduced in sufficient quantity to have significant reactivity effects. In water reactors the dominant concern is that the water in-leakage would have a lower neutron absorber concentration and/or temperature and therefore cause a reactivity insertion. The introduction of water into the coolant channels of gas-cooled reactors requires the most careful consideration. Although a possible combustion hazard if brought into contact with some materials, water is likely not to add reactivity in thermal reactors because it is not as good a moderator as heavy water or graphite. This, however, must be verified by a careful examination of each core configuration. The introduction of water in gas-cooled fast reactors may result in either decreased or increased reactivity depending on the amount. For small amounts the tendency will be to soften the neutron spectrum into the resonance energy range because of the increased moderation. This will result in a reactivity decrease, as explained in Chapter 4, caused by the decreased fission-to-capture ratio for neutrons in this energy range. If sufficient water is added to permit the slowing down of the neutrons past the capture resonances and into the thermal energy range, however, the consequence may be a large reactivity increase. In the latter situation the substantial enrichments of the fast reactor fuel may result in a water -moderated, thermal reactor configuration that is highly supercritical. The prevention of such an occurrence may be guaranteed by either of two approaches. First, the inventory of the steam generator or other possible water reservoirs can be limited so that even if all the water entered the primary system, it would not fill the reactor vessel to the point where it would begin to cover the core. Second, the fuel or other core components could be heavily doped with a thermal neutron absorber, so that even if the core were to be covered with water it would not become critical.

5-4 OVERPOWER TRANSIENTS

The transient behavior induced in power reactors by accidental insertions of reactivity is strongly dependent on the rate and magnitude of the insertion as well as the design characteristics of the reactor and the state of the system at the time of the accident. The examination of transients carried out in this and the following sections is facilitated by dividing reactivity accidents into broad categories according to the time-dependent properties of the initiating reactivity $\rho_i(t)$. Suppose we let the initiating reactivity be approximated by the ramp form

$$\rho_i(t) = \begin{cases} \dot{\rho}t, & 0 \le t \le t_0, \\ \dot{\rho}t_0, & t > t_0. \end{cases} \tag{5-19}$$

The insertion then is characterized by a ramp rate $\dot{\rho}$ and a magnitude $\dot{\rho}t_0$. If the initial neutron flux is orders of magnitude below the range where operation at power takes place, a very fast reactivity insertion may be completed before the power reaches the level where temperature and density changes begin to cause reactivity feedback. If, in addition, the insertion is completed before a power level that will cause the shutdown system to be actuated is reached, Eq. (5-19) may be approximated as an instantaneous step insertion. During the initial phase of the transient the reactor then may attain a constant period, as shown in Fig. 3-1, until the effects of feedback or shutdown reactivity become important.

For the preponderance of physically realizable accidents, the preceding step conditions are not met. Before the insertion is completed the reactor power is likely to rise sufficiently to cause reactivity feedback through the fuel temperature, to actuate the shutdown system, or to cause the control system to modify the course of the transient. Moreover, if the initial power level is within two or three decades of full power, prompt reactivity feedback will follow immediately from the power changes accompanying the insertion. In these situations it is the rate of reactivity insertion $\dot{\rho}$ that becomes of primary importance, since the transient may have been terminated—by either the shutdown system or reactivity feedback—before the insertion is completed. And, moreover, the maximum total reactivity $\rho(t)|_{max}$ that is reached during the accident is more closely correlated to $\dot{\rho}$ thann to $\dot{\rho}t_0$. If indeed the maximum reactivity is reached before the completion of the insertion, it is useful to categorize accidents broadly according to increasing ramp rates, since these correlate to accidents of increasing speed and severity, and if acceptable risk criteria are to be met, they also must correspond to accidents of decreasing probability. Going from small to large reactivity insertions, we classify transients as quasi-static, superdelayed critical, and superprompt critical.

When reactivity is added while the reactor is at power, the feedback reactivity and the control system tend to compensate for the insertion. If the reactivity addition rate is slow compared to the thermal time constant of the core, the delayed neutron behavior and the control system response time, the terms on the right of Eq. (5-3), may cancel one another, until a shutdown signal is generated and the reactor is scrammed. Therefore

$$\rho(t) \approx 0. \tag{5-20}$$

We refer to such transients as quasi-static, since the reactor slowly passes through a sequence of near-equilibrium conditions in which the power and temperature distributions are essentially the same as those that would occur during steady-state operation.

As consideration is given to increasingly faster rates of reactivity insertion, the quasi-static description is no longer valid, since the time scales become short enough that feedback and control reactivity can no longer respond quickly enough to compensate for the reactivity insertion. The properties of the delayed neutrons then become important in determining the neutron kinetics of the transient. Likewise, the transient is now likely to take place on a time scale that is comparable to the time required to remove heat from fuel to coolant. The fuel temperature is then governed by transient heat conduction and convection. This complex interaction of neutron kinetics and transient heat transfer is typical of transients for which the net reactivity rises above zero but does not reach prompt critical. We refer to reactivity accidents for which

$$0 < \rho(t)|_{max} < \bar{\beta} \qquad (5\text{-}21)$$

as superdelayed-critical transients.

We categorize as superprompt-critical transients those accidents for which reactivity is added so fast that the net reactivity rises above prompt critical,

$$\rho(t)|_{max} > \bar{\beta}, \qquad (5\text{-}22)$$

before reactivity reduction mechanisms become effective in reducing the reactivity. As illustrated by Fig. 3-3, as $\rho = \bar{\beta}$ is exceeded, the reactor period becomes so short that delayed neutrons no longer play an important role, and the neutronic behavior is governed by the prompt neutron generation. Moreover, heat transport out of the fuel is not very important on these short time scales, and the fuel may be assumed to exhibit an essentially adiabatic thermal behavior.

The mechanisms by which transients are terminated vary considerably as one passes from quasi-static to superdelayed-critical and from super-delayed-critical to superprompt-critical reactivity accidents. Since quasi-static transients are slow and mild, the preponderance of the transient time occurs between the initiating event and the time at which the protection system detects unacceptable changes in the state of the reactor core. On these relatively long time scales the additional power produced between the actuation of the shutdown system and the termination of the chain reaction by the control rods has negligible effect on the reactor core. In superdelayed-critical transients the chain reaction may again be expected to be terminated by driving the control rods into the core. With the more rapid rates of power increase characteristic of these transients, however, the power produced during inertial delays between shutdown system actuation and the move-ment of the control rods far enough into the core to bring the reactor subcritical may be a significant fraction of the total energy release. Superprompt-critical transients occur on such short time scales that prompt reactivity feedback that is inherent in the reactor fuel must be relied on to terminate the power burst. The inertial delays in inserting control rods or operating other active shutdown systems are too long for these systems to be effective against such power bursts.

The critical parameters on which fuel damage depends also change as one goes from very slow to very fast reactivity transients. In quasi-static tran-sients an essentially steady-state temperature distribution exists. Therefore, as seen from Eqs. (1-112) and (1-114), the maximum fuel temperature is determined by the maximum power level P_{max} that is reached before the transient is terminated. For superdelayed-critical accidents, transient heat transport equations such as Eqs. (5-10) and (5-11) must be solved to relate the maximum fuel temperature to the power history. With the very fast transients that characterize superprompt-critical reactivities, the adiabatic thermal behavior leads to fuel temperatures that are proportional to the total energy release: $\int_0^\infty P(t')\,dt'$.

If in any of the preceding situations the fuel melting point is reached, the fuel enthalpy rather than the temperature must be determined, since it determines what fraction of the fuel will melt. In extremely severe superprompt-critical transients the fuel may not only melt but vaporize. The high pressure may result in destructive blast effects. Since high pressures are generated only when the energy release rate is so fast that the pressure cannot be dissipated by expansion of the surrounding materials, the maximum power again become an indication of the destructiveness of the burst.

Although destructive superprompt-critical accidents represent the clearest threat to the integrity of the fission product barriers, the slower transients also have important safety implications. In the remainder of this section we examine those overpower transients for which prompt critical is not reached, and in the following sections the power bursts of superprompt-critical excursions are described.

Quasi-static Transients

As stated earlier, in quasi-static transients we are dealing with reactivity changes that are sufficiently slow that the reactivity insertion is continually negated by the effects of thermal feedback and/or the control system. Thus from Eq. (5-3) we have

$$0 = \rho_i(t) + \rho_{fb}(t) + \rho_c(t), \qquad (5\text{-}23)$$

where we assume that the shutdown system has yet to be acutated, and therefore $\rho_{sd} = 0$. The changes in temperature are assumed to be slow compared to the core time constant τ, and, consequently, steady-state temperature distributions exist. Under these conditions the reactivity feedback is determined from the power coefficient:

$$\rho_{fb}(t) - \rho_{fb}(0) = \int_{P(0)}^{P(t)} \frac{\partial \rho_{fb}}{\partial P}\,dP. \qquad (5\text{-}24)$$

We take the initial conditions to be

$$\rho_i(0) = \rho_{fb}(0) = \rho_c(0) = 0, \qquad (5\text{-}25)$$

and represent the reactivity insertion by Eq. (5-19). Then we approximate the power coefficient

$$\frac{d\rho_{fb}}{dP} = -\mu_P \tag{5-26}$$

as a negative constant. Over the range of interest we may write Eq. (5-23) as

$$0 = \dot{\rho}t - \mu_P[P(t) - P(0)] + \rho_c(t). \tag{5-27}$$

The reactor power is then given by

$$P(t) = P(0) + \frac{1}{\mu_P}[\dot{\rho}t + \rho_c(t)] \tag{5-28}$$

until the shutdown system is actuated. If the control system does not automatically modify the reactivity, we have the simple linear power increase

$$P(t) = P(0) + \frac{\dot{\rho}t}{\mu_P}. \tag{5-29}$$

A slow overpower transient, in a pressurized-water reactor, that reasonably approximates this behavior is shown in Figs. 5-3 and 5-4. The transient is initiated at full power, and the reactivity insertion is due to a postulated control rod withdrawal at the rate of $\dot{\rho} = 2 \times 10^{-5}$/sec. Because the core thermal time constant is the order of a few seconds in this system, the coolant temperature rise also approaches a linear time rise following an initial transient. Several plant parameters are sensed in this system to protect against damage from overpower transients. These include the neutron flux level, outlet temperature, and primary system pressure. For the slow power rise shown in Fig. 5-3 in which the delay between power and the sensed parameters is insignificant, the core outlet temperature shown in Fig. 5-4 provides the most sensitive indication of the overpower transient. The transient is terminated at 55 seconds by an outlet temperature trip set at 580.5°F. The neutron flux trip level is at 118% of full power, and therefore would not have occurred until much later. As discussed in Section 2-4, the flux level cannot be set as close to the full power rating because of uncertainties caused by changes in power distribution that then would cause an unacceptable frequency of spurious trip signals. After the trip signal is generated, the additional power produced before the control rods bring the reactor subcritical can be seen to have a negligible effect on the transient.

The most immediate consequence of slow transients such as that described is an increase in fuel and coolant temperature that is roughly proportional to the maximum power reached before shutdown. The protection system must function to terminate the transient before unacceptable

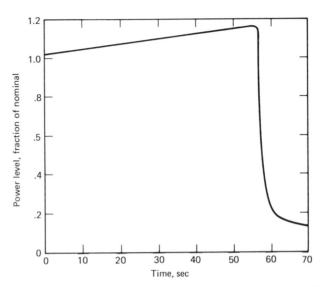

FIGURE 5-3 Power response to uncontrolled rod withdrawal at full power, $\dot{\rho} = 2 \times 10^{-5}$/sec. Adapted from Commonwealth Edison Company, "Final Safety Analysis Report—Zion Station," USAEC Docket No. 50-295, (1970).

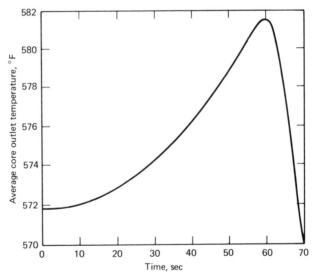

FIGURE 5-4 Temperature response to uncontrolled rod withdrawal at full power $\dot{\rho} = 2 \times 10^{-5}$/sec. Adapted from Commonwealth Edison Company, "Final Safety Analysis Report—Zion Station," USAEC Docket No. 50-295, 1970.

overheating occurs. There are two additional concerns that enter the safety analysis of slow transients: the possible loss of shutdown margin and the possibility that unforeseen threshold reactivity effects will cause a more serious nuclear excursion to develop.

If slow increases in reactivity are allowed to accumulate to the point where the shutdown margin is lost, the reactor can no longer be brought subcritical. Maximum worths on control rods and on control rod banks as well as stuck rod criteria can eliminate this concern with respect to rod withdrawal accidents. It is the more subtle long-term increases that may cause the primary concern: eroding away of control rod poisons, unexpectedly large depletion of burnable poisons, unanticipated buildup of plutonium-239, or other fissile transuranium elements, and so on. The difficulty with these slow reactivity buildups is that without proper precautions they may go undetected until it is too late.

It is not feasible to measure shutdown margin directly while the reactor is at power. Moreover, to maintain the reactor at the power level required by electrical demand, control rods or other control reactivity mechanisms may be continually adjusted, either automatically or by operator action, and, as a result, many of the plant parameters that are usually used to detect more rapid transients will remain within their acceptable range of values. A sufficiently large shutdown margin must be provided to accommodate uncertainties in reactivity buildup phenomena. In addition, vigilance requires that the physics of the reactor core be sufficiently well understood that the amount of poison control required for any anticipated operating state can be accurately predicted. For then the control poison can be monitored, so that if an unacceptably large poison requirement indicates that the shutdown margin is becoming dangerously small, the reactor can be immediately tripped.

Early termination of overpower transients is also desirable to preclude the possibility that accelerated or threshold reactivity effects will come into play, cause the power coefficient to become positive, and lead to an autocatalytic reactivity increase. When the slow reactivity increase is due to poorly understood causes, the danger that it will accelerate with increased power level cannot be ignored. Unanticipated increases caused by coolant boiling, fuel element bowing, erosion of control poisons, or a number of other more subtle effects may fall into this category. Threshold effects that are unrelated to the initial cause of reactivity insertion may also appear. Changes of phase as a result of overheating in some reactors may have such an effect. Coolant boiling in systems with positive void coefficients, or fuel or cladding melting and motion in fast reactors are two potentially serious threshold effects.

The transients caused by unexplained negative reactivity insertions are also cause for concern, since they may indicate the presence of unstable reactivity mechanisms that have the potential for suddenly becoming positive at a later time. For example, the buildup of deposits of neutron-absorbing corrosion products or other debris in the core may cause a slow reactivity decrease, but if a change in coolant flow rate, power level, or other dynamic condition should cause the deposits to be suddenly expelled from the core, a major reactivity increase would result. Likewise, flow obstructions may cause boiling in the coolant channels and result in a decreased reactivity if the void coefficient is negative. If the obstruction should then be dislodged, however, the cessation of the boiling would cause a sudden reactivity increase. As in positive reactivity insertion accidents, the presence of a control system, which adjusts the control poison in the core to maintain the reactor power level, will compound the consequences of the accident by tending to maintain the reactor at its original power level, thus increasing the overheating of the core components in question. The control system may also mask the accident from detection by neutron flux monitoring instruments. In such situations steps must be taken to monitor the control system for unexplained reactivity additions to the core, or to detect the accident in question by other means, as, for example, through the direct monitoring of component temperatures, or detection of fission products or delayed neutrons in the coolant stream.

Unexplained reactivity decreases also may be an indication of problems in the heat transport system that give rise to local fuel, coolant, or moderator overheating, since the presence of a negative temperature coefficient results in a reactivity decrease in these situations. For sufficiently slow deterioration, the subsequent behavior of the reactor power may be viewed using Eq. (5-28). Local core overheating causes a negative value of ρ_{fb} to appear. If no control action is initiated, the power level of the reactor decreases with decreasing reactivity. Although the decrease in power level is likely to mitigate the damage caused by the accident, the local overheating may still be sufficient to cause unacceptable damage to the ability of the fuel to retain fission products.

Superdelayed-Critical Transients

We now turn our attention to more rapid reactivity insertions that cause the net reactivity to rise significantly above zero and thereby cause neutron kinetics effects to become important. Provided the maximum reactivity occurring during the transient is small compared to $\bar{\beta}$, the delayed neutron fraction, the shortest reactor period occurring during the transient will still be long compared to Λ, the prompt neutron generation time. In this situation Λ may be considered to vanish completely since the kinetics of the transient are determined predominantly by the yields and half-lives of the delayed neutrons, as discussed in Chapter 3. The zero lifetime approximation given by Eq. (3-61) may then be adequate for determining the reactor power in terms of the reactivity. As more severe transients are considered, in which the maximum reactivity becomes a significant fraction of the prompt-critical condition $\rho = \bar{\beta}$, both the prompt neutron generation and the delayed neutron effects significantly influence the transient, and the full set of reactor kinetics equations, (5-1) and (5-2), is needed.

Since the superdelayed-critical transients occur over time scales that are often of the same order of magnitude as the core thermal time constant τ, the quasi-static power coefficient is no longer valid for determining the temperature feedback effects. Rather, transient thermal-hydraulic feedback models such as that obtained from Eqs. (5-10) and (5-11) are needed to describe the more complex coupling between power, temperatures, and reactivity.

Some of the characteristics that typically appear in superdelayed-critical transients can be highlighted by considering the behavior of the same pressurized-water reactor used to illustrate the quasi-static transient in the preceding subsection. Now, however, reactivity is inserted at a substantially higher rate. In Figs. 5-5 and 5-6 are shown the power and coolant outlet temperatures that result from a reactivity insertion at the rate of $\dot{\rho} = 8 \times 10^{-4}$ sec^{-1}, hypothesized to result from the uncontrolled withdrawal of two entire control rod banks. Comparing these figures to the corresponding $\dot{\rho} = 2 \times 10^{-5}$ sec insertion results in Figs. 5-3 and 5-4 reveals some interesting characteristics. The time scale in Figs. 5-5 and 5-6 is shorter, and the power rise is initially concave upward, indicating the positive reactivity, but its rate of increase begins to slow because of the effect of reactivity feedback from the fuel temperature. A flux trip is initiated at 118% of full power, but assumed delays in the protection and shutdown system allow the power to increase to 127% of full power before the control rods bring the reactor subcritical. Note that for this more rapid transient the trip on high outlet temperature would not be initiated until much later, since the several-second time constant of the reactor fuel effectively delays any significant rise in the coolant temperature until well beyond the point at which the flux trip is initiated. Because of the shorter time scales involved, the power generated after the trip is initiated is a significant part of the total energy release. The finite rate at which the shutdown rods can be driven into the core and the incremental worth of the rods as they enter the core become important considerations as more rapid transients are considered.

The preceding illustration indicates a number of plausible accident events that may give rise to overpower transients with a range of superdelayed-critical reactivites for most power reactors. The likelihood of many of these may be reduced to the point at which they are not to be expected within the lifetime of any one plant. For a given plant, however, there may be one or

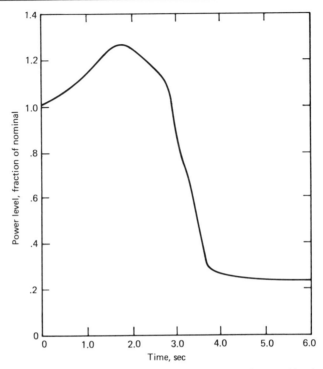

FIGURE 5-5 Power response to uncontrolled withdrawal of two rod banks at full power $\dot{\rho} = 8 \times 10^{-4}$/sec. Adapted from Commonwealth Edison Company, "Final Safety Analysis Report—Zion Station," USAEC Docket No. 50-295, 1970.

more events that would lead to such a transient, and at the same time may be reasonably anticipated to occur one or more times during the life of the plant. Transients arising from turbine trips, isolation valve closures, loss of off-site power to coolant pumps, for example, might fall into this category. Since the reactor protection and shutdown systems are relied on to terminate these transients before significant damage to the reactor fuel elements or primary system envelope can take place, it is imperative that the probability of a failure to shut down be extremely small. Otherwise, some of the so-called anticipated transients without scram might well progress to the point of causing a destructive power burst and become the dominant contributor to the risk from the nuclear reactor. A great deal of attention is given both to ensuring the reliability of the protection and shutdown systems and to providing backup methods for terminating such overpower transients (U.S. Atomic Energy Commission, 1973*b*).

Throughout the entire spectrum of potential overpower transients, great care must be taken to understand the safety ramifications of the control

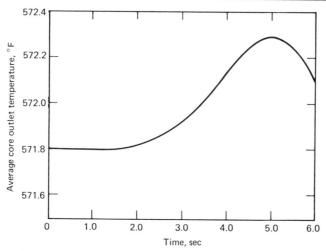

FIGURE 5-6 Temperature response to uncontrolled withdrawal of two rod banks at full power $\dot{\rho} = 8 \times 10^{-4}$/sec. Adapted from Commonwealth Edison Company, "Final Safety Analysis Report—Zion Station," USAEC Docket No. 50-295, 1970.

system design. As alluded to earlier, the control system may mask slower transients and delay their safe termination. Control systems may also increase the severity of overpower transients, or be their base cause. This is borne out by experience, for in several situations control actions have caused or significantly increased the core damage incurred from a reactor accident. For example, in the accidents in the Westinghouse test reactor and a Hanford production reactor, the negative reactivity effects of the over-heating and melting of fuel caused control rods to be withdrawn and therefore the damage to be extended. A nuclear excursion was caused in the HTRE-3 reactor because a faulty power level recorder indicated a fictitious decrease in power; this led to automatic control rod withdrawal, a consequent positive reactivity insertion, and therefore a nuclear excursion (Thompson, 1964).

5-5 SUPERPROMPT—CRITICAL EXCURSIONS

Dramatic changes take place in the nature of reactivity accidents as one passes from transients for which $\rho|_{max} < \bar{\beta}$ to superprompt-critical excursions for which $\rho|_{max} > \bar{\beta}$. As indicated in Fig. 3-3, the reactor period decreases by as much as several orders of magnitude as the reactivity passes through prompt critical. As a result the reactor power level rises so rapidly that the reactor shut down system is ineffectual in terminating the transient.

Rather, the negative Doppler coefficient of the fuel or other prompt inherent feedback mechanisms must be adequate to terminate the power burst if fuel damage is not to result.

   To illustrate some of the properties of superprompt-critical excursions, a power burst is shown in Fig. 5-7 for the same pressurized water reactor used to illustrate the quasi-states and superdelayed-critical transients in the preceding section. As in the superdelayed-critical case the reactivity insertion rate of $\dot{\rho} = 8 \times 10^{-4}$/sec is hypothesized to result from the uncontrolled withdrawal of two entire control rod banks. This accident is different, however, in that it is assumed to occur during a start up procedure, when the reactor is initially slightly subcritical, and at essentially zero power. The situation is thus analogous to the start up accident described in Section 3-3. Enough reactivity is inserted to exceed prompt critical before the power level rises to a high enough level either to cause a flux trip or to result in

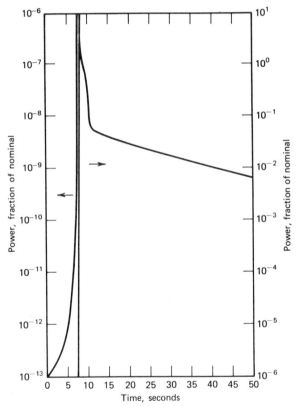

FIGURE 5-7 Uncontrolled withdrawal of two rod banks from a subcritical condition. Adapted from Commonwealth Edison Company, "Final Safety Analysis Report—Zion Station," USAEC Docket No. 50-295, 1970.

significant temperature feedback. Once prompt critical is exceeded, the rate of power increase is so rapid that the reactor is brought subcritical by the prompt negative reactivity feedback from the rising fuel temperature before the flux trip, set at 35% of full power, can become effective. Although the peak power is 885% of full power, the power burst is so narrow that the total energy released in the fuel in this particular excursion is not sufficient to cause damage.

There are mechanistic (albeit highly improbable) sequences of events that could lead to such excursions in other reactor types. For example, in some sodium-cooled fast reactor designs a power failure to all the coolant pumps followed by a failure of the reactor protection or shutdown system would lead to rapid sodium voiding capable of causing superprompt-critical reactivity insertions. Similarly, in some boiling water reactors multiple undetected operator errors during start-up procedures can lead to pathological control rod patterns in which there exist one or more rods with very large reactivity worths. A large reactivity would be rapidly added to the core if a gross mechanical housing failure caused such a rod to be ejected or if an uncontrolled withdrawal of the rod were coupled with a failure of the shutdown system.

So long as superprompt-critical excursions are physically possible, they must be analyzed to demonstrate that the resulting damage can be tolerated or to show that the excursions can be caused only by sequences of events that are too improbable to fall within the reactor's design basis. In what follows we examine these transients to understand their neutronic properties and to estimate the parametric dependences of the maximum power and energy release that result from the excursions. The problems associated with the blast damage and core meltdown that may result from catastrophic power bursts are addressed in Chapter 9.

The short time spans over which superprompt-critical power bursts take place allow substantial simplifications to the reactor dynamics model shown in Fig. 5-1. The reactivity contributions from the regulation and shutdown systems can be neglected during the initial phase of the excursion. In addition, the duration of the power burst may be expected to be short compared to the thermal time constant of the core. Therefore, the thermal-hydraulic feedback loop may be replaced by the assumption that the core behaves adiabatically, and $\rho_{fb}(t)$ is then given in terms of the prompt coefficients by Eqs. (4-102) through (4-106). Assuming that the prompt coefficient is negative, we have

$$\rho(t) = \rho_i(t) - |\mu| \int_0^t P(t')\, dt', \qquad (5\text{-}30)$$

which is often referred to as the linear energy model (Nyer, 1964).

For purposes of analysis, it is instructive to consider insertion reactivities as either steps or ramps. Although step insertions are not very realistic— except when extremely rapid insertions are made at very low power levels— they provide a good deal of insight into the nature of power bursts and serve as a basis for introducing the power bursts initiated by ramp insertions. In the two following subsections we utilize the linear energy model to examine superprompt-critical excursions induced by step and ramp reactivity insertions.

Step Insertions

In Fig. 5-8 are shown solutions to the point kinetics equations, (5-1) and (5-2), with the reactivity represented by the linear energy model given by Eq. (5-30). The reactivity insertion at $t = 0$ is a step of 1.2 dollars, $\rho_i = 1.2\bar{\beta}$. The prompt coefficient is taken as $|\mu| = 2 \times 10^{-5}$/joule, and delayed neutron data for uranium-235 is used. The power and energy release are plotted for prompt neutron generation times of $\Lambda = 10^{-4}$, 10^{-6}, and 10^{-8} sec.

Each of the power traces is characterized by a short burst followed by a slowly decaying tail. This characteristic may be understood in terms of the interaction of prompt and delayed neutron behavior. The delayed neutron emission rate at any time is determined by the precursor formation history over a time scale comparable to the precursor half-lives. Since the burst widths in Fig. 5-8 are short compared to these half-lives, the delayed neutron contribution is determined primarily by the rate of precursor formation before the initiation of the transient, and therefore it is quite small. The burst characteristics thus are determined by the prompt neutron behavior. Following the burst, the power decreases greatly, since the reactor becomes subcritical. Because large numbers of precursors are formed during the burst, however, a long-term tail results from the neutrons produced by the decay of these precursors.

For purposes of comparing results from different reactor types and from more severe accidents, it is desirable to have available simple relationships that relate the maximum power and the energy release to the insertion magnitude, prompt neutron generation time, and prompt coefficient. This can be done for step insertions by assuming that the burst is so rapid that the precursor half-lives appear to be essentially infinite, or, equivalently, $\lambda_i = 0$. The prompt neutron approximation that then results from combination of Eqs. (5-1) and (5-30) is

$$\frac{dP(t)}{dt} = \frac{1}{\Lambda}\left[\rho_0 - \bar{\beta} - |\mu| \int_0^t P(t')\,dt'\right]P(t), \qquad (5\text{-}31)$$

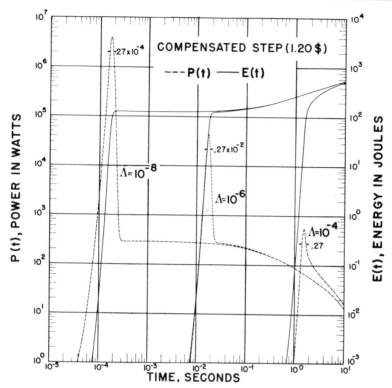

FIGURE 5-8 Compensated response to a step reactivity change of $1.20 ($\rho = 0.0078$) in U^{235} systems characterized by a negative prompt reactivity coefficient, $|\mu| = 2 \times 10^{-5}$ cm^3/J and representative prompt neutron generation times, $\Lambda = 10^{-8}$, 10^{-6}, and 10^{-4} sec. Dotted curves are power response and solid curves are energy response. Adapted from *Physics of Nuclear Kinetics* by G. R. Keepin with the permission of publishers, Addison-Wesley/W. A. Benjamin Inc., Advanced Book Program, Reading, Mass., U.S.A.

where ρ_0 is the magnitude of the step insertion. This equation was originally derived independently by Fuchs (1946), by Hansen (1952), and by Nordheim (1945).

Equation (5-31) can be solved for the time-dependent power and energy release for the step insertion, provided $\rho_0 > \bar{\beta}$. To do this, we first integrate the equation from zero to t:

$$P(t) = P(0) + \frac{1}{\Lambda} \left[(\rho_0 - \bar{\beta}) \int_0^t P(t')\, dt' - |\mu| \int_0^t P(t') \int_0^{t'} P(t'')\, dt''\, dt' \right].$$

$$(5\text{-}32)$$

Using integration by parts, it may be shown that

$$\int_0^t P(t') \int_0^{t'} P(t'')\, dt''\, dt' = \frac{1}{2}\left[\int_0^t P(t')\, dt'\right]^2.$$ (5-33)

Thus, using the definition in Eq. (4-105), we may write

$$\frac{dE(t)}{dt} = P(0) + \frac{(\rho_0 - \bar{\beta})}{\Lambda} E(t) - \frac{|\mu|}{2\Lambda} E(t)^2,$$ (5-34)

which may be integrated directly to yield

$$E(t) = \frac{\Lambda}{|\mu|}\left(R + \frac{\rho_0 - \bar{\beta}}{\Lambda}\right)\frac{(1 - e^{-Rt})}{(1 + A\, e^{-Rt})},$$ (5-35)

where

$$R = \sqrt{\frac{(\rho_0 - \bar{\beta})^2}{\Lambda^2} + \frac{2|\mu||P(0)|}{\Lambda}}$$ (5-36)

and

$$A \equiv \frac{\Lambda R + \rho_0 - \bar{\beta}}{\Lambda R - \rho_0 + \bar{\beta}}.$$ (5-37)

Differentiation of Eq. (5-35) then yields the time-dependent power

$$P(t) = \frac{2\Lambda R^2}{|\mu|}\frac{A\, e^{-Rt}}{(1 + A\, e^{-Rt})^2}.$$ (5-38)

The burst quantities of most interest are the total energy release, the maximum power, and the burst width at half maximum. The total energy release is just

$$E(\infty) = \frac{\Lambda}{|\mu|}\left(R + \frac{\rho_0 - \bar{\beta}}{\Lambda}\right).$$ (5-39)

The maximum power is obtained by setting the derivative of Eq. (5-38) equal to zero:

$$P_{\max} = \frac{\Lambda R^2}{2|\mu|}.$$ (5-40)

If the initial power is small so that $(\rho_0 - \bar{\beta})^2 \gg 2\Lambda|\mu||P(0)|$, and therefore

$$R \approx \frac{\rho_0 - \bar{\beta}}{\Lambda}$$ (5-41)

and

$$A \simeq \frac{2(\rho_0 - \bar{\beta})^2}{\Lambda |\mu| P(0)}, \tag{5-42}$$

we have

$$E(\infty) \simeq \frac{2(\rho_0 - \bar{\beta})}{|\mu|} \tag{5-43}$$

and

$$P_{\text{max}} \simeq \frac{(\rho_0 - \bar{\beta})^2}{2\Lambda |\mu|}. \tag{5-44}$$

Similarly, it may be shown that the burst width at half maximum is given by

$$\Gamma \simeq \frac{3.52\Lambda}{\rho_0 - \bar{\beta}}. \tag{5-45}$$

From this analysis we see that both peak power and energy release increase with the insertion strength and are inversely proportional to the prompt reactivity coefficient. The energy release is independent of the prompt neutron generation time, however, and none of the parameters depends on the initial power level of the reactor, provided it is small. Thus, although the differences in reactivity insertion mechanism and feedback effects in thermal and fast reactors are important to the characteristics of reactivity accidents, the difference in prompt generation time, per se, is not significant insofar as predicting the energy release is concerned. The small generation times in fast reactors do become detrimental, however, if transients are considered for which primary damage criteria depend on the maximum power. These conclusions apply only to excursions for which the initial reactivity is greater than one dollar. If the step insertion is less than one dollar, the distinct prompt burst followed by a delayed neutron tail disappears, and the resulting transient is much milder, occurring over a longer time span with a smaller power and energy release. These effects are illustrated in Fig. 5-9, where the kinetics equations solutions are plotted for the same reactor parameters but for a step insertion of just one dollar at $t = 0$.

Ramp Insertions

In Fig. 5-10 are shown the power traces and energy releases obtained from the point kinetics equations with linear energy feedback for a ramp insertion of 10 dollars/sec. The reactor characteristics are the same as those considered for the step insertion. As for step insertions, the results show an

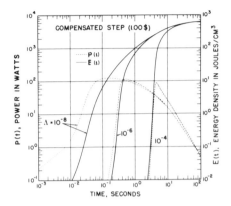

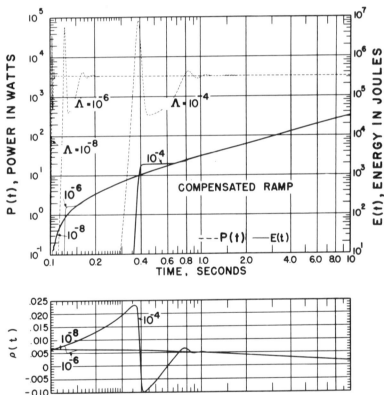

FIGURE 5-9 Compensated response to a step reactivity change of $1.00 ($\rho = 0.0065$) in U^{235} systems characterized by a negative prompt reactivity coefficient, $|\mu| = 2 \times 10^{-5}$ cm^3/J and representative prompt neutron generation times, $\Lambda = 10^{-8}$, 10^{-6}, and 10^{-4} sec. Adapted from *Physics of Nuclear Kinetics* by G. R. Keepin with the permission of publishers, Addison-Wesley/W. A. Benjamin Inc., Advanced Book Program, Reading, Mass., U.S.A.

FIGURE 5-10 Compensated response to ramp function reactivity rate of $10/sec for systems characterized by prompt neutron lifetimes $\Lambda = 10^{-4}$, 10^{-6} and 10^{-8} sec, and a negative prompt reactivity coefficient $|\mu| = 2 \times 10^{-5}$ cm^3/J. Adapted from *Physics of Nuclear Kinetics* by G. R. Keepin with the permission of publishers, Addison-Wesley/W. A. Benjamin Inc., Advanced Book Program, Reading, Mass., U.S.A.

initial sharp power burst. Now, however, the burst is followed by a series of oscillations that damp to a constant power. This oscillatory behavior may be understood as follows. Starting at low power, the ramp reactivity insertion causes the reactor to go on a period that decreases with time until sufficient power is developed for the feedback term to compensate for the ramp reactivity increase. At that time the reactor rapidly loses reactivity, becoming subprompt-critical, and the power rapidly drops. Eventually, the ramp increase overrides the negative reactivity resulting from the first burst, and the reactor period again becomes positive, starting the reactor on a second power burst. It has been demonstrated that in the absence of delayed neutrons these periodic power bursts would continue undamped (Hetrick, 1971). As sufficient time elapses for the buildup of delayed neutron precursors, however, the power oscillations are damped, and an equilibrium power level is reached. Setting the net reactivity to zero in Eq. (5-30), with $\rho_i = \dot{\rho}t$, this power may be shown to be

$$P(\infty) = \frac{\dot{\rho}}{|\mu|}. \tag{5-46}$$

It is usually the first burst that is most important in determining damage effects. As may be seen by comparing Figs. 5-8 and 5-10, the effect of prompt generation time on the maximum power and the energy release from this burst is not the same for step and ramp insertions. It is desirable to have available the parametric dependencies for these quantities in order to be able to predict the effect of design changes or ramp rate on accident severity. The starting point for estimating these relationships is the prompt neutron approximation, Eq. (5-31), with the step replaced by a ramp reactivity insertion:

$$\frac{dP(t)}{dt} = \frac{1}{\Lambda} \left[\dot{\rho}t - |\mu| \int_0^t P(t')\, dt' \right] P(t). \tag{5-47}$$

We assume in this equation that the reactor becomes prompt critical at $t = 0$, at a power level sufficiently small that the feedback effects up until $t = 0$ are negligible. A solution of this nonlinear differential equation (Hetrick, 1971) is more tedious than for the step reactivity insertion and is not pursued here. Using reasonable physical approximations, however, it is possible to obtain rough estimates for the maximum power and total energy release without solving for $P(t)$ and $E(t)$ (Nyer, 1964).

To estimate the maximum power and energy release we first rewrite Eq. (5-47) as

$$\frac{dP(t)}{dt} = \frac{\rho'(t)}{\Lambda} P(t), \tag{5-48}$$

where $\rho'(t)$ is defined as the reactivity in excess of prompt critical:

$$\rho'(t) = \dot{\rho}t - |\mu| \int_0^t P(t')\,dt'. \tag{5-49}$$

Observing that Eq. (5-48) can be expressed as

$$\frac{dP}{d\rho'}\frac{d\rho'}{dt} = \frac{\rho'}{\Lambda}P, \tag{5-50}$$

we insert the time derivative of ρ',

$$\frac{d\rho'}{dt} = \dot{\rho} - |\mu|P(t) \tag{5-51}$$

to obtain

$$\frac{dP}{P}(\dot{\rho} - |\mu|P) = \frac{\rho'\,d\rho'}{\Lambda}. \tag{5-52}$$

This expression may be integrated to yield ρ' in terms of the reactor power:

$$\frac{\rho'^2}{2} = \Lambda\dot{\rho}\ln\left(\frac{P}{P_0}\right) - |\mu|\Lambda(P - P_0), \tag{5-53}$$

where P_0 is the power at $t = 0$.

From Eq. (5-48) it is clear that the maximum power occurs when $\rho' = 0$. We thus have

$$P_{\max} = \frac{\dot{\rho}}{|\mu|}\ln\left(\frac{P_{\max}}{P_0}\right), \tag{5-54}$$

where we assume $P_{\max} \gg P_0$. Neglecting the slowly varying logarithmic dependence, we have

$$P_{\max} \propto \frac{\dot{\rho}}{|\mu|}. \tag{5-55}$$

To obtain a rough estimate of the energy release, we evaluate Eq. (5-49) at $t_{\max}$, the time at which $\rho'(t) = 0$:

$$\int_0^{t_{\max}} P(t')\,dt' = \frac{\dot{\rho}t_{\max}}{|\mu|}. \tag{5-56}$$

Physical arguments can be used to make two approximations. From the reactivity curves in Fig. 5-10 we see that no great error is introduced in the preceding equation by replacing $t_{\max}$ by t_m, the time at which $\rho'(t)$ is a maximum. Moreover, at t_m there is very little reactivity feedback, and therefore $\rho'(t'_m) \simeq \dot{\rho} t_m$. With these approximations

$$\int_0^{t_{max}} P(t)\, dt \simeq \frac{\rho'|_{max}}{|\mu|}. \tag{5-57}$$

The maximum reactivity is obtained by setting the derivative in Eq. (5-51) equal to zero, from which $P(t_m) = \dot{\rho}/|\mu|$, and substituting into Eq. (5-53), we have

$$\rho'_{max} = \sqrt{2\Lambda\dot{\rho}\left[\ln\left(\frac{\dot{\rho}}{|\mu|P_0}\right) - 1\right]}, \tag{5-58}$$

where P_0 is again considered to be small. Combining Eqs. (5-57) and (5-58) yields

$$\int_0^{t_{max}} P(t)\, dt \simeq \frac{1}{|\mu|}\sqrt{2\dot{\rho}\Lambda\left[\ln\left(\frac{\dot{\rho}}{|\mu|P_0}\right) - 1\right]}. \tag{5-59}$$

Finally, the energy release for the burst is estimated by assuming that the burst is approximately symmetric about $t = t_{max}$:

$$E \simeq \frac{2}{|\mu|}\sqrt{2\dot{\rho}\Lambda\left[\ln\left(\frac{\dot{\rho}}{\mu P_0}\right) - 1\right]}. \tag{5-60}$$

Neglecting the slowly varying logarithmic term, we have

$$E \propto \frac{\sqrt{\dot{\rho}\Lambda}}{|\mu|}. \tag{5-61}$$

The burst width is proportional to E/P_{max}, or

$$\Gamma \propto \sqrt{\frac{\Lambda}{\dot{\rho}}}. \tag{5-62}$$

From Eqs. (5-55) to (5-61) and (5-62), it is obvious that the transients become more severe with increasing ramp rate and decreasing prompt negative feedback coefficient. These dependencies are indicated by the power traces for varying ρ and $|\mu|$ in Fig. 5-11. In contrast to the step reactivity insertion, we find that when the prompt neutron generation time Λ is decreased, the maximum power does not change, and the energy release decreases as indicated in Fig. 5-10.

Although less important than the initial power burst, the additional energy released during the subsequent bursts often contributes significantly to the transients. This is particularly true when Doppler feedback is the

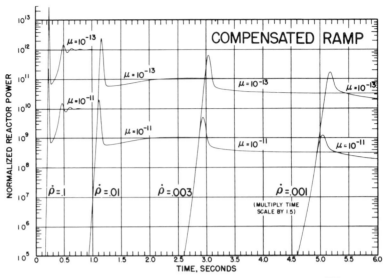

FIGURE 5-11 Compensated response to ramp function reactivity in U^{235} systems with prompt neutron generation time, $\Lambda = 5 \times 10^{-5}$ sec, and negative prompt reactivity coefficients in the range $|\mu| = 10^{-11}$ cm^3/sec to 10^{-13} cm^3/sec. Adapted from *Physics of Nuclear Kinetics* by G. R. Keepin with the permission of publishers, Addison-Wesley/W. A. Benjamin Inc., Advanced Book Program, Reading, Mass., U.S.A.

primary contributor to the prompt coefficients; because the Doppler coefficient decreases with increasing fuel temperature, the oscillations are less damped than indicated in Figs. 5-10 and 5-11. In contrast, the long time tail at constant power is of little interest for practical problems, since it would be expected to be clipped by the action of the shutdown system, and in any event heat conduction and coolant boiling and their associated reactivity effects that become important at longer times are not incorporated into the adiabatic fuel model.

Complex Shutdown Mechanisms

The linear energy feedback model given by Eq. (5-30) provides a basic understanding of superprompt-critical power bursts that are terminated solely by Doppler feedback, axial expansion, or other prompt mechanisms that are directly dependent on the fuel temperature. Increased accuracy, of course, requires that the temperature dependence of the reactivity coefficient be taken into account. For some reactor types more fundamental changes to the model are required, since the prompt coefficient is too small to provide the dominant shutdown reactivity, and more complex shutdown mechanisms must act to terminate the excursion.

Typical of systems relying on such complex shutdown mechanisms are the earlier of the SPERT-I cores, used to study reactivity accidents, and the SL-1 reactor, which was destroyed by a superprompt-critical power burst (Thompson, 1964). These cores were characterized by highly enriched metal fuels, cooled and moderated by water. Studies of the power bursts (Thompson, 1964; Nyer, 1964) indicated that in such cores the Doppler coefficient is very small and plays no significant role in terminating the power bursts. The primary shutdown mechanism is expulsion of the water moderator from the core. This may be caused both by the water boiling and by fuel expansion. The boiling is a delayed effect, since it requires that heat be transferred from fuel to coolant. Although the fuel expansion takes place more rapidly, even it may not be considered to be instantaneous with power production, since in very fast transients the inertial delay associated with the fuel expansion and coolant expulsion may become significant.

Whether the character of a superprompt-critical power burst is determined by the prompt coefficient or by one of the many possible complex reactivity feedback coefficients, it is to be expected that the burst energy release will increase with the magnitude and rate of the reactivity insertion. If the reactor is susceptible to very large reactivity insertions, the aforementioned shutdown mechanisms may be incapable of terminating the transient before a threshold is reached at which the fuel begins to melt or even to vaporize. With fuel vaporization, high pressure may be generated that acts to disassemble the reactor core explosively and thereby terminate the transient. For a number of reasons these most catastrophic of reactivity accidents have been intensely studied in conjunction with fast reactor development. We devote the following section to the discussion of such disassembly accidents.

5-6   FAST REACTOR DISASSEMBLY ACCIDENTS

The analysis of hypothesized nuclear excursions that are so severe as to cause fuel vaporization is given a great deal of attention in fast reactor design. To a great extent the concern with these is caused by a number of features that tend to be unique to fast reactor cores. Unlike most thermal reactors, a fast reactor is not in its most reactive configuration. A collapse of the core into a more compact configuration would lead to a substantial increase in reactivity, and in the larger cores of more recent design the boiling of the liquid metal coolant might also lead to rapid reactivity increases. Thus, in addition to the accidents caused by control rod ejection and other mechanisms shared with thermal reactors, accidents hypothesized to occur from sudden loss of coolant and/or fuel meltdown could lead to large reactivity insertions. Moreover, it is precisely for the latter types of accidents that a mechanistic analysis to place a realistic upper bound on the rate of reactivity insertion may be quite difficult.

Fast reactors also are characterized by prompt reactivity coefficients that are smaller than those in thermal reactors. As discussed in Section 4-4, a smaller Doppler coefficient stems from the fast spectrum and larger enrichments of fast reactors. In early fast reactor designs such as the EBR-II (Koch et al., 1957) and FERMI (Power Reactor Development Company, 1961) systems, the highly enriched metal fuels result in vanishingly small Doppler coefficients. In these systems the shutdown mechanisms are predominantly due to fuel expansion effects. These mechanisms, however, are somewhat delayed because of the inertia that must be overcome to produce fuel expansion. Moreover, any delay in the shutdown mechanism is more severe in fast than in thermal systems, since the small prompt generation times in fast reactors cause extremely rapid rates of power increase to take place in the event that the system becomes even slightly superprompt critical. This effect is graphically shown in Fig. 3-3, where the period is plotted versus reactivity for a range of prompt generation times.

To set an upper bound on the explosive energy release that could result from a catastrophic fast reactor power burst, Bethe and Tait (1956) analyzed the transient that would take place in a highly enriched, metal-fueled fast reactor if a sudden core collapse occurred. They hypothesized that all coolant is lost from the reactor and that the upper half of the core drops into the lower half with the acceleration of gravity. The result is an insertion rate of from 40 to 50 dollars per second. It was assumed that the reactivity increases beyond superprompt critical, but no reactivity shutdown mechanism appears until the fuel vaporizes and expands to fill the internal voids. When this threshold is reached, there is a rapid pressure surge from the vapor generation, which accelerates mass away from the center of the core and causes the reactor to disassemble, thus terminating the burst.

The Bethe–Tait analysis of the negative reactivity feedback caused by the fuel movement away from the core center has served as a basis for a series of increasingly refined hydrodynamic analyses of fast reactor core disassembly (Shaw and Hughes, 1970; Jackson and Nicholson, 1972). In the following subsection we derive a simplified expression for the disassembly reactivity feedback that shows the primary parametric relationships, while being consistent with the reactor kinetics model developed in Chapter 3. In the next subsection we follow arguments parallel to those of Bethe and Tait to obtain analytic estimates of the parametric dependence of the energy release from disassembly accidents when there is no Doppler feedback. Finally, we consider the more realistic situation for modern fast power reactors where a significant Doppler effect greatly mitigates the energy release from destructive power bursts.

Disassembly Reactivity Feedback

The reactivity effect in the disassembly model is due to the outward motion of mass near the center of the reactor core caused by the pressure generated by fuel vaporization. To calculate the reactivity effect we begin by writing the multiplication as in Eqs. (3-18), (3-20), and (3-22):

$$k = \frac{\langle k_\infty \psi_0^2 \rangle}{\langle \psi_0^2 \rangle} \times \left[1 + \frac{\langle M^2 (\nabla \psi_0)^2 \rangle + \langle \psi_0 M^2 (\nabla \ln l) \nabla \psi_0 \rangle}{\langle \psi_0^2 \rangle} \right]^{-1}. \tag{5-63}$$

In the Bethe-Tait analysis it is assumed that the material movements do not significantly perturb the power distribution, and therefore ψ is set equal to ψ_0. Owing to the subtle effects of the hydrodynamic redistribution of mass in the core, the last term in the denominator can no longer be neglected as it is throughout the reactivity feedback considerations in Chapter 4.

The reactivity feedback appears only from changes in the density m associated with material motions. Therefore, noting that $M^2 \propto m^{-2}$ and $l \propto m^{-1}$, and that k_∞ is independent of density, we may express the rates of change of these quantities in terms of the time derivative of the local density, m:

$$\frac{1}{M^2} \frac{\partial M^2}{\partial t} = -\frac{2}{m} \frac{\partial m}{\partial t}, \tag{5-64}$$

$$\frac{1}{l} \frac{\partial l}{\partial t} = -\frac{1}{m} \frac{\partial m}{\partial t}, \tag{5-65}$$

and

$$\frac{\partial k_\infty}{\partial t} = 0. \tag{5-66}$$

Using the continuity equation,

$$\frac{\partial m}{\partial t} = -\nabla \cdot m\mathbf{v}, \tag{5-67}$$

we may write the time derivatives in terms of the local velocity $\mathbf{v}$:

$$\frac{1}{M^2} \frac{\partial M^2}{\partial t} = \frac{2}{m} \nabla \cdot m\mathbf{v}, \tag{5-68}$$

and

$$\frac{1}{l} \frac{\partial l}{\partial t} = \frac{1}{m} \nabla \cdot m\mathbf{v}. \tag{5-69}$$

To estimate the disassembly reactivity ρ_d caused by material motions, we take the derivative

$$\frac{d\rho_d}{dt} \cong \frac{1}{k}\frac{dk}{dt} \tag{5-70}$$

of Eq. (5-63), assuming that the reactor volume remains unchanged. The latter assumption is reasonable since the reactor will become subcritical before the displacement becomes significant at the outer surface of the core. Utilizing Eqs. (5-66), (5-68), and (5-69) in the result, we obtain

$$\frac{d\rho_d}{dt} = -\frac{k}{\langle k_\infty \psi_0^2 \rangle}\left[\left\langle 2M^2\left(\frac{1}{m}\nabla\cdot m\mathbf{v}\right)(\nabla\psi_0)^2\right\rangle + \left\langle \psi_0 M^2\left(\frac{1}{m}\nabla\cdot m\mathbf{v}\right)\left(\frac{1}{l}\nabla l\right)\nabla\psi_0\right\rangle \right.$$
$$\left. + \left\langle \psi_0 M^2\left(\frac{1}{l}\nabla l\frac{1}{m}\nabla\cdot m\mathbf{v}\right)\nabla\psi_0\right\rangle\right]. \tag{5-71}$$

In the Bethe–Tait model it is also assumed that the material movement is sufficiently small that changes in the core density relative to the initial density can be ignored. Hence, if we assume that the reactor is initially uniform, we can replace m by $\bar{m}$, the core-averaged density, in Eq. (5-71). This then implies that k, k_∞, M^2, and l can be replaced by their core-averaged values, 1, $\bar{k}_\infty$, $\bar{M}^2$, and $\bar{l}$, respectively. Making these substitutions, we obtain

$$\frac{d\rho_d}{dt} = -\frac{\bar{M}^2}{\bar{k}_\infty\langle\psi_0^2\rangle}\langle 2(\nabla\cdot\mathbf{v})(\nabla\psi_0)^2 + \psi_0\nabla(\nabla\cdot\mathbf{v})\cdot\nabla\psi_0\rangle. \tag{5-72}$$

With the assumption that the flux at the core boundary vanishes, the volume integral on the right may be simplified further to yield

$$\frac{d\rho_d}{dt} = -\frac{\bar{M}^2}{\bar{k}_\infty\langle\psi_0^2\rangle}\langle\mathbf{v}\cdot\nabla[\psi_0\nabla^2\psi_0 - (\nabla\psi_0)^2]\rangle. \tag{5-73}$$

Ignoring changes in the density once again, we may use the constant density approximation to the momentum equation,

$$\frac{\partial\mathbf{v}}{\partial t} = -\frac{1}{\bar{m}}\nabla p, \tag{5-74}$$

to express the disassembly reactivity in terms of the pressure. Differentiating Eq. (5-73) with respect to time and using the momentum equation, we obtain, after applying the divergence theorem,

$$\frac{d^2\rho_d}{dt^2} = -\frac{\bar{M}^2}{\bar{m}\bar{k}_\infty\langle\psi_0^2\rangle}\langle p\nabla^2[\psi_0\nabla^2\psi_0 - (\nabla\psi_0)^2]\rangle. \tag{5-75}$$

The integral on the right may be evaluated for any number of appropriate flux distributions. The more important parametric dependences, however, may be estimated simply by assuming that the reactor is a bare, uniform sphere of radius $\tilde{R}$. The spatial distribution of the power is then given by (Lamarsh, 1966):

$$\psi_0(r) = \frac{\tilde{R}}{\pi r} \sin\left(\frac{\pi r}{\tilde{R}}\right), \tag{5-76}$$

where r is the radial distance. For this flux distribution it may be shown that

$$\nabla^2[\psi_0 \nabla^2 \psi_0 - (\nabla \psi_0)^2] = \frac{4}{3} \frac{\pi^4}{\tilde{R}^4} + O(r^2). \tag{5-77}$$

Since it is expected that the pressure buildup will be concentrated near the center of the core, where r is small, a rough first estimate results if we ignore terms of the order of r^2 and higher. Combining Eqs. (5-75) and (5-77) we obtain for the disassembly feedback

$$\frac{d^2 \rho_d}{dt^2} \simeq -\frac{8}{9} \frac{\pi^6 \bar{M}^2}{\bar{m} \bar{k}_\infty \tilde{R}^4} \frac{1}{V} \langle p \rangle. \tag{5-78}$$

By dropping higher-order terms in Eq. (5-77) we are neglecting the dependence of ρ_d on the details of the power distribution in the core. For more precise calculations it is necessary to carry at least some of this power shape information throughout the calculations. Aside from this power shape dependence, however, Eq. (5-78) correctly predicts the dependence of the disassembly reactivity on the other reactor parameters. A discussion of the effects of power shape on disassembly reactivity is given by Meyer et al. (1967), among others. Suffice it to point out that for a reactor of fixed size, power flattening causes a decrease in the magnitude of the disassembly reactivity.

The preceding rough estimate for the disassembly has two interesting properties. First, the acceleration of the feedback is related to the average pressure buildup in the core. Hence even with the rapid generation of large pressures, some time is required before ρ_d (as opposed to $d^2\rho_d/dt^2$) becomes significant. This is because inertial delays are encountered before the feedback becomes effective. Second, the presence of the $\bar{M}^2/\tilde{R}^4$ term indicates that as the reactor core becomes larger, either neutronically or physically, the effect of disassembly will decrease, and higher pressures will be required before the power burst is terminated.

In order to utilize expressions such as Eq. (5-78) in conjunction with the reactor kinetics equations, an equation of state must be specified to relate the pressure buildup to the energy density in the core. The energy density $e(r, t)$ here is taken to be the energy generated per unit mass of the core. Assume that $e(r, t)$ is approximately separable in space and time,

$$e(r, t) = E(t)N(r), \tag{5-79}$$

where

$$\langle N(r)m \rangle = 1. \tag{5-80}$$

The required equation of state must express the pressure in terms of the energy density and the density of the system:

$$p = p(e, \bar{m}). \tag{5-81}$$

Consistent with the foregoing assumptions, the density is assumed to retain its initial value $\bar{m}$ throughout the transient.

Increasingly sophisticated equations of state have come into use with refined models for hydrodynamic core disassembly. For purposes of the illustration that follows, the simple threshold model used in the original Bethe–Tait calculations suffices. In this model it is assumed that no pressure is generated below some threshold in energy density e_* and that the pressure increases linearly with energy density above this threshold. Thus we have

$$p = \begin{cases} 0, & e \leq e_* \\ (\gamma - 1)\bar{m}(e - e_*), & e \geq e_*. \end{cases} \tag{5-82}$$

where γ is the ratio of specific heats. If we insert this rudimentary equation of state into Eq. (5-78), we have

$$\frac{d^2 \rho_d}{dt^2} = -\frac{8}{9} \frac{\pi^6 \bar{M}^2 (\gamma - 1)}{\bar{k}_\infty \tilde{R}^4 V} \langle e - e_* \rangle_+, \tag{5-83}$$

where $\langle \cdot \rangle_+$ indicates that the integral is taken only over that part of the reactor volume for which the integrand is positive. Equation (5-83) may be rewritten in terms of $E(t)$, the total energy produced by the reactor between 0 and t, by noting that pressure will first start to be generated when $E = E_*$, where

$$E_* = \frac{e_*}{N(r)|_{max}}. \tag{5-84}$$

Thus combining Eqs. (5-79) and (5-84) with Eq. (5-83), we have

$$\frac{d^2 \rho_d}{dt^2} = -\frac{8}{9} \frac{\pi^6 \bar{M}^2 (\gamma - 1)}{\bar{k}_\infty \tilde{R}^4 V} \langle E(t)N(r) - E_* N(r)|_{max} \rangle_+. \tag{5-85}$$

Suppose attention is confined to very severe excursions in which the energy E becomes much larger than E_* before enough material motion can take place to result in a significant value of ρ_d. Under these circumstances the disassembly reactivity can be further approximated by neglecting the E_* term on the right of Eq. (5-85). Using the normalization condition, Eq. (5-80), we then have

$$\frac{d^2\rho_d}{dt^2} = -\frac{8}{9} \frac{\pi^6 \bar{M}^2(\gamma - 1)}{\bar{k}_\infty \tilde{R}^4 \bar{m} V} E(t).$$ (5-86)

Disassembly Accident Parameter Characterization

With the preceding expression for the disassembly reactivity, the linear energy feedback model used in the preceding section can be generalized to include superprompt-critical excursions that are so severe as to cause reactor disassembly. In the prompt neutron approximation, Eq. (5-47) becomes

$$\frac{dP(t)}{dt} = \frac{1}{\Lambda}[\dot{\rho}t - |\mu|E(t) + \rho_d(t)]P(t)$$ (5-87)

and may be numerically solved simultaneously with Eqs. (5-86) and (4-105) to yield the reactor power transient and energy release. The calculation again rests on the assumption that the reactor is brought prompt critical at $t = 0$ with a ramp rate $\dot{\rho}$, and at a power level $P(0)$ that is small enough that neither the prompt nor disassembly feedback terms have yet become significant.

Increasingly sophisticated improvements on the preceding model are being used to simulate destructive transients (Hummel and Okrent, 1970). For our purposes, however, it is of interest to follow arguments parallel to the original Bethe–Tait analysis to estimate the severity of transients governed by the disassembly phenomenon in terms of the reactivity insertion rate, prompt generation time, and other characteristic parameters.

Assume that the prompt coefficient is negligible. The superprompt-critical transients then consist of two distinct phases: a predisassembly phase, in which there is no reactivity feedback, and a disassembly phase. The disassembly begins when the energy release reaches a value of E_*, as indicated by Eq. (5-85). To estimate the energy release during the disassembly phase, the following information is needed from the pre-disassembly phase: the time t_* at the beginning of disassembly, the power $P(t_*)$, and the reactivity $\rho(t_*)$.

During the pre-disassembly phase the reactor power is determined from

$$\frac{dP(t)}{dt} = \frac{\dot{\rho}t}{\Lambda}P(t), \tag{5-88}$$

which has the solution

$$P(t) = P(0) \exp\left(\frac{\dot{\rho}}{2\Lambda}t^2\right). \tag{5-89}$$

At the initiation of disassembly, we then have

$$E_* = \int_0^{t_*} P(t)\,dt = P(0) \int_0^{t_*} \exp\left(\frac{\dot{\rho}t^2}{2\Lambda}\right) dt. \tag{5-90}$$

If the power $P(t_*)$ is much larger than $P(0)$, we may assume that $\dot{\rho}t_*^2/2\Lambda \gg 1$, and use the asymptotic form of the error function with imaginary argument (Abramowitz and Stegun, 1969) to write

$$E_* \simeq \frac{\Lambda P(0)}{\dot{\rho}t_*} \exp\left(\frac{\dot{\rho}t_*^2}{2\Lambda}\right). \tag{5-91}$$

The combination of this result and Eq. (5-89) then yields

$$P(t_*) = \frac{\Delta\rho_* E_*}{\Lambda}, \tag{5-92}$$

where $\Delta\rho_*$ is the prompt excess reactivity at the initiation of the disassembly phase. In the present case, where the prompt feedback coefficient vanishes,

$$\Delta\rho_* = \dot{\rho}t_*. \tag{5-93}$$

To estimate the time t_* we first rewrite Eq. (5-91) as

$$\frac{\dot{\rho}t_*^2}{\Lambda} = \ln\left\{\frac{\dot{\rho}}{\Lambda}\left[\frac{E_*}{P(0)}\right]^2\right\} + \ln\left(\frac{\dot{\rho}t_*^2}{\Lambda}\right), \tag{5-94}$$

and note that to a first approximation the ln term on the far right can be neglected compared to the term on the left, since $\dot{\rho}t_*^2/2\Lambda \gg 1$. Hence

$$t_* \simeq \sqrt{\frac{\Lambda}{\dot{\rho}} \ln\left\{\frac{\dot{\rho}}{\Lambda}\left[\frac{E_*}{P(0)}\right]^2\right\}}. \tag{5-95}$$

Equations (5-92), (5-93), and (5-95) are the relationships required to analyze the disassembly phase of the accident.

The foregoing analysis provides the initial conditions that permit the numerical integration of Eqs. (5-85) and (5-87) to obtain the energy release during the disassembly phase. A rough analytical estimate of the energy release following the initiation of the disassembly phase can be obtained by making some physical approximations. To do this we first note that at the initiation of the disassembly phase the reactor has a period of $\Lambda/\Delta\rho_*$, yielding

$$P(t) = P(t^*) \exp\left[\frac{\Delta\rho_*}{\Lambda}(t - t_*)\right], \qquad t \approx t_*. \tag{5-96}$$

In the absence of the prompt feedback, the excess prompt reactivity $\Delta\rho_*$ at t_* is given by Eqs. (5-93) and (5-95):

$$\Delta\rho_* \approx \sqrt{\dot{\rho}\Lambda \ln\left\{\frac{\dot{\rho}}{\Lambda}\left[\frac{E_*}{P(0)}\right]^2\right\}}, \tag{5-97}$$

and Eq. (5-92) can be used to express the power in terms of the threshold energy

$$P(t) = \frac{\Delta\rho_* E_*}{\Lambda} \exp\left[\frac{\Delta\rho_*}{\Lambda}(t - t_*)\right], \qquad t \approx t_*. \tag{5-98}$$

We now make three physical assumptions. First, assume that the disassembly phase is of such short duration that there is insufficient time for the reactivity insertion to increase significantly above $\Delta\rho_*$. Second, assume that the disassembly reactivity does not significantly affect the rate of power rise until it causes the reactor to be brought subprompt critical at t_{**}. Third, assume that the energy generated after the power peak that occurs at t_{**} can be ignored. The value of E obtained at t_{**} is a fair approximation to the total energy release, since by neglecting the energy after the power peak, a rough compensation is made for the overestimate of the energy generated during the power rise.

With these assumptions, both the power and energy transients during the disassembly phase are exponential. The energy release is obtained by integrating Eq. (5-98) between t_* and t:

$$E(t) = E_* \exp\left[\frac{\Delta\rho_*}{\Lambda}(t - t_*)\right]. \tag{5-99}$$

We confine our attention to severe transients in which the total energy release is very much larger than E_*. Then, over most of the disassembly phase, E_* can be neglected compared to E, and therefore the disassembly reactivity may be approximated by Eq. (5-86). Combining Eqs. (5-86) and (5-99), we have

$$\frac{d^2\rho_d}{dt^2} = -\frac{8}{9}\frac{\pi^6\bar{M}^2(\gamma-1)}{\bar{k}_\infty\tilde{R}^4\bar{m}V}E_* \exp\left[\frac{\Delta\rho_*}{\Lambda}(t - t_*)\right]. \tag{5-100}$$

To obtain $\rho_d(t)$ we integrate twice from t_* to t. Using Eq. (5-99), and assuming that the exponential term is large, we may approximate the disassembly reactivity by

$$\rho_d(t) \simeq -\frac{8}{9} \frac{\pi^6 \bar{M}^2 (\gamma - 1)}{\bar{k}_\infty \tilde{R}^4 \bar{m} V} \frac{\Lambda^2}{(\Delta \rho_*)^2} E(t). \qquad (5\text{-}101)$$

The reactor becomes subprompt critical when $\rho_i(t) = -\rho_d(t)$. Thus, with $\rho_i(t) = \dot{\rho} t_*$, we find the total energy release to be

$$E \simeq \frac{9}{8} \frac{\bar{k}_\infty \bar{m}}{\pi^6 \bar{M}^2 (\gamma - 1)} \tilde{R}^4 V \frac{(\Delta \rho_*)^3}{\Lambda^2}. \qquad (5\text{-}102)$$

Finally, substituting Eq. (5-97) for $\Delta\rho_*$,

$$E \simeq \frac{9}{8} \frac{\bar{k}_\infty \bar{m}}{\pi^6 \bar{M}^2 (\gamma - 1)} \left(\ln \left\{ \frac{\dot{\rho}}{\Lambda} \left[\frac{E_*}{P(0)} \right]^2 \right\} \right)^{3/2} \tilde{R}^4 V \frac{\dot{\rho}^{3/2}}{\Lambda^{1/2}}. \qquad (5\text{-}103)$$

This result indicates that in the absence of significant Doppler feedback, both the energy release and energy density, E/V, resulting from severe excursions may be expected to increase markedly with the size of the reactor, since $E \propto \tilde{R}^4 V$. Ignoring the mild effects of the logarithmic term, we also find that

$$E \propto \frac{\dot{\rho}^{3/2}}{\Lambda^{1/2}}. \qquad (5\text{-}104)$$

Thus one of the most important parameters is the reactivity insertion rate $\dot{\rho}$, but, unfortunately, this is also the parameter with the most uncertainty, since its accurate specification requires that the mechanism causing the insertion be amenable to precise analysis.

The burst energy is also proportional to $\Lambda^{-1/2}$. This dependence on Λ is in interesting contrast to the energy release from an excursion controlled by prompt feedback, which can be seen from Eq. (5-61) to be proportional to $(\dot{\rho}\Lambda)^{1/2}$. The reason for the difference in the Λ dependence is that in the excursion models controlled by the prompt feedback, the reactivity is assumed to change instantaneously with energy generation, without inertial effects being taken into account. In the Bethe–Tait model, however, inertial effects are dominant, and a reactivity reduction cannot take place until pressures build up and have time to move material. During this time additional energy will have been generated, the amount of which will increase with decreased Λ, since more neutron generation time will have elapsed. Bethe–Tait analyses of less severe transients indicate that the dependence on $\dot{\rho}$ and Λ is weaker (Jankus, 1962).

$$E - E_* \propto \frac{\dot{\rho}^{1/3}}{\Lambda^{1/9}}, \tag{5-105}$$

but still increases with increasing $\dot{\rho}$ and decreasing Λ.

Effect of Doppler Feedback on Explosive Energy Release

Not all the energy release from a nuclear excursion is available to do destructive mechanical work in the form of blast, shock, water hammer, or other mechanisms. Energy is required to heat, melt, and vaporize the fuel before destructive pressures are generated. Since in the energy threshold model e_* is the energy density that must be exceeded for significant pressures to be generated, only the energy density in excess of this threshold can be converted to work through isentropic expansion. More realistic treatments of the destructive work generated by the fuel vapor pressure, moreover, tend to lead to much smaller values of the explosive energy release than those obtained from the idealized isentropic expansion (Jankus, 1961). Even isentropic expansion of the fuel to low pressure, however, leaves the fuel in a state with considerable internal energy, and if this energy were rapidly transferred to a liquid coolant, with a much lower boiling point, additional destructive mechanical work might be produced as a result of a vapor explosion.

The mechanisms by which the thermal energy generated in a severe transient may cause damage to the fission product barrier, either as thermal energy or as destructive mechanical work, are varied. Consequently, it is often found useful to separate the analysis of the energy generation from that of the mechanisms causing damage to the system. Following this philosophy, in what follows we deal primarily with the explosive energy release, defined from the isentropic expansion of the fuel vapor to low pressure. The problem of determining the actual conversion efficiency to destructive mechanical work and the consideration of detailed damage mechanisms are taken up in Section 9-3.

The Bethe–Tait method for the analysis of explosive excursions was developed for moderate-sized metal-fueled reactors such as EBR-II (Koch et al., 1957) and FERMI-I (Power Reactor Development Company, 1961). Under pessimistic assumptions, explosive energy releases of the order of 1000 MW-sec were calculated. The short neutron lifetimes and lack of a significant Doppler coefficient indicated that the sharp burst would have pressurization characteristics that resembled those of chemical explosives.

Thus the practice came into use of equating the explosive energy release to an equivalent amount of TNT. Using the relationship 2 MW-sec(t) ≈ 1 lb of TNT, energy releases equivalent to several hundred pounds of TNT are calculated.

As indicated by Eq. (5-103) the energy yield resulting from Bethe-Tait analysis increases sharply with reactor volume. As an illustration (Hummel and Okrent, 1970), the upper limit of explosive energy release for a 50 dollar per second reactivity insertion is calculated to be 200 lb of TNT for the EBR-II core, but for the FERMI core with 5.5 times the volume, a 1000 lb TNT yield results when the same insertion rate is assumed, The cores of these reactors are relatively small. Those anticipated for 1000 MW(e) plants would be a factor of 10 or more larger in volume than the FERMI core. An obvious concern arises as to the magnitude of the energy release that could result from a severe excursion before termination by disassembly.

Because the severity of the disassembly accidents was predicted to increase as larger reactor cores came into being, a substantial effort has been expended toward lowering the upper limit of explosive energy release by refining the treatment of the excursion, and particularly of the disassembly model. The semianalytic model discussed in the preceding subsection has been replaced by numerical integration of the kinetics and disassembly feedback equations. More realistic geometrical representations and equations of state have been incorporated into the equations, and a series of computer codes have been written to treat the neutronic-hydrodynamic coupling in a more rigorous manner (Jackson and Weber, 1975). The single most important factor in determining the upper limit on the energy release is the magnitude of the negative Doppler coefficient (Meyer et al., 1967). This magnitude, moreover, has increased markedly as fast reactor design has evolved, since the large oxide-fueled cores now being proposed have softer spectra and therefore Doppler coefficients of larger magnitude.

The effect of the Doppler coefficient in decreasing the explosive energy yield is graphically illustrated in Fig. 5-12 for a proposed 1000 MW(e), sodium-cooled oxide-fueled fast reactor. Since in fast reactors the Doppler effect is inversely proportional to absolute temperature, $T(\partial k/\partial T)$ is taken as the abscissa. Even small values of the Doppler coefficient result in dramatic reductions of the explosive energy release. In the presence of increasingly large Doppler coefficients the energy yield versus the Doppler coefficient curve exhibits an oscillatory behavior. The increase of energy yield with reactivity insertion rate indicated for small values of the Doppler coefficient is expected. A more detailed examination must be made of the time-dependent behavior of the transient to gain some insight into the oscillatory behavior that sometimes results in smaller reactivity insertion rates leading to larger explosive energy releases.

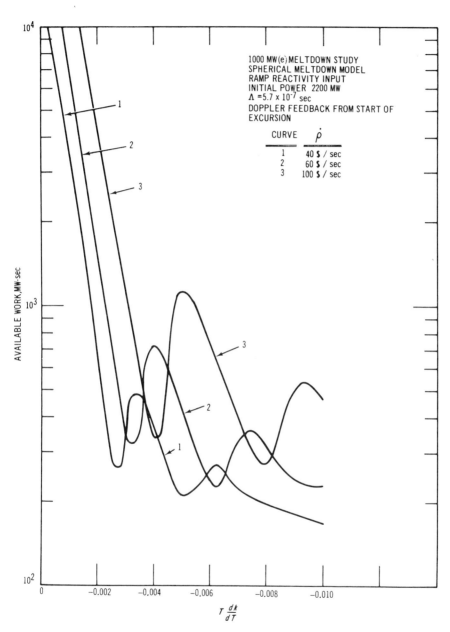

FIGURE 5-12 Influence of reactivity insertion rate on explosive energy release as a function of the Doppler coefficient. Adapted from R. A. Meyer, B. Wolf, N. F. Friedman, and R. Seifert, "Fast Reactor Meltdown Accidents Using Bethe–Tait Analysis," General Electric Report, GEAP-4809, 1967.

Clearly, Fig. 5-12 indicates that even a small Doppler coefficient results in a marked decrease in energy release. This rapid decrease in energy release with increasing Doppler coefficient may be understood as follows: Without the Doppler coefficient, power and energy build very rapidly because of the short prompt neutron generation time, and the reactor is on a very short period when the energy density threshold for disassembly is reached. The disassembly feedback is controlled by inertia, however, and even very large pressures must act over an appreciable time span before sufficient core expansion can take place to terminate the excursion. Thus a great deal of energy is generated during the disassembly phase. In the presence of Doppler feedback the reactivity is substantially reduced during the burst, causing the disassembly phase to be entered with less excess reactivity, and consequently a larger period. As indicated roughly by Eq. (5-102), the smaller value of $\Delta\rho_*$ at the initiation of disassembly causes less energy to be generated during the time required for pressure to build and create sufficient expansion to terminate the excursion.

Transients corresponding to the oscillatory part of Fig. 5-12 consist not of a single power burst, but rather of transients with two or more power peaks, as illustrated in Fig. 5-13. In such excursions the Doppler coefficient is strong enough to turn around the initial power rise and bring the reactor subprompt critical until the continuing ramp insertion causes the reactor to become superprompt critical once again. The power trace is characterized by repeated power peaks, just as in the case of the linear energy feedback model shown in Figs. 5-10 and 5-11. Now, however, there is little damping of the oscillations, since as the fuel temperature rises, the Doppler coefficient decreases. As shown in Fig. 5-13, the energy density rises to where fuel vapor pressure is generated, and after an inertial delay the transient finally is terminated by the disassembly reactivity.

The oscillatory behavior in the presence of larger Doppler coefficients, shown in Fig. 5-12, is due to the sensitivity of the energy release during disassembly to the phase of the power oscillation at which disassembly is initiated. For example, the disassembly in Fig. 5-13 is initiated near the second power peak, but the inertial delay does not allow sufficient disassembly reactivity to be generated to override the rapid reactivity addition taking place at the initiation of the third power burst. Disassembly shutdown then does not take place until the third burst has peaked, and the energy yield is near a maximum, as indicated in Fig. 5-12. If the Doppler coefficient had been somewhat smaller, more energy would have been generated in the first two bursts, and the disassembly would have been initiated earlier. In this case sufficient time would have elapsed from disassembly initiation for the inertial delay to be overcome and for shutdown to take place before the third power burst. Although the first two bursts then release more energy, the

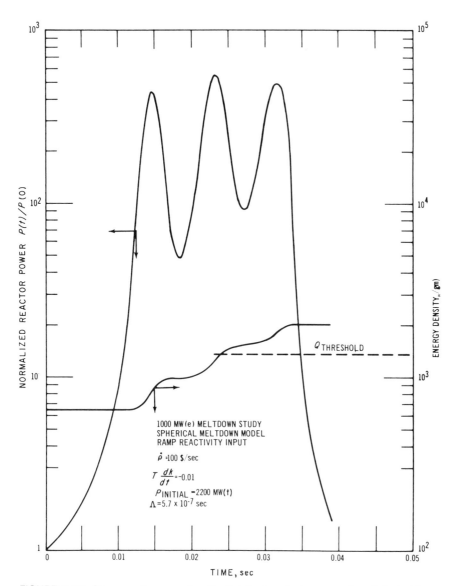

FIGURE 5-13 Power and energy density distribution during excursion. Adapted from R. A. Meyer, B. Wolf, N. F. Friedman, and R. Seifert, "Fast Reactor Meltdown Accidents Using Bethe–Tait Analysis," General Electric Report, GEAP-4809, 1967.

absence of the third burst results in a net decrease in the explosive energy release for the entire transient. Thus decreasing the Doppler coefficient may result in a smaller energy yield.

The interactions between prompt neutron generation time and Doppler coefficient are indicated in Fig. 5-14. When the Doppler coefficient is very small, a core with the shorter prompt neutron generation time will suffer the more severe explosive energy release. This is not surprising, since the Bethe–Tait analysis indicates that energy release increases with decreasing generation time in situations in which there is no significant prompt feedback. In contrast, when large Doppler coefficients are encountered, the energy release tends to increase with increased generation time. This is consistent with the analysis of the previous section, where it is shown that excursions dominated by prompt feedback yield energy releases that increase with the generation time.

The extensive parametric studies that have been carried out for fast reactors with significant Doppler coefficients (Meyer et al., 1967) lead to the conclusion that with the evolution of fast reactor design, the adverse effects of the larger core volumes on disassembly shutdown are largely compensated for by the increased Doppler coefficients. In the presence of a large Doppler coefficient, energy yields become less sensitive to reactivity insertion rates, the Doppler coefficient magnitude, and most other parameters (an exception is the fuel heat capacity). In design work, however, care must be taken to ensure that wide enough safety margins are included so that if uncertainty in the Doppler coefficient causes it to lie on a maximum instead of the calculated minimum on an energy release curve, such as that shown in Fig. 5-12, the safety criteria on explosive energy release still are not violated.

For survey calculations covering wide variations in design parameters, it is useful to have a criterion on the value of the Doppler coefficient that will cause the explosive release to become small, say, an order of magnitude less than that which would occur in the absence of prompt reactivity feedback. It has been estimated by Smith et al. (1965) that such a reduction would result if the Doppler coefficient provided just sufficient negative reactivity to balance the insertion rate at the initiation of the disassembly phase, so that at that instant the reactor would be just prompt critical. An estimate of the Doppler coefficient needed to meet this criterion is:

$$\bar{T}_{fe}\frac{d\rho_{fb}}{d\bar{T}_{fe}} = -\frac{[2\dot{\rho}\,\Lambda\,\ln\,(e_*\dot{\rho}\sqrt{(2/\pi\bar{\beta}P_0)})]^{1/2}}{\ln\,[1+0.73(e_*/c_vT_0)]},\qquad(5\text{-}106)$$

where P_0 and T_0 are the reactor power at delayed critical and the initial average fuel temperature and c_v is the specific heat at constant volume.

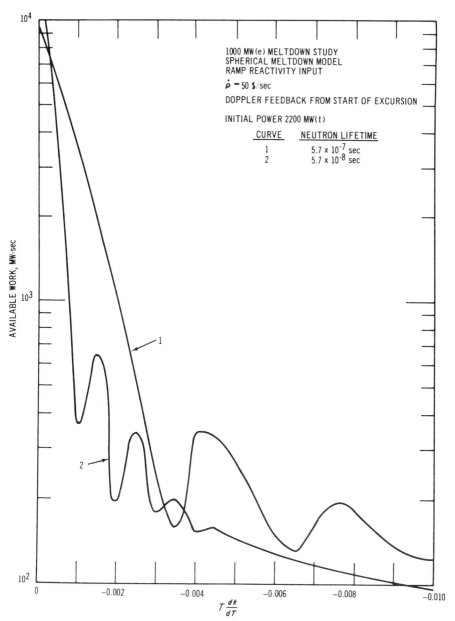

FIGURE 5-14 Influence of neutron lifetime on explosive energy release for excursions initiated by a ramp reactivity insertion. Adapted from R. A. Meyer, B. Wolf, N. F. Friedman, and R. Seifert, "Fast Reactor Meltdown Accidents Using Bethe–Tait Analysis," General Electric Report, GEAP-4809, 1967.

PROBLEMS

5-1. Assuming that the mass flow rate and core inlet temperature remain constant, show that Eqs. (1-118), (5-10), and (5-11) can be combined to yield the following equation for $\Delta T \equiv \bar{T}_0 - \bar{T}_i$, the coolant temperature rise through the core during a power transient:

$$a_1 \frac{d^2 \Delta T}{dt^2} + a_2 \frac{d \Delta T}{dt} + a_3 \Delta T = P(t),$$

where

$$a_1 = \tfrac{1}{2}(M_{fe}c_{fe}M_c c_p R),$$

$$a_2 = M_{fe}c_{fe}W c_p R + \tfrac{1}{2}(M_{fe}c_{fe} + M_c c_p),$$

$$a_3 = W c_p.$$

What are the initial conditions for a reactor that has been operating at steady state?

5-2. A sodium-cooled fast reactor has the following characteristics:

$P = 2400$ MW(t), $W = 76 \times 10^6$ lb$_m$/hr,

$\tau = 4.0$ sec, $c_p = 0.3$ Btu/(lb$_m$)(°F)$^{-1}$,

$M_{fe}c_{fe} = 7000$ Btu/°F, $M_c c_p = 1000$ Btu/°F.

$T_i = 800$°F,

Suppose the reactor undergoes a sudden trip, which may be approximated by setting the power equal to zero, and it is necessary to know the coolant outlet temperature transient in order to evaluate thermal shock effects. Assuming that the inlet temperature remains at its initial value, use the model of the preceding problem to evaluate the core outlet temperature transient and plot your results.

5-3. Assume that the reactor in the preceding problem suffers a control failure and undergoes a power transient

$$P(t) = P(0)[1 + 0.25t],$$

where t is in seconds.

a. Determine the outlet temperature transient and plot your results until T_0 has risen to 200°F above its initial value.

b. Suppose that the flux trip on the reactor fails and that the transient is terminated by a backup system that causes the reactor to trip if the core outlet temperature exceeds its full power value by more than 40°F. At what time will the trip occur? What will the power level be at the time of the trip?

c. Suppose that with the flux trip failed, the reactor power increases extremely slowly. What will the power level be at the time the outlet temperature trip causes the reactor to shut down?

5-4. From the power transient shown in Fig. 5-3 estimate the power coefficient of the 3250 MW(t) Zion Reactor.

5-5. Verify Eqs. (5-33) and (5-35).

5-6. A power reactor has core characteristics such that $(1/k)(\partial k/\partial \bar{T}_i) = 0$, and $(1/k)(\partial k/\partial \bar{T}_{fe})$ and $(1/k)(\partial k/\partial \bar{T}_c)$ are both negative. If the core undergoes a quasi-static transient induced by a slow change in the inlet temperature and/or the coolant flow rate, show that in the absence of a reactor trip:

a. The incremental change in feedback reactivity is given by

$$d\rho_{fb} = -\left|\frac{1}{k}\frac{\partial k}{\partial \bar{T}_{fe}}\right| d\bar{T}_{fe} - \left|\frac{1}{k}\frac{\partial k}{\partial \bar{T}_c}\right| d\bar{T}_c,$$

b. The incremental change in reactor power is

$$dP = \frac{\left|\dfrac{1}{k}\dfrac{\partial k}{\partial \bar{T}_{fe}} + \dfrac{1}{k}\dfrac{\partial k}{\partial \bar{T}_c}\right|\left(\dfrac{P}{2W^2 c_p}dW - d\bar{T}_i\right) + d\rho_c}{\left|\left(R + \dfrac{1}{2Wc_p}\right)\dfrac{1}{k}\dfrac{\partial k}{\partial \bar{T}_{fe}} + \dfrac{1}{2Wc_p}\dfrac{1}{k}\dfrac{\partial k}{\partial \bar{T}_c}\right|}$$

5-7. Consider a superprompt critical power burst that is adequately described by the prompt neutron approximation. Suppose that we have a reactivity feedback model such that

$$\rho(t) = \rho_i(t) - |\mu|\left[\int_0^t P(t')\,dt'\right]^n.$$

A power burst is initiated by a step reactivity insertion

$$\rho_i(t) = \rho^0 + \beta, \qquad t \ge 0,$$

from a very small initial power $P(0)$, at $t = 0$.

Show that the maximum power is given by

$$P_{\max} = \frac{n}{n+1}\left[\frac{(\rho^0/\Lambda)^{n+1}}{|\mu|/\Lambda}\right]^{1/n}$$

and the total energy release by

$$E(\infty) = (\rho^0/|\mu|)^{1/n}(n+1)^{1/n}$$

5-8. Verify Eqs. (5-71) through (5-78) for the disassembly reactivity.

5-9. Verify Eqs. (5-91) and (5-94).

5-10. Study the safety analysis report for a power reactor of your choice. Then discuss in some detail the design and procedural measures that are taken to prevent the sudden withdrawal of large amounts of control poisons from the core.

FUEL ELEMENT BEHAVIOR

Chapter 6

6-1 INTRODUCTION

One of the central objectives of the analysis of postulated accidents is to ascertain whether fuel elements will be damaged to the point where their integrity as fission product barriers is compromised. The details of the potential damage mechanisms vary considerably between fuel element designs. With a few exceptions, however, most present-day power reactor fuel elements consist of cylindrical pellets of ceramic—usually oxide—fuel housed in metal cladding. The general characteristics shared by elements of this type serve as a basis for understanding fuel element behavior and damage under accident conditions.

Under operating conditions the bulk of the fission product inventory remains trapped in the ceramic fuel. The fission products that escape from the fuel consist primarily of fractions of the inventories of noble gases and other species of more volatile fission products. These collect in the fuel-cladding gap and in other void regions within the cladding envelope. In some fuel designs a venting system is provided to remove the resulting fission product gas from these void regions on a continuous basis. Fuel failure most often is used to refer to the presence of holes, cracks, or other breaches of the cladding that permit the fission product gas that has collected in the void regions to escape to the coolant. Among the tens of thousands of fuel elements typically found in a large power reactor, at least some fuel failures of this type must be anticipated during normal plant operation. The number of failures that can be tolerated is determined primarily by the radioactive waste treatment system that is provided for removing the resulting radioactive materials from the primary system.

Under the conditions of some of the more severe postulated accidents, such failure of a substantial fraction of the fuel elements may be tolerable. For, although the radioactive release to the primary system will be of considerable magnitude, the preponderance of the inventory of the radioactive material nevertheless will remain entrapped in the ceramic fuel. The far greater danger to which much of safety assessment is addressed is that gross melting of the reactor core will result in the release of essentially all the noble gases and large fractions of the more volatile fission products. To prevent such destruction of fuel elements as a fission product barrier, the highly optimized heat transport geometry of the reactor core must be maintained through any anticipated accident transient. If fuel elements are damaged to the point where the large ratio of heat transfer surface to fuel mass provided by the small-diameter cylindrical pins is lost, or to where the flow channels through the core are plugged, the removal of decay heat becomes impossible. Increasing amounts of fuel will then overheat and melt. A likely consequence is that molten fuel will congeal into a larger mass, with a progressively decreasing heat transfer surface and cause blockage of additional flow channels. The process may then become autocatalytic and result in the meltdown of a substantial fraction of the reactor core.

The progression of events that may lead to the initiation of core meltdown varies considerably with the details of the accident transient and of the fuel design. Failure of the fuel element heat transport geometry may originate with molten or vaporized fuel, which then melts through or otherwise penetrates the cladding to plug the coolant channels. Alternately, over-stress, excessive temperatures or chemical reactions may cause the cladding to shatter or otherwise fail structurally before fuel starts to melt. The effect is then for the collapsing columns of fuel pellets to block the coolant channels. Other scenarios that are quite different may also appear, particularly when coated particles or other diverse fuel element designs are involved.

In this chapter we discuss the behavior of fuel elements under both the steady-state and transient conditions in order to understand the failure mechanisms leading to the loss of the fuel elements as a fission product barrier. In Chapter 9 the containment of meltdown and other core disruptive accidents is taken up. In the following section the steady-state temperature distributions are related to the rate of power production, fuel element properties, and the mechanisms for fuel-coolant heat transport. The thermal behavior of the fuel element during transients is then discussed. Finally, the interactions between temperature, stress, irradiation effects, and other phenomena necessary to the understanding of fuel element failure mechanisms are discussed.

6-2 STEADY-STATE HEAT TRANSPORT

Fuel element configurations vary significantly between power reactors. Their thermal behavior, however, usually may be understood in terms of the simple fuel pin model shown in Fig. 6-1. Heat is produced approximately uniformly across the fuel region. It is transported radially outward through

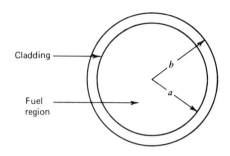

Cladding

Fuel region

FIGURE 6-1 Fuel pin cross section.

the fuel, across the fuel-cladding gap, through the cladding, and into the coolant channel. It is then carried away by coolant convection parallel to the fuel element axis. Axial heat transport along the fuel element is negligible because of the small diameter-to-length ratio of the element. The radial temperature distribution within the coolant channel is ignored, the convective heat transport being represented by Newton's law of cooling between the cladding surface temperature and the bulk coolant temperature.

As shown, the configuration is typical of the ceramic fuels with metal cladding used in most present-day power reactors. Although it is not strictly valid for fuel element configurations such as that shown for the HTGR in Fig. 1-12e, qualitatively correct temperature results may be obtained by artificially cylindrizing the fuel element and modifying the dimensions and material properties to conserve important thermal properties. For the HTGR configurations, one may represent the fuel element as a cylindrical pin where the "cladding" is now the graphite moderator through which heat is transported from fuel to coolant channel. In Table 6-1 are tabulated typical fuel element dimensions for a number of power reactor types, along with other thermal-hydraulic parameters to be defined in subsequent discussions.

In the foregoing chapters, steady-state fuel element heat transport is approximated in terms of $\tilde{T}_{fe}$, a weighted-average temperature across the cross section of the fuel element, and T_c the bulk temperature in the coolant channel. In earlier chapters these temperatures are written as $\tilde{T}_{fe}(r, z)$, $T_c(r, z)$ to indicate that they are associated with a fuel element lattice cell located a radial distance r from the core centerline and an axial distance z from the core midplane. The temperatures are related to the local linear heat rate through Eq. (1-100):

$$\tilde{T}_{fe}(r, z) - T_c(r, z) = R'_{fe}q'(r, z), \qquad (6-1)$$

where R'_{fe} is the thermal resistance per unit length of the fuel element. In order to evaluate R'_{fe}, and to determine the steady-state and transient

TABLE 6-1   Some Typical Values of Thermal Parameters

	PWR	BWR	HTGR	LMFBR	GCFR
$\bar{P}'''$, average core power density [kW/1.]	75	52	6.3	578	280
$\bar{q}'$, average linear heat rate [kW/ft]	5.8	7.1	1.0	9.6	12.7
$M_{fe}c_{fe}$, fuel element heat capacity [Btu/(°F)(ft)]	0.0446	0.079	0.196	0.019	0.029
h_c, coolant heat transfer coefficient [Btu/(hr)(ft)^2(°F)]	5300	5300	230	30,000	1880
τ, core time constant [sec]	4.12	7.08	17.3	2.79	4.04
τ', fuel cladding time constant [sec]	0.63	1.10	2.44	0.50	0.59
$\bar{q}'/(M_{fe}c_{fe})$, Adiabatic heat-up [°F/sec]	122	84	5.0	472	409

After Bellon, 1973.

behavior of a fuel element, we must determine the radial temperature distribution within the fuel element. We denote this radial distance from the fuel element centerline as ζ. For brevity, the (r, z) position of the fuel element in the reactor is deleted from the following discussions. This deletion causes no loss of generality, since the (r, z) coordinates appear only as fixed global parameters in the linear heat rate and the temperatures. We first treat heat conduction within the fuel element, assuming that the heat transfer coefficient between the cladding and coolant is known. Then the dependence of this heat transfer coefficient on the coolant thermodynamics and flow conditions is discussed, paying particular attention to the boiling crisis.

Fuel Element Temperature Distributions

With the axial heat transfer neglected, the steady-state heat conduction equation in the fuel region can be written in cylindrical geometry as (Kreith, 1965)

$$k_f \frac{1}{\zeta} \frac{d}{d\zeta} \zeta \frac{d}{d\zeta} T(\zeta) + q''' = 0, \tag{6-2}$$

where the fuel thermal conductivity k_f is assumed to be constant. In most cases no significant error is introduced by assuming the heat per unit volume q''' to be radially uniform over the fuel region. Thus it can be expressed in terms of the linear heat rate as

$$q''' = \frac{q'}{\pi a^2},\tag{6-3}$$

where a is the radius of the fuel region as indicated in Fig. 6-1. Equation (6-2) can be integrated twice to yield

$$T(\zeta) = C_2 - \frac{q'\zeta^2}{4\pi a^2 k_f} + C_1 \ln\zeta, \qquad 0 \leq \zeta \leq a,\tag{6-4}$$

where C_1 and C_2 are the constants of integration. Obviously, $C_1 = 0$, since otherwise the temperature would be infinite at the centerline; C_2 is determined by requiring that the temperature take on a value of $T(a^-)$ on the fuel side of the fuel-cladding interface. Performing the algebra yields,

$$T(\zeta) = T(a^-) + \frac{q'}{4\pi k_f}\left[1 - \left(\frac{\zeta}{a}\right)^2\right], \qquad 0 \leq \zeta < a.\tag{6-5}$$

Two quantities are of particular interest: The first is the temperature at $\zeta = 0$, the fuel centerline, which we denote as T_ϕ:

$$T_\phi = T(a^-) + \frac{q'}{4\pi k_f}.\tag{6-6}$$

The second is the radially averaged fuel region temperature,

$$\tilde{T}_f \equiv \frac{2}{a^2}\int_0^a T(\zeta)\zeta\,d\zeta,\tag{6-7}$$

which is found from Eq. (6-5) to be

$$\tilde{T}_f = T(a^-) + \frac{q'}{8\pi k_f}.\tag{6-8}$$

Unless an excellent thermal bond exists, there is likely to be a significant thermal resistance across the fuel-cladding interface. Such resistance is usually expressed empirically as a gap heat transfer coefficient h_g, defined in terms of the heat flux per unit area q''_g, and the temperature drop across an infinitely thin gap:

$$q''_g = h_g[T(a^-) - T(a^+)].\tag{6-9}$$

Noting that the total heat flux across the gap must be equal to the linear heat rate, we have

$$q' = 2\pi a q''_g,$$ (6-10)

or

$$q' = 2\pi a h_g [T(a^-) - T(a^+)].$$ (6-11)

The temperature distribution in the cladding is obtained from the source-free heat conduction equation in cylindrical geometry:

$$k_{cl} \frac{1}{\zeta} \frac{d}{d\zeta} \zeta \frac{d}{d\zeta} T(\zeta) = 0,$$ (6-12)

where k_{cl} is the cladding thermal conductivity. Integrating twice we have

$$T(\zeta) = C_1 + C_2 \ln \zeta, \qquad a < \zeta \le b.$$ (6-13)

The two constants of integration C_1 and C_2 may be determined in terms of the cladding surface temperature at b and heat flux at $\zeta = a^+$. The Fourier law of heat conduction is used along with Eq. (6-10) to give the heat flux per unit area as

$$\frac{q'}{2\pi a} = -k_{cl} \frac{dT(\zeta)}{d\zeta} \bigg|_{\zeta = a^+}$$ (6-14)

Evaluating the constants in Eq. (6-13), we obtain

$$T(\zeta) = T(a^+) - \frac{q'}{2\pi k_{cl}} \ln \left(\frac{\zeta}{a} \right), \qquad a < \zeta \le b.$$ (6-15)

The temperature drop across the cladding is then

$$T(a^+) - T(b) = \frac{q'}{2\pi k_{cl}} \ln \left(\frac{b}{a} \right),$$ (6-16)

and the volume-averaged cladding temperature, defined by

$$\tilde{T}_{cl} \equiv \frac{2}{b^2 - a^2} \int_a^b T(\zeta) \zeta \, d\zeta,$$ (6-17)

is

$$\tilde{T}_{cl} = T(a^+) - \frac{q'}{2\pi k_{cl}} \left[\frac{b^2}{b^2 - a^2} \ln \left(\frac{b}{a} \right) - \frac{1}{2} \right].$$ (6-18)

The temperature drop between the cladding surface and the bulk coolant temperature is also written in terms of the heat flux per unit area across this surface, q''_c, and a heat transfer coefficient h_c:

$$q''_c = h_c [T(b) - T_c].$$ (6-19)

Using the fact that the total heat flux into the coolant per unit length must be equal to the linear heat rate, we have

$$q' = 2\pi b h_c [T(b) - T_c]. \tag{6-20}$$

The coolant heat transfer coefficient h_c is a complex function of coolant properties and flow conditions. We presently treat it as a known constant, deferring the discussion of its properties to the next subsection.

At this point the temperature profile through the fuel element may be constructed in terms of the linear heat rate and the bulk coolant temperature. The temperature drops across the fuel, gap, cladding, and cladding–coolant boundary layer are given by Eqs. (6-6), (6-11), (6-16), and (6-20), respectively, and therefore $T_\frac{}{}$, $T(a^+)$, $T(b)$ may be determined in terms of q' and T_c. The temperature distributions within the fuel and cladding may then be determined from Eqs. (6-5) and (6-15). An example steady-state profile is given in Fig. 6-2 for a pin consisting of UO_2 pellets in a metal tube cladding and cooled by a liquid. For this composition the temperature drop in the fuel region is large because of the low conductivity of the oxide fuel. Likewise, a

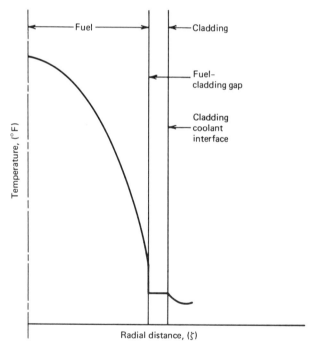

FIGURE 6-2   Typical steady-state temperature distribution for a liquid-cooled reactor with UO_2 fuel and metal cladding.

large temperature drop across the pellet-cladding gap may be expected because the thermal contact resulting from simply filling the tube with pellets is poor. This gap temperature drop is quite variable with operating conditions, because of the fuel expansion effects discussed in Section 6-4. The high conductivity of the metal cladding does not result in a large temperature drop, neither does the large coolant heat transfer coefficient normally present for forced convection in liquid-cooled systems. In a gas-cooled reactor the cladding-coolant temperature drop would be substantially larger.

In addition to the temperature profile, several additional relationships are useful in developing approximate lumped parameter models for treating both steady-state and transient conditions. First, from Eqs. (6-8), (6-9), and (6-18) the difference between average fuel and cladding region temperatures may be written as

$$q' = \frac{1}{R'_g}(\tilde{T}_f - \tilde{T}_{cl}),\tag{6-21}$$

where

$$R'_g = \frac{1}{8\pi k_f} + \frac{1}{2\pi a h_g} + \frac{1}{2\pi k_{cl}}\left[\frac{b^2}{b^2 - a^2}\ln\left(\frac{b}{a}\right) - \frac{1}{2}\right].\tag{6-22}$$

From Eqs. (6-16), (6-18), and (6-20), the temperature drop between the average cladding and the bulk coolant temperature may be expressed as

$$q' = \frac{1}{R'_c}(\tilde{T}_{cl} - T_c),\tag{6-23}$$

where

$$R'_c = \frac{1}{2\pi k_{cl}}\left[\frac{1}{2} - \frac{a^2}{b^2 - a^2}\ln\left(\frac{b}{a}\right)\right] + \frac{1}{2\pi b h_c}.\tag{6-24}$$

Combining these equations, we may also write

$$q' = \frac{\tilde{T}_f - T_c}{R'_g + R'_c}.\tag{6-25}$$

Finally, the total temperature drop from fuel centerline to coolant may be found from Eqs. (6-6), (6-11), (6-16), and (6-20) to be

$$T_\phi - T_c = R'_\phi q',\tag{6-26}$$

where

$$R'_\phi = \frac{1}{4\pi k_f} + \frac{1}{2\pi a h_g} + \frac{1}{2\pi k_{cl}}\ln\left(\frac{b}{a}\right) + \frac{1}{2\pi b h_c}.\tag{6-27}$$

In Section 1-4 the fuel element thermal resistance per unit length is defined either in terms of the fuel centerline temperature T_t or the averaged fuel element temperature $\tilde{T}_{fe}$. If the centerline temperature is used, then the thermal resistance per unit length is given by Eq. (6-27). To be consistent with the thermal transient model of Section 5-1, the average fuel element temperature $\tilde{T}_{fe}$ must be defined such that the specific heat per unit length of the fuel element multiplied by $\tilde{T}_{fe}$ is equal to the sensible heat stored in the element above the reference point $T = 0$. Assuming temperature independent specific heats, we have

$$\tilde{T}_{fe} \equiv \frac{M'_f c_f \tilde{T}_f + M'_{cl} c_{cl} \tilde{T}_{cl}}{M'_f c_f + M'_{cl} c_{cl}}. \tag{6-28}$$

Here M'_f and M'_{cl} are the mass of fuel and cladding, respectively, per unit length,

$$M'_f = \pi a^2 m_f,$$
$$M'_{cl} = \pi (b^2 - a^2) m_{cl}, \tag{6-29}$$

where m_f and m_{cl} are the fuel and cladding density and c_f and c_{cl} are the corresponding specific heats per unit mass. This definition of $\tilde{T}_{fe}$ may be combined with Eqs. (6-21) and (6-23) to yield

$$q' = \frac{1}{R'_{fe}} (\tilde{T}_{fe} - T_c), \tag{6-30}$$

if we take

$$R'_{fe} = \frac{M'_f c_f (R'_g + R'_c) + M'_{cl} c_{cl} R'_c}{M'_f c_f + M'_{cl} c_{cl}}. \tag{6-31}$$

Equation (6-30) is identical to the definition of the fuel element thermal resistance per unit length given in Eq. (1-100).

In the preceding derivations temperature-independent conductivities and specific heats are assumed. More precise results, of course, are obtainable using temperature-dependent thermal property data. If this is done, the thermal resistances R'_g, R'_c, R'_t, and R'_{fe} become temperature dependent. For survey calculations it is often adequate to use constant values of the thermal conductivity, specific heats, and gap heat transfer coefficient that are based on averages over a reasonable range of conditions, provided the errors introduced are no greater than the other uncertainties associated with the definition of the accident. Changes in the heat transfer coefficient h_c tend to be much more significant to the understanding of fuel element behavior during accidents. Coolant flow starvation, overheating, and/or depressurization—particularly when associated with a boiling crisis—may cause h_c to undergo large decreases in value and result in severe fuel element temperature transients. In what follows we take a closer look at the coolant conditions governing the behavior of the cladding-coolant heat transfer coefficient.

Convective Heat Transfer

The heat transfer coefficient is a function of the coolant flow rate, temperature, and pressure, as well as its conductivity, viscosity, and a number of other variables. Because of the difficulties in deriving theoretical expressions for it, the heat transfer coefficient is most often expressed as one of a number of semiempirical correlations between three dimensionless quantities:

$$Nu = Nu(Pr, Re), \qquad (6-32)$$

where Nu, Pr, and Re are the Nusselt, Prandtl, and Reynolds numbers, respectively. They are defined by

$$Nu = \frac{h_c D_e}{k}, \qquad (6-33)$$

$$Pr = \frac{c_p \mu}{k} \qquad (6-34)$$

and

$$Re = \frac{D_e \bar{v} m}{\mu}. \qquad (6-35)$$

The quantities m, k, μ, and c_p are the coolant density, thermal conductivity, viscosity and specific heat at constant pressure; $\bar{v}$ and D_e are the mean coolant speed and the equivalent diameter of the channel, respectively. Since m, k, and μ are temperature dependent, these dimensionless numbers are implicitly temperature dependent. Because the heat transfer correlations are not extremely accurate, however, it often is satisfactory simply to evaluate these thermal properties at an estimated average of the coolant temperature.

In Eq. (6-32) the Nusselt number is taken as the dependent variable because it is proportional to the heat transfer coefficient, the quantity that we wish to determine. The Prandtl number is a function only of the material properties of the fluid and not the flow conditions or channel geometry. As indicated in Table 6-2, Prandtl numbers for nonmetals are roughly unity, whereas those for liquid metals are much smaller because of their high conductivity. The Reynolds number is a measure of the ratio of inertial to

TABLE 6-2   Coolant Prandtl
Numbers

Water	4.52 @ 100°F
	0.87 @ 500°F
Gases	8.55 @ 500°F
	4.57 @ 800°F
Liquid metals	0.004 to 0.03

viscous forces. Large Reynolds numbers are associated with turbulent flow, and small numbers with laminar flow, the transition being at about Re = 3100.

A large number of correlations between Nu, Pr, and Re are available to correspond to particular geometries and specific flow conditions. For our purposes, two of the more generally applicable correlations for turbulent flow are sufficient. For single-phase turbulent flow in nonmetals—either liquids or gases—the Dittus–Boelter (1930) correlation is appropriate:

$$\text{Nu} = 0.023 \text{Pr}^{0.4} \text{Re}^{0.8}. \tag{6-36}$$

For liquid-metal coolants the Lyon–Martinelli correlation (Martinelli, 1947) finds the widest use,

$$\text{Nu} = 7 + 0.025 \text{Pr}^{0.8} \text{Re}^{0.8}. \tag{6-37}$$

The variation of the heat transfer coefficients with respect to coolant flow is easily seen by writing them in terms of a reference state denoted by the subscript 0. Thus, assuming the temperature dependence of the conductivity, viscosity, and specific heat can be neglected, we have for the Dittus–Boelter correlation

$$h_c = h_0 \left(\frac{G}{G_0}\right)^{0.8}, \tag{6-38}$$

where $G = \bar{v}m$ is the mass flow rate per unit area. For the Lyon–Martinelli correlation,

$$h_c = A + (h_0 - A)\left(\frac{G}{G_0}\right)^{0.8}, \tag{6-39}$$

where $A = 7k/D_e$. We see that for nonmetals the heat transfer coefficient rises not quite linearly with the mass flow rate. The same is true for liquid metals except at low flow rates, where a fairly large constant term dominates because of the heat transfer by conduction in the high conductivity metal. In either case, the adverse effect of flow starvation is apparent.

Typical values for the heat transfer coefficient for several reactor types are included in Table 6-1. In gas-cooled reactors, the low coolant density causes small values of h_c to occur relative to other reactor types. As a result, the temperature drop between cladding and coolant is large, and nominal errors in h_c may cause significant errors in the maximum cladding and fuel temperatures. For this reason a more accurate determination of h_c for the particular situation at hand may be warranted, especially if fins, surface roughening, or other devices to enhance gas cooling are utilized. In contrast, relatively large values of h_c are encountered in liquid-cooled or boiling systems. Thus a relatively large error in h_c may produce deviations in the temperature drop that are small compared, for example, to those resulting from uncertainties in the fuel conductivity and gap heat transfer coefficient.

The problem in liquid- or boiling-coolant systems is not that h_c cannot be calculated accurately enough under operating conditions. It is, rather, that the heat flux from the fuel may become sufficiently large or the coolant conditions may deteriorate enough to cause a boiling crisis and subsequent vapor blanketing of the fuel. Should this happen, the heat transfer coefficient would decrease by orders of magnitude, thus leading to near perfect insulation of the fuel pin from the coolant. We discuss boiling, and the boiling crisis, first with respect to water coolants, and then contrast the situation for liquid-metal coolants.

Boiling in Water

To examine the properties of the boiling crisis we first consider the situation in which a wire or other heated surface is placed in a pool of water, where no forced convection is present. The bulk temperature of the pool is maintained at T_{sat}, the saturation temperature of water, at a fixed, ambient pressure. As the wall temperature T_w of the heater undergoes a controlled increase, the heat flux q'' from heater to the pool exhibits the behavior shown in Fig. 6-3. The curve is divided into four regions, each of which has distinct physical characteristics.

In region I no boiling takes place, and the heat transfer is by single-phase natural convection. Even though the surface temperature is greater than the saturation temperature, no bubbles nucleate on the heated surface for the following reason. The bubbles would normally emanate from cavities and other imperfections on the heater wall. For the bubbles to nucleate, however, they must have enough internal pressure to overcome the fluid surface tension that acts to collapse them. The bubble pressure increases with its temperature. Since the inside of the bubble and the surrounding liquid are in thermal equilibrium, their temperatures are equal, and the only way the internal pressure may increase is by superheating the liquid at the heater

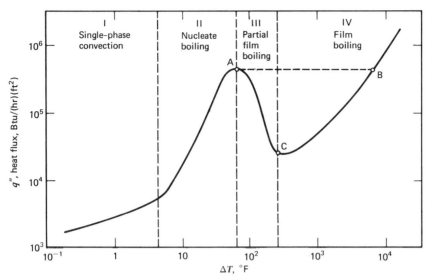

FIGURE 6-3 Heat flux versus surface temperature drop for a boiling system.

surface above the bulk saturation temperature. The amount of superheat required to initiate boiling is approximated by (Hsu, 1962):

$$T_w - T_{sat} = \frac{2\sigma T_{sat}}{H_{fg} m_g \tilde{r}_c}, \tag{6-40}$$

where σ is the surface tension, H_{fg} is the latent heat of vaporization, m_g is the vapor density, and $\tilde{r}_c$ is the radius at the mouth of an idealized conical cavity. As indicated in Fig. 6-3, water requires a few degrees superheat on a typical surface before boiling can be initiated.

In region II, where nucleate boiling takes place, the heat transfer is improved, since the formation and detachment of bubbles from the surface causes increased turbulence, which more effectively mixes the hot fluid at the wall surface with the cooler bulk of the fluid. The larger heat transfer coefficient that results is indicated by the steeper slope of the boiling curve in region II.

Point A on Fig. 6-3, which separates regions II and III, is of particular significance. It corresponds to the departure from nucleate boiling (or DNB). At this point the density of bubbles becomes so large that they coalesce, forming a vapor film that blankets the heated surface. Heat transfer must then take place by a combination of conduction and radiation across the vapor film. Neither of these two processes involved in film boiling is very effective; hence the heat flux decreases precipitously, even as the temperature difference is increasing. In region III the film blanket is not stable, giving rise to violent fluctuations between nucleate and film boiling. If the temperature difference exceeds that at point C, referred to as the Leidenfrost point, the liquid will no longer wet the wall. As a result, in region IV a stable film is formed, and the heat transfer is entirely by film boiling.

The significance of DNB is made clearer if one considers the boiling crisis that takes place if the power input to the heater is slowly increased until the critical heat flux q''_{cr}, corresponding to point A, is slightly exceeded. At this point a vapor blanket is formed, which results in near complete insulation of the heater element from the pool. The heater temperature than rapidly rises until a new equilibrium can be established between power input and heat transfer. The equilibrium does not occur until point B is reached, at which time the heat flux into the pool is once again equal to q''_{cr}. At this point the heat transfer across the film is predominantly by radiation, and the heater surface may be so hot that it melts or otherwise fails.

The analogy between the heater in the pool and a reactor fuel element cooled by water is obvious. In the flow boiling that takes place in a nuclear reactor, the heat flux versus temperature drop curve is similar to that in Fig. 6-3, even though the fuel element is cooled by forced convection under high-pressure conditions. As indicated earlier, nominal errors can be tolerated in the estimate of forced convection and nucleate boiling heat transfer coefficient, because they do not lead to excessive errors in the fuel element temperatures. It is imperative, however, to have available a means for determining a lower bound on the critical heat flux q''_{cr} that is extremely reliable under all operating conditions. For only then can the margin of safety between the heat fluxes occurring in the reactor and the corresponding critical heat fluxes be known. The minimum critical heat flux ratio (i.e., the minimum value of the critical heat flux to the actual heat flux anywhere in the core) plays a central role in water-reactor operations. In pressurized-water reactors it is often referred to as the departure from nucleate boiling ratio, or DNBR, and in boiling-water reactors it is called the minimum critical heat flux ratio, or MCHFR.

In order to predict accurately the critical heat flux under the flow boiling conditions found in water-cooled reactors, many parameters must be taken into account: pressure, temperature, flow rate, hydraulic diameter, and upstream power distribution, to name a few. Moreover, the critical heat flux has not been obtainable with sufficient accuracy from theoretical arguments; therefore the use of semiempirical correlations is relied on exclusively. To predict q''_{cr} with sufficient accuracy, many different correlations have been used, each quite lengthy, and often corresponding to a fairly narrow range of conditions. These correlations are not reproduced here, but rather the reader is referred to the extensive review of critical heat flux correlations by

Tong (1972). Qualitatively, the critical heat flux tends to increase with the mass flow rate. In saturated (sometimes called bulk) boiling, the critical heat flux exhibits a broad maximum in pressure at about 1000 psi. The critical heat flux increases with the amount of subcooling, defined as the difference $T_{sat} - T_c$ between the saturation and the bulk coolant temperature.

The dependence of the boiling crisis mechanism on the conditions that exist in the flow channel is illustrated by comparing the situations in pressurized-water reactors and boiling-water reactors as the heat flux rises to the critical value (Tong, 1972). The flow patterns are shown in Fig. 6-4. In a pressurized-water reactor the bulk coolant temperature is subcooled. Boiling therefore occurs in a bubbly boundary layer flow parallel to the hot cladding wall, where the local coolant temperature has risen sufficiently above the saturation temperature. A liquid core flows through the center of the channel where the temperature remains below the boiling point. When an excessive heat flux causes the bubbly boundary layer to separate from the wall, a pocket of stagnant vapor forms. The high heat flux at the surface causes the stagnant fluid to evaporate and hence results in a vapor blanket along the wall. This is the flow boiling crisis most often referred to as departure from nucleate boiling.

In a boiling-water reactor, where the bulk coolant temperature is at the saturation temperature and a substantial fraction of the volume of the coolant channel is occupied by steam, an annular flow pattern exists, as

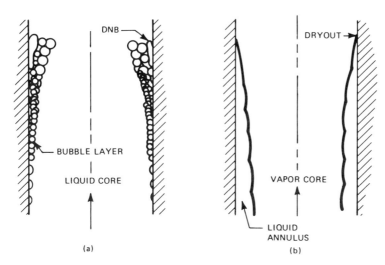

FIGURE 6-4 Boiling crisis mechanisms for PWR and BWR flow patterns. (*a*) PWR. (*b*) BWR. From L. S. Tong, *Boiling Crising and Critical Heat Flux,* U.S. Atomic Energy Commission, TID-25887, 1972.

shown in Fig. 6-4*b*. A liquid annulus flows along the cladding wall, with a vapor core flowing at a higher velocity through the center of the coolant channel. The boiling crisis in this flow regime is normally referred to as dry-out, since if an excessive heat flux causes the liquid film to become sufficiently thin, a dry spot forms on the cladding surface. The cladding temperature then rapidly rises, because of the poorer heat transfer characteristic of the steam vapor.

In a nuclear reactor a boiling crisis may be caused not only by an increase in the heat flux, but also by a number of changes in coolant conditions that may occur under accident conditions. A sufficient flow starvation or increased coolant inlet temperature leads to a decreased critical heat flux, and eventually a boiling crisis, whether of the vapor blanketing type shown in Fig. 6-4*a* or film dry-out as in 6-4*b*. If a sudden depressurization of the coolant channel should take place, the saturation temperature would drop sharply, and the bulk coolant temperature would remain about constant. Thus the coolant that was previously subcooled or at saturation temperature would suddenly become highly superheated, and a boiling crisis would take place.

Whether a boiling crisis is caused by excessive power production, flow starvation, depressurization, or any other cause, once it has occurred the cladding surface undergoes a severe increase in temperature, as indicated qualitatively by the transition from point A to point B in Fig. 6-3. Once the fuel element surface has reached these excessive temperatures, it becomes difficult to reestablish effective heat transfer, since water does not rewet the surface until its temperature first has been reduced below the Leidenfrost temperature, indicated as point C in Fig. 6-3. This reduction in temperature must be accomplished by radiation, vapor convection, and/or film boiling before the fuel elements can be quenched. The problem of rewetting overheated fuel elements occupies an important role in the behavior of the emergency core-cooling systems for water-cooled reactors discussed in Section 8-4.

Boiling in Sodium

In sodium-cooled fast reactors, boiling does not occur except under severe accident conditions. Even at the low operating pressures of these systems the saturation temperature of sodium is at about 1620°F, whereas the maximum coolant temperature is in the neighborhood of 1200°F. Thus a subcooled margin in excess of 400°F is present. Moreover, because the system is near atmospheric pressure, the danger of a sudden decrease in the saturation temperature because of a depressurization accident does not exist. When compared to water-cooled systems, the initiation of boiling in sodium is also hindered by the high surface tension and low Prandtl number.

These effects are illustrated in Fig. 6-5 by comparing the temperature profiles in a water- and a sodium-cooled system at the initiation of boiling. The larger surface tension results in the necessity for the sodium along the cladding surface to be superheated to temperatures substantially in excess of the saturation temperature. In operating systems, 30 to 50°F is likely to be required, although superheats as large as 500°F may be possible (Graham, 1971). The small Prandtl number, resulting from the high thermal conductivity in sodium, result in a smaller temperature gradient near the cladding surface than in the case of water; this, in turn, creates a flatter temperature profile across the channel, as indicated in Fig. 6-5. Thus, in contrast to water-cooled reactors, it is highly unlikely that sufficient superheat could be reached to initiate boiling at the cladding surface while the bulk sodium temperature is still substantially subcooled.

The same characteristics of sodium-cooled systems that tend to impede the initiation of coolant boiling contribute to a highly unstable situation should boiling be initiated. Under the flow and heat flux conditions found in operating fast reactors, there is no stable nucleate boiling region corresponding to that in water reactors. Rather, the initiation of boiling leads directly to a different form of boiling crisis, as may be seen from the following considerations.

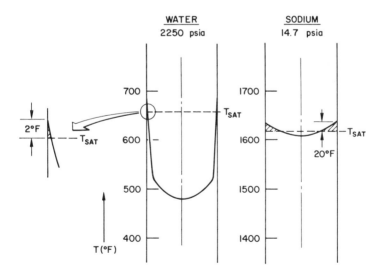

FIGURE 6-5 Comparative energy storage in water and sodium reactor channels. From D. Okrent and H. K. Fauske, "Lecture Background Notes on Transient Sodium Boiling and Voiding in Fast Reactors," Argonne National Laboratory Report, ANL/ACEA-101, 1972.

As indicated by Fig. 6-5, the large surface tension and small Prandtl number of sodium lead to a situation in which most of the sodium in the coolant channel is superheated before boiling is initiated. Hence, once a bubble does appear, all the superheated sodium may be instantaneously transformed into vapor, since thermodynamically, there is no need for any additional heat transport. The behavior following bubble nucleation is governed to a great extent by the large ratio of vapor to liquid specific volumes that exists in the low-pressure environment of a liquid-metal-cooled reactor. In contrast to water reactors, where many bubbles nucleate and grow, sodium boiling results in the formation of one or at most a few bubbles. The bubble formation results in a vapor that requires a much larger volume if it is to be at equilibrium pressure with the surrounding liquid. Thus high pressure is generated in the vapor, and the single bubble expands rapidly, pressurizing the coolant channel, and hence suppressing the nucleation of other bubbles. The single bubble becomes a high-pressure slug of vapor, which expels the liquid sodium from the ends of the coolant channel, typically within milliseconds (Okrent and Fauske, 1972).

This explosive expulsion results in a nearly instantaneous loss of flow through the coolant channel, although a thin film of sodium may momentarily adhere to much of the channel cladding surface and provide some cooling. Once the bubble has expanded outside of the core, the vapor may begin to condense and cause the bubble to collapse. Liquid sodium could then reenter the core, but this would be likely to give rise to a chugging motion in which the vapor bubble alternately grows and collapses until fuel failure results from the reduced heat transfer (Graham, 1971).

The fuel heat transfer considerations associated with the reentry of sodium into the core are likely to be overshadowed by the reactivity effects of the expulsion. As discussed in Section 4-4, the sodium void coefficient in the interior of a large fast reactor core will be positive, and the coolant expulsion will result in a rapid reactivity increase. Unless the coolant voiding can be confined locally to one or at most a few fuel assemblies, the coolant expulsion may give rise to a destructive superprompt-critical power burst.

6-3 TRANSIENT HEAT TRANSFER

The accurate description of the temperature distributions that appear in fuel elements under accident conditions requires that the time-dependent heat conduction equation be solved numerically, using a digital computer. The fuel element models incorporated into a wide variety of safety-related computer codes include detailed finite differencing of both space and time variables, and these often include provisions for temperature-dependent thermal properties, changes of phase, and other physical effects. For present purposes, we use simple, albeit less accurate, lumped parameter models to gain physical insight into the transient thermal behavior of fuel elements, while avoiding the introduction of the extensive mathematical apparatus required for finite difference solutions. The reader is referred to any one of the texts that cover the numerical solutions of such equations (Myers, 1971; Kreith, 1965).

Singly Lumped Parameter Model

A whole-core lumped parameter model for thermal transients is introduced in Section 5-1. In this model the core-averaged thermal behavior is described in terms of the core mass, heat capacity and steady-state thermal resistance. With the steady-state analysis of the preceding section completed, we are now able to evaluate these model parameters in terms of the fuel element geometry and material properties. The core-averaged fuel-element behavior is given in Eq. (5-12) by

$$\frac{d\bar{T}_{fe}}{dt} = \frac{P(t)}{M_{fe}c_{fe}} - \frac{\bar{T}_{fe} - \bar{T}_c}{\tau},\tag{6-41}$$

where M_{fe} is the core mass, c_{fe} is the core-averaged specific heat per unit mass and τ is the thermal time constant. The time constant is given by

$$\tau = RM_{fe}c_{fe},\tag{6-42}$$

where R is the core thermal resistance. Neglecting the effects of structural material and control poisons, the core mass may be written as

$$M_{fe} = NH(M_f' + M_{cl}'),\tag{6-43}$$

where the core consists of N identical fuel elements each of length H. The fuel and cladding masses per unit length are given by Eq. (6-29). The specific heat per unit mass of the fuel element is obtained by taking a mass average over the elements:

$$c_{fe} = \frac{M_f'c_f + M_{cl}'c_{cl}}{M_f' + M_{cl}'}.\tag{6-44}$$

To evaluate the thermal time constant we first note that R is related to R_{fe}', the fuel element thermal resistance per unit length, by Eq. (1-108): $R = R_{fe}'/NH$. We have evaluated R_{fe}' in terms of the fuel element properties, and therefore we may use this expression for R, along with Eq. (6-44), to obtain

$$\tau = R_{fe}'(M_f'c_f + M_{cl}'c_{cl}),\tag{6-45}$$

or, using Eq. (6-31),

$$\tau = M'_f c_f (R'_g + R'_c) + M'_{cl} c_{cl} R'_c. \tag{6-46}$$

Equation (6-41) may be expressed entirely in terms of the characteristics of an average fuel element by observing that a core-averaged linear heat rate may be defined by $\bar{q}' = P/NH$. Therefore, utilizing Eq. (6-43), we have

$$\frac{d\bar{T}_{fe}}{dt} = \frac{\bar{q}'(t)}{M'_{fe} c_{fe}} - \frac{(\bar{T}_{fe} - T_c)}{\tau}, \tag{6-47}$$

where

$$M'_{fe} = M'_f + M'_{cl}. \tag{6-48}$$

Typical values for τ, $\bar{q}'$, and $M'_{fe} c_{fe}$ are given in Table 6-1 for several reactor types. Note that the time constants are typically of the order of a few seconds, with the exception of the larger value for the high-temperature gas-cooled reactor (HTGR). In this system the heat must be transported through a large mass of graphite, as may be seen from Fig. 1-12e, which means the effective mass and thermal resistance per unit length of the fuel element are substantially larger than for the other cores. A wide range of values is seen to exist in the core-averaged linear heat rates, with the largest appearing in the fast reactor designs. A comparison of how rapidly the cores would overheat in the absence of cooling may be estimated from the quantity $\bar{q}'/M'_{fe} c_{fe}$, the adiabatic heat-up rate at full power. The most severe situation is seen to exist in the fast reactors, while the heat-up rate for the HTGR is the slowest, because of the large heat capacity of the graphite moderator. The quantity $\bar{q}'/M'_{fe} c_{fe}$ assumes that the reactor remains at full power during the course of the heat-up. The adiabatic heat-up from the decay heat following reactor shutdown would amount to a few percent of this value.

Doubly Lumped Parameter Model

The singly lumped parameter model previously discussed is useful in predicting average fuel element behavior, and in the form included in Section 5-2 it provides a simple model for thermal-hydraulic reactivity feedback. Equation (6-47) does not, however, give any indication of the temperature distribution within the fuel element. These distributions are particularly important for the analysis of individual fuel elements to determine the likelihood of failure as a result of a particular accident, and to understand the thresholds and modes of failure. Better estimates of the local fuel element behavior can be gained by proceeding from the singly lumped parameter model to the following doubly lumped parameter model.

To derive the doubly lumped parameter approximation, we first write the thermal energy balance separately per unit length of the fuel region and of the cladding region. We have, respectively,

$$M_f' c_f \frac{d\tilde{T}_f(t)}{dt} = q'(t) - q_g'(t) \qquad (6\text{-}49)$$

and

$$M_{cl}' c_{cl} \frac{d\tilde{T}_{cl}(t)}{dt} = q_g'(t) - q_c'(t), \qquad (6\text{-}50)$$

where $q_g'(t)$ is the rate per unit length at which heat is transferred from fuel to cladding and $q_c'(t)$ is the corresponding rate between the outer cladding surface and the coolant.

To eliminate $q_g'(t)$ and $q_c'(t)$ from the model, we require that the fuel and cladding energy balances reduce to the required steady-state temperature relationships when the time derivatives are set equal to zero. The steady-state relationships between the radial averages of the fuel, cladding, and coolant temperatures are given by Eqs. (6-21) and (6-23). Since $q' = q_g' = q_c'$ under steady-state conditions, these equations will be satisfied if we take

$$q_g'(t) = \frac{1}{R_g'} [\tilde{T}_f(t) - \tilde{T}_{cl}(t)] \qquad (6\text{-}51)$$

and

$$q_c'(t) = \frac{1}{R_c'} [\tilde{T}_{cl}(t) - T_c(t)]. \qquad (6\text{-}52)$$

Substituting these approximations into Eq. (6-49) and (6-50), we have

$$M_f' c_f \frac{d\tilde{T}_f(t)}{dt} = q'(t) - \frac{1}{R_g'} [\tilde{T}_f(t) - \tilde{T}_{cl}(t)] \qquad (6\text{-}53)$$

and

$$M_{cl}' c_{cl} \frac{d\tilde{T}_{cl}(t)}{dt} = \frac{1}{R_g'} [\tilde{T}_f(t) - \tilde{T}_{cl}(t)] - \frac{1}{R_c'} [\tilde{T}_{cl}(t) - \tilde{T}_c(t)], \qquad (6\text{-}54)$$

or alternately,

$$\frac{d\tilde{T}_f}{dt} = \frac{q'(t)}{M_f' c_f} - \frac{1}{\tau_1} [\tilde{T}_f(t) - \tilde{T}_{cl}(t)] \qquad (6\text{-}55)$$

and

$$\frac{d\tilde{T}_{cl}}{dt} = \frac{1}{\tau_2} [\tilde{T}_f(t) - \tilde{T}_{cl}(t)] - \frac{1}{\tau_3} [\tilde{T}_{cl}(t) - T_c(t)], \qquad (6\text{-}56)$$

where the new time constants are defined by

$$\tau_1 = M_f' c_f R_g', \tag{6-57}$$

$$\tau_2 = M_{cl}' c_{cl} R_g', \tag{6-58}$$

and

$$\tau_3 = M_{cl}' c_{cl} R_c'. \tag{6-59}$$

The initial conditions on Eqs. (6-55) and (6-56) are obtained by assuming that for $t = 0$; the reactor is operated at steady state with $q' = q_0'$. Setting the time derivatives equal to zero, we have

$$\tilde{T}_f(0) = T_c(0) + (R_g' + R_c')q_0' \tag{6-60}$$

and

$$\tilde{T}_{cl}(0) = T_c(0) + R_c' q_0'. \tag{6-61}$$

Overpower Transients

Some general characteristics for behavior of the fuel and cladding temperatures under accident conditions of various types may be obtained by solving Eqs. (6-55) and (6-56) for highly idealized situations. First, we consider the case where the reactor that has been operating at steady state goes on a positive period. For simplicity, we approximate the power production by

$$q'(t) = q_0' e^{t/\mathcal{T}}, \tag{6-62}$$

where $\mathcal{T}$ is the period. If the coolant temperature remains constant, it is easily shown that after the initial transients die out, the asymptotic behavior of both the fuel and cladding temperature is proportional to $\exp(t/\mathcal{T})$. We therefore assume that Eqs. (6-55) and (6-56) have solutions of the form

$$\tilde{T}_f(t) \simeq T_c + A_f e^{t/\mathcal{T}} \tag{6-63}$$

and

$$\tilde{T}_{cl}(t) \simeq T_c + A_{cl} e^{t/\mathcal{T}}. \tag{6-64}$$

Substituting these expressions into Eqs. (6-55) and (6-56), and solving for A_f and A_{cl}, we obtain

$$\tilde{T}_f(t) = T_c + \frac{\left[1 + \left(\dfrac{1}{\tau_2} + \dfrac{1}{\tau_3}\right)\mathcal{T}\right]}{\left[1 + \left(\dfrac{1}{\tau_1} + \dfrac{1}{\tau_2} + \dfrac{1}{\tau_3}\right)\mathcal{T} + \dfrac{\mathcal{T}^2}{\tau_2 \tau_3}\right]} \frac{\mathcal{T}}{M_f' c_f} q_0' e^{t/\mathcal{T}} \tag{6-65}$$

and

$$\tilde{T}_{cl}(t) = T_c + \frac{1}{\left[1 + \left(\dfrac{1}{\tau_1} + \dfrac{1}{\tau_2} + \dfrac{1}{\tau_3}\right)\mathcal{T} + \dfrac{\mathcal{T}^2}{\tau_1\tau_3}\right]} \frac{\mathcal{T}^2}{\tau_2 M'_f c_f} q'_0\, e^{t/\mathcal{T}}. \quad (6\text{-}66)$$

These solutions have distinctly different behaviors, depending on whether the period is large or small compared to the thermal time constants τ_1, τ_2, and τ_3. For mild overpower transients, in which the reactor period is long, we can assume that $\mathcal{T} \gg \tau_1, \tau_2, \tau_3$. Therefore the preceding solutions reduce to

$$\tilde{T}_f(t) \simeq T_c + (\tau_2 + \tau_3)\frac{1}{M'_f c_f} q'_0\, e^{t/\mathcal{T}} \quad (6\text{-}67)$$

and

$$\tilde{T}_{cl}(t) \simeq T_c + \frac{\tau_1\tau_3}{\tau_2 M'_f c_f} q'_0\, e^{t/\mathcal{T}}. \quad (6\text{-}68)$$

Using the definition of the thermal time constants, we find that the fuel and cladding obey the quasi-steady-state relationships

$$\tilde{T}_f(t) \simeq T_c + (R'_g + R'_c)\frac{M'_{cl} c_{cl}}{M'_f c_f} q'(t) \quad (6\text{-}69)$$

and

$$\tilde{T}_{cl}(t) \simeq T_c + R'_c q'(t). \quad (6\text{-}70)$$

If more detailed finite-difference calculations are made, the temperature distribution for such slow power transients is found to resemble closely the steady-state distribution. Figure 6-6a is indicative of such a situation.

If a power transient with a very short period, such as might occur in a superprompt-critical excursion, is considered, we have $\mathcal{T} \ll \tau_1, \tau_2, \tau_3$. Equations (6-65) and (6-66) then reduce to

$$\tilde{T}_f(t) \simeq T_c + \frac{\mathcal{T}}{M'_f c_f} q'_0\, e^{t/\mathcal{T}} \quad (6\text{-}71)$$

and

$$\tilde{T}_{cl}(t) \simeq T_c + \frac{\mathcal{T}^2}{\tau_2 M'_f c_f} q'_0\, e^{t/\mathcal{T}} \quad (6\text{-}72)$$

Noting from Eq. (6-62) that

$$\int_0^t q'(t)\, dt \simeq \mathcal{T} q'_0\, e^{t/\mathcal{T}} \quad (6\text{-}73)$$

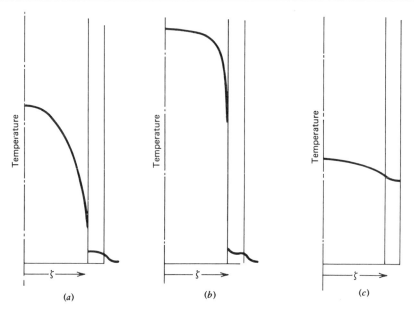

FIGURE 6-6 Fuel pin temperature profiles for transient conditions. (*a*) Quasi-steady state. (*b*) Nuclear excursion. (*c*) Cooling failure.

when $t \gg \mathcal{T}$, we have

$$\tilde{T}_f(t) \simeq T_c + \frac{1}{M'_f c_f} \int_0^t q'(t)\, dt \qquad (6\text{-}74)$$

and

$$\tilde{T}_{cl}(t) \simeq T_c + \left(\frac{\mathcal{T}}{\tau_2}\right) \frac{1}{M'_f c_f} \int_0^t q'(t)\, dt. \qquad (6\text{-}75)$$

Thus we see that if the period becomes short compared to τ_2, the time required to transport heat from fuel to cladding, the fuel behaves approximately adiabatically, and only a small fraction, $\mathcal{T}/\tau_2$, of the heat produced is stored in the cladding. In this situation more detailed treatment of the spatial dependence would indicate that the heat has very little time to move, and thus there results a nearly flat temperature distribution within the fuel region. This situation is indicated by the temperature distribution shown in Fig. 6-6*b*.

Cooling-Failure Transients

Thus far we have assumed that both the coolant temperature T_c and the coolant heat transfer coefficient h_c remain constant during the course of the transient. Since h_c is dependent on the mass flow rate and T_c is dependent on both mass flow rate and core inlet temperatures, cooling-failure accidents in particular are likely to cause major changes in these quantities. In coolant-induced accidents that progress slowly, such as a slow rise in inlet temperature or slow flow ·starvation, the quasi-state relationships given by Eqs. (6-69) and (6-70) roughly describe the situation, provided time-dependent values of $T_c(t)$ and $h_c(t)$ are used.

In faster transients it is necessary to account more accurately for the changes in the coolant conditions. In gas-cooled systems the large cladding-to-coolant temperature drops may necessitate the use of explicit representation of the heat transfer coefficient in terms of the mass flow rate even for rather mild flow transients. Because they tend to be much smaller in liquid-cooled systems, changes in the cladding-to-coolant temperature drops induced by variation in mass flow rate tend to be less significant in determining cladding failure than in gas-cooled systems. As a result, it may be acceptable to neglect variations in h_c and therefore in τ_3 during flow transients, provided a boiling crisis does not occur. In such situations the fuel element temperature profile may closely resemble that at steady state, in which case the singly lumped parameter model provides a qualitatively correct estimate of the fuel conditions.

In the event that excessive heat flux or deteriorated coolant conditions cause a boiling crisis, the heat transfer coefficient h_c may be expected to decrease by orders of magnitude, as indicated by the discussions of the preceding section. Should this occur, near-perfect insulation between fuel element and coolant occurs and the singly lumped parameter model becomes completely inadequate, since much of the cladding overheating results from the redistribution of heat within the fuel element.

The effect of a boiling crisis on the redistribution of heat within the fuel element may be illustrated by a simple example. Suppose that at $t = 0$ a reactor is instantaneously shut down. Then, if decay heat is neglected, $q'(t) = 0$ for $t > 0$. We assume that at the same instant of time a boiling crisis causes a complete cooling failure such that $h_c = 0$ for $t > 0$. It may be seen from Eqs. (6-24) and (6-59) that $\tau_3 = \infty$ for $t > 0$. With these conditions Eqs. (6-55) and (6-56) reduce to

$$\frac{d\tilde{T}_f(t)}{dt} = -\frac{1}{\tau_1}[\tilde{T}_f(t) - \tilde{T}_{cl}(t)] \qquad (6\text{-}76)$$

and

$$\frac{d\tilde{T}_{cl}(t)}{dt} = \frac{1}{\tau_2}[\tilde{T}_f(t) - \tilde{T}_{cl}(t)], \qquad (6\text{-}77)$$

where the initial conditions are given by Eqs. (6-60) and (6-61).

The nature of the solution of these equations is easily seen as follows. First, multiply Eqs. (6-76) and (6-77) by τ_1 and τ_2, respectively, and then add the two equations to obtain

$$\frac{d}{dt}[\tau_1 \tilde{T}_f(t) + \tau_2 \tilde{T}_{cl}(t)] = 0. \tag{6-78}$$

Using the definitions of τ_1 and τ_2, this is seen to be equivalent to stating that the average fuel pin temperature

$$\tilde{T}_{fe} = \frac{M'_f c_f \tilde{T}_f(t) + M'_{cl} c_{cl} \tilde{T}_{cl}(t)}{M'_f c_f + M'_{cl} c_{cl}} \tag{6-79}$$

is time independent. If we now subtract Eq. (6-77) from Eq. (6-76), we obtain

$$\frac{d}{dt}[\tilde{T}_f(t) - \tilde{T}_{cl}(t)] = -\left(\frac{1}{\tau_1} + \frac{1}{\tau_2}\right)[\tilde{T}_f(t) - \tilde{T}_{cl}(t)]. \tag{6-80}$$

This equation has the solution

$$\tilde{T}_f(t) - \tilde{T}_{cl}(t) = [\tilde{T}_f(0) - \tilde{T}_{cl}(0)]e^{-t/\tau'}, \tag{6-81}$$

where

$$\tau' = \frac{\tau_1 \tau_2}{\tau_1 + \tau_2}. \tag{6-82}$$

The preceding results have some important implications. Even if decay heat can be ignored, and the reactor is shut down at the same time that the fuel element becomes insulated from the coolant, a major temperature transient is introduced by the redistribution of stored heat between fuel and cladding. Although the average fuel element temperature remains constant, the cladding temperature rises, and the fuel temperature falls until thermal equilibrium between the two is reached. Whereas the initial temperature distribution is dependent only on the thermal resistance between fuel, cladding, and coolant, the final equilibrium temperature is strongly dependent on the relative heat capacities as well. The temperature profile collapses with a time constant τ', which is therefore a measure of how fast the cladding temperature will rise in such an accident; estimated values of τ' are given for several reactor types in Table 6-1.

In Fig. 6-7 are plotted the fuel and cladding temperatures for this hypothesized behavior. The data used is for a boiling-water reactor, but the curves qualitatively characterize most ceramic fuels with metal cladding. Initially, the cladding and coolant temperature are close together, with the average fuel temperature being much higher due to the poor conductivity of

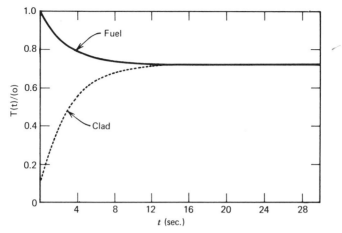

FIGURE 6-7   Fuel pin temperature response to an instantaneous loss of the heat source and cooling capacity. Adapted from E. E. Lewis, "A Transient Heat Conduction Model for Reactor Fuel Elements," *Nucl. Eng. Des.,* **15**, 33–40 (1971).

the UO_2. Because the fuel has a substantially larger specific heat than the cladding, the equilibrium temperature is much closer to that of the fuel than of the cladding. The temperature at which metals lose their strength or otherwise fail is likely to be well below the operating temperature of the ceramic fuels. Thus the temperature transient shown would likely bring the metal cladding to a point well above its failure temperature.

The redistribution of heat from fuel to cladding tends to dominate the early stages of cooling-failure accidents. Provided the reactor is shut down, moreover, it is the interaction of heat redistribution and decay heat production that determines the nature of the thermal transients. The coupling of these effects is illustrated in Fig. 6-8, where temperature traces obtained from a doubly lumped parameter model are plotted. In all the curves it is assumed that the reactor is tripped at $T = 0$, but power production continues at reduced levels because of decay heat production. The boiling crisis and complete insulation are assumed to take place simultaneously with the reactor shutdown in the curves labeled $T = 0$. In this situation the rapid cladding temperature increase is similar to that of Fig. 6-7. Now, however, thermal equilibrium with the fuel is not achieved. Rather, the ongoing production of decay heat results in the fuel temperature remaining somewhat above that of the cladding, with both cladding and fuel temperatures increasing with time.

If the reactor protection system can detect the cause of the cooling failure and shut down the reactor even a few seconds before the time of the boiling crisis T, the character of the thermal transient is dramatically altered. As

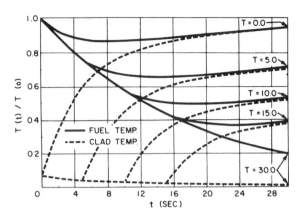

FIGURE 6-8 Fuel pin behavior during idealized loss-of-coolant accidents with boiling crises at 0, 5, 10, 15, and 30 seconds. Adapted from E. E. Lewis, "A Transient Heat Conduction Model for Reactor Fuel Elements," *Nucl. Eng. Des.*, **15**, 33–40 (1971).

indicated by Fig. 6-8, the time delay permits a part of the heat stored in the fuel to be dumped to the coolant. Thus when the boiling crisis does occur, there is less stored heat to be redistributed from fuel to cladding, and the average temperature toward which fuel and cladding temperatures converge is substantially lower. The decay heat eventually causes fuel and cladding to reach excessive temperatures. Now, however, these high temperatures are not reached until a substantially later time, providing additional time for emergency measures to be taken to reestablish cooling.

After the heat redistribution in the fuel element has taken place and the fuel element heat-up has begun, the radial temperature profile has the properties indicated in Fig. 6-6c. In such situations where the cladding temperature is only slightly less than that of the fuel, failure and even melting of metal cladding are likely to occur before the ceramic fuel approaches its melting point. Once again, although the qualitative behavior in HTGR's is similar, the quantitative results are quite different, since not only is the redistribution time constant τ' much larger, but the graphite moderator can withstand much higher temperatures than metal cladding.

6-4 FAILURE MECHANISMS

The steady-state and transient temperature distributions discussed in the preceding sections provide a basis for understanding the modes by which fuel elements may fail. Fuel failure mechanisms arise from interactions of the thermal-mechanical properties of the fuel and cladding, the buildup and migration of fission products, radiation damage, temperature-sensitive chemical reactions and a host of other phenomena whose relative importance varies not only between classes of fuel elements but also between accident types. For clarity of discussion, the present treatment is limited to the two most widely used classes of fuel elements: mixed oxide pellets consisting of UO_2 or UO_2 and PuO_2, clad in metal tubing, and coated-particle carbide fuels. Discussions of both the metallic and dispersion fuels popular in early reactors and of recent developments in carbide and nitride fuels may be found elsewhere (Gurinsky and Issevow, 1973; Washburn and Scott, 1971).

The analysis of fuel failure is facilitated by first placing considerable emphasis on the steady-state behavior. The reasons for this are twofold. First, failure modes under accident conditions are strongly dependent on the state of the fuel elements at the time of accident initiation. Second, many of the failure modes that result from driving fuel too hard under steady-state conditions are also applicable to transient situations, particularly slower transients such as those pictured in Fig. 6-6a, where the fuel element temperature profile does not deviate greatly from steady-state behavior. We consider first the steady-state behavior of metal-clad oxide fuel. For these fuel elements we then discuss failure under conditions of severe overpower transients having temperature profiles resembling Fig. 6-6b, and for cooling failure accidents, with temperature profiles as in Fig. 6-6c. We conclude with a brief outline of the properties of coated-particle carbide fuel.

Mixed-Oxide Fuel

Fuel pellets consisting of UO_2 and/or PuO_2 have a relatively low thermal conductivity and therefore exhibit large temperature differences between fuel centerline and surface. The large temperature gradients and high centerline temperature result in fuel cracking and restructuring. Under steady-state conditions, Eq. (6-26) indicates that the temperature drop across the fuel is proportional to the linear heat rate q', but independent of the fuel radius a. As a result, the temperature-dependent cracking and restructuring phenomena can be correlated to the linear heat rate without reference to the fuel dimensions, provided the fuel surface—and therefore coolant—temperature changes with increased power level are small compared to those at the fuel centerline. The effects of increasing linear heat rate and therefore of increasing interior temperature for a typical fast reactor fuel element of mixed UO_2 and PuO_2 in sintered pellets are shown in Fig. 6-9.

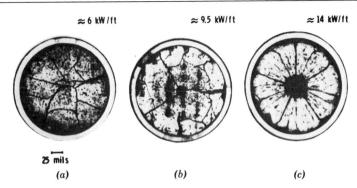

25 mils

(a)   (b)   (c)

FIGURE 6-9 Comparison of mixed oxide fuel structure with varying heat rates. Adapted from J. H. Scott, G. E. Gulley, C. W. Hunter, and J. E. Hanson, "Microstructural Dependence of Failure of Oxide LMFBR Fuel Pins," *Proc. Fast Reactor Safety Meeting*, Beverly Hills, Calif., CONF-740401, April 1974.

Cracking and Restructuring. The low conductivity of oxide fuels causes the temperature drop between fuel centerline and surface to be large even at nominal linear heat rates. From Eq. (6-6),

$$\Delta T = \frac{q'}{4\pi k_f}. \tag{6-83}$$

In the absence of geometrical constraints, the hot fuel interior would expand and the cool exterior contract relative to the density at the average fuel temperature. Because the free expansion induced by the temperature profile in the fuel is not compatible with the cylindrical shape of the pellets, however, internal thermal stresses arise. These may be determined analytically for the idealized case of an infinitely long cylinder containing a uniform heat source, provided the fuel is assumed to be linearly elastic with no external force applied (Freudenthal, 1957). It may be shown that the tangential (φ or hoop), radial (ζ), and axial (z) thermal stresses are given by

$$\sigma_\varphi = \frac{\alpha E \Delta T}{4(1-\nu)}\left[3\left(\frac{\zeta}{a}\right)^2 - 1\right], \tag{6-84}$$

$$\sigma_\zeta = \frac{\alpha E \Delta T}{4(1-\nu)}\left[\left(\frac{\zeta}{a}\right)^2 - 1\right] \tag{6-85}$$

and

$$\sigma_z = 0. \tag{6-86}$$

In these equations E, ν, and α are Young's modulus, Poisson's ratio, and the coefficient of linear thermal expansivity, respectively.

Oxide fuels are quite brittle. As a result, they tend to fracture from thermal stresses before significant plastic deformation takes place. Equations (6-84) and (6-85) indicate that the cracks are likely to originate at the outer fuel surface, where the maximum value of the hoop stress occurs. If we define σ_{cr} as the stress for which cracking is initiated, we find that cracking occurs when

$$\Delta T \geq \frac{2\sigma_{cr}(1 - \nu)}{\alpha E}, \tag{6-87}$$

or utilizing Eq. (6-83),

$$q' \geq \frac{8\pi k_f \sigma_{cr}(1 - \nu)}{\alpha E}. \tag{6-88}$$

The value of σ_{cr} may be substantially smaller than the yield stress of the oxide, since, as we discuss in Section 8-1, cracking in brittle materials normally propagates from imperfections on the surface, where local stress concentrations exceed the yield stress even while the nominal stress level is still well below the elastic limit. For this reason, σ_{cr} is often referred to as a rupture modulus. In oxide fuels such cracking occurs to relieve thermal stresses, even though the linear heat rate amounts to only a few kilowatts per foot, as indicated in Fig. 6-9a.

If the fuel centerline temperature is kept below about 1500°C, the oxide microstructure remains unchanged. However, as higher interior temperatures result from increased linear heat rates, dramatic effects take place in the fuel microstructure, as indicated by Figs. 6-9b and c. Between about 1500 and 1800°C, sintering takes place, causing grain growth and recrystallization in an equiaxed structure. If the interior temperatures rise above 1800°F, corresponding to about 9 kW/ft for the fuel element shown in Fig. 6-9b, a central void appears along with the formation of a columnar grain region. These appear because the pores in the oxide migrate up the temperature gradient, forming high-density radially oriented columnar grains. The pore migration and increased density of the columnar region cause the void region to form about the fuel centerline.

The grain growth and crystallization causes the cracks induced by thermal stress to anneal away while the fuel pin is at power. When the temperature profile collapses with reactor shutdown, however, new thermal stresses arise from the inability of the interior, restructured fuel region to contract freely. Thus thermal stress cracking, such as that visible in Figs. 6-8b and 6-9c, propagates radially from the central void region outward along the columnar grains.

Fission Product Behavior. In fresh fuel cracking, restructuring and central void formation take place over a relatively short time span, before there is a significant buildup of fission products. With the buildup of fission products, new thermal-mechanical effects appear that are closely tied to the ability of the fuel to retain its radioactive inventory. Each time a fission reaction takes place, one atom is replaced by two, and additional volume is required to accommodate the increasing number of atoms. The result is that the fuel swells as the fission product inventory builds. This swelling may be divided into two components. Fission products that remain in solid form at operating temperatures remain relatively immobile and make a small but significant contribution to the fuel swelling. The more varied and pronounced effects arise from the fission products that are in gaseous form at operating temperatures.

The gaseous fission products are insoluble in the fuel and therefore tend to form bubbles at a rate determined by their rate of diffusion through the oxide lattice to the grain boundaries, and by their mobility along the grain boundaries. The nucleation, growth, and migration of these bubbles determine both the amount of swelling in the fuel and the amount of fission gas that is released from the fuel. The behavior of the bubbles in the columnar, equiaxed, and unrestructured grain regions of the fuel is quite different (Okrent et al, 1972; Scott et al, 1974). Nearly all the fission gas formed in the columnar grain region forms bubbles at the grain boundaries, migrates up the thermal gradient, and is released to the central void. In the equiaxed region the bubbles are less mobile, and only between 25 and 50% are released. Nearly all the gaseous fission products are retained in the unrestructured regions, except for those released from the grain boundaries along which thermal stress cracking has occurred.

In oxide fuels it is not uncommon for the centerline temperature to be quite close to 2800°C, the melting temperature, in the highest power density regions of the core. As the linear heat rate increases and the temperature approaches melting, increasing fractions of the fission product gases are released from the fuel, since more of the fuel forms in columnar grains. Thermal expansion and swelling also become more pronounced as the centerline temperature increases, and if melting occurs, the fuel undergoes an approximately 10% decrease in density, causing further volume increase.

Metal Cladding

Two types of failure are considered in the analysis of cladding behavior. If the cladding ruptures it will fail to retain the gaseous fission product inventory within the fuel element. If it deforms excessively so as to block the flow of coolant, or loses its capability to support the column of fuel pellets structurally, a second, more serious failure occurs, since the ability to transport heat out of the reactor core then is seriously impeded. The discussion of fuel element behavior relevant to cladding failures is conveniently structured by first examining the sources of the forces on the cladding and then relating these to the cladding properties as they appear in the reactor environment.

The pressure on the cladding exterior is due to the reactor coolant and may range from near atmospheric, for the stainless steel cladding used in liquid-metal-cooled systems, to well over 2000 psi, for the zircaloy used in pressurized-water reactors. The pressure on the interior of the cladding arises from at least three sources: differential thermal expansion, fuel swelling, and fission gas pressure (Scott et al., 1974).

Temperature profiles such as that shown in Fig. 6-6a cause the radial expansion of the fuel to be greater than that of the cladding, even though the fuel may have the smaller coefficient of expansivity. The radial growth of the fuel region is further enhanced by the cracking and restructuring of the oxide pellets, and by the swelling that increases with irradiation time because of the accumulation of solid and gaseous fission products. The temperature increase in the cladding between zero and full power is much smaller than that in the fuel, and therefore there is a much smaller radial expansion. The net effect of differential expansion and swelling is to narrow, and in some cases close, the annular gap between the fuel pellet and cladding wall. If the gap is completely closed, the cladding must deform either elastically or plastically to accommodate the fuel expansion. If the cladding is not sufficiently ductile to accommodate the increased fuel radius, rupture occurs.

Assuming that a sufficient fuel–cladding gap is maintained to accommodate the dimensional increases in the fuel, we must consider the pressure on the internal cladding wall resulting from the buildup of fission product gas. Heuristically, the pressure on the cladding may be viewed as that of a perfect gas that occupies the central pellet voids, the fuel–cladding gap and any plenum regions located above or below the pellet stack. Let V_1, V_2, V_3 be the volumes of each of these void regions. If the pellets are not stacked so as to block all gas flow completely between void regions, we may assume that these volumes are in pressure equilibrium. We may then write

$$p = \frac{NR}{V}\bar{T},\tag{6-89}$$

where $V = V_1 + V_2 + V_3$, N is the number of moles of gas filling the voids, and R is the gas law constant. The temperature in the central void is much higher than that at either the fuel–cladding gap or the plenum regions. Thus $\bar{T}$ must be taken as the appropriate average over the void spaces:

$$\bar{T} = \frac{V}{\sum_i (V_i/T_i)}.\tag{6-90}$$

Utilizing Eq. (6-89), the effects of increased linear heat rate and of fuel exposure in increasing the pressure on the cladding are easily seen. The temperature of the central void region rises linearly with the fuel linear heat rate, thereby increasing the average temperature on the right of Eq. (6-89). Likewise, the differential expansion increases with the fuel power rating, effectively causing the gap component of V to decrease. Finally, the fission products produced with increasing fuel exposure cause the pressure to increase for the following reason: Most void regions are initially occupied by an inert gas to equalize the coolant pressure and thereby prevent cladding collapse about the fuel pellets. As fission reactions take place, some of the gaseous products are released from the columnar and equiaxed grain regions and increase N, the number of moles of gas in the void regions. Those fission products not released to the void contribute to fuel swelling, which in turn decreases the gap volume. The internal pressure may also be increased if high vapor pressure impurities, such as moisture, are trapped within the cladding, or if impurities interact with fuel or cladding constituents to produce additional gaseous material.

The maximum cladding stress is in the tangential direction. If we approximate the cladding as a thin shell, this stress may be related to the interior and cooling pressures, p and p_c, by the hoop stress formula

$$\sigma_\varphi = \frac{b}{b-a}(p - p_c), \qquad (6\text{-}91)$$

where a and b are defined in Fig. 6-1. If the internal pressure is due to the gap closure accompanying fuel expansion, cladding deformation that is sufficient to accommodate the fuel expansion relieves the stress. Small cladding deformations, however, do not relieve the stress caused by fission gas pressure, since the pressure is relieved significantly only if the cladding deformation is so large as to cause a substantial increase in the total void volume V. In the event that the cladding is brittle, failure would be by rupture, since the ultimate stress would be reached before sufficient cladding expansion could take place. For very ductile materials, sufficient pressure relief might take place, but the danger would then be that excessive cladding ballooning would block coolant flow through the core or hinder the structural function of the cladding.

Whether the cladding pressure arises from mechanical fuel contact or internal gas pressure, the cladding temperature must be taken into account in determining whether failure takes place. Since both the cladding yield and ultimate stress deteriorate at elevated temperatures, excessive cladding temperature contributes to the mechanical failure. This is illustrated by the stress temperature curves shown in Fig. 6-10 for a stainless steel used as a fast reactor cladding. For this cladding, properties are seriously deteriorated

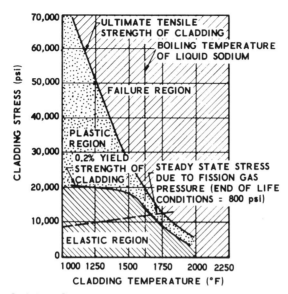

FIGURE 6-10 Stainless Steel cladding properties versus temperature. From C. A. Anderson, Jr., R. A. Markely and J. F. P. Henson, "Fuel Rod Design Basis," *Proc. Fast Reactor Fuel Technology*, New Orleans, La., April 1971. Used by permission.

by the time the coolant boiling temperature is reached. Included in the figure is an indication of stress increase with temperature caused by fission gas pressure at the end of life for the particular fuel element. It is seen that failure of the cladding would occur at 1800°F, well below the cladding melting point.

Figure 6-10 does not indicate the presence of several other phenomena that cause deterioration of cladding properties with fuel exposure and that at the same time are enhanced with increasing temperatures (Okrent et al., 1972). Possibly the most widely studied of these is the swelling of stainless steel cladding that takes place under irradiation of fast neutrons. At high temperatures, creep, or viscoelastic behavior, may also cause excessive deformations, even though the yield stress is not exceeded. Radiation-induced creep may be a particular problem at high temperatures and long irradiation times. Finally, high temperature increases the rate of fuel-cladding and cladding-coolant chemical reactions, which in some cases may degrade the mechanical behavior of the cladding.

Transient Behavior

The discussion of metal-clad oxide fuels under the transient conditions induced by reactor accidents is made difficult by the wide variety of circumstances that may be encountered during any accident and by the complexity of the interacting physical phenomena. Different accident types lead to vastly different fuel element temperature profiles, as indicated by Fig. 6-6. Moreover, for any one accident the fuel element behavior is strongly dependent on the power density, the burnup to which the element has been subjected, the coolant conditions, possible fuel–coolant or cladding–coolant chemical interactions, and a host of other variables. There are some general considerations, however, that tend to structure the discussion of this variety of phenomena. In major, albeit improbable, accidents the primary concern is that the cladding maintain its integrity in the sense that it neither fracture, causing the fuel element coolant geometry to be lost, nor deform excessively, causing coolant flow to be hindered. So long as the fuel does not fail in the preceding sense, some fuel melting, cladding cracking, chemical reaction, or other damage may be tolerated during major accidents, since with the exception of the release of some fraction of the gaseous fission product inventory, the function of the element as a fission product barrier is not destroyed.

The nature of the thermal-mechanical transients that may lead to structural failure of the cladding differs significantly between accident types. This may be illustrated by considering accidents associated with the quasi-static, nuclear excursion and cooling-failure temperature profiles of Fig. 6-6. As stated earlier, the failure mechanism in slow transients often apears as aggravation of the temperature-dependent phenomena present under steady-state conditions. Thus cladding failure under the quasi-static transient would be expected to arise from increased transient pressure on the cladding, caused by the increased fuel temperature, coupled with the deterioration of cladding strength with excessive temperature. The relative importance of cladding stress level and temperature tends to vary with the type of accident. In fast overpower transients, represented by Fig. 6-6b, the energy input to the fuel causes excessive hoop stress in the cladding, causing it to fail even though the cladding is not at an excessive temperature. The extreme example of this is the classic Bethe–Tait analysis, discussed in Section 5-6, where the fuel is vaporized, causing such high fuel pressures to be generated that the mechanical strength of the cladding may be neglected altogether. At the other end of the spectrum, a cooling-failure transient, such as that seen in Fig. 6-6c, may result in some fuel region cooldown and possibly even a reduction in cladding stress. In this case, however, the cladding may reach such high temperatures that it loses its strength, to the point where structural failure, or even melting, takes place. The extreme example of this situation is in some simplified cooling-failure accident analysis where it is assumed that the cladding melts and falls from the core before the fuel reaches excessive temperatures.

Overpower Transients. Overpower transients, and particularly the fuel behavior induced by superprompt-critical power bursts, have received substantial attention in conjunction with both thermal and fast reactor projects. Since such transients may be completed before appreciable heat transfer can take place, probably the most important parameter in determining fuel damage is the energy input per unit mass of fuel. Figure 6-11 indicates the effect of various energy inputs to fresh uranium dioxide fuel clad in zircaloy and placed in water coolant. The data on failure and metal–water reactions between cladding and coolant are accumulated from extensive testing in the TREAT reactor (Baker and Liimalainen, 1969). Relative to room temperature, fuel melting takes place between 200 and 300 cal/g, and vaporization occurs above 350 cal/g. The most critical parameter that must be determined for design criteria is the amount of energy that can be absorbed without causing structural cladding failure.

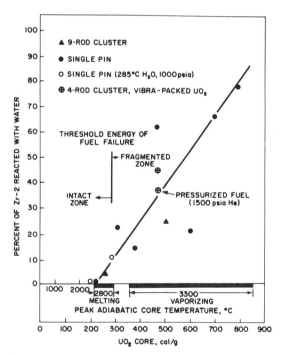

FIGURE 6-11 Results of meltdown experiments in TREAT on Zircaloy-2 clad, UO_2-core, simulated fuel elements submerged in water (water at 30°C and 30 psia, except as noted). Reprinted from "Chemical Reactions" by L. R. Baker, Jr. and R. C. Liimatainen, in *The Technology of Nuclear Power Reactor Safety*, Vol 2, T. J. Thompson and J. G. Beckerly (Eds.), 1973, by permission of the M.I.T. Press, Cambridge, Mass.

Figure 6-11 indicates that fragmentation occurs after only a fraction of the fuel becomes molten. This fragmentation is caused by fuel expansion, particularly the approximately 10% increase in volume that takes place with melting. The expansion, with possible direct contact of molten fuel with the cladding, causes stress and temperature conditions that result in catastrophic cladding failure. Following fragmentation, molten fuel and/or cladding may interact with the coolant. Of particular concern is the reaction of the zircaloy cladding with water at elevated temperatures. Figure 6-11 indicates that this cladding–coolant interaction increases rapidly with the energy input to the fuel.

If the fuel element has previously been brought to a sufficient power density to cause the cracking, restructuring, and central void formation discussed earlier, the overpower transient behavior may be significantly modified. Since the maximum temperature occurs at the periphery of the central void, melting is initiated near the void. If the heating rate is low, the fuel tends to run down the central void as it melts, since the viscosity of molten uranium dioxide is quite low. If this occurs the molten fuel accumulates near the bottom of the fuel element and tends to melt through the cladding at that point. In the event of a superprompt-critical excursion, however, the heating is so rapid that the inertial delay prevents downward fuel slumping from becoming significant over the time scale of the accident. The tendency then is for fuel melting to progress until the 10% volume increase of the molten fuel is sufficient to fill the central void region. When this happens the pressure in the fuel interior builds rapidly, causing the remaining solid annulus of fuel to expand against the cladding, and causing molten fuel to be forced through the interpellet gaps and thermal stress cracks to make direct cladding contact, which results in cladding failure.

The fuel exposure, and therefore the accumulation of fission products, is also an extremely important variable in determining failure thresholds under transient overpower conditions (Culley et al., 1971; Scott et al., 1974; Weber et al., 1974). Now, in addition to the volume increase that the fuel undergoes as a result of thermal expansion and melting, transient fission gas release and solid fuel swelling, caused by rapid precipitation and growth of fission gas bubbles within the solid fuel, become important factors. In irradiated fuel, significant fission gas pressure is already likely to be present while the reactor is at steady-state power. Increasing fuel temperature increases the pressure, and therefore adequate plenum volume or venting must be incorporated into the fuel element design to prevent excessive pressure from developing in the element during either steady-state operation or mild anticipated transients. For more severe overpower transients, the transient fission gas release and fission product induced swelling of the fuel pose a more serious problem than the gas already present in the void regions.

Under transient overpower conditions, fission gas entrained in the fuel causes rapid swelling and fission gas release as the melting temperature is approached. This is due to the accelerated precipitation and growth of fission gas bubbles along intergranular boundaries. Rapid heating of fuel that contains a high concentration of fission gas may result in such a large pressure buildup in the intergranular bubbles as to cause intergranular breakup and the fragmentation of the fuel before the melting point is reached. When melting occurs the remaining fission gases in the fuel material are released, causing a frothing behavior. The fuel breakup and/or frothing caused by melting causes the fuel to expand to fill the available void space and increase the pressure on the cladding. Thus failure thresholds tend to occur at substantially lower energies when large inventories of gaseous fission products are entrained in the solid fuel.

The effect of the fission product inventory also strongly interacts with the linear heat rate at which the fuel element has operated in determining the failure threshold. To illustrate, we consider fuel pins that have been exposed to the same total number of fissions, but have operated at differing linear heat rates in a fast reactor.

Suppose we first examine a section of fuel element, such as that shown in Fig. 6-9c that has sufficiently high linear heat rate that the central void region is present and the columnar grain region occupies a significant part of the cross-section area. It will be recalled that essentially all the fission gas is released from the columnar region, and 25 to 50% is released from the equiaxed region. However, essentially all the gaseous fission products are retained in the unrestructured region. Thus if an overpower transient causes melting only in the columnar grain region, there is little swelling, frothing, or fuel breakup resulting from entrained fission products. Moreover, the central void provides space into which the molten fuel can expand without causing excessive pressures to be generated. Only as the equiaxed and unstructured regions approach melting temperatures does the fission gas become an effective driving force toward pressurizing the fuel element and driving molten fuel into contact with the cladding.

This behavior may be contrasted to that of the fuel element section pictured in Fig. 6-9a, which has suffered a comparable number of fissions, but for which the linear heat rate has been sufficiently small that restructuring and central void formation are absent. As the fuel is rapidly heated in this situation, the gaseous fission products retained throughout the fuel region cause fuel swelling and pressurization as the centerline temperature approaches melting. Furthermore, there now is no central void region to accommodate fuel swelling and thermal expansion. Thus the fuel may be expected to fail at substantially lower energy inputs.

Models have been developed to take linear heat rates and other parameters into account in setting fuel failure criteria. Shown in Fig. 6-12 are the results of one such model for a sodium-cooled fast reactor fuel. As indicated in the foregoing discussion, fuel that has operated at high linear heat rates can sustain larger energy inputs than that for which no restructuring or central void formation has taken place. As might be expected, pellets that have sustained sufficient linear heat rates to undergo equiaxed growth, but for which columnar grains have not been found, have a behavior between the two extreme cases.

In Fig. 6-12 it is clear that the internal energy that the fuel can sustain without failure decreases as the cladding temperature increases. This is in accordance with our previous arguments that the decreased strength of the cladding at elevated temperatures means that less stress from fuel expansion can be withstood. The accidents with high fuel energy and low cladding temperature correspond to very fast power bursts over time spans in which

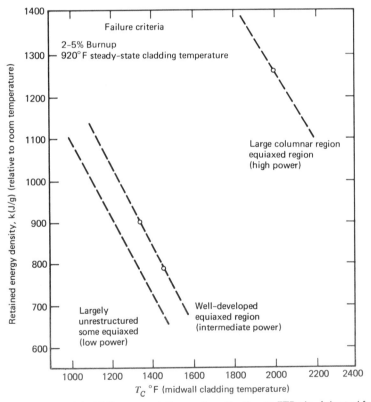

FIGURE 6-12 Empirical failure correlation and application to FTR pin. Adapted from J. H. Scott, G. E. Gulley, C. W. Hunter, and J. E. Hanson, "Microstructural Dependence of Failure of Oxide LMFBR Fuel Pins," *Proc. Fast Reactor Safety Meeting*, Beverly Hills, Calif., CONF-740401, April 1974.

there has been little time for heat transfer. Later, in these transients, or in transients with slower power increases, heat is transferred to the cladding, causing situations where the energy density in the fuel becomes smaller but the cladding temperature increases.

Thus far our attention has been directed toward describing cladding failure during overpower transients. The fuel element behavior immediately following such failure must also be carefully considered. The violent metal–water reactions that may take place when hot cladding and the coolant make contact are but one example of such concerns. Consideration must be given not only to chemical reactions, but to other types of reactions between fuel element debris and the coolant. Of particular concern in both water- and liquid-metal-cooled systems are the vapor explosions that may be possible if molten fuel causes rapid vaporization of coolant. The events following fuel failure and, in particular, the possibility that failure may propagate from one fuel element to another have been of particular concern in liquid-metal fast reactors (Van Erp et al., 1974). In these systems we must consider the specter of rapid fuel failure propagation causing coolant voiding or fuel motion to the point where a superprompt-critical configuration is reached.

Cooling-Failure Transients. To recapitulate earlier arguments, cooling-failure transients are most often characterized by a collapse of the temperature gradient between fuel and cladding, and a rapid rise in the cladding temperature. The net power input to the fuel element may be much less than that for nuclear excursions, but may vary widely, depending on whether the reactor is tripped, and on the reactivity effects induced by the accident. In either situation the cladding may fail or even melt while the oxide fuel is still intact. A central concern in fast reactor accidents, in which the trip is assumed to fail, is in the positive reactivity effects that may be associated with the movement of the molten cladding out of the interior regions of the core. In the event that the reactor is successfully shut down, the primary concern is that the cladding maintain the fuel elements in a coolable geometry so that emergency cooling measures can be utilized for the removal of decay heat.

Cladding considerations associated with cooling-failure accidents are best illustrated with an important example: In water-cooled reactors, major loss-of-coolant accidents are often considered to place the severest performance criteria on the safety systems. In these accidents the temperature at which zircaloy cladding can maintain the fuel element structural integrity in a low-pressure steam environment determines performance requirements on the emergency core-cooling systems. Thus a great deal of attention has been given to the understanding of cladding behavior in the high-temperature, low-pressure steam environments expected to appear during such accidents (Baker and Liimatainen, 1973; Rittenhouse, 1971).

The cladding deterioration becomes more pronounced with increasing temperature. About 700°F is the maximum operating temperature for water reactor cladding. Between 1200 and 1800°F swelling and creep cause the zircaloy to rupture, releasing gaseous fission products to the coolant. Above about 1800°F the metal–water reaction

$$Zr + 2H_2O \rightarrow ZrO_2 + 2H_2 \qquad (6\text{-}92)$$

becomes an increasingly important consideration. If an adequate supply of steam is available, the reaction rate is a rapidly increasing function of temperature as indicated by the parabolic rate law (Baker and Just, 1962):

$$\frac{d\zeta}{dt} = -\left(\frac{0.0615}{\zeta_0 - \zeta}\right) \exp\left(\frac{-41,200}{T}\right), \qquad (6\text{-}93)$$

where

ζ = radius of the reacting interface (in.),

ζ_0 = initial radius (in.),

T = absolute interface temperature (°R),

t = time (sec).

In water-cooled reactors this exothermic reaction becomes significant above about 2000°F, and following a major loss-of-coolant accident it may become autocatalytic above about 4800°F. However, design criteria are set to prevent the cladding from reaching 2000°F based on the following rationale. For this improbable accident, rupture of cladding that permits escape of some gaseous fission products can be tolerated. If, however, its temperature is allowed to exceed 2000°F for several minutes, a significant fraction of the cladding will react with the steam, causing the cladding to be coated with brittle ZrO_2. It has been shown that if more than about 18% of the cladding reacts to form this oxide, it becomes susceptible to fragmentation from thermal shock. Thus an unacceptable situation would exist, since the eventual quenching of the cladding by emergency flooding of the core would result in the destruction of the cladding as a structural support for the fuel, and thereby lead to an uncoolable configuration of core debris. Note that the temperature criterion is well below the melting temperature. At the same time it is independent of the stress level in the cladding, since the stress would be relieved at lower temperatures by swelling and creep rupture.

Coated-Particle Fuel

The foregoing discussions are strictly applicable only to oxide pellet fuel in metal cladding, although many analogies can be drawn to the behavior of other types of fuel. Although the metal-clad oxide fuel elements are by far the most widely used in present-day reactors, there is one other fuel type in use that deserves consideration because of its unique features. That is the coated-particle fuel used in high-temperature graphite-moderated gas-cooled reactors (Kaae et al., 1971).

Unlike more conventional fuel elements in which a metallic cladding provides for both gaseous fission product containment and structural rigidity, coated-particle fuel elements contain only ceramics that can withstand high temperatures, and the fission product retention and structural functions are not provided by a single cladding material. As shown in Fig. 1-12e, the structural support for the fuel is provided by the graphite moderator block that contains both fuel and coolant holes. The fuel holes are filled with thousands of fertile and fissile coated particles embedded in a carbide matrix material. The fission product retention is provided primarily by the coating on the many small particles. The matrix material and graphite moderator provide structural support while at the same time providing an additional barrier to the diffusion of the fission products into the coolant channels.

The fuel particles are typically a few hundred microns in diameter and are contained in the design illustrated in Fig. 6-13. Each carbide or oxide particle is clad with a multilayer ceramic coating. The coating structure is more elaborate for the fissile material because the preponderance of the fission products are produced in these particles. For the particular design shown, the central kernel of fissile material is 93% enriched uranium carbide. The inner coating is a buffer region to absorb recoil energy of the fission fragments; it also contains voids to provide volume for fuel swelling and the generation of fission product gases. The two remaining carbon

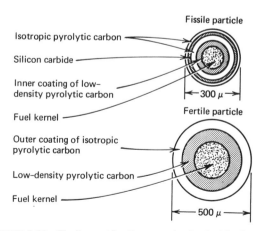

Isotropic pyrolytic carbon

Silicon carbide

Inner coating of low-density pyrolytic carbon

Fuel kernel

Outer coating of isotropic pyrolytic carbon

Low-density pyrolytic carbon

Fuel kernel

Fissile particle

$\leftarrow$ 300 μ $\rightarrow$

Fertile particle

$\leftarrow$ 500 μ $\rightarrow$

FIGURE 6-13   Fissile and fertile coated spherical fuel particles.

coatings act as a pressure vessel and provide support and protection for the coating of silicon carbide that is the primary fission product barrier. This barrier is necessitated by the high diffusion rate of metallic fission products, particularly Cs and Sr, through the carbon coatings. The silicon carbide coating is not used in the particles containing only fertile materials because the number of fission products produced in these is much smaller.

Some of the failure mechanisms for coated-particle fuels are similar to those discussed previously for metal-clad pins. The buildup of fission products in the central kernel leads to swelling and to the generation of pressure from the fission product gases. The resulting pressures may cause cracking of the particle coating. Thermal stress must also be considered because the expansivities of the carbon and the silicon carbide coatings are not identical. Determining the fraction of fission products escaping these particles is a statistical problem, because millions of independent particles are present in a reactor core.

There is an additional failure mechanism present that is unique to coated-particle fuels. This is the so-called amoeba effect. At high temperatures there is an interaction between fuel kernel and coating in the presence of a thermal gradient across the particle. The pyrocarbon coating material goes into solution in the carbide fuel kernel at its hot surface and is discharged at its cool surface. This is due to the decreased solubility at lower temperatures. Thus there is mass transport of the fuel kernel through the coating toward the hot side of the particle, eventually causing the particle to fail. Experimental correlations have indicated that the migration speed is proportional to the temperature difference across the pellet and is a strongly increasing function of pellet temperature (Stansfield et al., 1975). The amoeba effect also occurs in oxide particles, but here it is less well understood.

Since there is no metal in cores in which the graphite moderator blocks support the fuel pins, which consist of coated particles, much higher temperatures can be withstood under both steady-state and transient conditions. The moderator block design must take into account dimensional changes due to creep and radiation damage at high temperatures. Likewise, care is required to eliminate air and moisture from the coolant channels, since at elevated temperatures the graphite reacts with both oxygen and water. Finally, if graphite is allowed in high neutron irradiation fields at low temperatures, precautions must be taken to avoid uncontrolled release of Wigner energy. This is the energy stored in the form of lattice defects formed by neutron bombardment at temperatures below about 300°C. If graphite temperature is then elevated, these defects anneal out in an exothermic process that may become autocatalytic if the defect density is great enough. With these phenomena properly taken into account, however, the structural support provided by the graphite retains its strength to very high temperatures. And even then it does not melt, but sublimates at temperatures above 6600°F, a temperature greater than that at which the carbide and oxide fuel melt.

As a result of the structure of high-temperature gas-cooled reactor cores, fuel failures leading to the release of fission products due to conditions of excess temperature are quite different from that in systems with metal-clad fuel elements. Excessive core temperatures are accompanied by higher internal pressure, aggravated amoeba effects and chemical attack of certain fission products on the coatings of the fuel particles. There is thus an increased rate of coating failure leading to more rapid release of fission products to the carbide matrix material. The diffusion of fission products through the graphite moderator blocks to the coolant channels is also accelerated by the elevated temperature. In the event of an accident, however, there is another source of radionuclides that may rival the direct release from the fuel elements. Because of the migration of fission products out of the small percentage of failed or deteriorated coated particles during normal operation, there may be a significant amount of the metallic fission products deposited on the coolant surfaces of the primary system. Should sudden overheating take place, these would become resuspendent in the coolant stream and therefore be available for escape in the event of a primary system breach.

PROBLEMS

6-1. A reactor has slab fuel elements of fuel thickness and cladding thickness as shown. Find the fuel element time constant in terms of a, d, k_f, k_{cl}, m_f, c_f, m_{cl}, c_{cl}, h_g, and h_c.

6-2. A BWR cylindrical fuel element has a UO_2 fuel radius of 0.246 in and a cladding thickness of 0.030 in. Using the following data:

a. Calculate $M'_f c_f$, $M'_{cl} c_{cl}$, R_g, R_c, R'_{fe}, τ.

b. Calculate $T(0)$, $T(a^-)$, $T(a^+)$, $T(b)$, $\tilde{T}_f$, $\tilde{T}_{cl}$, $\tilde{T}_{fe}$ for a linear heat rate of $q' = 7$ KW(t)/ft and a coolant temperature of 550°F.

c. Draw the temperature profile through the fuel element. Data:

$m_f c_f = 50.8 \text{ Btu}/(°F)(ft^3)$, $h_c = 2400 \text{ Btu}/(hr)(°F)(ft^2)$,

$m_{cl} c_{cl} = 75.9 \text{ Btu}/(°F)(ft^3)$, $k_f = 1.15 \text{ Btu}/(hr)(ft)(°F)$,

$h_g = 1000 \text{ Btu}/(hr)(°F)(ft^2)$, $k_{cl} = 10 \text{ Btu}/(hr)(ft)(°F)$.

6-3. Suppose that the fuel element in the preceding problem suffers a complete cooling failure at $t = 0$ (i.e., $h_c = 0$, $t > 0$), but the power remains at its full value $q'(0) = 7 \text{ kW}(t)/ft$. Using the doubly lumped parameter model,

a. Derive an analytical expression for $\tilde{T}_f(t)$ and for $\tilde{T}_{cl}(t)$ in terms of $q'(0)$, $\tilde{T}_f(0)$, $\tilde{T}_{cl}(0)$, $T_c(0)$, and the thermal properties of the fuel element $M_f' c_f$, $M_{cl}' c_{cl}$, R_g', R_c'.

b. Plot $\tilde{T}_f(t)$ and $\tilde{T}_{cl}(t)$ from $t = 0$ to $t = 3\tau$.

c. How long after the failure would the cladding reach the failure temperature of 1600°F?

6-4. Verify the overpower temperature transients given by Eqs. (6-65) and (6-66).

6-5. Suppose that the fuel element specified in problem 6-2, is located in a reactor with a coolant pressure of 1100 psia. The pressure in the fuel-cladding gap is 1500 psia, and the cladding may be approximated as a thin shell.

a. Calculate the hoop stress in the cladding when $q' = 7 \text{ kW/ft}$.

b. Calculate the change in hoop stress if the primary system should suddenly depressurize to atmospheric pressure.

c. If gap suddenly closes, causing the fuel instantaneously to make excellent thermal contact with the cladding (i.e., $h_g \rightarrow \infty$) estimate the maximum value of the thermal hoop stress induced by the ensuing temperature transient (see Eq. 8-12).

Cladding properties:

$k = 10 \text{ Btu}/(hr)(ft)(°F)$, $E = 12 \times 10^6 \text{ psi}$,

$\alpha_l = 4 \times 10^{-6}/°F$, $\nu = 0.38$.

6-6. The zirconium cladding on the fuel pin described in problem 6-2 undergoes a metal–water reaction that is governed by the parabolic rate law.

a. How long will it take for 20% of the cladding to react if the cladding is at 1000°F, at 2000°F, and at 3000°F?

b. How long will it take all the cladding to react at each of these three temperatures?

c. Using the data from problem 6-2 estimate the average fuel element temperature increase that results if all the heat of reaction remains in the fuel element. (Assume that the specific heat per unit volume of the Zr and of the ZrO_2 are equal and that the heat of reaction = 2880 cal/g of Zr.)

COOLANT TRANSIENTS

Chapter 7

7-1 INTRODUCTION

As outlined in Section 5-1, reactor transients revolve about imbalances between heat production and removal. The transients may be caused by reactivity insertions that cause excessive power production, on the one hand, or heat transport system failures that hinder heat removal, on the other. Chapter 5 emphasizes accidents in which excessive power production constitutes the primary hazard. In this and the following chapter, accidents in which the failure of heat transport systems poses the primary threat are considered. It must be continually emphasized that the strong feedback coupling between reactivity effects and heat transport system behavior often makes a sharp distinction between these accident categories impossible. In the case of cooling-failure accidents, the behavior of the reactor power during the course of the transient plays an important role. Even after the reactor has been successfully shut down, the afterheat from the decay of fission products is sufficient to melt the core and cause penetration of the primary system if a path for orderly heat transport out of the system is not reestablished.

For purposes of analysis, we divide cooling-failure accidents into two categories. In loss-of-coolant accidents, treated in the next chapter, the primary system envelope is breached and some or all of the primary coolant escapes. In what we refer to as coolant transients are included accidents in which the coolant inventory of the primary system is maintained, but for which the orderly heat transport from the core and through the primary system envelope nevertheless is disrupted.

The global effects of coolant transients may be seen by considering quasistatic changes in the core conditions. For such slow changes, the effects of the core and coolant heat capacities can be ignored, and the static relationships for the core-averaged temperatures given by Eqs. (1-115) and (1-117) can be written in the time-dependent forms:

$$\bar{T}_{fe}(t) = \bar{T}_i(t) + \left[R + \frac{1}{2W(t)c_p} \right] P(t) \tag{7-1}$$

and

$$\bar{T}_o(t) = \bar{T}_i(t) + \frac{1}{W(t)c_p}P(t). \tag{7-2}$$

The fuel element and coolant outlet temperatures are seen to increase with decreases in core mass flow rate W and with increases in the inlet temperature $\bar{T}_i$.

Flow failures may be divided into two types by expressing the flow in terms of the core pressure drop Δp and hydraulic loss coefficient $\mathcal{K}$, as in Eq. (1-125):

$$W = A_T\sqrt{\frac{2m\,\Delta p}{\mathcal{K}}}, \tag{7-3}$$

where A_T and m are the flow area and coolant density. The average core and coolant outlet temperatures then become

$$\bar{T}_{fe}(t) = \bar{T}_i(t) + \left[R + \frac{1}{2A_Tc_p}\sqrt{\frac{\mathcal{K}}{2m\,\Delta p}}\right]P(t) \tag{7-4}$$

and

$$\bar{T}_o(t) = \bar{T}_i(t) + \frac{1}{A_Tc_p}\sqrt{\frac{\mathcal{K}}{2m\,\Delta p}}P(t). \tag{7-5}$$

In the event of the failure of one or more coolant pumps, or of the closure of a coolant loop containing a coolant pump, the pressure drop across the core decreases, causing decreased mass flow rate and increased core temperatures as indicated in Eqs. (7-3), (7-4), and (7-5). Flow starvation may also take place if there is an increase in the core hydraulic loss coefficient or a decrease in the effective flow area. Accidents affecting the latter parameters are most likely to be localized blockages in which flow through one or a few fuel assemblies is starved as a result of debris impacted on the core inlet structure or of deterioration of the flow geometry within the coolant channels caused by foreign debris, cladding deformation, fuel melting, or any number of other causes.

Increases in the core inlet temperature $\bar{T}_i$ usually are attributable to various forms of the loss-of-heat-sink accident. If the heat exchangers coupling the primary and secondary systems fail, or if there is a failure anywhere in the secondary side of the heat exchangers that blocks the orderly transport and dissipation of energy, the energy produced by the reactor will be bottled in the primary system, and will be sensed in the reactor core as a rising inlet temperature.

The effects of cooling-failure accidents on fuel element integrity are indicated by the transient heat transfer formulation of the preceding chapter. There it is shown that fuel and cladding temperatures increase with the local coolant temperature as well as with the linear heat rate. Equally important are the effects of the coolant conditions on the heat transfer across the cladding-coolant surfaces. Flow decreases result in deterioration of this heat transfer coefficient, and in particular, the deterioration of coolant flow and/or elevation of the coolant temperature in some situations lead to a boiling crisis and subsequent insulation of the fuel elements from the coolant. Flow failures and/or heat-sink losses also lead in general to rising primary system temperatures and therefore to thermal expansion of the reactor coolant. If the primary system pressure control system is not adequate to cope with this coolant expansion, excessive pressures may result and place undue stress on the primary system envelope.

To assess the impact of flow- and heat-sink-failure accidents, dynamic models for coolant flow rates and temperatures must be used, since such accidents may be expected to occur on time scales that are too short for the quasi-static relationships of Eqs. (7-1) through (7-5) to be valid. To this end, transient hydraulic models are formulated in the next section. In Sections 7-3 and 7-4 these are applied in conjunction with the thermal models developed in Chapters 5 and 6 to the various flow-failure accidents. In Section 7-5 the modeling of the thermal and pressure transients induced in the primary coolant by loss-of-heat-sink accidents is discussed. The chapter is concluded with a discussion of the events outside the primary system that may result in such heat sink failures.

Since the transients treated in this chapter do not involve the large changes in primary system coolant inventory associated with loss-of-coolant accidents, most of the phenomena may be discussed in terms of incompressible flow models that are applicable to all reactor coolants, whether they be liquids or gases. Exceptions to this generic approach, however, must be made in the discussion of pressure transients where gases and low and high vapor pressure liquids behave quite differently. Likewise, the properties of the particular coolant become very important, as discussed in Section 6-2, should boiling occur, and consequently, direct-cycle boiling-water reactors often must be modeled quite differently from other reactor systems.

7-2 HYDRAULIC MODELING

The determination of both the steady-state and transient flow behavior associated with reactor accidents requires that we formulate hydraulic models for the reactor core and primary system. In this section we utilize the continuity and mechanical energy balance equations to arrive at an incompressible flow model for a flow channel. We then summarize the empirical relationships between mass flow rate and pressure drop for channels, fittings, and other configurations likely to be found in reactor systems. Finally, Kirchhoff's laws for hydraulic systems, which allow the expressions for individul flow channels to be linked into a primary system model, are stated.

Mass and Mechanical Energy Balances

We begin our formulation of hydraulic transient models by writing the macroscopic, or control volume, form of the continuity and mechanical energy balance equation in a Eularian (i.e., stationary) coordinate system (Bird, Stewart, and Lightfoot, 1960):

$$\frac{\partial}{\partial t} \int_V m \, dV + \int_{\mathscr{S}} m\hat{n} \cdot \mathbf{v} \, d\mathscr{S} = 0, \tag{7-6}$$

and

$$\frac{\partial}{\partial t} \int_V \tfrac{1}{2}mv^2 \, dV + \int_{\mathscr{S}} \tfrac{1}{2}mv^2 \hat{n} \cdot \mathbf{v} \, d\mathscr{S}$$

$$= -\int_V \mathbf{v} \cdot \boldsymbol{\nabla} p \, dV + \int_V m\mathbf{v} \cdot \mathbf{g} \, dV - E_v + E_p, \tag{7-7}$$

where

m = fluid density,

V = fluid volume,

$\mathscr{S}$ = surface of volume V,

$\hat{n}$ = outward normal to $d\mathscr{S}$,

v = fluid speed,

$\mathbf{v}$ = fluid velocity,

p = pressure,

$\mathbf{g}$ = gravitational force vector,

E_v = rate of mechanical energy dissipation due to viscous forces,

E_p = rate of mechanical energy generation due to a pump.

We now consider a specific flow channel such as that shown in Fig. 7-1. If ζ is the distance along the channel centerline, we may write $dV = A(\zeta) \, d\zeta$, where $A(\zeta)$ is the channel area at ζ. At a macroscopic level the fluid velocity perpendicular to the centerline direction $\hat{\zeta}$ can be neglected. Moreover, in

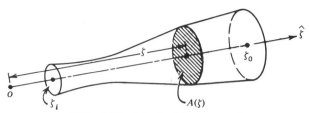

FIGURE 7-1 Idealized flow channel.

most engineering calculations where single-phase turbulent flow is involved, the velocity profile parallel to $\hat{\zeta}$ is fairly flat across the area of the channel. We thus may approximate $\mathbf{v}$ and v by single quantities that are a function only of ζ and time:

$$v = v(\zeta, t), \qquad \mathbf{v} = v(\zeta, t)\hat{\zeta}. \tag{7-8}$$

Under these conditions variations in the fluid density and pressure perpendicular to the direction of flow also may be neglected, and we have

$$m = m(\zeta, t) \tag{7-9}$$

and

$$p = p(\zeta, t). \tag{7-10}$$

With these assumptions we may take the volume V between an inlet ζ_i and an outlet ζ_o, where v is in the positive ζ direction. Equations (7-6) and (7-7) then may be written as

$$\frac{\partial}{\partial t} \int_{\zeta_i}^{\zeta_o} m(\zeta, t) A(\zeta) \, d\zeta + m(\zeta_o, t) v(\zeta_o, t) A(\zeta_o) - m(\zeta_i, t) v(\zeta_i, t) A(\zeta_i) = 0 \tag{7-11}$$

and

$$\frac{\partial}{\partial t} \int_{\zeta_i}^{\zeta_o} \tfrac{1}{2} m(\zeta, t) v(\zeta, t)^2 A(\zeta) \, d\zeta + \tfrac{1}{2} m(\zeta_o, t) v(\zeta_o, t)^3 A(\zeta_o)$$
$$-\tfrac{1}{2} m(\zeta_i, t) v(\zeta_i, t)^3 A(\zeta_i)$$
$$= -\int_{\zeta_i}^{\zeta_o} v(\zeta, t) A(\zeta) \frac{\partial}{\partial \zeta} p(\zeta, t) \, d\zeta - g \int_{\zeta_i}^{\zeta_o} m(\zeta, t) v_z(\zeta, t) A(\zeta) \, d\zeta$$
$$-E_v + E_p, \tag{7-12}$$

where g is the magnitude of the gravitational vector and v_z is the vertical component of v.

For liquid systems and for gases in which the pressure is not rapidly changing, the density of the fluid along the flow channel is not likely to change rapidly with time. Thus in the absence of boiling, it is usually adequate to treat primary system flow transients by assuming incompressible flow:

$$\frac{\partial m}{\partial t} = 0, \tag{7-13}$$

which means that Eq. (7-9) is hereafter replaced by

$$m = m(\zeta). \tag{7-14}$$

It is easily seen that the time independence of m causes the first term in Eq. (7-11) to vanish, and we have

$$m(\zeta_o)v(\zeta_o, t)A(\zeta_o) = m(\zeta_i)v(\zeta_i, t)A(\zeta_i). \tag{7-15}$$

Moreover, since ζ_o may be chosen arbitrarily in Eq. (7-11), it is clear that $m(\zeta)v(\zeta, t)A(\zeta)$, the mass of fluid that passes past ζ per unit time, is independent of ζ. Thus we define the mass flow rate at any point along the channel as

$$W(t) = m(\zeta)v(\zeta, t)A(\zeta). \tag{7-16}$$

The mechanical energy balance equation may be reduced to a more manageable form by using the incompressible flow approximation contained in Eqs. (7-14) and (7-16). Combining Eq. (7-12) with these approximate relationships, we have

$$W(t) \int_{\zeta_i}^{\zeta_o} \frac{d\zeta}{m(\zeta)A(\zeta)} \frac{dW(t)}{dt} + \frac{W(t)^3}{2} \left(\frac{1}{m_o^2 A_o^2} - \frac{1}{m_i^2 A_i^2} \right)$$

$$= -W(t) \int_{\zeta_i}^{\zeta_o} \frac{1}{m(\zeta)} \frac{\partial p}{\partial t}(\zeta, t)\, d\zeta - gW(t)(z_o - z_i) - E_v + E_p, \tag{7-17}$$

where the subscripts o and i designate quantities evaluated at ζ_o and ζ_i. Now dividing by $W(t)$ and rearranging terms, we obtain

$$\frac{W(t)^2}{2} \left(\frac{1}{m_o^2 A_o^2} - \frac{1}{m_i^2 A_i^2} \right) + \int_{p(\zeta_i)}^{p(\zeta_o)} \frac{dp}{m(p)} + g(z_o - z_i)$$

$$+ \frac{L}{\bar{m}\bar{A}} \frac{dW(t)}{dt} + \tilde{E}_v - gH_p = 0 \tag{7-18}$$

where

$$L = \zeta_o - \zeta_i, \tag{7-19}$$

$$\bar{m} = \frac{1}{V} \int_{\zeta_i}^{\zeta_o} m(\zeta)A(\zeta)\, d\zeta, \tag{7-20}$$

$$\frac{1}{\bar{A}} = \frac{\bar{m}}{L} \int_{\zeta_i}^{\zeta_o} \frac{d\zeta}{m(\zeta)A(\zeta)}, \tag{7-21}$$

$$\tilde{E}_v = \frac{E_v}{W(t)}, \tag{7-22}$$

$$H_p = \frac{E_p}{gW(t)}. \tag{7-23}$$

The first three terms in Eq. (7-18) are recognized as the widely applied Bernoulli equation for steady state flow in frictionless fluids. The addition of the last three terms permits us to account approximately for inertial and viscous effects and to treat pipe links in which pumps are located. The term $\tilde{E}_v$ accounts for the viscous energy loss per unit mass of fluid. It is usually written in the form

$$\tilde{E}_v = \tilde{e}_l \frac{v^2}{2}, \tag{7-24}$$

where $\tilde{e}_l$ is an empirical loss coefficient, and $v^2/2$ is the fluid kinetic energy per unit mass. A closer examination of the loss coefficients is included in the following subsection.

The term H_p in the foregoing equations is the pump head. Since gH_p in Eq. (7-18) is the amount of mechanical energy imparted to each unit mass of the fluid by the pump, and g is the gravitational acceleration, H_p has units of length. The head of a pump of a given design can be correlated to the pump size, rotational velocity, fluid density, and mass flow rate through correlations of dimensionless numbers (Kay, 1963). For fixed coolant properties, pump size, and speed these relationships are often reduced to curves relating the normalized pump head and normalized flow rate. The normalized head is just $H_p(W)/H_p(W_R)$, where W_R and $H_p(W_R)$ are just the rated flow and head, respectively, and the normalized flow is W/W_R. The correlations between normalized head and flow typically take the form (Stepanoff, 1957)

$$\frac{H_p(W)}{H_p(W_R)} = \frac{1 - C_1(W/W_R) - C_2(W/W_R)^2}{1 - C_1 - C_2}. \tag{7-25}$$

A representative pump characteristic, with $C_1 = 0.222$ and $C_2 = 0.253$, is plotted in Fig. 7-2.

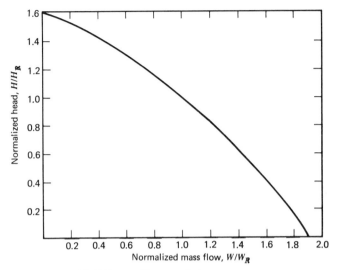

FIGURE 7-2 Characteristic pump curve.

For primary system transients in which the coolant inventory remains constant, the density gradients along the flow paths tend to be small, provided no boiling is taking place. It is reasonable under these conditions to simplify Eq. (7-18) further by making the assumption that

$$\int_{p_i}^{p_o} \frac{dp}{m(p)} \simeq \frac{p_o - p_i}{\bar{m}}. \tag{7-26}$$

If we then define the channel pressure drop as

$$\Delta p = p_i - p_o, \tag{7-27}$$

we can rewrite Eq. (7-18) as a sum of contributions to the pressure drop from inertial, acceleration, viscous, hydrostatic, and pump energy terms, respectively:

$$\Delta p = \frac{L}{\bar{A}} \frac{dW}{dt} + \left[\left(\frac{\bar{m}\bar{A}}{m_o A_o} \right)^2 - \left(\frac{\bar{m}\bar{A}}{m_i A_i} \right)^2 \right] \frac{W^2}{2\bar{m}\bar{A}^2} + \tilde{e}_l \frac{W^2}{2\bar{m}\bar{A}^2} + \bar{m}g(z_o - z_i)$$
$$- \bar{m}gH_p(W). \tag{7-28}$$

Before proceeding, it is necessary to emphasize that the preceding lumped parameter treatment is not appropriate for very fast transients where acoustical phenomena become important or to situations where transient boiling or rapid depressurization of gaseous systems cause rapid time changes in the fluid density to occur. For a discussion of the application of mass, momentum, and energy balance equations to such situations see Meyer (1960, 1961).

Pressure Drop Relationships

Before the flow model given by Eq. (7-28) can be used, expressions for the loss coefficient $\tilde{e}_l$ must be known not only for straight, uniform channel lengths, but also for pipe bends, valves, orifices, and other fittings. In addition, where contractions or expansions in channel diameter or where a significant density change take place over the length of the channel, the acceleration term must be evaluated. For brevity, we lump the acceleration and viscous pressure drop contributions into a single coefficient

$$\mathscr{K} = \tilde{e}_l + \left(\frac{\bar{m}\bar{A}}{m_o A_o}\right)^2 - \left(\frac{\bar{m}\bar{A}}{m_i A_i}\right)^2, \qquad (7\text{-}29)$$

which allows us to write Eq. (7-28) in the contracted form

$$\Delta p = \frac{L}{\bar{A}}\frac{dW}{dt} + \mathscr{K}\frac{W|W|}{2\bar{m}\bar{A}^2} + \bar{m}g\,\Delta z - \bar{m}gH_p(W), \qquad (7\text{-}30)$$

where

$$\Delta z = z_o - z_i. \qquad (7\text{-}31)$$

In the loss term W^2 is replaced by $W|W|$ to account approximately for the situation where flow reversal occurs. Since the channel inlet, outlet, and pressure drop are defined in terms of an assumed initial flow direction, a flow reversal causes the sign of the viscous losses to change. These losses usually dominate acceleration effects; hence the loss term on the right of Eq. (7-30) has the correct sign in the event of a flow reversal. If accurate calculations are to be made, it may be necessary to recalculate the magnitude of $\mathscr{K}$ too, since the size of viscous flow losses often depends on the flow direction. In general $\mathscr{K}$ also may vary with transient conditions. Equation (7-30) is not applicable to fast transients where acoustical or transient boiling cause the deviations of $\mathscr{K}$ from values measured at steady state to be largest.

We hereafter restrict our discussion to flow channels consisting of uniform cross-section area. Each channel, however, may contain bends, orifices, or other localized flow resistances. For purposes of analysis, these uniform area flow channels are considered to be linked together by expansions or contractions in flow area between channels or by large volume plenums that may form the hydraulic junction between two or more coolant channels. A reactor coolant system, for example, might be analyzed by considering the reactor coolant channels, the heat exchanger tubing, and the hot and cold legs of the coolant loops each to be flow channels that are linked by the inlet and outlet plenums of the core and heat exchanger.

Since each flow channel has viscous—and possibly acceleration—losses due to the local flow resistances as well as to the channel walls, it is convenient to write the hydraulic resistance $\mathscr{K}$ as a sum,

$$\mathscr{K} = \sum_l \mathscr{K}_l, \tag{7-32}$$

where $\mathscr{K}_1$ is the flow resistance due to the channel walls and the remaining terms correspond to the localized flow resistances. The pressure drop arising from the contractions and expansions in flow area between channels must also be taken into account. For this purpose we follow the convention of associating these terms with the channel having the smaller of the two flow areas (i.e., expansion pressure drops are associated with the inlet channel and contraction pressure drops with the outlet channel).

For incompressible single-phase flow there are no acceleration losses added to the wall friction in uniform channels, since the flow area $\bar{A}$ is constant and $m_i = m_o = \bar{m}$. We therefore have only wall losses. These are estimated in terms of the Darcy–Weisback friction factor f:

$$\mathscr{K}_1 = f\left(\frac{L}{D_h}\right), \tag{7-33}$$

where L is the channel length and D_h is the hydraulic diameter, defined as four times the flow area divided by the wetted perimeter of the channel. Alternately, the losses are sometimes stated in terms of the Fanning friction factor, which is equal to $f/4$. In either case, the friction factor is a dimensionless quantity that is related to the Reynolds number, defined by Eq. (6-35), and the pipe roughness by empirical correlations. The functional dependence of f on the Reynolds number is shown in Fig. 7-3; the roughness of the pipe wall is given by ε/D_h, where ε is a measure of the projection of irregularities into the flow area.

As indicated in Fig. 7-3, when flow rates are low, corresponding to Reynolds numbers of less than about 2,000, a laminar flow region exists, and the friction factor is inversely proportional to the flow rate:

$$f = \frac{64}{\text{Re}}. \tag{7-34}$$

For most reactor coolant calculations the flow rates are high enough that the flow will have passed through the critical zone, and turbulent flow will exist. For smooth flow channels the friction factor is only mildly dependent on the Reynolds number:

$$f = 0.184\text{Re}^{-0.2}. \tag{7-35}$$

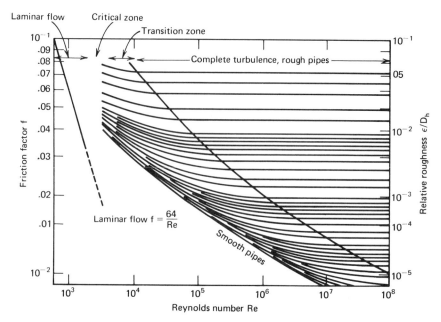

FIGURE 7-3 Friction factor for single-phase channel flow. From L. F. Moody, "Friction Fraction for Pipe Flow," *Trans. ASME,* **66,** 671 (1944). Used by permission.

For a sufficiently rough pipe—or high Reynolds number—the flow becomes independent of the Reynolds number and dependent only on the surface characteristics of the channel.

Although the incompressible turbulent flow friction factor given by Eq. (7-35) is adequate to describe the pressure drop in liquid-cooled systems, it may be desirable to make corrections for acceleration losses when gas-cooled systems are considered. For steady-state behavior an approximately corrected pressure drop expression may be obtained for uniform channels by taking (Waters and Walker, 1963)

$$\mathcal{K}_1 = f \cdot \frac{L}{D_h} + 2 \frac{(T_o - T_i)}{T_o + T_i} + \ln \left(\frac{p_o}{p_i} \right), \tag{7-36}$$

where T_o, T_i, p_o, p_i are the core outlet and inlet temperatures and pressures.

When flow in boiling-water reactors is considered, the foregoing single-phase pressure drop formulas are no longer adequate. In these systems the volume ratio of vapor to liquid becomes large enough that an annular flow pattern, such as that shown in Fig. 6-4b, develops. In such an annular flow pattern the vapor core at the channel center travels faster than the liquid annulus. Thus both viscous and acceleration terms must be modified to account for this vapor-liquid slip as well as for the acceleration loss due to the decreasing density of the two-phase fluid as it passes through the channel. The changes are accounted for in the formulation of Martinelli and Nelson (1948). We have

$$\mathcal{K}_1 = \frac{f}{D_h}(L_1 + F_m L_2) + \frac{2r_m}{m}, \qquad (7\text{-}37)$$

where f and m are calculated from the single-phase liquid flow at the channel inlet and L_1 and L_2 are the partial channel lengths over which single-phase and annular flow exist. The friction multiplier F_m and the acceleration correction r_m are determined from Figs. 7-4 and 7-5. The exit quality appearing in these figures is the ratio of steam to total mass flow rates out of the channel.

In addition to wall and acceleration losses associated with uniform channels, account must be taken of the localized pressure losses. In the incompressible approximation, the entire pressure drop is due to the viscous loss, since both flow area and density are the same up- and downstream from the resistance. The values of $\mathcal{K}_i$ for some of the more common localized flow resistances are given in Table 7-1. For the orifice, β is the ratio of the orifice opening to the flow channel area. Also included in Table 7-1 are the $\mathcal{K}$ values for contractions and expansions. For these, both acceleration and viscous terms contribute; we follow the previously stated convention by

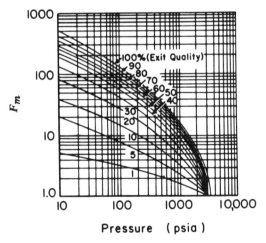

Pressure (psia)

FIGURE 7-4 Friction multiplier F_m as a function of exit quality and absolute pressure. From R. C. Martinelli and D. B. Nelson, "The Prediction of Pressure Drop During Forced Circulation Boiling Water," *Trans. ASME*, **70**, 695–702 (1948). Used by permission.

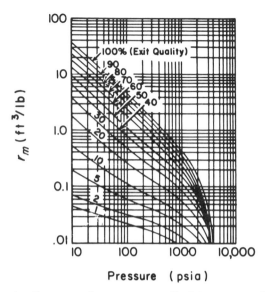

Pressure (psia)

FIGURE 7-5 Acceleration correction r_m versus absolute pressure for various exist qualities. From R. C. Martinelli and D. B. Nelson, "The Prediction of Pressure Drop During Forced Circulation Boiling Water," *Trans. ASME*, **70**, 695–702 (1948). Used by permission.

TABLE 7-1 Approximate Loss Factors for Turbulent Flow

Obstacle	$\tilde{e}_v$
Sudden changes in flow area*	
Rounded entrance to pipe	$1.05 - \beta^2$
Sudden contraction	$1.45 - 0.45\beta - \beta^2$
Sudden expansion	$2\beta(\beta - 1)$
Sharp-edged orifice	$2.7(1 - \beta)(1 - \beta^2)\beta^{-2}$
Fittings and valves	
Rounded 90° elbow	0.4–0.9
Square 90° elbow	1.3–1.9
45° elbow	0.3–0.4
Open globe valve	6–10
Open gate valve	0.2

* $\beta =$ (smaller flow area)/(larger flow area)
Data from Kramers, 1958.

taking β to be the ratio of the smaller to the larger flow areas. More detailed expressions may be found for a great variety of valves and fittings elsewhere (Crane, 1957).

The importance of the hydrostatic pressure drop compared to the viscous and acceleration terms in Eq. (7-30) varies greatly with the situation. In gas-cooled systems the hydrostatic term may be ignored entirely. Likewise, in liquid-cooled systems it tends to amount to a small correction, which often may be ignored when the reactor is operating at power. The hydrostatic pressure drop is, of course, of utmost importance in liquid-cooled systems that rely on natural convection for decay heat removal.

Kirchhoff's Laws

Heretofore we have dealt with only a single flow channel in order to relate the mass flow rate W and pressure drop Δp. To analyze the flow in reactor coolant systems, we often must treat part or all of the entire piping network that constitutes the primary system. We therefore require Kirchhoff's laws as applied to hydraulic systems. These may be stated simply as follows. Since the pressure at any point in the system is unique , the potential law states that the sum of the pressure drops around any loop of the channels must be equal to zero:

$$\sum_l \Delta p_l = 0. \tag{7-38}$$

The pressure drops Δp_l here are defined by assuming that the flows are all in the direction in which the loop is traversed. If the initial flow direction assigned to a channel is in the direction opposite to that of the loop traversal, then a negative sign must be assigned to the pressure drop, since in Eq. (7-30) we defined the pressure drop in terms of the assumed initial flow direction. Thus, for example, in terms of the flow directions assigned in Fig. 7-6a, we have

$$\Delta p_1 + \Delta p_2 - \Delta p_3 + \Delta p_4 = 0. \tag{7-39}$$

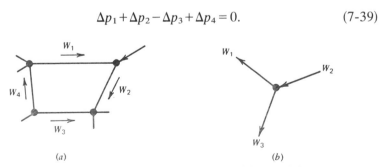

(a) (b)

FIGURE 7-6 Hydraulic circuit loops and junctions. (a) Loop. (b) Junction.

The junction law states that at any junction in the hydraulic circuit the sum of the flows W_j away from the junction must vanish:

$$\sum_j W_j = 0. \tag{7-40}$$

Again, care is required in assigning the signs to the terms. For example, in terms of the flow directions assigned in Fig. 7-6b, we have

$$W_1 - W_2 + W_3 = 0. \tag{7-41}$$

The junction law is valid for plenum volumes as well as for simple junctions so long as the mass of fluid in the plenum does not vary significantly with time. The latter condition is usually met with sufficient accuracy in a reactor primary system, provided there is no breach of the primary system envelope.

The rules for pressure drop Δp and flow W for systems in parallel and in series flow follow immediately from Kirchhoff's laws. Suppose we have n channels with pressure drops Δp_i and mass flow rates W_i. If the channels are in parallel:

$$\Delta p_p = \Delta p_1 = \Delta p_2 = \cdots \Delta p_n, \tag{7-42}$$

$$W_p = W_1 + W_2 + \cdots W_n. \tag{7-43}$$

If they are in series:

$$\Delta p_s = \Delta p_1 + \Delta p_2 + \cdots \Delta p_n, \tag{7-44}$$

$$W_s = W_1 = W_2 = \cdots W_n. \tag{7-45}$$

If, as often is the case, we have parallel flow through identical channels, we may use the fact that

$$W_p = nW_l \tag{7-46}$$

$$A_p = nA_l \tag{7-47}$$

to write Eq. (7-30) for the parallel components as

$$\Delta p = \left(\frac{L}{A}\right)_p \frac{dW_p}{dt} + \frac{\mathcal{K}_p}{2\bar{m}}\left(\frac{W_p}{A_p}\right)^2 + g\bar{m}\,\Delta z - g\bar{m}H_p\left(\frac{W_p}{n}\right), \tag{7-48}$$

where $\mathcal{K}_p = \mathcal{K}_l$, and $(L/A)_p = L_l/(nA_l)$. Similar expressions for series flow in identical channels follow from the fact that $\Delta p_s = n\Delta p_l$.

Before proceeding with the analysis of specific classes of flow failures, it should be pointed out that in cases where flow reversal occurs, the magnitude of the loss term $\mathcal{K}$ does not in general remain constant, since flow resistance through pumps, some valves, and other fittings may change markedly with flow reversal. If no check valves are present, however, it is often found convenient to a first approximation to neglect these changes. Likewise, the mild dependence of the friction factor on the Reynolds number is often neglected when only turbulent flow is present.

7-3 PARTIAL FLOW FAILURES

We now consider the partial core flow failures that arise from two sources: blockages in the core or its inlet structure, which cause localized flow starvation in one or a few fuel assemblies, and single pump failures or coolant loop blockages, which cause a more or less uniform flow reduction across the core. The neutronic implications of such accidents are stressed in earlier chapters. We here concentrate on the hydraulic consequences.

Flow Blockages

The multitude of potential mechanisms for causing flow blockages over a localized region of a reactor core may be divided roughly into two groups; debris swept against the core inlet structure by the force of the coolant flow, and failure of fuel or other components within the core. The debris may consist of nuts, bolts, broken pump impeller blades, or other fragments of failed components within the primary system envelope. Similarly, tools, scrap materials, or other items may be left in the primary system at the time of construction or refueling. Fuel pin bowing, cladding ballooning, and structural materials deformation tend to block coolant channels. Molten fuel that penetrates the cladding and is injected into the coolant channel may cause plugging. Most seriously, if extensive fuel melting or structural collapse of cladding occurs, the flow channel geometry may be destroyed, leading to complete localized cooling failure and massive plugging of the fuel assemblies.

Flow blockages differ from many other reactor core accidents because of their localized nature. Usually one or at most a few fuel assemblies are initially affected. This often makes the prompt detection of the accident difficult, since the localized changes in power density and coolant conditions may not have sufficient effect on the core flux levels, coolant pressure drop, or average coolant outlet temperature to cause the protection system to shut the reactor down. As a result, in some systems it is deemed necessary to instrument the core to monitor outlet temperatures or fission product concentrations of individual groups of fuel assemblies, delayed neutrons in the coolant stream, or other variables that are more sensitive to localized conditions.

In many reactor designs flow blockage and resulting fuel failure within a single assembly may be tolerated with rather minor consequences as compared to other reactor accidents. The more serious safety-related questions then center on the likelihood of a flow blockage occurring over a significant part of the core, and particularly on the potential for the reactivity and/or thermal effects of a blockage that originates in one fuel assembly propagating to cause a global core disruption before the accident can be detected and the reactor shut down. The propagation of such accidents is of particular concern in large sodium-cooled fast reactors, where flow starvation and consequent coolant boiling may lead to large reactivity additions if the overheated area propagates laterally across the core.

The design of the core-cooling configuration can accomplish a great deal in the prevention of massive coolant blockages. This may be seen by examining the hydraulic aspects of the core more carefully. To demonstrate the effects of partial core inlet blockages, suppose we consider a coolant channel with a hydraulic resistance $\mathscr{K}_0$ and a flow area A_0. If the pressure drop across the core is Δp, the mass flow rate is determined from

$$\Delta p = \frac{\mathscr{K}_0}{2\bar{m}}\left(\frac{W_0}{A_0}\right)^2. \tag{7-49}$$

Now if a blockage occurs that affects only a small fraction of the many coolant channels through the core, the hydraulic resistance of the core as a whole is not affected appreciably, and therefore the core pressure drop Δp does not change as a result of the blockage. The flow blockage, rather, appears as an additional flow resistance in series with $\mathscr{K}_0$. Thus

$$\Delta p = \frac{(\mathscr{K}_0 + \mathscr{K}_b)}{2\bar{m}}\left(\frac{W}{A_0}\right)^2. \tag{7-50}$$

Of course, if the blockage were complete, $\mathscr{K}_b$ would be infinite, causing the flow to vanish completely. Partial blockages are difficult to characterize, since even if the geometry of the flow were known precisely only empirical correlations would provide accurate values of $\mathscr{K}_b$. Often, however, it is adequate to characterize the obstruction simply by the fraction of the flow channel area that is blocked, and make a rough estimate of $\mathscr{K}_b$ by assuming that the obstruction is in the form of a sharp-edged orifice with an aperture area $A_0 - A_b$. If we take $\gamma = A_b/A_0$ as the fraction of the flow area that is blocked, we have from Table 7-1:

$$\mathscr{K}_b = 2.7\gamma\frac{(2-\gamma)}{(1-\gamma)^2}. \tag{7-51}$$

Utilizing Eqs. (7-49) and (7-50) we then find the flow reduction to be

$$\frac{W}{W_0} = (1-\gamma)\left[(1-\gamma)^2 + \frac{2.7}{\mathscr{K}_0}\gamma^2(2-\gamma)\right]^{-1/2}. \tag{7-52}$$

The large ratio of heat transfer surface to flow area in a reactor core causes $\mathscr{K}_0$ to be relatively large. This in turn means that a large fraction of the flow area must be blocked before there is a serious flow reduction. For example, $\mathscr{K}_0 = 15$ is a typical value for a pressurized-water reactor (Tong and Weisman, 1970). As indicated in Fig. 7-7, the fraction of the flow area blocked must be quite large before a substantial flow reduction takes place. Similarly, in a typical sodium-cooled fast reactor lattice, it has been found that 90% of the flow area must be blocked before a 50% reduction in flow takes place (Teague, 1970).

   In the foregoing analysis, the question of determining the effective flow area A_0 of the channel is not addressed. It is important, however, that this area be as large as possible, since if a coolant channel has an extremely small effective flow area, complete blockage may be possible even from a small piece of debris. The evaluation of the area over which a blockage must occur before fuel elements may be affected by flow starvation is dependent both on the structure of the reactor fuel assembly lattice and on the design of the core inlet structure.

   For purposes of discussion we divide the fuel assembly lattices into three types: closed lattices, open shrouded lattices, and open unshrouded lattices. In a closed lattice, such as that shown for a hgh-temperature gas-cooled reactor in Fig. 7-8a, each coolant channel is separated from the other channels by the presence of the solid moderator. In such a configuration a

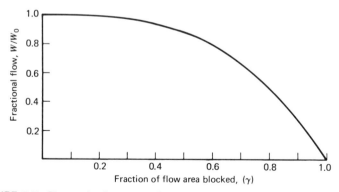

FIGURE 7-7   Flow reduction versus fractional area blocked for a reactor lattice.

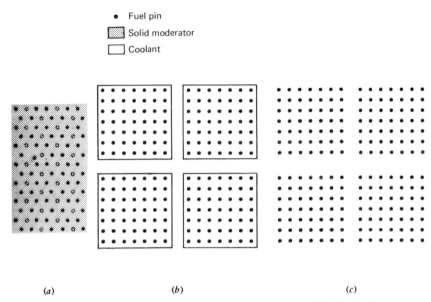

- Fuel pin
- Solid moderator
- Coolant

(a)   (b)   (c)

FIGURE 7-8 Reactor lattice configuration (a) Closed lattice (HTGR). (b) Open-lattice shrouded fuel assembly. (c) Open-lattice unshrouded fuel assembly.

relatively small obstruction can result in a plugging of the channel. In such situations it is highly desirable to ensure that heat can be transferred from the fuel pins to the unimpeded cooling channels without excessive core heating, even though one or more channels are completely blocked.

In open lattices, cross flow can take place between the flow paths adjacent to the individual fuel pins. Hence if the channel between two fuel pins is obstructed, flow from adjoining lattice cells tends to mix and cool the affected fuel pins, except for a stagnant wake area immediately downstream from the blockage. If a fuel assembly consisting of such an open lattice is shrouded, as shown, for example, in Fig. 7-8b, then the cross flow can only take place within the lattice cells of a single assembly. A blockage covering the inlet area of an entire assembly would thus terminate flow in that assembly. In a pressure tube reactor the situation is similar to that in a shrouded open lattice, since cross flow can take place between the fuel elements within a single pressure tube, and the blockage of the cross-section area of the tube results in flow starvation to the entire assembly of fuel elements within that tube. In an unshrouded open lattice, such as that shown in Fig. 7-8c, cross flow may take place between as well as within assemblies of fuel elements. Consequently, it may be possible to sustain blocked flow areas at the core inlet that are substantially larger than the area of a single fuel assembly. In this situation the area of the blockage that can be tolerated is dependent on the flow mixing characteristics of the lattice, and particularly on the nature of the wake of reduced flow directly downstream from the blockage.

In considering the effects of flow blockages in reactor lattices, the inlet structure of the core must be taken into account, since most core inlet structures are designed to eliminate or at least mitigate the effects of many potential flow blockage problems. The geometrical configurations of inlet structures vary greatly between reactor designs, and often they are included as an integral part of the fuel assembly structure. The basic idea, however, is to place a protective grid structure upstream from the fuel assemblies. The grid then stops large pieces of debris that would otherwise block the core-coolant channels. The grid openings are made small enough so that debris that can pass through them is not large enough to block the major part of a fuel assembly. Between the protective grid structure and the entry to the fuel element lattice there must be sufficient room for cross flow to take place in the wake of the obstruction, so that each fuel assembly receives an adequate coolant flow even though a large area of the protective grid is blocked.

Flow starvation caused by core inlet blockage manifests itself as increased coolant outlet temperature. In gas-cooled reactors this in turn may lead to excessive cladding temperatures. In liquid-cooled reactors the core outlet may reach the coolant boiling point in the obstructed channel, or alternately, a boiling crisis may be caused by exceeding the critical heat flux in a higher power density region of the obstructed channel. Once boiling is initiated it becomes increasingly difficult to reestablish flow in a damaged channel. As indicated for water in Eq. (7-37), the hydraulic resistance of the channel increases rapidly with the amount of boiling in the channel, since both acceleration losses and the effects of wall friction increase sharply with the exit void fraction. As a result, an unstable thermal-hydraulic coupling may come into play, in which reduced flow rate causes boiling and a further increase in flow resistance. The increased flow resistance leads to a yet smaller mass flow rate, which leads to a larger fraction of coolant vaporization, and so on. In boiling-water reactors the flow channels and thermal-hydraulic coupling must be designed to permit stable steady-state operation under conditions in which the acceleration and friction losses due to the presence of vapor in the coolant channels are always present.

As pointed out in Chapter 6, the initiation of boiling in a sodium-cooled system leads to the very rapid expulsion of coolant from the entire flow channel, and in large systems to a reactivity increase caused by the positive void reactivity coefficient. From this effect originates an interesting safety trade-off between the use of unshrouded and shrouded fuel assemblies. If unshrouded assemblies are used, the likelihood of a blockage large enough to produce sodium boiling is decreased, since cross flow can take place between assemblies, and several assemblies would have to be blocked before the outlet temperature would reach the boiling point. On the other hand, should boiling be initiated, there would be no mechanical restraint to the lateral spreading of the sodium void through many fuel assemblies. The potential would then exist for a superprompt-critical excursion. In contrast, if the fuel assembly shrouding is designed to withstand the pressures associated with sodium boiling, the blockage will result only in sodium voiding of the blocked channels. With careful neutronic design it may be possible to void one or more assemblies without causing the reactor to approach superprompt critical. Thus the use of shrouding may increase the probability of sodium boiling but decrease its consequences.

If a flow blockage occurs within the core proper instead of in the core inlet structure, more serious situations may result (Fontana et al., 1975). If the fuel assembly is not completely blocked, in an open lattice flow mixing may prevent the core outlet temperature from becoming excessive. However, if the obstruction appears in a high power density region of the core, as is likely to be the case if the obstruction's cause is cladding deformation of molten fuel motion, the local hot spot will appear at and immediately downstream from the obstruction.

Coolant Loop Failures

Reductions in core flow result whenever the pressure drop across the core is reduced. This pressure drop decreases if one or more of the coolant pumps fails or if a coolant loop is blocked, as would be the case if an isolation valve were inadvertently closed. The valve closure problem can be reduced to manageable proportions either by eliminating valves from the primary system loops or by designing the valves such that the maximum speed of closure is so slow as to allow time for detection of the failure and orderly shutdown of the plant before the core flow is substantially decreased. Pumps, however, are normally in an active state of operation, and they are subject to rapid failure either from mechanical seizure or from power failure. A catastrophic mechanical seizure of a pump may occur nearly instantaneously, causing a precipitous drop of the pump head to zero. Pump failure due to loss of power does not result in an instantaneous loss of head, since even if the pump motor suddenly fails, the rotational inertia of the pump impeller and fly wheel will cause the pump to coast down gradually over a period of time that is often on the order of several seconds or more.

In postulating reactor accidents, the simultaneous seizure of more than one pump is not considered, since the probability of two or more random mechanical failures—each of which is highly improbable—occurring at the same instant may be considered to be vanishingly small. Simultaneous failure of all the pumps because of power failure, however, would not be an incredible event, since common mode failure could cause failure of the supply of power to all the coolant pumps.

Flow Determination. The dynamics of a loss-of-flow accident in which power to all pumps is lost simultaneously is taken up in the next section. Presently, we consider the more probable event in which a single pump is lost. In such a situation it is necessary to determine the equilibrium flow rate that will be established following the hydraulic transient. From this flow it can be determined whether the reactor can remain at full power without causing excessive temperatures or whether a rapid power cutback or shutdown is required to prevent core damage. Since at least one coolant pump remains in operation, in the following analysis we assume that there is adequate circulation to remove decay heat in the event that the reactor is successfully shut down. Moreover, for brevity we make two additional assumptions that do not cause significant errors for most power reactors. First, the hydrostatic pressure drops both before and after the failure transient are neglected. Second, the mild dependence of the flow resistances on the Reynolds number is neglected for the highly turbulent flow conditions that exist both before and after the transient.

Suppose we consider the single pump failure in a primary system consisting of a core and N identical coolant loops. The core is characterized by a loss coefficient $\mathcal{K}_c$ and a flow area A_T. Likewise, each of the loops has a loss coefficient $\mathcal{K}_l$ and flow area A_l. In order to eliminate the flow area from the equations, we define the normalized flow resistances

$$\tilde{\mathcal{K}}_c = \frac{\mathcal{K}_c}{A_T^2} \qquad (7\text{-}53)$$

and

$$\tilde{\mathcal{K}}_l = \frac{\mathcal{K}_l}{A_l^2}. \qquad (7\text{-}54)$$

Before the pump failure we have the initial conditions:

$$\Delta p_c = \frac{\tilde{\mathcal{K}}_c W_0^2}{2\bar{m}} \qquad (7\text{-}55)$$

and

$$\Delta p_l = \frac{\tilde{\mathcal{K}}_l W_l^2}{2\bar{m}} - \bar{m}gH_p(W_l), \qquad (7\text{-}56)$$

where Δp_c and Δp_l are the core and loop pressure drops, and W_0 and W_l are the corresponding mass flow rates. Applying Kirchhoff's potential and junction laws, we have

$$\Delta p_c + \Delta p_l = 0 \qquad (7\text{-}57)$$

and

$$W_0 = NW_l. \qquad (7\text{-}58)$$

Therefore the initial core flow rate is

$$W_0 = \bar{m}\left[\frac{2gH_p(W_0/N)}{\tilde{\mathcal{K}}_c + \tilde{\mathcal{K}}_l/N^2}\right]^{1/2}. \qquad (7\text{-}59)$$

Now suppose that the pump in one of the loops fails, a hydraulic transient ensues, and a new flow equilibrium is established. Let W_∞, W_a, and W_b be the flows through the core, the loop with the failed pump, and each of the intact loops, respectively, following the transient. The pressure drops across the core and each of the intact loops are then

$$\Delta p_c = \tilde{\mathcal{K}}_c \frac{W_\infty^2}{2\bar{m}} \qquad (7\text{-}60)$$

and

$$\Delta p_b = \tilde{\mathcal{K}}_l \frac{W_b^2}{2\bar{m}} - \bar{m}gH_p(W_b). \qquad (7\text{-}61)$$

If we assume that the pump failure has no effect on the flow resistance in this disabled loop, then the pressure drop across this loop is given by

$$\Delta p_a = \tilde{\mathcal{K}}_l \frac{|W_a|W_a}{2\bar{m}}, \qquad (7\text{-}62)$$

where the absolute value signs are included here to ensure that the flow reversal in loop a is properly treated.* Now applying Kirchhoff's laws, we have

$$\Delta p_c + \Delta p_a = 0, \qquad (7\text{-}63)$$

$$\Delta p_c + \Delta p_b = 0 \qquad (7\text{-}64)$$

and

$$W_\infty = W_a + (N-1)W_b. \qquad (7\text{-}65)$$

---

* Failure to account for the flow reversal leads to a contradiction when Kirchhoff's potential law is applied.

Combining Eqs. (7-60) through (7-65), we find the flow rate through the core to be

$$W_\infty = \bar{m}\left\{\frac{2gH_p[(1+\sqrt{\tilde{\mathscr{H}}_c/\tilde{\mathscr{H}}_l})(N-1)^{-1}W_\infty]}{\tilde{\mathscr{H}}_c+\tilde{\mathscr{H}}_l(N-1)^{-2}(1+\sqrt{\tilde{\mathscr{H}}_c/\tilde{\mathscr{H}}_l})^2}\right\}^{1/2}. \tag{7-66}$$

Finally, this expression may be combined with Eq. (7-59) to yield

$$W_\infty = W_0\left\{\frac{1+N^{-2}\tilde{\mathscr{H}}_l/\tilde{\mathscr{H}}_c}{1+(N-1)^{-2}(1+\sqrt{\tilde{\mathscr{H}}_l/\tilde{\mathscr{H}}_c})^2}\right\}^{1/2}\left\{\frac{H_p[(1+\sqrt{\tilde{\mathscr{H}}_c/\tilde{\mathscr{H}}_l})(N-1)^{-1}W_\infty]}{H_p[W_0/N]}\right\}^{1/2}. \tag{7-67}$$

From this expression it is seen that for a fixed number of coolant loops the ratio of the loop to the core flow resistance $\tilde{\mathscr{H}}_l/\tilde{\mathscr{H}}_c$, and the pump characteristic curve $H_p(W)$ determine the core flow. The functional dependence of the flow on N and $\tilde{\mathscr{H}}_l/\tilde{\mathscr{H}}_c$ is illustrated graphically in Fig. 7-9; the solution of Eq. (7-67) is plotted for the pump characteristic illustrated in Fig. 7-2 where the rated pump flow is taken as W_0/N.

Effect of Check Valves. Primary systems typically have small ratios of loop to core flow resistance if there is only a small flow resistance associated with the primary side of the heat exchanger. This occurs, for example, in some fast reactor systems where the low-resistance shell side of the heat exchanger is in the primary coolant loop and the high-resistance tube side is in the intermediate loop. In these cases Fig. 7-9 indicates that even the loss of a single pump may cause a large decrease in the core flow, since a large

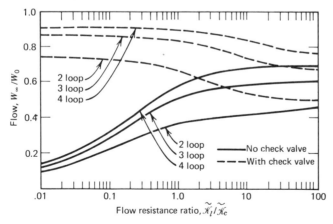

FIGURE 7-9 Flow reduction following a single cooling-pump failure.

part of the coolant pumped through the intact loops bypasses the core through the disabled loop. Granted, the preceding simplified analysis assumes that the magnitude of the flow resistance does not change with flow reversal, whereas, in fact, the reversal is likely to increase the flow resistance through the pump and therefore decrease the bypass flow. Even then it may be found that the resulting core flow is too small following the transient. As a result, when the ratio $\tilde{\mathcal{H}}_l/\tilde{\mathcal{H}}_c$ is small, check valves may be placed in the coolant loops to eliminate flow reversals. The loss of a pump then has the same effect as blocking the coolant loop, and the analysis is as follows.

If the flow through loop a is blocked completely, then Kirchhoff's laws reduce to

$$\Delta p_c + \Delta p_b = 0 \qquad (7\text{-}68)$$

and

$$W_\infty = (N-1) W_b. \qquad (7\text{-}69)$$

Equations (7-60) and (7-61), relating the pressure drop and flow rates in the core and intact loops, are still valid. Combining these equations with Kirchhoff's relationships yields

$$W_\infty = \bar{m} \left\{ \frac{2gH_p[(N-1)^{-1}W_\infty]}{\tilde{\mathcal{H}}_c + (N-1)^{-2}\tilde{\mathcal{H}}_l} \right\}^{1/2}. \qquad (7\text{-}70)$$

Normalizing this result to W_0, we obtain

$$W_\infty = W_0 \left\{ \frac{1 + N^{-2}(\tilde{\mathcal{H}}_l/\tilde{\mathcal{H}}_c)}{1 + (N-1)^{-2}(\tilde{\mathcal{H}}_l/\tilde{\mathcal{H}}_c)} \right\}^{1/2} \left\{ \frac{H_p[(N-1)^{-1}W_\infty]}{H_p[N^{-1}W_0]} \right\}^{1/2}. \qquad (7\text{-}71)$$

The ratio W_∞/W_0 is plotted in Fig. 7-9 as a function of N and $\tilde{\mathcal{H}}_l/\tilde{\mathcal{H}}_c$, using, once again, the pump characteristic from Fig. 7-2. The use of check valves to prevent the core-bypass flow reversal in the disabled loop is seen to cause a significant increase in W_∞ in systems for which the ratio of loop-to-core flow resistance is small. Significant benefit is not derived, however, from the check valves in systems in which the loop-to-core flow resistance ratio is large, since even in the absence of the valves there is little core-bypass flow.

Temperature and Pressure Consequences. Analyses such as the preceding yield the core flow rate following a partial flow failuure. If the power level following the transient is given, the corresponding core outlet temperature can be determined from Eq. (7-2). Conversely, the criteria for the maximum permissible core outlet temperature can be used to determine if the reactor power can be allowed merely to adjust to the reactivity feedback effects induced by the decreased coolant flow, or whether protective action is required to trip the reactor or at least to reduce the power level.

The considerations made in determining the allowable coolant outlet temperatures are similar in some respects to those for the localized flow blockage accident. Now, however, the outlet temperature over the entire core, rather than over one or a few fuel assemblies in the case of the flow blockage accident, is affected. Consequently, the effect on the temperature in the core outlet plenum and the hot leg of the coolant loop is more pronounced. This increased temperature may lead to overpressurization of the primary system, if, as discussed in Section 7-5, there is insufficient means for compensating for the thermal expansion of the coolant. In addition, the overheating and reactivity effects on the core are global for pump failure accidents, whereas they are confined to the disabled assemblies in a flow blockage that does not propagate.

In the event that the temperature rises in the core become large, the flow failure may be aggravated by a thermal hydraulic coupling between rising coolant temperature and flow resistance that tends to increase the core flow resistance and hence further decrease the flow. This effect may be illustrated for gas-cooled systems through Eq. (7-36): As the temperature difference across the core increases the acceleration loss increases, because of the decrease in outlet coolant density, and therefore the flow resistance increases as $2(\bar{T}_0 - \bar{T}_i)/(\bar{T}_0 + \bar{T}_i)$. This is somewhat compensated by the decrease in the average coolant density $\bar{m}$ that tends to decrease the friction losses as indicated by Eq. (7-30).

In liquid-cooled systems the effect of coolant temperature on flow resistance is much milder unless the liquid begins to boil. If boiling occurs a strong feedback effect takes place. In sodium systems the energy stored as superheat and the large ratio of liquid to vapor density cause the coolant to be rapidly expelled from the core, as discussed in Section 6-2. In the presence of a positive void reactivity coefficient, the voiding may then lead to a severe nuclear excursion. In light-water systems the inception of boiling may also lead to a vapor blanketing in isolated cooling channels if the local power density is large enough. This effect may be understood as follows.

In water systems there is relatively little superheat at the initiation of boiling. Therefore the extent to which vapor blanketing of the channel takes place depends critically on the thermal-hydraulic coupling between heat input and flow resistance following the inception of boiling (King, 1964; Boure et al., 1973). The situation may be understood from the plots of pressure drop versus flow (Fig. 7-10). The dashed lines represent the pressure drop of channels filled only with vapor or only with liquid. The solid lines represent the pressure drops that result from boiling in two channels that might be typical of an average core channel and a hot channel. Consider now a water-cooled core operating with an initially high pressure drop and mass flow rate at point A. The mass flow rate through all the channels is identical, being equal to W_A. With a pump failure, however, the core

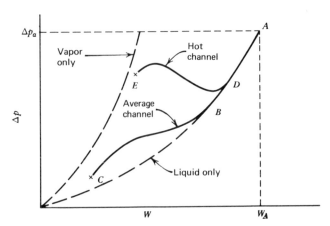

FIGURE 7-10 Two-phase pressure drop versus mass flow rate.

pressure drop decreases. If point B is reached the average channel outlet temperature reaches boiling, and further decreases in pressure drop cause much larger decreases in mass flow rate as indicated by the flattening of the curve in Fig. 7-10. Eventually the decreased flow rate and the ensuing vapor buildup in the channel causes a boiling crisis to occur at point C.

As channels with higher heat input are considered, the boiling portion of the curve becomes flatter and may cause an unstable situation if the slope becomes negative. The latter situation is illustrated for the hot channel in Fig. 7-10. If the pressure drop is decreased to the point where boiling begins in the hot channel at point D, the flow rate begins to drop more rapidly because of the acceleration and two-phase friction losses. The high rate of heat input, however, causes the boiling to increase rapidly with decreased flow rate, and the flow rate further deteriorates with increased boiling. The result is an unstable situation in which, once boiling is initiated at point D, an autocatalytic vapor binding of the channel takes place, resulting in a boiling crisis at point E. Thus for our hypothetical core, a sufficient reduction in the pressure drop will result in the rapid vapor blanketing of the hot channel even though coolant at the outlet of the average channel has not yet reached the boiling point. The instability between points D and E can be removed by reducing or flattening the power distribution in the reactor sufficiently so that the negative slope of the curve is removed.

7-4 THE LOSS-OF-FLOW ACCIDENT

In single pump failure accidents, such as that treated in the preceding section, the determination of the flow rate following the transient allows one to determine whether a power cutback or reactor trip is necessary to prevent excessive overheating of the coolant. When such protective action is necessary a dynamic analysis of the flow transient may be required to ensure that the power cutback can be carried out fast enough that excessive coolant temperatures are not reached during the early stages of the transient. In the loss-of-flow accident, caused by a power failure to all the coolant pumps, the equilibrium flow following the transient is quite small, since it is driven only by natural convection. Therefore the reactor must be shut down. The central question then becomes whether the power can be reduced rapidly enough that the power flow mismatch during the transient does not lead to unacceptable coolant temperatures. A determination must be made of the mass flow rate versus time, which can then be incorporated into a thermal model along with the power trace to determine the coolant temperature. Following the flow transient—usually referred to as the flow coastdown—the reactor trip will have reduced the power level to the decay heat level. It then must be ascertained whether natural convection is adequate for the removal of decay heat. If it is not, means must be incorporated to bring emergency systems for forced circulation into operation before excessive core heat-up takes place.

In what follows we first discuss the hydraulic analysis leading to a flow coastdown curve. We then examine the temperature behavior of the coolant during such an accident. Finally, we analyze the flow and temperature behavior during natural convection removal of decay heat, and we discuss some possible emergency core-cooling measures.

Flow Transient

To determine the flow transient, we apply pressure drop flow relationships along with Kirchhoff's laws in a manner reminiscent of the preceding section. Now, however, we must include the effect of the fluid inertia in the pressure drop relationships, and take into account the pump inertia following the loss of power. The analysis is simplified by the assumption that all the pumps behave identically following the power failure, thus permitting the accident to be treated in terms of the reactor core and N identical coolant loops.

We first rewrite Eq. (7-30) for the core and for the coolant loops, respectively:

$$\Delta p_c = \left(\frac{L}{A}\right)_c \frac{dW}{dt} + \tilde{\mathcal{K}}_c \frac{W^2}{2\bar{m}} + g(\bar{m}\,\Delta z)_c \tag{7-72}$$

$$\Delta p_l = \left(\frac{L}{A}\right)_l \frac{dW_l}{dt} + \tilde{\mathcal{K}}_l \frac{W_l^2}{2\bar{m}} + g(\bar{m}\,\Delta z)_l - \bar{m}g\left(\frac{\omega}{\omega_0}\right)^2 H_p, \tag{7-73}$$

where we take $\tilde{\mathcal{H}}_x = \mathcal{H}_x / A_x^2$. In these equations $(L/A)_x$, $\tilde{\mathcal{H}}_x$, and $(\bar{m}\ \Delta a)_x$ indicate that the quantities are appropriately summed over the series and parallel flow paths within the core and the loops, respectively. For example, the loop typically consists of series of paths through the hot and cold legs and the heat exchanger, and the heat exchanger consists of many parallel flow paths. The term $(\omega/\omega_0)^2$ multiplies the initial pump head in order to approximate the effect of the decreasing speed of the pump during the transient.

For a reactor with N identical loops Kirchhoff's laws are

$$\Delta p_c + \Delta p_l = 0 \qquad (7\text{-}74)$$

and

$$W = NW_l. \qquad (7\text{-}75)$$

Combining these expressions with Eqs. (7-72) and (7-73), we find that the core flow is determined from

$$\left(\frac{L}{A}\right)_{pr}\frac{dW}{dt} + \tilde{\mathcal{H}}_{pr}\frac{W^2}{2\bar{m}} + g(\bar{m}\ \Delta z)_{pr} - \bar{m}g\left(\frac{\omega}{\omega_0}\right)^2 H_p = 0, \qquad (7\text{-}76)$$

where for the entire primary system

$$\left(\frac{L}{A}\right)_{pr} = \left(\frac{L}{A}\right)_c + \frac{1}{N}\left(\frac{L}{A}\right)_l, \qquad (7\text{-}77)$$

$$\tilde{\mathcal{H}}_{pr} = \tilde{\mathcal{H}}_c + \frac{1}{N^2}\tilde{\mathcal{H}}_l, \qquad (7\text{-}78)$$

and

$$(\bar{m}\ \Delta z)_{pr} = (\bar{m}\ \Delta z)_c + (\bar{m}\ \Delta z)_l. \qquad (7\text{-}79)$$

To simplify the analysis, we divide the accident into two stages. In the first stage the pump head and flow rate are large compared to the hydrostatic term, and therefore it is neglected. In the second stage, the transient is completed, the pump head has vanished, and the hydrostatic term drives the steady-state natural convection. In reality there is a period of time during the latter part of the flow coastdown when both pump and hydrostatic head contribute significantly. However, we obtain a pessimistic underestimate of the flow if we simply assume that there is no hydrostatic head until the flow decays to the value given by steady-state natural convection, and thereafter assume that pump head and fluid inertia make no contribution to the flow.

During the flow coastdown the speed of the pump must be modeled, since it appears in Eq. (7-76). Following the loss of power to the pump, the pump torque immediately drops to zero. The pump then coasts down at a rate determined by the fly wheel inertia and the windage torque within the pump. If I is the moment of inertia of the pump and C is a loss coefficient relating the windage torque to the square of the pump speed ω,

$$I\frac{d\omega}{dt} = -C\omega^2. \tag{7-80}$$

Solving in terms of the initial pump speed ω_0, we obtain

$$\omega = \frac{\omega_0}{1 + t/t_p}, \tag{7-81}$$

where the pump half-time t_p is defined by

$$t_p = \frac{I}{C\omega_0}. \tag{7-82}$$

We are now prepared to treat the flow transient. After substituting Eq. (7-81) into Eq. (7-76), and neglecting the hydrostatic terms in accordance with our previous arguments, we obtain

$$\left(\frac{L}{A}\right)_{pr}\frac{dW}{dt} + \tilde{\mathcal{K}}_{pr}\frac{W^2}{2\bar{m}} = \frac{\bar{m}gH_p}{(1 + t/t_p)^2}. \tag{7-83}$$

The initial condition,

$$\tilde{\mathcal{K}}_{pr}\frac{W_0^2}{2\bar{m}} - \bar{m}gH_p = 0 \tag{7-84}$$

allows us to write this nonlinear differential equation in the normalized core flow

$$t_l\frac{d}{dt}\left(\frac{W}{W_0}\right) + \left(\frac{W}{W_0}\right)^2 = \frac{1}{(1 + t/t_p)^2} \tag{7-85}$$

where t_l is the loop half-time, given by

$$t_l = \frac{2\bar{m}}{W_0\tilde{\mathcal{K}}_{pr}(A/L)_{pr}}. \tag{7-86}$$

In examining the flow transient, it is helpful to consider two extreme cases. First consider the hypothetical, if incredible, situation where there is a simultaneous seizure of all the coolant pumps causing their pressure heads to drop instantaneously to zero. The right-hand side of Eq. (7-85) then vanishes, whereupon

$$t_l \frac{d}{dt}\left(\frac{W}{W_0}\right) + \left(\frac{W}{W_0}\right)^2 = 0. \qquad (7\text{-}87)$$

or

$$W = \frac{W_0}{1 + t/t_l}. \qquad (7\text{-}88)$$

In this case the flow coastdown is governed by the inertia of the fluid as represented by the loop half-time. If t_l is expressed in terms of flow velocities and the more elementary loss coefficient, $\mathcal{K}_{pr}$, it is readily seen that in an average sense, the loop half-time increases with the time required for the coolant to traverse the primary system loop and decreases with increasing loss coefficients. Flow deterioration such as this takes place most rapidly in systems with short loop traversal times and high flow resistance.

Although the foregoing situation is not considered credible, the same influences are present in the determination of the speed at which the flow deteriorates to a new equilibrium in the event of a single pump seizure. The quantitative treatment of the flow transient, however, is more difficult because of the lack of symmetry between the multiple coolant loops.

In the event of a power failure to all pumps, Eq. (7-85) is applicable as it stands. If, however, the coolant pumps are designed with sufficiently large flywheels, the pump inertia may be made much larger than that of the fluid, and hence $t_p \gg t_l$. If this is the case it is reasonable to ignore the first term in Eq. (7-85) and approximate the flow coastdown by

$$W = \frac{W_0}{1 + t/t_p}. \qquad (7\text{-}89)$$

In this situation the pump design determines the speed of the flow coastdown. In fact, power reactors are sometimes designed so that this approximation is applicable, and the inertia of the flywheel is set so that the loss-of-flow accident does not lead to unacceptable results.

In the intermediate range, where t_l and t_p are of the same order of magnitude, an analytical or numerical solution to Eq. (7-85) must be found (Burgreen, 1959). The results of such calculations are shown in Fig. 7-11, where the normalized flow is plotted versus time. The curves are given for several different values of the ratio $\alpha = t_l/t_p$. For pumps of perfect efficiency this parameter may be shown to be the ratio of the kinetic energy of the fluid in the loop to the stored kinetic energy associated with the flywheel and other rotating parts of the pump (Burgreen, 1959). Since Fig. 7-11 is plotted in terms of loop half-times, the effect of changing the pump stored energy in

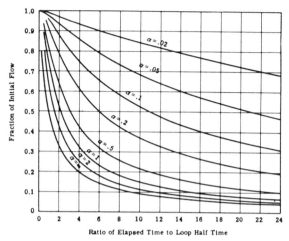

FIGURE 7-11   Analytical flow coastdown curves. From D. Burgreen, "Flow Coastdown in a Loop After Pumping Power Cutoff," *Nucl. Sci. Eng.* **6**, 306–312 (1959). Used by permission.

a coolant loop of fixed properties is graphically illustrated. The small α curves represent situations where the energy stored in the pump flywheel is large. Consequently, the pump will continue to circulate the fluid for a substantial length of time, and the flow coastdown will be slow. A large value of α indicates that little energy is stored in the pump relative to the fluid in the coolant loop. In this situation flow resistance becomes a prominent factor in causing a much faster flow coastdown.

Coolant Temperature Transient

Once a determination of the flow coastdown has been made, it is necessary to evaluate the resultant transient in the coolant temperature. For only then can it be determined whether a boiling crisis, cladding overheating, primary system overpressure, or other potentially damaging event will produce an unacceptable threat to the core or primary system integrity. In general, the evaluation of the transient must take into account the reactivity effects through temperature feedback, shutdown, and control systems. As indicated in Fig. 5-1, the flow deterioration now drives a transient in which thermal and neutronic behavior are coupled through these mechanisms. The following simple model will suffice to illustrate a few of the more rudimentary features of such transients.

Suppose we consider a reactor in which the ratio of fuel element temperature drop to coolant temperature increase is quite large so that the assumptions leading to the decoupled thermal model given by Eqs. (5-17) and (5-18) are met. Writing the coolant temperature increase across the core as

$$\Delta T_c = 2[\bar{T}_c(t) - T_i(t)], \tag{7-90}$$

we may utilize Eq. (5-18) to describe the coolant temperature behavior during the transient:

$$\frac{d\,\Delta T_c}{dt} = \frac{P(t)}{Wc_P\tau} - \left(\frac{1}{\tau} + \frac{d}{dt}\ln W\right)\Delta T_c. \tag{7-91}$$

Using the integrating factor,

$$\exp\left[-\int_0^t \left(\frac{1}{\tau} + \frac{d}{dt'}\ln W\right)dt'\right], \tag{7-92}$$

we may solve this equation for ΔT_c in terms of the reactor power and flow transient:

$$\Delta T_c(t) = \Delta T_c(0)\frac{W(0)}{W(t)}\left[e^{-t/\tau} + \frac{1}{\tau}\int_0^t \frac{P(t)}{P_0}e^{-[(t-t')/\tau]}\,dt'\right]. \tag{7-93}$$

Thus if the power and flow transients are known, the ΔT_c transient caused by the mismatch can be evaluated. This can then be used in conjunction with peaking factors, or an assumed spatial coolant temperature distribution, to find the local conditions necessary to determine critical heat flux ratios and other damage criteria.

In the absence of a detailed calculation of the power transient associated with the loss-of-flow accident, suppose we consider two highly idealized situations. First, if the reactor remains at its initial power, Eq. (7-93) reduces to just

$$\Delta T_c(t) = \Delta T_c(0)\frac{W(0)}{W(t)}. \tag{7-94}$$

Second, neglect decay heat and assume that the reactor power drops immediately to zero. Equation (7-93) then becomes

$$\Delta T_c(t) = \Delta T_c(0)\frac{W_0}{W(t)}e^{-t/\tau}. \tag{7-95}$$

Suppose that we now assume that the flow coastdown is governed by the pump inertia so that $W(t)$ may be represented by Eq. (7-89). Equations (7-94) and (7-95) then become, respectively,

$$\Delta T_c(t) = \Delta T_c(0)\left(1 + \frac{t}{t_p}\right) \tag{7-96}$$

and

$$\Delta T_c(t) = \Delta T_c(0)\left(1+\frac{t}{t_p}\right)e^{-t/\tau}.$$ (7-97)

From the first of these relationships we see that if the reactor were to remain at power there would be a linear temperature rise, with the doubling time being equal to the pump half-time t_p. From the second equation it is clear that even if there were no power production following the initiation of the flow coastdown, if the pump half-time is shorter than the core time constant, there is a peak in $\Delta T_c(t)$ that is larger than its initial value. This phenomenon occurs because in a core where $t_p < \tau$, most of the energy stored in the fuel elements as sensible heat is not transferred to the coolant until after the flow deterioration has become large.

Natural Convection Cooling

Following the total loss of pumping power only natural convection remains to circulate the coolant through the primary system. Provided the reactor has been shut down fast enough to prevent damage during the flow coast-down, a central question then becomes whether an equilibrium state can be established for the natural convection removal of decay heat without the occurrence of excessive temperatures. In what follows we assume that an equilibrium has been reached and determine the mass flow rate and coolant temperature rise through the core.

If steady-state natural convection exists in the primary system, both the inertial and pump head terms vanish from Eq. (7-76). Thus the equilibrium flow W_∞ through the core must satisfy

$$\mathcal{H}_{pr}\frac{W_\infty^2}{2\bar{m}}+g(\bar{m}\,\Delta z)_{pr}=0.$$ (7-98)

To evaluate the hydrostatic pressure head we must determine the elevation change and coolant density for each segment of the primary loop. Suppose we model the reactor primary system as indicated in Fig. 7-12. The path of the coolant now consists of the core, the upper plenum, the hot leg, the heat exchanger, the cold leg, and the lower plenum. We denote these by the subscripts c, up, hl, hx, cl, and lp, respectively. Now if each of these is characterized by an inlet and an outlet elevation, z_i and z_o, the hydrostatic head in Eq. (7-98) may be determined from

$$(\bar{m}\,\Delta z)_{pr} = m(z_o-z_i)|_c + m(z_o-z_i)|_{up}+m(z_o-z_i)|_{hl}+m(z_o-z_i)|_{hx}$$

$$+m(z_o-z_i)|_{cl}+m(z_o-z_i)|_{lp}.$$ (7-99)

Ideally, heat is exchanged with the coolant only in the core and heat exchangers. If we can also ignore the small changes in coolant density due to the pressure drops around the loop, we can assign a density of m_o to the fluid between the core outlet and the heat exchanger inlet and a density of m_i to that between the heat exchanger outlet and the core inlet. If we also assume that the density variation in the fluid is approximately linear as it passes through the core and heat exchanger, we can reduce Eq. (7-99) to

$$(\bar{m}\,\Delta z)_{pr} = m_o[\tfrac{1}{2}(z_o - z_i)|_c + (z_o - z_i)|_{up} + (z_o - z_i)|_{hl} + \tfrac{1}{2}(z_o - z_i)|_{hx}]$$
$$+ m_i[\tfrac{1}{2}(z_o - z_i)|_{hx} + (z_o - z_i)|_{cl} + (z_o - z_i)|_{lp} + \tfrac{1}{2}(z_o - z_i)|_c]. \tag{7-100}$$

To reduce this expression further, we make use of the fact that the outlet elevation of each segment of the flow path is equal to the inlet elevation of the next segment: $z_o|_{up} = z_i|_{hl}$, $z_o|_{hl} = z_i|_{hx}$, $z_o|_{hx} = z_i|_{cl}$, $z_o|_{cl} = z_i|_{lp}$, $z_o|_{lp} = z_i|_c$ and $z_o|_c = z_i|_{up}$. Hence

$$(\bar{m}\,\Delta z)_{pr} = -(m_i - m_o)(\bar{z}_{hx} - \bar{z}_c), \tag{7-101}$$

where $\bar{z}_{hx}$ and $\bar{z}_c$ are the heat exchanger and core midplane elevations.

Equation (7-101) may now be used to evaluate the core mass flow rate and the coolant temperature increase as follows. Suppose that we assume that the volumetric coefficient of thermal expansion

$$\beta = -\frac{1}{m}\frac{\partial m}{\partial T}\bigg|_p \tag{7-102}$$

can be considered to be constant over the temperature range of interest. Then the density decrease across the core can be written as

$$m_o - m_i = -\bar{m}\beta\,\Delta T_c \tag{7-103}$$

where $\bar{m}$ is the average coolant density in the core. This expression can be combined with Eq. (7-101), and the result may be substituted into Eq. (7-98), whereupon

$$\tilde{\mathcal{H}}_{pr}\frac{W_\infty^2}{2\bar{m}} - \bar{m}g\beta(\bar{z}_{hx} - \bar{z}_c)\,\Delta T_c = 0. \tag{7-104}$$

To determine the flow and temperature increase we need an additional relationship between ΔT_c and W. For steady-state flow we see from Eq. (1-117) that

$$\Delta T_c = \frac{P_d}{W_\infty c_p}, \tag{7-105}$$

where P_d is the power level produced by decay heat. Eliminating either W_∞ or ΔT_c yields

$$W_\infty = \left[\frac{2\bar{m}g\beta P_d}{\tilde{\mathcal{K}}_{pr}c_p}(\bar{z}_{hx} - \bar{z}_c) \right]^{1/3} \qquad (7\text{-}106)$$

for the core flow, and

$$\Delta T_c = \left(\frac{P_d}{\bar{m}c_p} \right)^{2/3} \left[\frac{\tilde{\mathcal{K}}_{pr}}{2g\beta(\bar{z}_{hx} - \bar{z}_c)} \right]^{1/3} \qquad (7\text{-}107)$$

for the coolant temperature rise.

Clearly, if natural convection is to be relied on to remove decay heat, it is necessary to elevate the heat exchangers well above the core, assuming the flow is upward through the core. The foregoing analysis assumes that flow resistance of the primary system is a constant. In reality it may be necessary to take the Reynolds number dependence of $\tilde{\mathcal{K}}_{pr}$ explicitly into account, particularly if situations are encountered where laminar flow causes the loss coefficients to display an inverse Reynolds number dependence. In water-cooled systems natural convection boiling may sometimes take place, thereby causing both an increase in the density change $m_i - m_o$ and in the flow resistance $\tilde{\mathcal{K}}_{pr}$. The net effect, however, can be expected to enhance the efficiency of the natural convection heat removal, unless the heat fluxes are so large as to cause a boiling crisis and/or vapor binding.

In gas-cooled systems the fluid density is too small to allow decay heat removal by natural convection. Thus emergency measures must be taken to provide forced convection removal of the decay heat before excessive temperatures are reached. Such cooling systems most often take the form of auxiliary coolant loops that are set into operation on detection of the loss of flow. One of the most important criteria in the design of such loops is that their power supply be independent of the main coolant pumps so that they also are not disabled by the mechanism initiating the loss-of-flow accident. Such auxiliary loops might be driven, for example, by small turbines that utilize directly the steam produced in the steam generators by the decay heat. Conversely, in gas-cooled reactors, the primary coolant pumps are often driven directly by steam produced from the reactor in order to eliminate those failure modes that originate in the electrical power supply. The auxiliary loop might then be driven by electrical pumps, with a provision for the power being supplied by on-site diesel generators.

Our discussion of the loss-of-flow accident has been in terms of the coolant temperature rise through the core. Provided the inlet temperature is known, the evaluation of ΔT_c allows one to determine whether excessive core temperatures, primary system overpressure, or thermal shock occur. If it can be assumed that the heat exchanger is intact and removes heat at the same rate as it is produced, the assumption of a constant inlet temperature may be reasonable. If there is a failure in the secondary system that causes the removal of heat from the primary system to be impaired, the inlet temperature will rise. Should such a failure occur either alone or in conjunction with a loss-of-primary-system flow, a loss-of-heat-sink accident is said to occur. These are discussed in the following section.

7-5 LOSS OF HEAT SINK

The preponderance of the preceding chapters is devoted to accidents that originate within the reactor core or primary system. Any interruption of the orderly flow of energy from the reactor fuel to its ultimate disposition as electricity and waste heat, however, can lead to excessive overheating of nuclear power plant components and to undue stress on the fission product barriers. In particular, if an accident takes the form of a disruption of energy transport through the steam cycle or other systems beyond the primary system envelope, it is reflected first back into the primary system as rising coolant temperature and pressure because of the inability of the heat exchangers to remove heat. We therefore refer to these accidents as a loss of heat sink. Excessive stress to the primary envelope and damage to the reactor fuel may result from the primary system temperature and pressure transients that accompany such accidents. In examining such loss-of-heat-sink accidents we consider only indirect- and intermediate-cycle systems. We first model the primary coolant temperature transients that result from the energy imbalance in the primary system. We then examine the pressure transients associated with uncontrolled rises in coolant temperature. Finally, we categorize the myriad of secondary system faults that may lead to losses of heat sink and discuss the safety systems that are used to deal with such accidents.

Temperature Transients

To approximate the coolant temperature transients that may appear as the result of a heat-sink failure, we consider a primary system consisting of core and lower and upper plenums, and identical coolant loops consisting of hot legs, heat exchangers, cold legs, and pumps, as shown in Fig. 7-12.

A Primary System Model. If the heat losses from the primary system can be neglected, then during steady-state operation the coolant between the core outlet and the heat exchanger inlet will be at the same temperature T_o

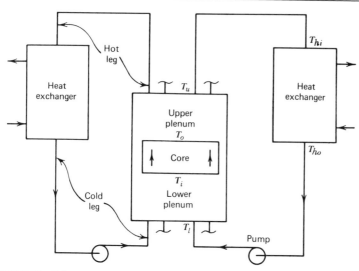

FIGURE 7-12 Primary system diagram (two of four identical coolant loops shown).

and the coolant between the heat exchanger outlet and the core inlet will be at T_i. Under transient conditions an energy balance can be written for the coolant within each of these regions by assuming that the flow is incompressible. For the upper plenum and hot leg,

$$M_1 c_p \frac{d\bar{T}_1(t)}{dt} = W c_p [T_o(t) - T_{hi}(t)], \qquad (7\text{-}108)$$

and for the lower plenum and cold leg,

$$M_2 c_p \frac{d\bar{T}_2(t)}{dt} = W c_p [T_{ho}(t) - T_i(t)]. \qquad (7\text{-}109)$$

In these expressions M_1 and M_2 are the masses of coolant located, respectively, between core outlet and heat exchanger inlet, and between heat exchanger outlet and core inlet, and $\bar{T}_1$ and $\bar{T}_2$ are the mass-averaged coolant temperatures over these volumes. The temperatures T_{hi} and T_{ho} are those of the heat exchanger inlets and outlets, respectively, and it is assumed that flow transients are not so rapid as to cause the mass flow rate $W(t)$ to vary significantly in space around the coolant loops.

Equations (7-108) and (7-109) may be used to arrive at a simple lumped parameter model for the average coolant temperature in the primary system. To do this we assume that the coolant mass within the core and the heat exchanger contributes only a small part of the primary system heat capacity. Then we simply include these small contributions in M_1 and M_2, and assume that coolant heat capacities can be neglected in the core and heat exchanger models. Adding Eqs. (7-108) and (7-109) yields

$$M_{pr}c_p\frac{d\bar{T}_{pr}}{dt} = Wc_p(T_o - T_i) - Wc_p(T_{hi} - T_{ho}), \qquad (7\text{-}110)$$

where the primary system coolant mass is

$$M_{pr} = M_1 + M_2, \qquad (7\text{-}111)$$

and the mass-averaged primary system coolant temperature is

$$\bar{T}_{pr} = \frac{M_1\bar{T}_1 + M_2\bar{T}_2}{M_{pr}}. \qquad (7\text{-}112)$$

The core may be modeled using Eqs. (5-17) and (5-18), provided the coolant heat capacity is neglected. With the assumption that $\bar{T}_c = 0.5 \times (T_o + T_i)$, Eq. (5-18) can be solved to yield

$$Wc_p(T_o - T_i) = \frac{1}{\tau}\int_{-\infty}^{t} P(t')\, e^{-(t-t')/\tau}\, dt'. \qquad (7\text{-}113)$$

The heat exchanger may be roughly modeled as follows. The rate at which energy is removed by the heat exchanger can be written as

$$Wc_p(T_{hi} - T_{ho}) = hA\, \Delta T_m. \qquad (7\text{-}114)$$

For a heat exchanger consisting of thin-walled tubes with total area A, the effective heat transfer coefficient is determined from

$$\frac{1}{h} = \frac{1}{h_p} + \frac{\delta}{k_{hx}} + \frac{1}{h_s}, \qquad (7\text{-}115)$$

where h_p and h_s are the coefficients associated with the primary and secondary side of the exchanger, and δ and k_{hx} are the thickness and conductivity, respectively, of the heat exchanger walls. The quantity ΔT_m is the mean temperature drop between primary and secondary sides of the heat exchanger and may be defined differently, depending on whether a steam generator, cross-flow, multiple-flow, or counterflow system is under analysis. If we consider the heat exchanger to be a steam generator in which the secondary fluid is at or near the saturation temperature along the heat transfer surface, we may assume that the secondary side of the steam generator is at a single temperature T_s. For this configuration the mean temperature drop may be shown to be (Kay, 1963)

$$\Delta T_m = \frac{T_{hi} - T_{ho}}{\ln\left(\dfrac{T_{hi} - T_s}{T_{ho} - T_s}\right)}. \qquad (7\text{-}116)$$

Strictly speaking, the foregoing relationships are valid only for steady-state operation. Usually, however, the heat capacity of the primary fluid and tubing within the steam generator can be neglected when compared to that of the core and primary system coolant inventory. Therefore a reasonable approximation to transient behavior is obtainable simply by assuming that the temperatures, and possibly in some situations the heat transfer coefficient, in Eq. (7-114) are time dependent. An additional approximation results in further simplification of the heat exchanger model. Suppose we write Eq. (7-114) as

$$W c_p (T_{hi} - T_{ho}) = h A \Gamma (\bar{T}_{pr} - T_s),$$ (7-117)

where

$$\Gamma = \frac{\Delta T_m}{\bar{T}_{pr} - T_s}.$$ (7-118)

In these expressions it is assumed that the effective temperature drop across the heat exchanger walls is proportional to $\bar{T}_{pr} - T_s$. This should be acceptable to a first approximation since $\bar{T}_{pr}$ will lie between T_{hi} and T_{ho}. The proportionality factor Γ should remain relatively constant, provided the temperature profile along the heat exchanger walls does not change drastically. To illustrate, suppose we consider the particular case where

$$\bar{T}_{pr} \simeq \tfrac{1}{2}(T_{hi} + T_{ho}).$$ (7-119)

The definition of ΔT_m may be utilized to express Γ as

$$\Gamma = \frac{2}{\ln R_T}\left(\frac{R_T - 1}{R_T + 1}\right),$$ (7-120)

where

$$R_T = \frac{T_{hi} - T_s}{T_{ho} - T_s}.$$ (7-121)

Thus if the ratio of heat exchanger inlet to outlet temperature drops between primary and secondary systems remains constant, the value of Γ is fixed, and the lumped parameter treatment is valid.

Primary System Energy Balance. At this point we may substitute the simple core and heat exchanger models given by Eqs. (7-113) and (7-117) into Eq. (7-110) to obtain a primary system energy balance in terms of the core power and the steam temperature in the secondary system:

$$\frac{d\bar{T}_{pr}}{dt} = \frac{1}{M_{pr}c_p \tau} \int_{-\infty}^{t} P(t')\, e^{-(t-t')/\tau}\, dt' - \frac{1}{\tau_{pr}}(\bar{T}_{pr} - T_s).$$ (7-122)

Observe that in addition to the core time constant τ, the average coolant temperature is governed by a primary system time constant

$$\tau_{pr} = \frac{M_{pr}c_p}{hA\Gamma}. \tag{7-123}$$

This energy balance may be expected to give reasonable results provided the transients are not fast compared to τ or τ_{pr}, and where the time scales involved are larger than the loop transit time $t_{tr} = M_{pr}/W$, which is a measure of the time required for the coolant to complete one cycle through the primary system.

The primary system time constant τ_{pr} is a measure of the time required to remove heat from the primary to secondary system. Its significance is made clearer by assuming, as often is the case in liquid-metal-cooled systems, that the core time constant is much shorter than that of the primary system. Equation (7-122) then reduces to

$$\frac{d\bar{T}_{pr}}{dt} = \frac{P(t)}{M_{pr}c_p} - \frac{1}{\tau_{pr}}(\bar{T}_{pr} - T_s), \tag{7-124}$$

indicating that in this situation the power produced enters into the coolant stream instantaneously. If we now hypothesize that there is an instantaneous termination of the power production—say at $t = 0$—the primary system temperature would decay exponentially toward T_s as

$$\bar{T}_{pr}(t) = T_s + [\bar{T}_{pr}(0) - T_s]e^{-t/\tau_{pr}}. \tag{7-125}$$

In some systems, such as those gas-cooled reactors in which the heat must pass through large masses of graphite moderator before reaching the low-density coolant, the core time constant may be larger than that of the primary system. In the extreme case where the small mass of the coolant gas causes $\tau \gg \tau_{pr}$, the derivative term in Eq. (7-122) can be dropped, and we have simply

$$\bar{T}_{pr}(t) = T_s + \frac{hA\Gamma}{\tau} \int_{-\infty}^{t} P(t')\, e^{-(t-t')/\tau}\, dt'. \tag{7-126}$$

In this situation, a sudden termination of power at $t = 0$ would result in the primary system cooling down to T_s as

$$\bar{T}_{pr}(t) = T_s + [\bar{T}_{pr}(0) - T_s]e^{-t/\tau}. \tag{7-127}$$

In a more realistic setting, it is likely that Eq. (7-122) will require solution when τ and τ_{pr} are of comparable magnitude. Moreover, in the loss-of-heat-sink accident, one may expect a rising value of T_s, or an increasing value of

τ_{pr}. In particular, if vapor blanketing on the secondary side of the heat exchanger causes h to become very small, then τ_{pr} may become so large that the last term in Eq. (7-122) is insignificant. In this situation of complete insulation, the adiabatic heatup of the primary system is governed by

$$\frac{d\bar{T}_{pr}}{dt} = \frac{1}{M_{pr}c_p}P(t), \tag{7-128}$$

where for simplicity we again have assumed that τ is small.

It is likely that a loss-of-heat-sink accident will be in the form of impairment rather than a total loss of the primary system heat-removal capability described by the adiabatic heatup. Such impairment may result from several situations. For the case in which the steam generator pressure increases, the saturation temperature T_s increases. If, on the other hand, there is a substantial decrease in pressure, causing T_s to decrease, a boiling crisis is likely to occur, causing h_s and therefore h to decrease sharply. Other difficulties can also be envisioned: Decreases in primary or secondary flow cause smaller values in hA to result; loss of inventory from the secondary side of the heat exchanger may cause the tubes to be uncovered, causing the heat transfer area to become smaller. The causes in the secondary system that can result in such steam generator problems are taken up in a subsequent subsection.

For the case in which the primary system heatup rate is slow, such as would be the case following reactor shutdown, other heat capacities can come into play to slow the rate of temperature rise. The most significant of these is likely to be that of the reactor vessel walls. At the same time, the penetration of the transient temperature distribution into vessel walls can add thermal stresses to the increasing pressure stress level that accompanies the rising coolant temperature. Furthermore, in prestressed concrete vessels, temperatures can eventually be reached that cause decomposition of the concrete.

The average primary system temperature provides a useful indication of the severity of heat-sink failure accidents, and in particular it points out the importance of an early reactor trip in slowing the heatup rate. Nonetheless, if faster transients occur, such as those in conjunction with flow-failure accidents, the onset of core damage and primary system overpressure are more accurately indicated by the temperatures at the core outlet and in the hot leg. Unfortunately, the foregoing model gives no indication of temperature distribution within the primary system.

In general, the modeling of the transient temperature distribution within the primary system is an arduous task. A reasonably simple model for numerical computations can be formulated by coupling the core and heat exchanger described previously to models of the plenums and loops that involve the extreme flow patterns of complete mixing or pure transit delays (Grace, 1964). A simple model for the system shown in Fig. 7-12 is as follows.

Suppose that the mass of fluid in the hot and cold legs is small compared to that in the plenums. Then the masses M_1 and M_2 in Eqs. (7-108) and (7-109) can be associated with the average plenum temperatures. If the flow mixing within each of the plenums is idealized to be complete and instantaneous, the plenum outlet temperatures are the same as the average plenum temperatures. The plenum energy balances can be written as

$$M_1 c_p \frac{dT_u(t)}{dt} = W c_p [T_o(t) - T_u(t)] \qquad (7\text{-}129)$$

and

$$M_2 c_p \frac{dT_i(t)}{dt} = W c_p [T_l(t) - T_i(t)], \qquad (7\text{-}130)$$

where T_u and T_l are the upper plenum outlet and lower plenum inlet temperatures as indicated in Fig. 7-12, and T_i and T_o are again the core inlet and outlet temperatures. If we now assume that there is no flow mixing in the hot and cold legs, the inlet and outlet temperatures in these flow channels may be represented by pure transit delays:

$$T_{hi}(t + t_h) = T_u(t) \qquad (7\text{-}131)$$

and

$$T_l(t + t_c) = T_{ho}(t), \qquad (7\text{-}132)$$

where t_h and t_c are the transit times for the fluid to pass through the hot and cold legs, respectively.

Equations (7-129) through (7-132) can be combined to give equations that can be solved for the heat exchanger inlet temperature in terms of the core outlet temperature

$$M_1 c_p \frac{dT_{hi}(t)}{dt} = W c_p [T_o(t - t_h) - T_{hi}(t)] \qquad (7\text{-}133)$$

and for the core inlet temperature in terms of the heat exchanger outlet temperature

$$M_2 c_p \frac{dT_i(t)}{dt} = W c_p [T_{ho}(t - t_c) - T_i(t)]. \qquad (7\text{-}134)$$

With these two equations the primary system temperature transient can be followed, provided a core model, such as Eq. (7-113), is available to relate T_o to T_i and the heat exchanger can be modeled, as in Eq. (7-117), to relate T_{ho} to T_{hi}.

Pressure Transients

With increasing temperature, the thermal expansion of the coolant leads to increasing pressure unless sufficient space is available to accommodate the increased coolant volume. Large increases in primary system pressure due to this thermal expansion can occur during accidents involving excessive coolant temperatures unless protective features are incorporated into the system design. These pressures can result in excessive stress on the primary system envelope, and eventually in a loss-of-coolant accident emanating from the rupture of that envelope. Therefore an integral part of each reactor design must be a system of expansion volumes and/or relief and safety valves that allow for control of pressure transients and prevention of excessive stress on the primary system envelope.

Gas-Cooled Systems. The control of pressure transients strongly depends on whether the primary system coolant is a gas or a low or a high vapor pressure liquid. In gas-cooled systems, the perfect gas law may be used to relate the primary system pressure to the average coolant temperature,

$$p = \bar{m}R\bar{T}_{pr}, \tag{7-135}$$

where $\bar{T}_{pr}$ here is in absolute units. Since the primary system volume remains constant during overpressure transients, $\bar{m}$, the average coolant density, is also fixed. Therefore the pressure and temperature are related by

$$p(t) = p(0)\frac{\bar{T}_{pr}(t)}{\bar{T}_{pr}(0)}. \tag{7-136}$$

If $p(t)$ approaches the design limits of the primary envelope, protective action must be taken to discharge some of the gas from the primary system and prevent further pressure increases. The preferred method for discharging coolant is through motor-operated relief valves that carry out a controlled relief of just enough coolant to prevent further pressure increases.

A backup system must be provided for the possibility that the transient is of such speed that the relief valves cannot be opened fast enough or that their flow rates are not large enough to stem the pressure rise. Such a system consists of safety valves that blow open as a direct result of the internal pressure and do not reseat until the pressure is relieved. These are set to open at a higher pressure than the relief valves. Since the gas coolant is noncondensible, even at room temperature, a large reservoir must be provided into which the relief and safety valves can discharge without causing an undue back pressure. Often only the containment vessel is sufficiently large to serve this purpose. Hence the effects of such pressure relief must be taken into account both in setting the design temperature and pressure of the containment and in treating the fission product activity of the coolant that will be released to the containment.

Sodium-Cooled Systems. The volumetric change undergone by liquid coolants during temperature transients is so small that usually it can be neglected in flow and temperature transients. Nevertheless, it plays an important part in pressure transients. Because the liquid is nearly incompressible, a vapor volume must be provided to accommodate the expansion; otherwise the expansion will be felt as deformations in the primary system envelope, which will lead eventually to rupture if the temperature increase is sufficiently large. The mechanism by which expansion volumes are provided differs between systems cooled by low and by high vapor pressure liquids. We therefore treat them separately and take sodium-cooled and water-cooled reactors as examples.

Since the saturation temperature of sodium is high, even at atmospheric pressure, there is no need to pressurize the primary system. Rather, as shown in Fig. 8-11, the reactor vessel is filled with sodium to a prescribed level, with the remainder of the vessel being occupied by an inert cover gas. The compressibility of this cover gas permits the sodium to expand with increasing temperatures without causing an undue pressure increase. The magnitude of the pressure increase can be estimated as follows.

Suppose β, given by Eq. (7-102), is the coefficient of thermal expansion. Neglecting the temperature dependence of β, the increase in the sodium volume V_{Na} can be shown to be given by

$$V_{Na}(t) - V_{Na}(0) = \beta[\bar{T}_{pr}(t) - \bar{T}_{pr}(0)]V_{Na}(0), \qquad (7\text{-}137)$$

and this must be just equal to the decrease in the cover gas volume. Since the mass of cover gas is fixed, its density increase is

$$\bar{m}_{cg}(t) = \bar{m}_{cg}(0)\frac{1}{1 - \dfrac{V_{Na}}{V_{cg}}\beta[\bar{T}_{pr}(t) - \bar{T}_{pr}(0)]}, \qquad (7\text{-}138)$$

where V_{Na}/V_{cg} is the initial ratio of sodium to cover gas volume. Now the perfect gas law may be used along with this expression to estimate the cover gas pressure transient:

$$p(t) = p(0) \frac{T_{cg}(t)}{\overline{T}_{cg}(0)} \frac{1}{1 - \dfrac{V_{Na}}{V_{cg}} \beta [\overline{T}_{pr}(t) - \overline{T}_{pr}(0)]}. \tag{7-139}$$

The cover gas temperature will vary, depending on the rate of coolant expansion. For slow transients it may be expected to remain in thermal equilibrium with the sodium, and hence we may take $T_{cg}(t) = \overline{T}_{pr}(t)$ in the preceding expression. For very fast transients the compression of the cover gas may be nearly adiabatic. In this limiting case Eq. (7-139) is replaced by

$$p(t) = p(0) \frac{1}{\left\{ 1 - \dfrac{V_{Na}}{V_{cg}} \beta [\overline{T}_{pr}(t) - \overline{T}_{pr}(0)] \right\}^{\gamma}}, \tag{7-140}$$

where γ is the ratio of cover gas specific heat at constant pressure to constant volume.

To limit pressure increases in sodium systems it is desirable to make the cover gas volume V_{cg} as large as practicable. To prevent the sodium hammer effects that may arise in the event of a core disruptive accident (see Section 9-3), however, it is desirable to minimize the height of the gas column between the sodium surface and the reactor vessel head. For this reason the reactor vessel is often connected to an auxiliary tank containing enough cover gas volume to achieve an acceptable ratio V_{Na}/V_{cg}. With this configuration, the expansion of sodium over large temperature ranges can be accommodated, provided its boiling point is not reached. If the sodium in the core boils, a large increase in V_{Na} and a more severe pressure transient may result. Such an event would only be expected to occur, however, in the case of a core disruptive accident, such as those discussed in Sections 5-6 and 9-3. The effects of blast and sodium hammer then become the primary determinants of damage to the primary system envelope.

Water-Cooled Systems. The increase in pressure with coolant temperature is more difficult to predict when water or any other high vapor pressure liquid is the coolant because transient two-phase behavior must be modeled (Grace, 1964). If the primary system operates below the boiling temperature, as in the case of a pressurized-water reactor, the liquid is maintained in equilibrium with a volume of its own vapor in a special pressurizer chamber, such as that shown in Fig. 8-20. During normal operation the system pressure is controlled by the pressurizer through the use of a system of electric heaters and sprays that maintain the pressurizer temperature at a fixed value; consequently, a constant pressure is maintained at the vapor liquid interface. The pressure in the pressurizer is the same as that of the primary system at the point where the connecting surge line enters the coolant loop. The pressurizer temperature, however, must be higher than that in the coolant loops, since the pressurizer is at saturation temperature while the remainder of the primary system is subcooled.

Rapid increases in the primary system temperature cause the coolant to expand into the vapor volume of the pressurizer. As in the case of sodium, the volume increase of the primary system coolant is given approximately by

$$V_{H_2O}(t) = V_{H_2O}(0) + \beta [\bar{T}_{pr}(t) - \bar{T}_{pr}(0)] V_{H_2O}(0), \qquad (7\text{-}141)$$

from which the surge rate in the pressurizer can determined. So long as the pressurizer does not fill completely with liquid, the pressure transient is tempered by the condensation of the vapor phase that accompanies the increase of saturation temperature with pressure. For this reason it is desirable to have a large vapor volume initially present in the pressurizer. Further pressure relief is provided through the use of motor-operated relief valves that open when excessive pressures are present, and vent the excess pressurizer inventory to a dump tank in a controlled manner. The relief valves, however, may not be fast acting enough or of large enough capacity to prevent the pressurizer from filling and causing excessive pressures in the event of a severe transient. For this reason safety valves venting directly from the primary system to the containment are set at a pressure that is higher than the relief valves, but that does not produce damaging stress to the primary system envelope. With the opening of fast-acting safety valves the pressurizer transient is terminated, but as in the case of gas-cooled reactors one must then consider the effect of the release of coolant that may be quite radioactive to the containment; in particular, the effects of the release on containment pressure and temperature must be examined.

Direct-Cycle Boiling-Water Reactors.   In direct-cycle boiling-water reactors, such as that shown in Fig. 8-22, both liquid and vapor phases are present within the reactor vessel, and the temperature-pressure transients induced by the disruption of the orderly transport of heat out of the reactor vessel are compounded by the following effects. Mismatches between steam and feedwater flow can lead to uncovering of the core, on the one hand, or filling of the reactor vessel with liquid, on the other. If the core is uncovered, the reactor will automatically shut down because of the negative void coefficient, but the fuel will overheat from the resulting vapor binding of the fuel. For this reason, high-pressure coolant injection systems must be available to supplement the feedwater system in maintaining the water inventory under accident conditions. If the excessive coolant inventory is added to the vessel, the pressure will rise, and the core may undergo a nuclear excursion because of the collapse of the core voids with increasing pressure. To prevent overstress of the reactor vessel, it is equipped with both safety and motor-operated relief valves that operate to terminate the pressure transient, again at the cost of discharging coolant into the containment.

Initiating Events

By and large, heat-sink-failure accidents originate outside the primary system envelope. Therefore, to examine the mechanisms leading to such accidents and the protective actions that are taken to mitigate their consequences, we must examine the transport of energy from the primary system heat exchangers to its ultimate disposition as electricity and waste heat. As indicated by Fig. 1-15, the systems by which this is accomplished can differ substantially. In direct-cycle systems steam is generated within the reactor vessel, as shown in Fig. 1-15b. Direct cycle gas turbine configurations are also being proposed. In the more common indirect cycle, the steam is produced in separate steam generators, as shown in Fig. 1-15a. An intermediate coolant loop, such as that shown in Fig. 1-15c, is added in liquid-metal-cooled systems. Most of the considerations that appear in conjunction with loss-of-heat-sink accidents, however, can be illustrated using the indirect-cycle system.

The Secondary System. The principal features of a typical secondary system associated with an indirect or intermediate-cycle nuclear power plant are shown schematically in Fig. 7-13. Water is pumped into the steam generators by feedwater pumps. There, high-pressure and high-temperature steam is generated. The steam then passes through steam lines to the turbine. In the turbine the steam loses pressure and temperature as its thermal energy is converted to mechanical energy. The steam leaves the turbine at reduced temperature and subatmospheric pressure and enters the condenser. There it loses heat to the circulating water through a heat exchanger and condenses to the liquid phase. The cycle is completed by the water being pumped out of the condenser, through the feedwater lines, and back into the steam generators.

A steam cycle must be designed to meet variable demand for electrical power production. The turbine generator frequency is fixed by the synchronization of the electrical power network. The inlet temperature and pressure conditions for the turbine are normally stipulated by turbine design to within a relatively narrow range. Therefore the power production is controlled by regulating the mass flow rate of the steam by throttle valves at the turbine inlet, and by the feedwater flow through control valves in the

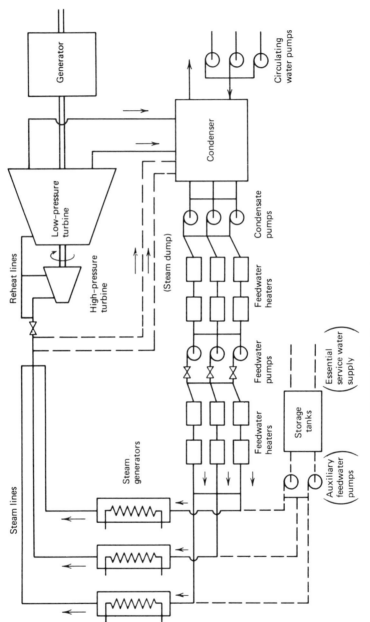

FIGURE 7-13 Steam cycle diagram.

feedwater lines. The reactor power must, of course, also be regulated to match the demand of the turbine generator.

Many causes for the failure of the orderly transport of energy through the secondary system can be enumerated: mechanical failure of pumps, valves, or other active components, ruptures of lines carrying the steam or water, failure in the steam or feedwater flow control systems, power failures, and so on. As shown in Fig. 7-13, the provisions for parallel heat transport paths through the feedwater system, steam generators,and steam lines minimize the consequences of many of the more common single-component failures. Moreover, appropriate valving can further reduce both the likelihood and consequences of such failures: Isolation valves permit the maintenance and repair of equipment while the reactor is operating at partial power, and check valves prevent the flow reversals that aggravate many accident transients.

Even with these precautions a number of accidents can entirely distrupt the energy transport system. There is only one turbine condenser and therefore problems with this system affect all heat transport paths. There are typically a number of headers common to the parallel transport paths whose rupture would lead to a general breakdown of the orderly energy flow. Likewise, failure of the feedwater or steam flow controls or of the supply of A.C. power to the various pumps could incapacitate all the parallel heat transport paths. Since the probability of all these accidents is not vanishingly small, it is necessary to provide alternate paths by which heat can be removed from the primary system.

In the event of an interruption of the flow of energy away from the primary system, it is essential that the malfunction be sensed and the protective system promptly function to trip the reactor. This having been accomplished, the energy input to the plant is reduced to the level of decay heat. Several options may then be used for emergency heat dissipation. Included in all plants is the steam dump, as shown in Fig. 7-13, which allows the turbine to be bypassed and the heat to be dissipated in the condenser. Although this system provides a viable method for dealing with turbine failures, it is not adequate if the failure is in another part of the secondary system. To deal with these other failures, one of two methods is normally used. First, a completely independent heat transport system may be provided. This would consist of an auxiliary primary system coolant loop and heat exchanger, and a heat dissipation system on the secondary side of the auxiliary heat exchanger. Alternately, the steam generators may be used, as indicated in Fig. 7-13, as the basis for an open cycle heat dissipation system. As indicated in the diagram, auxiliary feed pumps inject water from a reservoir into the steam generators. The water then boils and is vented to the atmosphere through relief and/or safety valves in the steam generators.

Secondary System Disturbances. The use of the auxiliary heat removal systems is best illustrated by considering some of the more common accidents that may originate in the secondary systems. Often it is useful to group these in pairs, since for many of the accidents in which heat removal through the steam generators is impaired there is a converse situation in which excessive heat removal takes place. Although the loss of heat sink results in rising primary system coolant temperature and pressure, excessive heat removal may result in a sudden cooldown of the primary system. The latter situation can result in excessive thermal stresses to the reactor vessel and other components. Moreover, for reactors with large negative coolant temperature coefficients, the decrease in core inlet temperature associated with the cooldown induces a reactivity increase, and the reactor then undergoes a nuclear transient analogous to the cold-water accidents discussed earlier.

The measures to protect a nuclear plant from adverse occurrences invariably include a reactor trip, to eliminate the heat source, and a turbine trip, to protect the turbine generator from damage. The turbine trip constitutes the most common interruption of the secondary system energy flow, because it is used not only in the event of an accident within the primary system, but also to prevent damage from causes arising within the turbine generator or from adverse demand conditions on the electrical network. Trip signals, for example, are originated by loss of electrical load, overload, excessive load changes, turbine overspeed, bearing overheating, loss of lubricating fluid, and so on.

In the event of a turbine trip, the energy stored in the primary system and steam generators is dissipated by closing the turbine intake valves and opening valves to the steam dump system, thereby providing a bypass path to the condenser. In some systems the steam dump may not have the speed and flow capacity to handle trip from full power without resulting in excessive pressure on the secondary side of the steam generator. Thus in this case a momentary opening of steam generator relief valves is required to stop the pressure buildup in the secondary system.

A loss-of-heat-sink accident will occur if in conjunction with a turbine trip the steam dump valves do not open. This event may take place because of failure in the steam dump valves or their control system, or it may occur as a result of deliberate action taken in conjunction with some condenser failure accidents: If the condenser fails, either as a result of a loss of condenser vacuum or from failure of the circulating water system, the continued supply of steam through either the turbine or steam dump causes the condenser pressure to increase. To prevent the massive rupture of the condenser from internal pressure, some protective systems are designed to cause turbine trip with no steam bypass through the dump system in the event of this accident.

If, for whatever reason, a turbine trip without steam bypass does occur, there will be a steam line flow stoppage. This in turn will prevent heat removal from the steam generators and cause a rapid pressure buildup. The protective action in this event consists of opening steam generator relief and safety valves to reduce the pressure and provide an atmospheric heat sink. Since the water supply to the condenser is cut off, the reservoir for the feedwater system may be inadequate for long-term heat removal. It therefore may be necessary to utilize the auxiliary feedwater system to provide water to the steam generators.

The converse situation to the turbine trip without bypass occurs when there is an uncontrolled opening of the turbine throttle or steam dump valves, or a steam line rupture. Increased steam flow then removes heat from the steam generator at an accelerated rate, causing an initial cooldown of the primary system. At a later time the steam generator may become ineffectual, since a continuation of the steam flow in excess of what can be provided by the feedwater system results in the loss of steam generator water inventory and subsequent vapor binding of the heat transfer surfaces. An important part of safety analysis is to demonstrate that the plant can survive such a transient until the auxiliary heat removal path can be established without incurring unacceptable damage.

Complete disruption of feedwater flow can be caused by loss of feedwater control, loss of power to the feedwater pumps, or rupture of the headers in the feedwater system. The flow loss causes the steam generator inventory to be exhausted, whereupon vapor binding of the heat transfer surface and thermal insulation of the primary system takes place. The accident is countered by rapid trip of the reactor, so that the water inventory in the steam generators is sufficient to provide cooling until the auxiliary feedwater system can be brought into operation for decay heat removal. For these actions to be successful in the event of a rupture in the feedwater system, check valves must be present downsteam of the rupture to prevent the drainage of the steam generator inventory through the break.

Conversely, control failures that result in increased feedwater flow, or loss of feedwater heaters, give rise to excessive heat removal transients from the resultant cooling of the steam generator inventory. Likewise, the inadvertent start-up of the auxiliary feedwater system results in steam generator cooldown. In either event it must be shown that the resulting transient can be sustained without plant damage.

In a direct-cycle boiling-water reactor, the reactor vessel takes the place of the steam generators, and the strong negative reactivity feedback resulting from voids in the core causes the foregoing accidents to have a strong and immediate effect on the reactor power. In particular, the pressure rise associated with turbine trip without bypass causes the bubbles in the core to

collapse and a nuclear excursion to result. Increases in feedwater or recirculation flow also may cause reactivity increases as a result of the sweeping of
bubbles from the core. Conversely, steam line feedwater line breaks cause
increased void formation and reactivity decreases.

Power-Failures.   In some nuclear plants the loss of off-site A.C. power can
result in the most serious of heat sink failures, since if all the pumps are
electrical, simultaneous turbine trip, loss of primary flow, loss of feedwater
flow, and loss of circulating water flow will take place. The control and
protection system will continue to operate using D.C. power stored in
on-site battery banks, and reactor shutdown will be driven by gravitational
or some form of stored energy. Nevertheless it is necessary to have diesel
generators or other reliable sources of on-site power that can be quickly
brought into operation. If primary system natural circulation is adequate for
the removal of decay heat, the emergency power is still required to drive the
auxiliary feedwater pumps. If, as in the case of gas-cooled reactors, forced
convection is required to remove decay heat from the primary system,
emergency power also must be supplied to auxiliary primary system coolant
pumps. In the latter situation a common design option is to provide a
completely separate path for decay heat removal rather than use the same
coolant loops and steam generators as during normal operation.

    To reduce the dependence of the heat transport system on A.C. power,
and in some cases to improve the plant economics, it is often desirable to
utilize some of the steam from the generators to drive feedwater or other
pumps through the use of small steam turbines. If coolant or feedwater
pumps are powered in this manner, the use of electrically driven emergency
equipment adds valuable diversity to the safety systems. Conversely, if
coolant or feedwater pumps are electrically driven, the effects of loss of A.C.
power can be minimized by providing an auxiliary decay heat removal path
that can be operated entirely by natural convection and/or from the steam
produced by decay heat. Finally, further increase in safety system reliability
may be provided by permitting the option of operating auxiliary feedwater
pumps and other essential equpment either by on-site emergency power or
from steam produced by the decay heat, auxiliary boilers, or other devices.

    The foregoing discussion is centered about the immediate problem of
establishing a stable path for decay heat removal following heat-sink-failure
accidents. It also is essential that both the power supply and the cooling-
water reservoir be capable of providing for this heat removal ad infinitum.
Furthermore, in these or any other accident types, provisions must be made
for cooling down the primary system, and, if necessary, dismantling the
reactor core. In low-pressure sodium systems a single type of auxiliary heat
removal system may be adequate over the entire range of coolant tempera-

tures encountered during such procedures. Likewise, variable-speed circulators may allow the auxiliary heat removal system in a gas-cooled reactor to operate over the entire range of pressures and temperatures. Often, however, a second auxiliary system must be provided for use following reactor cooldown. For example, in pressurized-water reactors, the steam generators are inoperative following the cooldown of the reactor and removal of the vessel head. Thus, the auxiliary feedwater system shown in Fig. 7-13 can no longer be used. Rather a low-pressure residual heat removal system must be provided to extract decay heat directly from the pressure vessel. Finally, in all reactors reliable heat removal systems must be provided for the fuel after it is removed from the core, for it is only by ensuring that reliable decay heat removal is provided for all eventualities that fuel melting and subsequent releases of unacceptable amounts of fission products can be avoided.

PROBLEMS

7-1. Equation (7-48) relates the flow rate and pressure drop for n identical channels placed in parallel. Derive the analogous expression for n identical channels placed in series.

7-2. An incompressible coolant of density m flows through a reactor channel of length L and area A with a mass flow rate $W(0)$. A partial flow blockage at the channel entrance causes the pressure drop to fall instantaneously to one-half of its initial value Δp_0. Assuming that the friction factor, fluid density, and hydraulic diameter D remain constant, determine the mass flow rate $W(t)$ as a function of time following the blockage.

7-3. Verify the following equations that relate to pump failures:
a. Eqs. (7-66) and (7-67).
b. Eqs. (7-70) and (7-71).

7-4. A four-loop pressurized-water reactor operates at 2200 psi. The core pressure drop is 45 psi at the rated core flow of 120×10^6 lb_m/hr. The pump head at rated flow is 280 ft. Assume the pump characteristic is given by Eq. (7-25) with $C_1 = 0.20$ and $C_2 = 0.30$. If the pump in one of the identical loops should fail, estimate the fractional reduction in core flow rate
a. if the coolant loops contain check valves.
b. if the coolant loops contain no check valves.

7-5. Work the preceding problem for a three-loop sodium-cooled reactor (Na density 52 lb_m/ft^3) with a core pressure drop 105 psi at the rated flow of 42×10^6 lb_m/hr. The pump head at rated flow is 282 ft (of Na), and the characteristic is again defined by Eq. (7-25) with $C_1 = 0.20$ and $C_2 = 0.30$.

7-6. Verify Eq. (7-93).

7-7. A pressurized-water reactor has two identical coolant loops. The hydraulic parameters for the system are as follows:

$$\left(\frac{L}{A}\right)_c = 5 \text{ ft}^{-1}, \qquad \left(\frac{L}{A}\right)_l = 100 \text{ ft}^{-1},$$

$$\Delta p_c = 50 \text{ psia}, \qquad \Delta p_l = 50 \text{ psia},$$

$$W_c = 7 \times 10^6 \text{ lb}_m/\text{hr}, \qquad m = 62.0 \text{ lb}_m/\text{ft}^3.$$

At time zero power is lost to the pumps in both loops. Neglecting natural convection estimate the time for the flow rate to be reduced to one-half of its initial value

a. if the inertia of the pumps can be neglected

b. if the pumps have half-times of 5 sec.

7-8. Verify Eqs. (106) and (107). Then find the counterparts to these equations under the assumption that the flow is laminar and therefore that $\mathcal{H}_{pr} = C/W_\infty$, where C is a constant.

7-9. Decay heat is to be removed from a three-loop 1000 MW(t) sodium-cooled fast reactor by natural convection. At full power the coolant pump heads are 300 ft (of Na) and the coolant flow rate is 40×10^6 lb_m/hr. The design requirement is that at 2% of full power natural convection must prevent the coolant temperature rise in the core from exceeding the value at full power. Assuming that the core inlet temperature and hydraulic resistances remain constant, and that the flow is upward through the core, calculate the distance that the heat exchangers must be elevated above the core. [Assume that the sodium has the following properties: $\beta = 1.6 \times 10^{-4}/°F$, $c_p = 0.3$ Btu/(lb_m) $(°F)$, $m = 54$ lb_m/ft^3.]

7-10. A 3250 MW(t) PWR has a primary coolant volume of 13,000 ft^3 and a core flow rate of 135×10^6 lb/hr. The core inlet and outlet temperatures are 530 and 590°F. The steam generator has an effective heat transfer coefficient of $h = 1000$ Btu/$(hr)^{-1}(°F)(ft)^2$, the effective area of each of the four steam generators is 50×10^3 ft^2, and the temperature on the secondary side of the steam generators is 500°F.

a. Estimate the primary system time constant τ_{pr}.

b. Estimate the loop transit time t_{tr}.

c. Suppose the loss of the feedwater heaters causes the steam generator temperature suddenly to drop by 100°F. Plot the primary system temperature transient under the assumption that the reactor is immediately tripped and that the core time constant is $\tau = 5$ sec.

7-11. A 3500 MW(t) direct cycle BWR has a primary system coolant volume of 18,000 ft^3. At full power the reactor produces 10.5×10^6 lb$_m$ of steam per hour at a pressure of 1025 psia; approximately 60% of the coolant volume is filled with liquid and the remaining volume is filled with steam. Suppose the reactor is tripped; the steam line valves close instantaneously, but a control failure causes feedwater to continue to enter the reactor vessel at the full power rate. Roughly, how many seconds will elapse before the vessel overpressurizes causing relief and/or safety valves to open. (Hint: to a first approximation, overpressurization will occur when the vessel becomes completely filled with liquid.)

7-12. Study the safety analysis report for a power reactor of your choice. Then for each of the following accidents discuss in some detail the method by which the accident is detected and the provisions for the orderly removal of the stored energy in the primary system and for the short- and long-term removal of decay heat:

a. excessive electrical load.

b. loss of electrical load.

c. loss of off-site power.

d. condenser vacuum failure.

e. control system failures.

LOSS-OF-COOLANT ACCIDENTS

Chapter 8

8-1 INTRODUCTION

The cooling-failure accidents treated in Chapter 7 consist of flow blockages, coolant-pumping failures, and heat-sink losses of various forms. All of these accidents have the common feature that the coolant inventory is maintained, at least to the point where pressure relief is required through the use of relief and/or safety valves. Since the total coolant mass remains approximately constant in these cases, many of the features of the accidents are amenable to study through the use of relatively simple incompressible flow models, regardless of the reactor type.

We now turn our attention to accidents in which the primary system envelope is breached, and a substantial loss of the coolant inventory is incurred. Under these situations the various reactor types show little similarity in behavior, owing to the diverse properties of the coolants. Liquid-metal-cooled systems are operated near atmospheric pressure, and therefore breaches in the primary system result in a pouring of coolant out of the reactor vessel. Under these conditions, hydrostatic heads, siphon effects, and pump suctions play dominant roles in determining if adequate cooling of the core is available following the accident. Gas- and water-cooled systems, in contrast, operate under high pressures, which means that primary envelope rupture results in violent blowdown of the primary system. In gas-cooled systems the core must be cooled following the accident by forced circulation of the low-pressure gas remaining in the system. In water-cooled systems the high-temperature coolant flashes to steam during the blowdown, and cooling following the accident must be reestablished by refilling the vessel with cool water at near-atmospheric pressure.

For each of these coolants the spectrum of accidents falling under the heading of loss of coolant is wide, both in likelihood and in consequences. The most likely failures of the primary system envelope are expected to be in the form of slowly growing cracks or other imperfections caused by fatigue, corrosion, or other deleterious mechanisms. With adequate instrumentation for monitoring the coolant inventory and a program of periodic inspection of primary system components, however, most such gradual deterioration should be detected early enough for an orderly shutdown of the plant for repair to be carried out. Although they are much less probable, larger breaches of the primary system envelope caused by rapid failure of valve casing, coolant piping, vessel penetrations, or other components must also be considered. The worst of these breaches, usually hypothesized as either the guillotine rupture of the largest-diameter coolant pipe or the failure of the largest vessel penetration, often provide the design basis for the emergency core-cooling system and for the containment strength.

In the event of the worst conceivable loss-of-coolant accident, the massive rupture of the reactor vessel, it is unlikely that a meltdown of the reactor core could be prevented, and an analysis to demonstrate that the containment structure would not be penetrated would be problematic. As a result, the preponderance of effort has been directed toward assuring that the probability of vessel rupture is extremely small; the probability of reactor vessel fracture is estimated to be less than one failure per 10^6 reactor years of operation in present day water-cooling reactors (Advisory Committee on Reactor Safeguards, 1974). Less study has been devoted to detailed analysis of the sequence of events that would follow such a failure.

In this chapter we first examine the integrity of primary system envelopes, discussing both systems consisting of steel vessels and coolant loops and systems in which the primary coolant envelope is enclosed entirely within a massive prestressed concrete vessel. Then, in Sections 8-3 through 8-5, we discuss the loss-of-coolant accident and address the problem of emergency core cooling for liquid-metal-, gas-, and water-cooled reactors. Questions of containment response as well as of core meltdown are deferred to Chapter 9.

8-2 PRIMARY SYSTEM INTEGRITY

The materials and structure of a primary system envelope are strongly dependent on the pressure and temperature at which the reactor operates and on the volume of the primary system. Because of their high-temperature strength and ability to resist corrosion, stainless steel alloys are invariably selected for the piping in primary system coolant loops of liquid-cooled systems. Different materials, however, are used in the reactor vessel.

Light-water-cooled reactors are operated at high pressures, in the neighborhood of 1000 psia for boiling-water reactors and 2200 psia for pressurized-water reactors. However, because coolant temperatures are in the neighborhood of 600°F it is acceptable to fabricate the reactor vessels from low-alloy carbon steels that are then lined with stainless steel for corrosion resistance. In heavy-water moderated reactors, however, the reactor vessel is replaced by many pressure tubes that carry coolant to individual fuel assemblies. Since these pressure tubes are exposed to the same intense thermal and fast neutron bombardment as the fuel element cladding, they are usually constructed of zirconium alloys. Sodium-cooled reactors operate at near atmospheric pressure and therefore the pressure vessels are thinner than in water-cooled systems. However, since sodium-cooled systems may be required to operate at temperatures approaching 1000°F, the high-temperature strength of stainless steel is required in the vessel construction.

Steel has also been used in the design of high-pressure gas-cooled reactors. However, with increasing power ratings, the volumes required to house the low-power density cores become large. As a result of these large volume requirements, and the desirability of increased design pressures and temperatures, the use of prestressed concrete vessels that are insulated and lined with steel has become increasingly attractive. In such vessels the entire primary system can be housed internally as a series of chambers and ducts, thereby eliminating the need for exposed coolant loops; and with such vessels the volume requirements of the primary system no longer present a serious design limitation.

In what follows we first discuss some of the primary considerations in ensuring that steel vessels of the type used in water-cooled reactors will not fail. We then examine mechanisms by which the integrity of primary system piping may be lost, and then address the phenomena that are unique to high-temperature sodium-cooled systems. Finally, an outline is given of some of the principal considerations relating to the integrity of prestressed-concrete reactor vessels.

Steel Vessels

In discussing the use of steel reactor vessels, a distinction must be made between systems that operate at temperatures of less than about 800°F and those that operate at higher temperatures. In the United States the design of vessels in the first group is governed by Section III, "Nuclear Vessels," of the American Society of Mechanical Engineers (ASME) Boiler and Pressure Vessel Code. This code is based on the long experience gained in the use of steel vessels in high-pressure systems (Whitman et al., 1967), as well as the extensive research efforts that have been directed toward understanding the behavior of such vessels. The code is continuously revised to assure its applicability to evolving design practices. In vessels operating above about 800°F the time-dependent deformations caused by the viscoelastic or creep properties of the steel become important to the reliable design of the vessel. Because much less experience is available at these elevated temperatures, design criteria are not included in Section III of the ASME Boiler and Pressure Vessel code. Rather, the evolving criteria for high-temperature vessel design are contained in code case 1592.

Since boiling-water and pressurized-water reactors operate with coolant temperatures well under 800°F, they are governed by Section III. The vessels of sodium-cooled fast reactors may be required to withstand temperatures in excess of 1000°F, and therefore the creep criteria in code case 1592 must be incorporated into their design. In this subsection we discuss the integrity of the mild steel vessels used in light-water-cooled reactors. A brief discussion of the creep phenomena that must be considered in the design of the stainless steel vessels used in sodium-cooled reactors is combined with the discussion of the high-temperature piping systems in the following subsection.

Section III of the ASME code is much more stringent than the other nonnuclear sections because a massive failure of the primary system envelope would present the threat of large releases of radioactive materials. The stringent design and fabrication practices endorsed by the ASME code ensure that the primary stresses in the vessel are well below yield and that plastic flow will relieve localized stresses that may be beyond the elastic limit without endangering the integrity of the vessel. Particular modes of failure must be considered in the analysis. These include ductile rupture and buckling caused by short-term overloads, and loss of function caused by gross distortion or racheting. The possibility of fatigue failure because of either high-cycle vibrations or low-cycle high-strain thermal cycling on the life of the vessel also must be eliminated.

The long design life, often in excess of 40 years, and the adverse environmental conditions to which a nuclear pressure vessel is subjected also require that a great deal of attention be given to the choice of materials. The high-intensity radiation emanating from the reactor core precludes the routine inspection of important sections of the vessel. At the same time fast neutron irradiation may gradually alter material properties and cause the vessel integrity to deteriorate, particularly under cyclic loading conditions. In addition, a poor matching of material choice to operating criteria may lead to long-term deterioriation from corrosion, erosion, or stress corrosion cracking.

Brittle Fracture. Perhaps the greatest concern addressed in the pressure vessel code is in avoiding the possibility of loss of ductility of the steel. For with such a loss, a vessel may become subject to a catastrophic failure by brittle fracture. The effect of the loss of ductility on a stress–strain curve is illustrated in Fig. 8-1. A ductile material such as a mild steel may not have an extremely large yield stress. However, even if the yield stress is exceeded,

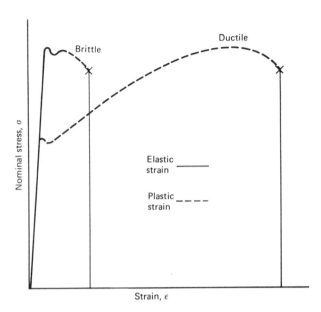

FIGURE 8-1 Stress–strain curves for ductile and brittle materials with the same ultimate stress.

there is a substantial margin between the yield stress and ultimate strength of the material. Moreover, beyond the yield stress the steel can absorb large amounts of strain energy per unit volume—as indicated by the area under the curve—and therefore yield to create a more favorable stress distribution. In contrast, if the material becomes brittle, it may well have a higher yield strength. However, if the yield strength is exceeded there is likely to be a much smaller margin between yield and ultimate strength. Moreover, stress relief by plastic flow now is severely limited by the small amount of energy that the material is capable of absorbing, as indicated by the small area under the curve.

If a vessel becomes brittle, fracture can take place even though the nominal stress level in the vessel is well below yield. Cracks or other flaws may exist in the vessel because of poor machining, bad welds, and so on. Since the stress concentration at the tips of these cracks exceeds the yield stress, cracks may propagate rapidly causing fast fracture. The theory for such fast fracture is due to Griffith (1920), Irvin et al. (1964), and others. Griffith postulated that the stability of a cracklike defect depended on the balance between total energy available to propagate the crack and energy absorbed by the propagating crack. The mechanism driving the crack propagation is the release of elastic strain energy from the vicinity of the crack as it propagates. At the tip of the crack there is a stress concentration, resulting in a small region of plastic behavior. As the crack propagates, energy is absorbed in the plastic work at the crack tip and in forming the free surface of the crack. Thus if the rate of elastic strain energy released is greater than the rate at which work is done at the crack tip, the crack is unstable and will propagate.

This process can be simply viewed through a heuristic example. Consider the flat plate in Fig. 8-2a that is subjected to the stress σ perpendicular to the crack of length $2a$. In the absence of the crack, the strain energy per unit volume of the plate would be $\sigma^2/2E$, where σ is the stress and E is Young's modulus. The normal component of the stress at the crack surfaces must vanish. The resulting decrease in the elastic strain energy surrounding the crack is roughly approximated by the total strain energy that would exist in the circular volume of radius a if the crack were not present. Thus for a plate of thickness τ_w the elastic energy released in crack formation is

$$\mathcal{U}_e = \frac{\pi a^2 \tau_w \sigma^2}{2E}. \tag{8-1}$$

The release of elastic energy with increased crack length is

$$\frac{d\mathcal{U}_e}{da} = \frac{\pi a \tau_w \sigma^2}{E}. \tag{8-2}$$

Now suppose that p is the plastic work required at the crack tip per unit increase in the cross-section crack area, and γ is the material's surface energy. The cross-section crack area is $2a\tau_w$. Hence the total energy absorbed in the crack formation is

$$\mathcal{U}_p = 2\tau_w a(p + \gamma), \tag{8-3}$$

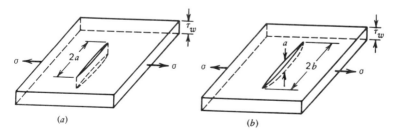

FIGURE 8-2 Idealized crack geometries.

whereupon the rate at which energy is absorbed with crack formation is

$$\frac{d\mathcal{U}_p}{da} = 2\tau_w(p+\gamma).$$ (8-4)

From Eqs. (8-2) and (8-4) it is apparent that the net energy release with increased crack length is

$$\frac{d}{da}(\mathcal{U}_e - \mathcal{U}_p) = \frac{\pi a \tau_w \sigma^2}{E} - 2\tau_w(p+\gamma);$$ (8-5)

note that the net release is an increasing function of crack length. If the net energy release from increasing the crack length becomes positive, the crack is unstable and will propagate. For a given stress level, the critical crack length $2a_c$ for propagation is established by setting the derivative equal to zero:

$$\pi a_c \sigma^2 = 2(p+\gamma)E.$$ (8-6)

The quantity p and γ are functions of the properties of the vessel material. For steels $p \gg \gamma$, and therefore γ may be neglected. Material-testing procedures such as Charpy impact testing (Whitman et al., 1964) can be used to measure the fracture toughness $\mathcal{G}_c$, which is simply related to p by $\mathcal{G}_c = 2p$. To treat accurately more realistic geometrical configurations, the foregoing Griffith theory energy arguments have been supplemented by the techniques of linear elastic fracture mechanics (American Society for Testing and Materials, 1965). In using these more advanced methods, it is convenient to express results in terms of the critical stress concentration factor K_c, instead of p or $\mathcal{G}_c$. This stress concentration factor is a function of both the material properties and the state of the stress in the absence of the flaw. For the plane stress situation in Fig. 8-2a, $K_c^2 = E\mathcal{G}_c$. Therefore Eq. (8-6) may be written as

$$K_c = \sigma\sqrt{\pi a_c}.$$ (8-7)

In pressure vessels the flaws are not expected to take the form of the simple through crack shown in Fig. 8-2a. They are more likely to penetrate the vessel wall only partially and to have a more complex geometry. For example, many flaws can be characterized in terms of the elliptical crack geometry shown in Fig. 8-2b, where the critical dimension is now the crack depth a. If the crack length $2b$ is large compared to the depth, the limiting state of plane strain is approached. In plane strain the plastic work is related to the critical stress concentration factor by $K_c^2 = E\mathcal{G}_c/(1-\nu^2)$, where K_c is also referred to as the plane strain fracture toughness. Using the techniques of stress analysis, relationships can be derived that relate the critical flaw dimension a to the nominal stress level (through σ), the material properties (through K_c), and the crack geometry. For example, the elliptical crack shown in Fig. 8-2b may be characterized by (Whitman et al., 1967)

$$K_c = \frac{1.12\sigma\sqrt{\pi a_c}}{\Phi_0}, \qquad (8\text{-}8)$$

where Φ_0 is the geometrical correction factor tabulated in Table 8-1.

TABLE 8-1 Geometric Correction for an Elliptical Crack

a/b	Φ_0	a/b	Φ_0
0	1.0	0.6	1.277
0.1	1.016	0.7	1.345
0.2	1.051	0.8	1.418
0.3	1.097	0.9	1.403
0.4	1.151	1	1/2
0.5	1.211		

From Whitman et al., 1967.

Temperature Dependence of Brittle Fracture. Since K_c decreases with the amount of plastic work done at the crack tip, a transition from ductile to brittle behavior will cause a precipitous drop to take place in K_c and therefore in a_c. This is indicated by the decrease in the areas under the stress–strain curves in Fig. 8-1. The steels used in reactor vessels are ductile at room temperature and above. Hence they have values of K_c that are so large that even if σ is set equal to the yield stress in Eq. (8-8) even large cracks will not propagate. If the temperature is decreased sufficiently, however, the steel will undergo a nil-ductility transition (NDT) that transforms it into a brittle metal similar to cast iron. Below the nil-ductility transition temperature (NDTT), where the steel is brittle, K_c is small enough that cracks of nominal length may propagate even though σ is well below the yield strength. For mild steels the NDTT ranges from 0°F to −20°F. However, the NDTT increases as the steel is exposed to the intense neutron radiation from a reactor core, and increases of as much as 300°F have been observed. Moreover, reactor vessels are subject not only to intense neutron irradiation over long periods of time, but to cyclic loading resulting from reactor start-up, shutdown, and power changes. Such cyclic loading causes cracks initially present to grow in size during the life of the vessel through low cycle fatigue. To ensure that the integrity of a reactor vessel is never jeopardized the designer must adequately account for coupled effects of the initially present flaws and their growth from cyclic loading, for the inability to inspect the vessel thoroughly after operations have begun, and for a NDTT that increases with neutron exposure and therefore with vessel life.

The effect of temperatures on allowable stresses and flaw sizes is graphically represented in the generalized fracture analysis diagram given in Fig. 8-3. The family of curves gives the combination of flaw size and nominal stress that will result in fracture as a function of temperature. The nil-ductility transition, or NDT, is the point below which the steel plate will fracture at the yield stress even though only very small flaws are present. The reason for this is indicated in the brittle curve of Fig. 8-1: Below the NDT temperature the ultimate strength is no larger than the yield strength. The crack arrest temperature, or CAT, curve represents the nominal stress level below which propagating brittle fracture is arrested. Thus, for stress–temperature combinations lying to the lower right of the CAT curve, brittle fracture is not expected regardless of flaw size. The FTE, or fracture transition for elastic loading, is the temperature above which cracks will propagate only if the plate is subjected to loading beyond the yield stress. Even then the crack will propagate only in the deformed, plastically loaded region, and not to regions of the vessel where the stress level is below the

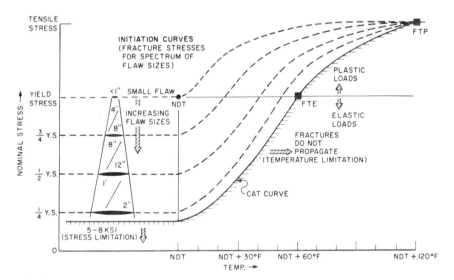

FIGURE 8-3 Fracture analysis diagram. From G. D. Whitman, G. C. Robinson, Jr., and A. W. Sarolainen, "Technology for Steel Pressure Vessels for Water-Cooled Reactors," Oak Ridge National Laboratory Report, ORNL-NSIC-21, 1967.

yield point. Above the FTP, or fracture transition for plastic loading, fracture can no longer take place by crack propagation.

As indicated by the figure, the fracture transition for elastic loading (FTE) may be expected to occur at about NDT + 60°F and the fracture transition for plastic loading at about NDT + 120°F. Therefore, pressure vessel safety may be approached by specifying a minimum temperature at full vessel pressure based on a measured NDT. For example, a common practice has been to reduce pressure and thereby limit stresses whenever the vessel is cooled below NDT + 60°F.

Increasingly, the fracture analysis diagram is being supplanted by the application of fracture mechanics (Mager et al., 1972) in determining operating restrictions and safety margins. In theory, this technique consists of the following. Exhaustive inspections are made to determine the length, location, orientation, and other relevant characteristics of all significant flaws as they exist in a vessel. A thorough stress analysis and an experimental determination of material properties as a function of temperature then permit the actual crack length to be compared to the corresponding critical crack length for specified vessel pressure and temperature. From this comparison, quantitative operating specifications can be established to meet required safety margins.

Transient Loadings. In using either the fracture analysis diagram or fracture mechanics to set vessel specifications, both changes in the critical stress intensity with time and slow crack growth must be accounted for. Fast neutron bombardment tends to translate the curves in Fig. 8-3 to higher temperatures. Such irradiation embrittlement is often expressed in terms of the change in NDT, as shown, for example, in Fig. 8-4. This gradual deterioration of the material properties through plant life is most pronounced in the belt-line region of the vessel, directly opposite the core midplane, where the radiation is most intense.

The crack growth during the life of the vessel resulting from cyclic stresses accompanying the start-up and shutdown of the nuclear plant is best correlated to the stress intensity factor K, which describes the stress condition at the tip of the advancing crack. For the crack in Fig. 8-2b,

$$K = \frac{1.12\sigma\sqrt{\pi a}}{\Phi_0}. \tag{8-9}$$

The most common relationship for crack growth is (Whitman et al., 1967)

$$\frac{da}{dN} = C_0(\Delta K)^4, \tag{8-10}$$

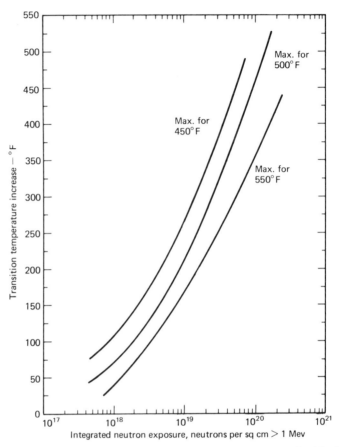

FIGURE 8-4 Radiation-induced increase in transition temperature for Mn–Mo steel. Adapted from Commonwealth Edison Company, "Final Safety Analysis Report—Zion Station," USAEC Docket No. 50-295, 1970.

where ΔK is the change in K occurring during the cyclic loading, N is the number of stress cycles, and C_0 and n are determined by experimental measurements.

To assure protection from brittle fracture, the preceding methods are used to generate curves of maximum pressure versus temperature for vessel operation. An example is shown in Fig. 8-5. In addition to specifying steady-state limitations it is necessary to set conditions on the rate of vessel cooldown. The reason for this may be understood in terms of Fig. 8-6, where temperature and stress distributions are shown for a vessel wall in contact with the reactor coolant on the left side and with a perfect insulator on the

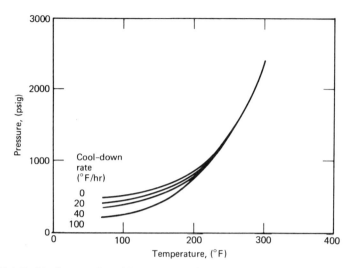

FIGURE 8-5 Maximum allowable pressures for reactor coolant system cool-down. Adapted from Commonwealth Edison Company, "Final Safety Analysis Report—Zion Station," USAEC Docket No. 50-295, 1970.

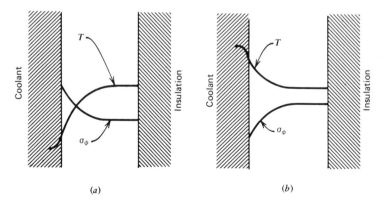

FIGURE 8-6 Temperature and hoop stress distributions in reactor vessel wall during cool-down and during heat-up. (*a*) Cooldown. (*b*) Heat-up.

right. For simplicity the vessel is assumed to be a thin shell. Under steady-state conditions the temperature and stress are space independent. The largest nominal stress is in the tangential direction and is given by the "hoop" stress formula

$$\sigma_\varphi = \frac{R}{\tau_w} p, \tag{8-11}$$

where p is the coolant pressure, R the vessel radius, and τ_w the wall thickness. If the temperature is nonuniform, a thermal stress is superimposed on the pressure stress. The thermal stress is given approximately by (Freudenthal, 1957)

$$\sigma_\varphi(r) = \frac{\alpha E}{1-\nu}[\bar{T}_w - T_w(r)], \tag{8-12}$$

where α is the coefficient of linear thermal expansion and $\bar{T}_w$ and $T_w(r)$ are the average and local wall temperature.

During vessel cooldown the minimum temperature will be at the surface in contact with the coolant, Fig. 8-6a, with a zero temperature gradient at the insulated surface. From Eqs. (8-11) and (8-12) we see that the maximum stress will occur at the inner surface as shown. The difference between surface and average temperature increases with cooldown rate; therefore the stress becomes very intense if there is a sudden drop in coolant temperature. During vessel cooldown the situation is further aggravated by neutron irradiation. Since the neutron flux is attenuated as it passes outward through the vessel wall, the maximum radiation embrittlement occurs at the same place as the maximum stress—at the coolant surface. For this reason the limiting case is often considered to occur during cooldown for a crack on the inner surface of the vessel at the vessel belt line. During vessel heat-up the situation is quite different. Because the temperature distribution now appears as in Fig. 8-6b the hoop stresses at the inner surface tend to cancel. At the outer surface they are additive, but the maximum value is smaller than in the cooldown case because of the zero temperature gradient at the insulated surface. This, taken together with the fact that the radiation damage is not as intense at the point of maximum stress, results in the vessel heat-up transient being substantially less severe.

Primary System Piping

In general, breaches of the primary system envelope outside of the reactor vessel lead to loss-of-coolant accidents that are substantially less severe than those that would result from the massive rupture of the vessel itself. Moreover, in a properly arranged primary system, emergency cooling systems can be designed to cope with the entire spectrum of such accidents up to and including the guillotine rupture of the largest coolant pipes in the system. Nevertheless the likelihood of a major loss-of-coolant accident must be minimized by ensuring that the probability of breaches of the primary system envelope is small, and that those that may occur will progress gradually so that coolant leakage can be detected and the plant shut down before a large rupture can take place. The design of highly reliable coolant loops and primary system components outside a vessel is made somewhat easier by the fact that they are not subject to intense neutron bombardment, and that they are likely to be more accessible to inspection and repair. At the same time, the nature of primary system piping requires not only that many of the same phenomena appearing in pressure vessel design be carefully considered, but that a number of additional potential causes of possible failure be addressed (Palladino, 1973; Freudenthal, 1974).

A significant difference in quality control may exist for fabrication of the reactor vessel and the remainder of the primary system envelope. Fabrication of the vessel may be carried out in the controlled environment of the manufacturer's shop, with inspection of welds being performed from both inside and outside the vessel. In contrast, much of the coolant pipe welding and inspection must be done under field conditions. Significant flaws in the welding can be expected to occur more frequently, since only exterior welds can be used in piping; furthermore, the heat-affected regions in the vicinity of welds may be particularly susceptible to certain failure modes such as the stress corrosion cracking discussed later.

Other challenges also arise in the fabrication of highly reliable coolant envelopes. Dissimilar metals must be joined, for example, in connecting stainless steel piping to mild steel vessels. Not only the coolant piping but the casings, stems, and seals associated with pumps and valves must be designed, fabricated, and joined to the primary coolant loops with such care that the overall integrity of the system is not compromised.

Cyclic Loading and Fatigue.   Steel piping systems must be designed in accordance with the ASA code for Pressure Piping and the ANS-I Nuclear Power Piping Code. In addition to the loading caused by primary system pressure and the weight of supported components, the design and layout of piping systems must carefully account for the thermal expansions and contractions that take place at start-ups, shutdowns, power level changes, and other transients within the design basis. If sufficient flexibility is not incorporated into the coolant loops, local thermal stresses may give rise to severe cyclic loading, possibly into the plastic region. The possibility of failure by low-cycle high-strain thermal fatigue must then be addressed. In systems, such as liquid-cooled reactors, where the core outlet temperature may undergo rapid changes of hundreds of degrees because of trips and other power level changes, such thermal fatigue is aggravated by thermal shock at coolant loop inlets, heat exchanger tubing, and other points within the primary system.

Although a primary system piping layout must be sufficiently flexible to withstand thermal cycling without suffering fatigue damage, it also must be properly anchored in order to withstand the severe shaking it may undergo in the event of a major earthquake. Restraints on piping movement also are necessary to control the destructive piping whip that could endanger vital safety system components if a coolant loop rupture should occur.

In general, the rupture of small-diameter piping used in the primary system for a variety of auxiliary and safety functions will not lead to as severe a loss-of-coolant accident as the rupture of a main coolant line. At the same time there are a number of reasons why it is difficult to maintain the same degree of reliability in these systems as in the major components and primary coolant lines. Because of its lighter weight and smaller dimensions, small-diameter piping is more susceptible to damage during construction. It may also be more vulnerable to fatigue damage resulting from thermal cycling, particularly if it is anchored too rigidly between more massive components. Flow vibrations or vibrations originating in pumps or other components may also give rise to high cycle fatigue in these systems. Finally, auxiliary systems often are intermittently isolated from the primary system for substantial periods of time; thus when valves are opened or pumps started following isolation, the entry of coolant into these lines may cause more severe swings in temperature and fluid chemistry than are likely to take place in the main coolant loops.

Corrosion.   Corrosive attack by impurities in the primary coolant and/or working fluid may lead to deterioration of the primary system envelope. Fortunately, corrosive mechanisms tend to lead to failure by gradually increasing leakage rather than by sudden rupture. Therefore damage can be detected and repaired before the possibility of a major loss-of-coolant accident arises. In water-cooled reactors, where the greatest experience has been gained, the most difficult corrosion problems occur on surfaces that are in contact with the working fluid. This is attributable to the ingress of impurities into the secondary system from a number of sources, most particularly the leaking in of air and condenser cooling water.

In direct-cycle boiling-water reactors, of course, the working fluid circulates through the entire primary system. In these systems intergranular stress corrosion cracking has been found to originate in the inside of small-diameter core spray and recirculation bypass lines (Castro, 1976). The intergranular cracks are formed from the selective attack of oxygen and other coolant impurities on areas of high surface stress such as appear in heat-affected areas near welds.

In indirect-cycle pressurized-water reactors the more stringent control of oxygen and other impurities in the primary coolant prevents stress corrosion cracking. However, on the secondary side of the steam generator tubing, where the primary system envelope is in contact with the working fluid, pitting, wastage, and intergranular stress corrosion cracking have occurred frequently (Martel, 1976). In recirculating-type steam generators, the corrosive attack is driven by the chemical residues that are concentrated in the liquid phase as steam is produced. Damage occurs in crevices, at high-stress surfaces, and in other areas where local conditions further enhance corrosive attack. Improved water treatment techniques for the secondary system are currently being applied to alleviate these problems.

Creep. The steel primary system envelopes for sodium-cooled reactors must be designed to meet criteria that are made more demanding than those of water-cooled reactor systems by the high operating temperature. With sodium temperatures of 1000°F or higher often being specified, stainless steels appear to be most capable of providing the high-temperature strength and corrosion resistance necessary to ensure the integrity of the envelope. As indicated in ASME code case 1592, careful attention must be given to the viscoelastic or creep behavior that becomes important in stainless steels at temperatures in excess of 800°F.

Primary systems are normally designed so that following the plant shakedown all stress levels are below yield, with adequate margins to account for yield stresses that decrease with increasing temperature. At sufficiently elevated temperatures, however, metals are found to undergo continuous elastic straining even though they are subjected to stresses well below the yield point. This so-called creep behavior is illustrated schematically in Fig. 8-7 in which strain is plotted versus time for several values of

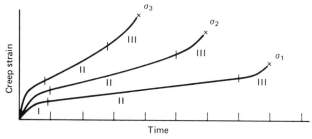

FIGURE 8-7 Creep diagram at constant temperature for three stress levels $\sigma_1 < \sigma_2 < \sigma_3$.

stress and a fixed value of the temperature. After the initial transient in stage I, the specimen undergoes a constant rate of strain in stage II. This strain rate increases with the stress level as well as with temperature. In stage III the strain rate increases markedly, necking occurs, and the specimen ruptures. The end of the creep life of a material can be defined as the transition point between stages II and III in Fig. 8-7. For constant temperatures this creep life can be measured as a function of stress, as indicated in Fig. 8-8.

The reactor vessel and primary system coolant loops in liquid-metal-cooled systems must be carefully designed to ensure that the stress levels are low enough that the creep life of the steel will not be exceeded during the life of the plant. A number of factors complicate these considerations, however, and make the straightforward use of a diagram such as Fig. 8-8 inappropriate. First, functional failures of critical components such as pumps, baffles, or valves are likely to take place from excessive deformation substantially before the end of creep life takes place. Second, plant transients that result in increased temperatures and the induction of large thermal stresses may lead to rapidly accelerated stage II creep rates and significantly reduce the life of the component even though the transient conditions may last only a very short time. Finally, cyclic loading may lead to damage from the interaction of creep and fatigue. This may be understood as follows: Following thermal or mechanical loading at high temperatures, localized stresses will be relaxed by creep deformation even though they are below yield. When the load is then removed, the deformed state of the component causes new stresses to arise, and these in turn are relaxed by creep deformation. If the loading is repeated, cyclic plastic creep deformation will take place. The result is similar to the low-cycle high-strain fatigue that may occur at lower temperatures. It differs, however, in that low-cycle fatigue at lower temperatures becomes pronounced only if the cyclic loading exceeds the yield strength of the metal. In contrast, high-temperature creep fatigue

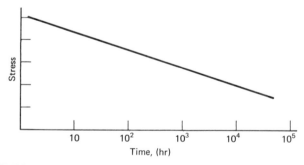

FIGURE 8-8   Transition to stage III creep at a constant temperature.

damage may occur at an accelerated pace even though the stress levels never approach the yield strength.

Prestressed-Concrete Vessels

Prestressed-concrete vessels are considered for use in nuclear power plants because they offer some significant advantages (Landis and Watson, 1974). The vessels can be built of larger dimensions than is possible for steel systems. This is a particularly important factor in the economic design of the large, low-power-density, gas-cooled thermal reactors. In addition, considerable freedom is gained in the choice of vessel geometry, which is advantageous in a structural and functional sense. For example, in recent designs—such as that illustrated for a proposed gas-cooled fast reactor in Fig. 8-9—the need for external coolant loop piping is eliminated by arranging the primary system to consist of a series of component chambers, linked by ducts for coolant flow.

The safety features of prestressed-concrete vessels as compared to steel vessel and piping systems are most pertinent to our discussion. As pointed out by the proponents of concrete systems, the exposed primary steel piping is eliminated and replaced by coolant loops that are embedded deeply within a massive structure. Thus a great deal of protection from external forces is an inherent feature of the systems. More important, the potential of catastrophic failure without warning from brittle fracture may be eliminated by using prestressed-concrete vessels. The prestressed-concrete failure mode is progressive with signs of the vessel deterioration being present long enough before the impending failure to allow shutdown of the plant. To understand these differences in behavior it is necessary first to outline the design of the prestressed-concrete vessels and then to discuss the failure modes.

A concrete structure has a substantial compressive strength but very little tensile strength. Prestressing the structure consists of creating a state of compression in those regions where the actions of the loads otherwise would cause tensile stresses to arise within the concrete. The prestress is introduced by the use of high-strength steel tendons. The pattern of tendon placement may be quite intricate. For example, in Fig. 8-9 the vessel shown has complexes of axial and transverse tendons in addition to those circumferentially wrapped around the vessel. After the concrete is formed, the steel tendons are carefully tightened between their anchor points in the concrete. This places the steel in tension and the concrete in compression. When the vessel chambers are pressurized, the net tension on the vessel is increased. This results in a decrease in the concrete compression and an increase in the steel tension. In properly designed vessels, however, the number, placement, and size of the steel tendons is such that some compression remains in

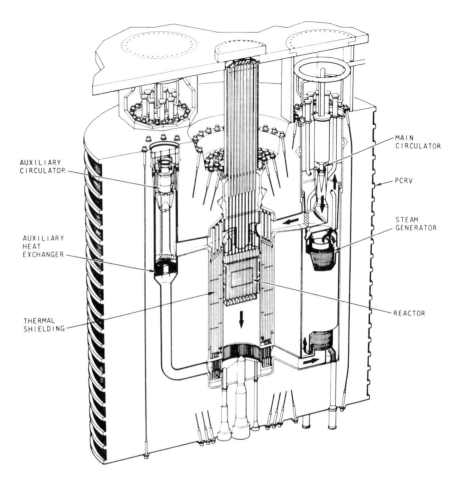

FIGURE 8-9 Gas-cooled fast reactor in a prestressed concrete primary system. From Gulf General Atomic, "Gas-Cooled Fast Breeder Reactor Preliminary Safety Information Document," GA-10298, 1971. Used by permission.

the concrete, and the tension in the steel tendons remains below the yield strength.

In contrast to steel vessels, the pressure load is borne not by a monolithic shell, which may become subject to fracture, but by the numerous independent steel tendons. These are sufficiently redundant that the sudden fracture of one or even several of the tendons would not cause overstress on those remaining. Furthermore, it is impossible to perform detailed inspections of steel vessels once the reactor has been in operation because of the high radiation field from fission products. In contrast, the tendons in concrete vessels can usually be placed at locations sufficiently remote from the reactor core that they can be monitored by strain gauges or other means. Thus there is warning of impending loss of prestress or other deterioration. In some designs it is also possible to replace faulty tendons.

With the possibility of fast fracture under nominal operating conditions eliminated, behavior under accidental overpressurization must be considered. Like steel systems, the concrete vessel must be protected by a system of relief and safety valves. It is nevertheless instructive to examine the ultimate overpressure failure of such a vessel in the absence of these devices (Rockenhauser, 1969). The vessel is characterized by the pressure-deformation curve shown in Fig. 8-10. With no load there is a compressive stress in the concrete vessel imposed by tensioning of the tendons. In regime 1, the vessel strain is linear elastic. In this regime, minor concrete cracking may occur at points of localized stress concentration, but the vessel behaves basically as a monolithic structure. As the internal pressure increases beyond the design pressure and into regime 2, compression is lost in the concrete, and the tension reaches a point where major cracking commences. The steel tendons remain in the elastic region throughout regime 2, although the concrete cracking becomes progressively worse. The net behavior is still

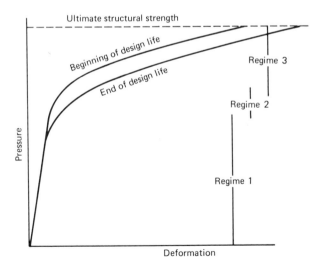

FIGURE 8-10   Idealized general load–deformation relationship for a prestressed concrete vessel. From W. Rockenhauser, "Structural Design Criteria for Primary Containment Structures (Prestressed Concrete Vessels)," *Nucl. Eng. Des., 9*, 449–466 (1969). Used by permission.

elastic, but not necessarily linear. That is to say, the vessel deformation is more or less reversible, because as the pressure is relieved the tendons will cause the cracks to close. At even higher internal pressure, in regime 3, the concrete becomes extensively cracked and the steel tendons are stressed into the plastic range. Increased pressure then causes very rapid vessel deformation until the ultimate strength of the structure is reached. At that point failure results from the ductile fracture of the steel tendons. From this short outline it is evident that even if the vessel is overpressurized into regime 2, leakage through the concrete cracks will occur only during the overpressure transient, because the elastic behavior of the steel tendons closes the cracks once the pressure is relieved. Moreover, a vessel liner of mild steel is invariably used to prevent leakage even though cracks in the concrete may open.

The positive safety features of prestressed-concrete pressure vessels notwithstanding, there is substantially less experience in their use than with steel systems. Thus codes that set minimum standards for their design and construction are at a much earlier stage of development. Prestressed concrete vessels for nuclear power plants are covered at present under division 2 of Section III of the ASME Boiler and Pressure Vessel Code and in ACI standard 359-44. Moreover, a number of specific problems must be addressed if concrete vessels are to have the extreme reliability required for the primary system envelope. Among these are the difficulty in assuring that the concrete is of a uniformly high quality, and that vessels with intricate chamber and ducting geometry can be formed without permitting voids within the concrete volume. The vessel ducting and penetrations required for refueling and maintenance also complicate the placement of the steel tendons in a configuration such that localized areas with inadequate prestressing do not occur. Much remains to be learned about long-term deterioration of the vessel from radiation damage to the concrete and from corrosion of the steel cables. Finally, the enclosure of the primary system within the massive vessel makes inspection and repair of internal components exceedingly difficult.

Probably the most serious limitations on the use of concrete in pressure vessels result from the poor properties of concrete at elevated temperatures and in thermal gradients (McAfee et al., 1976). As indicated in Fig. 8-11, the strength of concrete deteriorates rapidly with temperature. It also is susceptible to increasing rates of creep deformation at elevated temperatures, and it virtually disintegrates at about 850°C. The high-temperature disintegration, moreover, may take the form of violent spalling, caused by the high-pressure boiling of water trapped in pores in the concrete. As a result of these poor properties at high temperatures, nominal pressure vessel operating temperatures are typically limited to 150°F, with localized temperatures not exceeding 250°F (Gulf General Atomic, 1971).

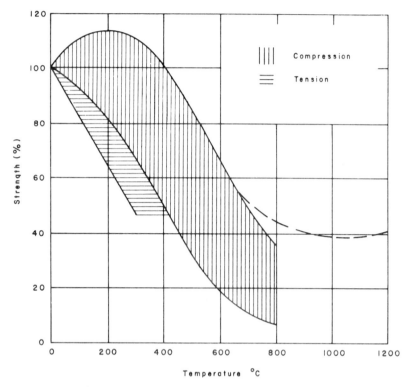

FIGURE 8-11 Relative "hot" strengths of concrete to strength at room temperature. From D. S. Schweitzer, "HTGR Safety Evaluation Division Quarterly Report," Brookhaven National Laboratory Report, BNL-50450, 1975.

Since concrete cannot directly withstand the temperatures of the reactor coolant, insulation must be placed inside the steel vessel liner, and often a forced-convection cooling system may be required between the liner and the concrete. As a result, it is necessary to consider potential accidents that may lead to the breakdown of this thermal barrier because of the loss-of-liner cooling or the rupture of the leaktight liner. Pumping failures, for example, could cause a loss-of-liner cooling. The rupture of the liner itself can result if the liner anchoring to the concrete is not compatible with the deformations of the concrete, particularly if the concrete is subject to cracking under overpressure conditions. Furthermore, such accidents would be aggravated if the liner were allowed to become brittle, from neutron irradiation or other causes, and thereby susceptible to brittle fracture.

8-3 LIQUID-METAL-COOLED REACTORS

Owing to its excellent heat transfer properties, the liquid metal sodium is the dominant coolant choice for fast breeder reactors. The thermal property that has the most direct bearing on loss-of-coolant accidents is the high boiling point of 1620°F at atmospheric pressure. Since other material constraints normally limit coolant temperatures in breeder reactors to about 1200°F, the primary system can be operated at or near atmospheric pressure while retaining a wide margin between operating and coolant boiling temperatures. Operation near atmospheric pressure is desirable both in reducing the probability of the occurrence of a loss-of-coolant accident and in minimizing the consequences should such an accident occur. The low system pressure permits the design of primary system envelopes in which the stress levels are quite low, and hence the probability of catastrophic vessel or pipe rupture is minimized. Furthermore, if the primary system envelope is breached, a rapid coolant blowdown from the primary system that occurs in high-pressure systems will not take place. The accident, rather, will be more accurately characterized by a pouring of the fluid from the breach that is driven only by the hydrostatic and/or pump heads.

Although the low operating pressures are an attractive feature, liquid-metal-cooled fast reactors have a number of characteristics that are less desirable when the possibility of a loss-of-coolant accident is considered. First, changes of power level often entail swings of hundreds of degrees in the coolant outlet temperature. As a result, large thermal stresses may appear in the reactor vessel, coolant piping, or other components, and care must be taken to ensure that thermally induced shock, creep, or fatigue do not reduce the integrity of the primary system envelope. Second, should sodium escape and come into contact with either water or oxygen, a fire will result, thereby increasing the thermal and mechanical loading on the containment structures. Normally, accidental contact of the primary system sodium with the water in the secondary system is eliminated by using the intermediate loop design shown in Fig. 1-15c. The possibility of sodium-oxygen contact is minimized by placing the primary system in a vault with a nitrogen atmosphere. Finally, and possibly most important, if a loss of coolant results in the uncovering of the fast reactor core before the reactor trip becomes effective, a major nuclear excursion may result, since the net sodium void coefficient of reactivity is likely to have a large positive value. Even if shutdown is accomplished, the high power densities in fast reactors result in only short time spans before core meltdown begins to take place. As a result, enphasis is placed on ensuring that the core will always remain covered with sodium even in the event of a primary system breach.

Primary System Leakage

The parameters governing the rate at which fluid will leak from a low-pressure primary system may be simply illustrated by considering the situation shown in Fig. 8-12a, where a pipe has ruptured at an elevation that is a distance $\Delta z(0)$ below the fluid level in the vessel. The mass flow rate W_0 out of the break may be determined in a quasi-static manner using the Bernoulli and continuity equations as follows. First apply the pressure drop relationship, Eq. (7-30) between the point a inside the vessel and at point 0 at the break. There is no pump or change in elevation. Hence with $\tilde{A} = A_0$ we may write Eq. (7-30) as

$$p_a - p_{\text{at}} = (1 + \tilde{e}_l)\frac{W_0^2}{2\bar{m}A_0^2},$$ (8-13)

where the inertial effects are neglected, and the cross-section flow area at point a is large enough that the A_i^{-2} term in Eq. (7-29) can be neglected. Since the velocity of the fluid between the pool surface and the pipe entry is very small, only the hydrostatic pressure head contributes significantly to the pressure difference:

$$p_a - p_{cg} = \bar{m}g\,\Delta z.$$ (8-14)

Combining these equations, we have

$$W_0 = A_0\bar{m}\left[\frac{2g\,\Delta z + 2\bar{m}^{-1}(p_{cg} - p_{\text{at}})}{1 + \tilde{e}_l}\right]^{1/2}.$$ (8-15)

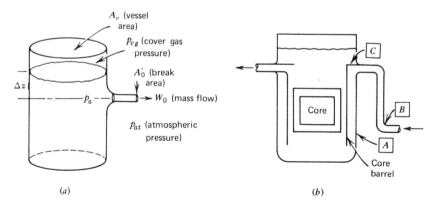

A_v (vessel area)

p_{cg} (cover gas pressure)

A_0' (break area)

Δz

p_a

W_0 (mass flow)

p_{at} (atmospheric pressure)

C

Core

B

A

Core barrel

(a)

(b)

FIGURE 8-12 Reactor vessels.

Of course as coolant flows out of the vessel, Δz will decrease. From mass conservation,

$$\frac{d\,\Delta z(t)}{dt} = -\frac{1}{A_v \bar{m}} W_0(t), \tag{8-16}$$

and therefore combining equations yields

$$\frac{d\,\Delta z(t)}{dt} = -\frac{A_0}{A_v}\left\{\frac{2g\,\Delta z + 2\bar{m}^{-1}(p_{cg} - p_{at})}{1 + \tilde{e}_l}\right\}^{1/2}. \tag{8-17}$$

If the cover gas is isolated from the exterior of the vessel, as it would be in a pressurized system, p_{cg} would be expected to decrease as the liquid level falls. If, for simplicity, we assume that the cover gas is initially at atmospheric pressure and remains there throughout the transient, then $p_{cg} = p_{at}$ and Eq. (8-17) is easily solved to yield

$$\Delta z(t) = \Delta z(0)\left[1 - \frac{A_0}{A_v}\sqrt{\frac{g}{2\,\Delta z(0)(1 + \tilde{e}_l)}}\,t\right]^2. \tag{8-18}$$

The length of time elapsed before the break is uncovered is then

$$t_0 = \frac{A_v}{A_0}\sqrt{\frac{2(1 + \tilde{e}_l)\,\Delta z(0)}{g}}. \tag{8-19}$$

From this it is seen that the area ratios of the break to the vessel, the initial height of liquid above the break level, and the flow resistance $\tilde{e}_l$ of the discharge line determine the rate of coolant level decrease. The flow rate is increased if $p_{cg} > p_{at}$ or if the break is on the downstream side of a pump. In the latter case the pump head term gH_p would be added to the numerator on the right of Eq. (8-17).

Clearly, if the core is to remain covered with sodium following a loss-of-coolant accident, no breaches can be allowed at elevations that are below the top of the reactor core. To ensure that this criterion is met, the reactor vessels are built so that all the piping penetrations are above the top of the core. Moreover, it is often necessary to maintain not only vessel penetrations but the entire primary loop at elevations above the core. Otherwise a situation such as that shown in Fig. 8-12b might develop because of the arrangement of the vessel internals and coolant piping. The isolation of the core inlet from the core outlet in this configuration causes siphoning action to take place between points A and B should a break occur at B. Thus the hydrostatic pressure head expelling coolant from the vessel is the same whether the break is in the coolant pipe at point B or in the vessel at point A.

This problem may be eliminated by including a siphon breaker, which in this case could be a hole at point C allowing some bypass flow between the ruptured line and the upper vessel plenum. With this direct communication between core inlet and outlet, the sodium level will drop only to level C, because the cover gas will then enter the inlet pipe and break the siphoning effect.

Two methods are often used to ensure the near-absolute protection required against system breaches at elevations below the top of the core. First, the reactor vessel, as well as any piping at low elevations, is made of double-walled construction. The gap between the inner vessel and the outer or guard vessel is small enough that the rupture of the interior vessel does not result in the sodium level dropping far enough to uncover the core. Alternately, the vessel may be located in a leaktight vault or cavity that is small enough that rupture of the reactor vessel will result in the cavity filling either to the point where the core will remain covered or to the point where an external reservoir of sodium can be used to fill the cavity rapidly to the required height. The vault configuration has an additional desirable feature. The vault walls may be designed to withstand and absorb the blast effect generated by hypothesized nuclear excursions. Thus, even if the vessel ruptured through such an event, the vault would act to maintain sodium in the primary system.

Decay Heat Removal

Sodium-cooled reactors must be designed to ensure that in the event of a breach of the primary system not only does the reactor remain covered with coolant, but there is a reliable system for decay heat removal. Such emergency core cooling can be accomplished either through the use of the remaining intact coolant loops or by a separate, auxiliary heat removal system.

Additional geometrical considerations become important when the post accident removal of decay heat is examined. In the vessel configuration in Fig. 8-12b, for example, the remaining intact coolant loops (not shown) could not be relied on for decay heat removal. A break in either the hot or the cold leg at the elevation at which the legs penetrate the reactor vessel causes the liquid level to fall below the intakes of the remaining intact coolant loops, causing them to become vapor bound. Conceivably, an auxiliary coolant loop of similar design could be placed at a slightly lower elevation. But if this were done, a breach of the auxiliary loop could have the effect of vapor-binding both the main and auxiliary coolant loops. Alternately, sodium could be injected into the vessel from an elevation at or higher than the break level and could be allowed to overflow through the break and onto the floor of the cavity in which the vessel is located. It then could be recirculated through a heat exchanger and back into the vessel.

Primary systems can be designed to make accidents that lead to an uncovering of the core nearly incredible, while at the same time allowing the intact heat transport loops to be used for postaccident heat removal. These may be classified either as loop or pot systems. An example of a loop system is shown in Fig. 8-13a. The entire heat transport system outside the reactor vessel is elevated above the core midplane, and the vessel is protected by a guard vessel as well as a minimum volume vault. In addition, the coolant loops are designed so that a path for the removal of decay heat remains intact regardless of where the primary system breach takes place. This latter property is most clearly recognized by considering the consequences of pipe ruptures at three strategic locations: the core outlet, the core inlet and the pump outlet, points A, B, and C, respectively.

If a pipe rupture takes place at the core outlet, point A in Fig. 8-13a, the fluid level falls only to the elevation of the break, and backflow through the pipe is prevented by the check valve. With the intake to the outlet pipe constructed as shown, the inlets of the intact coolant loops will remain submerged, and therefore these loops can be used to remove heat. With the intermediate heat exchanger elevated as shown, this removal is facilitated by natural convection, as discussed in the preceding chapter. Thus, provided the reactor is promptly tripped, the loops remaining are adequate for decay heat removal, even if the coolant pumps are no longer operative. There is, however, a limitation on the height of the heat exchangers, since zero pressure at the highest elevation would cause the liquid volume to separate, breaking the siphoning action through the loop.

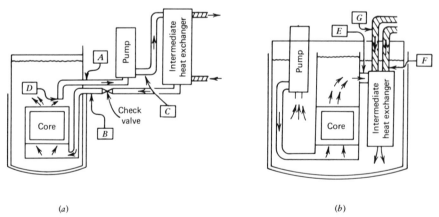

(a)   (b)

FIGURE 8-13 Primary systems with liquid-metal coolants (only one coolant loop shown). (a) Loop system. (b) Pot system.

In Fig. 8-13a a pipe rupture in the inlet piping at point B is more serious for a number of reasons. First, the check valve, while preventing flow reversal, does not prevent flow from the core outlet from passing around the loop and then discharging from the rupture. Thus coolant is expelled from both ends of the ruptured pipe. Second, the drop in pressure at the inlet plenum tends to cause a precipitous flow decrease through the core, since the rupture creates a low flow resistance path in between the inlet plenum and the atmosphere. In any event, the sodium level drops to B, at which time the backflow out of the break will terminate. If no more sodium is lost, the intact loops still are capable of removing heat from the reactor. Keeping these loops submerged, however, constitutes a third problem.

If the pump in the ruptured loop continues to operate, it will suck sodium out of the vessel and expel it to the vault until the liquid level falls below the intake of the outlet pipe at point D. At this time, flow will end because of vapor binding not only in the broken loop, but in the intact loop as well. Thus while the core remains covered with sodium, there is no path for decay heat removal. To prevent vapor binding of the intact loops, it is necessary to shut down the pump rapidly in the broken loop, since only when the pump pressure head becomes less than that required to lift the sodium from the intakes to the pump level will flow terminate. The pump cannot be stopped instantaneously, but must coast down in the same half-time t_p as described in the preceding chapter. In contrast to the pump power failure accident described earlier, it is here desirable to make the pump half-time as small as possible, to ensure that once the trip signal is given, the pump speed and therefore the flow out of the loop will decay very rapidly. The pump half-time must be weighed against the inventory of sodium that may be lost before the loop intakes uncover as well as the rates at which coolant is being lost through the vessel inlet break. Examples of parameter studies relating these parameters are given elsewhere (Harper and Hart, 1969). The result must be that the drawing of sodium out of the vessel by the pump in the broken loop must be terminated before the loop intakes are uncovered.

In the event that the pipe rupture is located directly at the pump outlet, at point C, the situation is somewhat different because the check valve prevents flow out of the vessel inlet. However, the broken loop pump will still cause all the heat transfer loops to become vapor bound if it is not tripped soon enough.

From the foregoing discussion it is apparent that a pipe break may require not only that the reactor be tripped, but that the pump in the broken loop be shut down. The latter requirement may cause difficult protection system design problems, since it is necessary to determine which loop has been breached in order to trip the proper pump. Tripping a pump in an intact loop would only aggravate the situation. An alternative is to trip all the coolant pumps simultaneously with the reactor, since then it is not required to know which loop has been breached. This action is viable only if the heat exchangers are sufficiently elevated to allow decay heat removal by natural convection in the intact loops following the reactor and pump trip.

Thus far the accidents considered have been for loop configurations of liquid metal primary systems. Many of the potential accidents are eliminated if a second configuration is used: the pot (or pool) primary system shown schematically in Fig. 8-13b. In contrast to the loop configuration, the core, primary coolant piping, pumps, and intermediate heat exchangers are submerged in a single pool containing the entire inventory of primary system sodium. The vessel containing this pool is typically of double-walled construction, to minimize the probability of failure, and is placed in a minimum volume vault to ensure that in the event of failure the pool level is retained at a high enough elevation to ensure orderly heat transport of decay heat.

With the pool level ensured, detailed analysis of many of the pipe break accidents, important in loop systems, is obviated. This may be seen by considering the consequences of breaks at various places in the reactor piping. If a pipe is broken on the primary side of the intermediate heat exchanger, at a point E, only forced convection through a single loop is impaired, and there is no loss of sodium from the vessel. Similarly, if the pipe is broken at point F, mixing occurs between the primary sodium and that of one of the intermediate loops, possibly impairing heat transport through that loop, but again resulting in no loss of sodium from the primary vessel. This effect is similar to that which would occur from ruptured tubes in the intermediate heat exchanger. A break in the intermediate loop outside the vessel, say at point G, results only in the loss of sodium from that loop. Finally, in the event of a loss-of-heat-sink accident, the large sodium inventory results in a slow primary system temperature rise, providing a longer time interval before damage occurs to reestablish cooling.

Although the inherent safety features of pool systems are substantial, there are also a number of disadvantages. It is difficult to elevate the intermediate heat exchangers sufficiently above the core to give satisfactory natural convection flow requirements under accident conditions, because the cost of the increased height of the vessel is likely to be prohibitive. Maintenance of coolant pumps, heat exchangers, and other equipment necessary for the safe operation of the plant is also made more difficult by their immersion in the sodium pool.

8-4 GAS-COOLED REACTORS

The initiating events that typically are postulated to cause loss-of-coolant accidents in gas-cooled reactors tend to be quite different from the pipe ruptures postulated in sodium- or water-cooled systems. In contrast to the steel primary systems of most liquid-cooled systems, the large primary system volume requirements of gas-cooled reactors have increasingly led to the use of prestressed concrete vessels. In these the primary system volume consists of chambers and ducts deeply embedded in the massive concrete structure. Hence exposed coolant loop piping that may be subject to fracture is eliminated. Large penetrations through the concrete vessel are required, however, for refueling, steam generator maintenance, and other purposes, and the failure of such a penetration could lead to the rapid depressurization of the primary system. During operation concrete plugs are secured in the penetrations through the use of elaborate hold-down devices. As a result, the plug removal required for complete opening of the penetration is assumed to be incredible, and loss-of-coolant accidents are postulated to be caused only by the failure of the seal between plug and vessel. Such accidents have relatively small break areas when compared to the large primary system volume, and lead to rather long depressurization times, typically of the order of minutes (Baxter and Swanson, 1974; Buttemer and Larrimore, 1974).

Loss-of-coolant accidents in gas-cooled and in liquid-cooled systems differ in another respect. The boiling of a liquid coolant may bring about large reactivity changes, because of the coolant density coefficient, and a precipitous drop in the heat removal capability may result from boiling crises and vapor binding of the reactor coolant channels. In contrast, the coolant density in a gas-cooled reactor is so small that it has no appreciable effect on reactivity. Gradual reactivity changes are brought about only from the fuel and/or solid moderator temperature coefficients as the core overheats because of cooling deterioration. Likewise, the cooling deterioration varies continuously as the coolant density decreases with depressurization.

Before a reactor shutdown is accomplished, the reactor core temperature will rise, since the loss of cooling capability far exceeds the nominal decrease in power that results from the negative reactivity effects. The rate of core heat-up varies substantially between fast and thermal gas-cooled systems. The high-power-density fast reactor core will tend to have relatively little heat capacity, and therefore the temperature rise may be of the order of 400°F/sec, as indicated in Table 6-1. In contrast, thermal gas-cooled reactor designs often require the heat to pass through large masses of graphite moderator before reaching the coolant channels. In these cases the core heat capacity is much larger, and as indicated in Table 6-1, the full-power adiabatic heat-up rate may be of the order of only 5.0°F/sec. Since these heat-up rates determine how much time can elapse before fuel damage and eventually core meltdown occur in the absence of a trip, they indicate that the reactor shutdown must be much faster in the fast systems than in the thermal reactor core.

Once the shutdown is accomplished and the depressurization is complete, there remains the problems of decay heat removal. Natural convection of a gas coolant is completely inadequate for decay heat removal, and the addition of a liquid coolant to the system is likely to lead to many materials, reactivity, and thermal shock problems. Means therefore must be provided for sufficient forced convection of the coolant gas at or near atmospheric pressure to accomplish the orderly removal of decay heat.

In what follows, we examine the gas-cooled reactor depressurization accident in more detail. First the critical flow that governs the coolant leak rate is discussed. It is then incorporated into a model for determining the rate of primary system depressurization. Finally, the modeling of the core flow rate and the methods used to ensure decay heat removal following the transient are outlined.

Depressurization Transient

The salient features of a depressurization of a gas-cooled system may be illustrated by considering the blowdown of a simple vessel such as that shown in Fig. 8-14. Let V be the vessel volume and A_0 the break area of the discharge nozzle. The fluid in the vessel obeys the perfect gas law

$$p = mRT. \tag{8-20}$$

Since the pressure in the vessel is initially much higher than that of the surrounding environment, the compressibility of the gas must be taken into account in determining the discharge flow through the break.

If the break area is much smaller than the vessel volume, as is usually the case for the range of accidents considered, substantial simplification can be made. Under these circumstances the rate at which the average vessel pressure is decreasing will be sufficiently slow that the change in vessel pressure during the time required for fluid to traverse the distance between a stagnation point just inside the nozzle and the point of discharge (points i and 0, respectively, in Fig. 8-14) can be ignored. As a result, the discharge flow can be determined assuming that the pressure and temperature p_i and T_i as well as the conditions outside the vessel are time independent. Once the mass flow rate is determined as a function of the vessel conditions, a mass balance for the gas in the vessel volume can be made to determine the depressurization transient.

To determine the flow, we assume that the wall friction in the nozzle can be neglected. With this and the foregoing assumptions, the inertial, wall friction, and pump terms can be eliminated from Eq. (7-18) to yield the

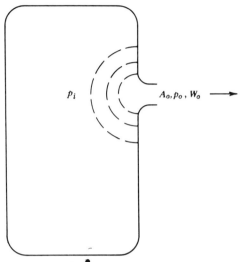

FIGURE 8-14   Simple vessel.

Bernoulli equation:

$$\frac{v_0^2}{2} - \frac{v_i^2}{2} + \int_{p_i}^{p_0} \frac{dp}{m(p)} + g(z_0 - z_i) = 0. \qquad (8\text{-}21)$$

Now, since the inlet to the flow path is taken at a point where the fluid velocity can be neglected, $v_i = 0$. There is no change in elevation in the flow path and therefore $z_0 = z_i$. Finally, the mass flow rate out of the vessel is just $W_0 = A_0 m_0 v_0$, and therefore from the Bernoulli equation we have

$$W_0^2 = 2m_0^2 A_0^2 \int_{p_i}^{p_0} \frac{dp}{m(p)}. \qquad (8\text{-}22)$$

To evaluate the mass flow rate, we make the further, reasonable assumption that heat transfer through the nozzle walls is insignificant, and therefore the perfect gas undergoes an adiabatic expansion in discharging from the vessel. It is well known that in an adiabatic process the temperature and pressure of a perfect gas are related by

$$\frac{T}{T_i} = \left(\frac{p}{p_i}\right)^{(1-1/\gamma)}, \qquad (8\text{-}23)$$

where γ is the ratio of specific heats. Using the perfect gas law to eliminate T and T_i, we have

$$\frac{m_i}{m} = \left(\frac{p_i}{p}\right)^{1/\gamma}, \tag{8-24}$$

which may be substituted into Eq. (8-22) for m. The integration over pressure then yields

$$W_0 = A_0\sqrt{2m_ip_i}\,\psi\!\left(\frac{p_0}{p_i}\right), \tag{8-25}$$

where

$$\psi\!\left(\frac{p_0}{p_i}\right) = \sqrt{\frac{\gamma}{\gamma-1}}\left[\left(\frac{p_0}{p_i}\right)^{2/\gamma} - \left(\frac{p_0}{p_i}\right)^{(1+1/\gamma)}\right]^{1/2}. \tag{8-26}$$

The flow rate is written as a product of two factors. The first depends only on the break area and the gas conditions inside the vessel. The second, ψ, is a function of (p_0/p_i) the ratio of pressure at the nozzle exit to that in the vessel. Obviously, when $p_0 = p_i$, the flow is zero. If the conditions in the vessel remain constant, it may be seen that decreasing p_0—or correspondingly, increasing the pressure drop across the flow channel—increases the flow rate as expected. However, if one continues to decrease the ratio p_0/p_i it is easily shown that $\psi(p_0/p_i)$ passes through a maximum at

$$\frac{p_0}{p_i} = \left(\frac{2}{\gamma+1}\right)^{\gamma/(\gamma-1)}, \tag{8-27}$$

with a value of

$$\psi_{max} = \sqrt{\frac{\gamma}{2}\left(\frac{2}{\gamma+1}\right)^{(\gamma+1)/(\gamma-1)}} \tag{8-28}$$

and then decreases for smaller ratios of p_0/p_i. When ψ_{max} is reached the fluid speed at the nozzle exit has reached its sonic velocity at pressure p_0. Physically, the pressure in the nozzle exit will go no lower than this critical value although the pressure on the outside of the vessel may have a smaller value. The nozzle therefore is said to be choked. So long as this condition exists, the mass flow rate is a function only of the flow area and the pressure and density of the gas within the vessel, and choked or critical flow is said to exist.

For gas-cooled reactor systems the ratio of containment to primary system pressure is much smaller than the critical value of Eq. (8-27). Thus critical flow exists over most of the depressurization transient, and the conditions in the primary system are uncoupled from those in the containment. Although the vessel pressure will eventually decrease and the containment pressure increase to a point where the flow is no longer choked, it is often adequate simply to assume that critical flow exists until the primary system comes to equilibrium with the containment. We may further assume that the mixing within the vessel is thorough enough that p_i and m_i can be replaced by the vessel-averaged quantities p and $\bar{m}$. Then

$$W_0 = A_0 \sqrt{2\bar{m}p}\ \psi_{max}. \tag{8-29}$$

The vessel depressurization rate can now be estimated from the mass balance

$$\frac{d}{dt}(mV) = -W_0, \tag{8-30}$$

Using the perfect gas law to eliminate the density from the preceding equations, we obtain

$$\frac{d}{dt}\left(\frac{p}{T}\right) = -\frac{A_0}{V}\sqrt{\frac{2R}{T}}\ \psi_{max}p. \tag{8-31}$$

The energy conservation equation must also be solved in order to determine the rigorous relationship between p and T during the transient. If no heat transfer were to take place between the coolant and the core or vessel walls, an adiabatic depressurization would result, with a corresponding cooling of the coolant because of the work done in accelerating the fluid through the break. With the large amounts of stored heat in the primary system, however, it is reasonable to assume that at any time the heat stored in the core and other components causes the coolant in the primary system to remain at approximately its original temperature. Moreover, this assumption results in a more rapid blowdown than the adiabatic approximation. We therefore take T as a constant. Equation (8-31) thus reduces to

$$\frac{dp}{dt} = -\frac{p}{\tau^*} \tag{8-32}$$

and

$$\tau^* = \frac{V}{A_0\sqrt{2RT}\ \psi_{max}}. \tag{8-33}$$

The solution is just

$$p(t) = p(0)\ e^{-t/\tau^*}, \tag{8-34}$$

where the pressures are measured above atmospheric. If R and ψ_{max} are evaluated for common coolants,

$$\tau^* = \begin{cases} 0.302\dfrac{V}{A_0\sqrt{T}} & \text{(He)} \\[2ex] 1.06\dfrac{V}{A_0\sqrt{T}} & \text{(CO}_2\text{)} \end{cases} \tag{8-35}$$

Where V is in liters, A_0 in cm^2, T in °K, and τ^* in seconds.

Coolant Flow Transient

The rate of power production and the mass flow through the primary system have the greatest effect on the core temperature during the depressurization transient. Since the loss of coolant has relatively little effect on reactivity, the power level may be expected to remain roughly constant until the reactor trip causes it to be reduced rapidly to the level of decay heat. The mass flow rate, on the other hand, is more difficult to describe, since it is affected both by the decreasing pressure in the primary system and the transient in circulator performance following the accident.

If depressurization is rapid, the fluid dynamic description of the primary system may be quite complex, requiring the simultaneous solution of mass, momentum, and energy balance equations (Ludewig and Epel, 1974). The technique used may then be similar to those used in water reactor depressurization calculations, except that the two-phase equations of state for water are replaced by the perfect gas law.

The relatively small break areas hypothesized for many depressurization accidents, however, often permit a number of simplifications to be made. In particular, with slow depressurization rates a compressible model is needed only for the coolant discharge through the break, and the flow within the primary system can be treated using incompressible flow models (Buttemer and Larrimore, 1974). Some of the more important phenomena affecting the flow transient can be highlighted through the following heuristic model.

Suppose we compare the depressurization time constant τ^* to the time $t_{tr} = M_{pr}/W$ required for the coolant to make one cycle through the primary system. If τ^* is smaller or of the same order of magnitude as t_{tr}, a detailed compressible flow model is needed to obtain an acceptable approximation to the flow transient. Over the range of postulated accidents, however, the leak rate often is small enough that $t_{tr} \ll \tau^*$. Under these circumstances the leak rate W_0 may be expected to be small compared to the mass flow rate W through the core. The average coolant pressure will decrease very little during the time in which the coolant makes one cycle around the primary system. As a result, the relative magnitudes of the pressure drops around the coolant loops will be affected only indirectly through the gradually decreasing system pressure. Under these conditions an adequate flow approximation may be obtained by uncoupling the flow and depressurization calculations as follows.

We first perform a system-averaged depressurization calculation using the simple model of the preceding subsection. From this, the time dependence $\bar{m}(t)$ of the average coolant density can be determined. The incompressible flow model of Chapter 7 then is simply modified to account for the small leak by assuming the fluid density to have the time dependence $\bar{m}(t)$, wherever it appears. Equation (7-76) for the core flow then becomes

$$\left(\frac{L}{A}\right)_{pr}\frac{dW(t)}{dt} + \tilde{\mathcal{K}}_{pr}\frac{W(t)^2}{2\bar{m}(t)} + g[\bar{m}(t)\,\Delta z]_{pr} = \bar{m}(t)g\left[\frac{\omega(t)}{\omega_0}\right]^2 H_p. \quad (8\text{-}36)$$

Moreover, in gas-cooled reactors the fluid density is so small that the inertial and hydrostatic terms have little effect on the flow transient. Eliminating these from Eq. (8-36) yields

$$W(t) = \sqrt{\frac{2gH_p}{\tilde{\mathcal{K}}_{pr}}}\,\bar{m}(t)\frac{\omega(t)}{\omega_0}. \quad (8\text{-}37)$$

Finally, if we normalize to $W(0)$ and eliminate $\bar{m}(t)$ through the use of the perfect gas law, we have

$$\frac{W(t)}{W(0)} = \frac{T(0)p(t)\omega(t)}{T(t)p(0)\omega_0}. \quad (8\text{-}38)$$

A rough assessment of the flow transient now may be made by assuming that the absolute temperature T following the pressure transient is not appreciably different from its initial value. Gas-cooled reactors typically operate at from 500 to 1000 psi. Assuming that the containment remains at atmospheric pressure, a flow reduction to between 1.5 and 3.0% of the full-power value will result from depressurization if the circulators can continue to operate at their initial speed.

The finite size and increased temperature of the containment, however, may cause the system pressure to stabilize at two atmospheres or more following the transient. Even with a 1000 psi initial pressure the core flow then could be maintained at about 3% of the full power level, assuming the circulators remain at their initial speed. This 3% figure is significant for the following reason. The temperature rise through the core is roughly proportional to the ratio of power to mass flow rate as indicated by Eqs. (7-1) and (7-2). If the reactor is·promptly tripped following the onset of the accident, then by the time the depressurization is completed, the decay heat level will also be of the order of 3%. Thus the circulators operating at their initial speed should be capable of preventing excessive temperatures from occurring in the core.

In reality, the circulator speed is not likely to remain constant during a depressurization transient, since its power requirements decrease with depressurization. Moreover, it is desirable to control the circulator speed so that neither core overheating before reactor trip nor thermal shock following the trip takes place. The central relationship between reactor power, circulator speed, and the temperature increase of the coolant as it passes through the core can be understood as follows.

Assume that the pressure is given by Eq. (8-34) until equilibrium is reached with the containment. Then, combining Eqs. (8-34) and (8-38), we have for the flow

$$\frac{W(t)}{W(0)} = \frac{T(0)}{T(t)} e^{-t/\tau^*} \frac{\omega(t)}{\omega_0}. \tag{8-39}$$

The transient in the coolant temperature rise through the core can be obtained by inserting this expression into the transient heat transfer model of Eq. (7-93). We obtain

$$\frac{\Delta T_c(t)}{T(t)} = \frac{\Delta T_c(0)}{T(0)} e^{t/\tau^*} \frac{\omega_0}{\omega(t)} \left\{ e^{-t/\tau} + \frac{1}{\tau} \int_0^t \frac{P(t)}{P_0} \exp\left[-\frac{(t-t')}{\tau} \right] dt' \right\}, \tag{8-40}$$

where τ is the core time constant.

Now consider the two idealized cases. Before reactor trip, we assume that the reactor remains at full power. Thus the foregoing equation reduces to

$$\frac{\Delta T_c(t)}{T(t)} = \frac{\Delta T_c(0)}{T(0)} e^{t/\tau^*} \frac{\omega_0}{\omega(t)}. \tag{8-41}$$

During this time period we would like to increase the circulator speed as much as possible to compensate for the decreasing pressure. Since the power input requirement of the circulator decreases with decreased coolant density, there will be a tendency for $\omega(t)$ to increase. Typically, the circulator is limited to less than about 50% speed increase over its initial value to prevent damage from excessive centrifugal forces. As a result, the circulator can compensate for the leak only for a time of substantially less than τ^*, and thus the reactor must be tripped within a short time span to avoid the core overheating associated with flow starvation.

To emphasize the effect of the sudden power decrease associated with reactor shutdown, suppose we now ignore the time-dependent effects of delayed neutrons and decay heat and simply assume that at $t = 0$ the reactor power instantaneously drops to P_d, a time-independent approximation to the decay heat level. Equation (8-40) reduces to

$$\frac{\Delta T_c(t)}{T(t)} = \frac{\Delta T_c(0)}{T(0)} \frac{\omega_0}{\omega(t)} \left\{ \frac{(P_0 - P_d)}{P_0} \exp\left[-\left(\frac{1}{\tau} - \frac{1}{\tau^*}\right)t \right] + \frac{P_d}{P_0} e^{t/\tau^*} \right\}.$$

(8-42)

The decay heat level is only a few percent of full power, and hence initially the behavior is dominated by the first bracketed term. In fast reactors the fuel time constant τ may be expected to be much smaller than τ^*. As a result, if the circulator speed remains constant or increases following the trip there would be a precipitous drop in ΔT_c. The resulting decrease in the core outlet coolant temperature could then cause unacceptable thermal shock damage to the fuel-element cladding, steam generator tubing and other components. The problem is not so severe in the graphite-moderated thermal reactors, since the long core time constants slow the cool-down rate, and moreover only ceramic materials are included in the reactor core. In both fast and thermal systems, however, a power input to the circulators is automatically cut back in order to obtain a better power to flow match during the ensuing shutdown transient. At times that are long compared to τ, the second bracketed term in Eq. (8-42) dominates. It now is necessary to compensate for the decreasing pressure by gradually increasing the circulator speed to where the decay heat can be removed at near-atmospheric pressure without causing the core outlet temperature to become excessive.

Emergency Core Cooling

The foregoing examples indicate the importance of being able to control the circulator speed during the depressurization accident both in executing sharp cutback to avoid thermal shock problems at the time the reactor is tripped and in affecting a gradual increase in circulator speed to provide for decay heat removal as the primary system pressure deteriorates. The details of how this is accomplished depend strongly on the circulator design and particularly on its power supply. For large nuclear plants, power requirements for the gas circulators are large, typically 5% or more of the plant's electrical output. Consequently, they usually consist of axial compressors driven by steam extracted directly from the steam generators. Since the steam generators contain large masses of water, they are both excellent heat sinks for the primary system and are capable of supplying steam to drive the circulators for a substantial period of time during the depressurization transient. Moreover, following the transient, an equilibrium condition can be established in which energy is supplied to the steam generator from the transfer of decay heat out of the primary system. A part of the steam produced from this decay heat is supplied to drive the circulators at an adequate speed to prevent excessive core temperatures. Such a scheme for emergency core cooling is feasible, however, only if highly reliable instrumentation and controls can closely regulate the feedwater flow, steam generation inventory and pressure, and the flow to the circulator turbine.

The interactions between the reactor power decay, circulator, and steam generator transients are likely to result in rather complex time-dependent interaction between the circulator speed, cladding temperature, and other variables. For example, in Fig. 8-15 the results from the analysis of a gas-cooled fast reactor depressurization accident are shown (Buttemer and Larrimore, 1974). The 25 in^2. leak in the reactor inlet plenum causes a depressurization with a time constant of $\tau^* = 60$ sec. The reactor is tripped at 4.5 sec, after the accident initiation, at which time the sharp cutback in circulator and feedwater flow to the steam generator also is initiated. The circulator speed then gradually increases as the system depressurizes, and is thereafter controlled to establish a long-term equilibrium. The steam generator inventory and pressure gradually decrease during the transient until a new equilibrium is established between the reduced levels of feed-water flow, decay heat input, and steam supply to the circulators. Following the reactor trip, the maximum cladding temperature first falls but then increases somewhat as the stored heat in the fuel elements is transferred to the coolant under conditions of a rapidly decreasing flow rate. Since the core

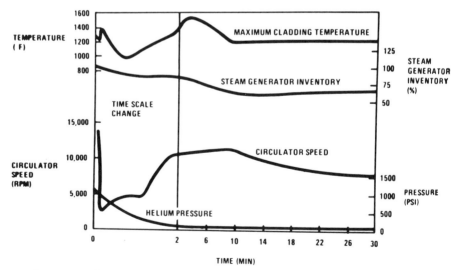

FIGURE 8-15   Gas-cooled fast reactor depressurization transient. From D. R. Buttermer and J. A. Larrimore, "Response of Gas-Cooled Fast Breeder Reactors to Depressuriza-tion Accidents," *Nucl. Eng. Des.*, **26**, 195–200 (1974). Used by permission.

time constant τ is the order of a few seconds, this stored heat is removed from the fuel within about 10 sec following the trip. Thereafter the cladding temperature undergoes more gradual variations that are controlled primarily by the core flow transient during and after the depressurization.

The foregoing results assume the proper operation of both the reactor trip and the core-cooling systems. As in other accidents, backup measures are provided both for shutdown and for the emergency core cooling. The reliability of the core cooling is maximized, for example, not only by providing parallel coolant loops and helium circulators, but by providing more than one power supply for each circulator. For example, the turbines driving the circulators may be supplied with steam either from steam generators or from auxiliary boilers. In some cases pelton wheels utilizing high-pressure service water may also be capable of driving the circulators. Finally, a common practice is to have completely independent auxiliary coolant loops with electrically powered circulators as a backup for the main heat transport path. These must be capable of fast start-up and of decay heat removal for indefinitely long periods of time following the reactor trip.

8-5 WATER-COOLED REACTORS

The loss-of-coolant accident, or LOCA, occupies a central position in the safety analyses of light-water-cooled reactors. The high-pressure and high-temperature water represents a large inventory of stored energy that may be released over a short period of time in the event of a major breach of the primary system envelope. Should such a breach occur, the water would rapidly flash to steam. The flashing would in most cases cause the reactor to be shut down, because of the negative coolant void coefficient, and halt the transfer of heat out of the core, because of vapor binding of the coolant channels. The containment would be pressurized by the escaping steam, and in the case of pressurized-water reactors, the core support structures would undergo a severe shaking over the short period of time during which sonic decompressurization waves passed through the primary system. In general, the accident would be qualitatively similar in a water-cooled pressure tube reactor, with the exception that a positive void coefficient might in some cases result in a mild power excursion accompanying the loss-of-coolant transient. Our attention hereafter is directed predominantly to pressurized- and boiling-water cores housed in steel vessels.

The problem of cooling the core to prevent meltdown following a loss of coolant is qualitatively different from that in a gas- or liquid-metal-cooled reactor. Since the vessel depressurization results in the liquid coolant being replaced by steam vapor, the heat transport paths for the liquid coolant become vapor bound and therefore inoperable. Moreover, emergency cooling by forced convection of the vapor remaining in the primary system is precluded, since the requirements for large gas circulators, high gas flow velocities, and so on, are incompatible with the design of the water-reactor primary systems. Consequently, emergency core cooling is provided by reflooding the core with water following the depressurization at near atmospheric conditions. After a transient flooding period, the core is cooled by nucleate boiling, with the resulting steam escaping to the containment, and cool water being continually supplied to the reactor vessel.

The spectrum of possible loss-of-coolant accidents is wide. The accidents range from small leaks, which can be compensated for through normal coolant inventory control systems, to the massive fracture of the reactor vessel. Current licensing criteria generally require that safety systems be provided to ensure that neither core meltdown nor breach of the containment envelope result from a wide range of breaches up to and including the rupture of the largest coolant pipes in the primary system. In light-water reactors, the massive rupture of the steel reactor vessel is thought to be too improbable to warrant consideration with regard to the design of the safety systems. In heavy-water reactors the problem of vessel rupture is eliminated by containing the coolant in many separate pressure tubes instead of a single vessel.

Safety systems must be capable of responding to small primary system breaks as well as large ones. The most stringent conditions on their design, however, most often are encountered following the double-ended rupture of the largest coolant loop pipes in pressurized water reactors or of the recirculation and steam lines in direct-cycle boiling-water reactors. Hence design bases for core structures, emergency core-cooling systems, and containment envelopes are to a great extent derived from the necessity of mitigating the consequences of these large-break accidents. As a result, a detailed understanding of the nature of the depressurization transient is required.

In the following subsections we first examine the depressurization of a simple vessel filled with high-pressure and high-temperature water. From this, several of the more important characteristics of the depressurization transient become apparent. We then examine in more detail the sequence of events that is likely to occur should a loss of coolant occur in a pressurized- or boiling-water reactor. In the final subsection the problem of emergency core cooling is taken up in some detail.

Blowdown Modeling

The primary system of a water-cooled reactor constitutes a network of flow paths and junctions the details of which must be taken into account in the modeling of the course of a loss-of-coolant accident. Some of the more important features of the depressurization transient can be more simply illustrated, however, by ignoring the flow resistances within the primary system and considering the behavior of a simple vessel, such as shown in Fig. 8-14, that is filled with high-temperature subcooled water. Simple analytical models do not exist even for the depressurization transient of a simple vessel, since neither the incompressible flow approximation nor the perfect gas law are applicable. Instead, mass and energy balance equations must be solved in conjunction with an equation of state that describes the subcooled, two-phase, and superheated behavior of water. Since the equation of state is not available in analytical form, fluid properties must be utilized in tabular form from steam tables (e.g. Keenan and Keyes, 1959).

Before proceeding to the quantitative formulation of vessel blowdown, it is useful to consider the qualitative features of such transients by considering the thermodynamic behavior of the coolant as the transient proceeds. Figure 8-16 shows a p-v diagram for water. At the outset of the blowdown the coolant will be in a compressed liquid state as indicated by point A. For a pressurized-water reactor a temperature of 550°F and a pressure of 2200°F is typical. As the blowdown takes place, the vessel pressure will decrease and the specific volume will increase. Thus the trajectory of fluid properties must

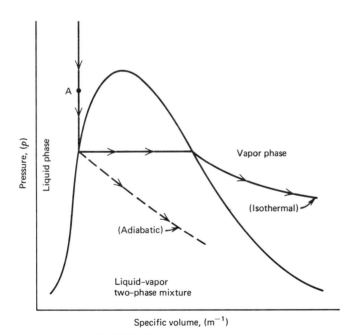

FIGURE 8-16 Idealized p-v diagram.

proceed from the upper-left-hand corner to the lower-right-hand corner of the p-v diagram. If we assume that phase separation can be ignored, the thermodynamic trajectory can be expected to lie between two idealized limits. If the thermal contact with the vessel walls and other heat reservoirs is sufficient to retain the coolant at its initial temperature, the fluid properties will follow the isotherm as indicated. If, on the other hand, there is no heat addition to the coolant, the blowdown will be adiabatic and follow the trajectory of constant entropy.

Within the envelope encompassed by these two extremes, the blowdown may be divided into two, or sometimes three, phases: (1) the subcooled blowdown is characterized by a precipitous drop to the saturation pressure with very little change in density. Here there is no significant difference between the isothermal and adiabatic trajectories. In practice the subcooled blowdown lasts only a very short time—usually a small fraction of a second—since only a small amount of fluid must be expelled before the saturation condition is reached. Subcooled blowdown is characterized by the emission of sonic decompression waves that propagate from the break and are reflected from the vessel walls. (2) The two-phase blowdown results in either no pressure change or a more gradual pressure decrease as coolant is expelled. The large decrease in density is required before the saturated steam curve can be reached, and the lower break flow rates that occur under two-phase conditions cause the two-phase blowdown to last an appreciably longer period of time—typically, several seconds or more for pipe ruptures in reactor vessels. (3) Superheated blowdown appears at the end of the transient if enough heat is contained in the fluid to raise it above the saturation temperature. It is characterized by a more rapid decrease of pressure as steam is expelled from the vessel.

Mass and Energy Balance. The explicit time dependence of a depressurization transient is predicted by solving the macroscopic forms of the mass and energy transport equations. We make the assumption that no significant pressure gradients exist within the vessel volume and neglect the small effects of potential and kinetic energy. The macroscopic or control volume, forms of the mass and energy conservation equations may be written as (Bird, Stewart, and Lightfoot, 1960)

$$V\frac{d\bar{m}}{dt} = -W_0 \tag{8-43}$$

and

$$V\frac{d}{dt}\bar{m}\bar{u} = -h_0 W_0 + Q(t), \tag{8-44}$$

where the internal energy and enthalpy are related by

$$\bar{u} = \bar{h} + \frac{p}{J\bar{m}}, \tag{8-45}$$

and

V = vessel volume,

$\bar{m}$ = average coolant density,

$\bar{u}$ = average coolant internal energy,

$\bar{h}$ = average coolant enthalpy,

h_0 = coolant stagnation enthalpy at break entry,

p = vessel pressure,

J = mechanical to thermal energy conversion constant,

W_0 = break mass flow rate,

Q = heat transfer rate into coolant.

Equation (8-45) may be used to eliminate the internal energy from the energy balance. The mass and energy balance may then be rewritten as

$$\frac{d\bar{m}}{dt} = -\frac{W_0}{V} \tag{8-46}$$

and

$$\frac{d\bar{h}}{dt} = \frac{1}{V\bar{m}}\left[(\bar{h} - h_0)W_0 + \frac{V}{J}\frac{dp}{dt} + Q(t) \right]. \tag{8-47}$$

The steam tables may be used to relate the enthalpy to the pressure and density, provided thermodynamic equilibrium is assumed within the vessel:

$$\bar{h} = \bar{h}(\bar{m}, p). \tag{8-48}$$

We now have three equations, (8-46), (8-47), and (8-48), but five time-dependent unknowns: $\bar{m}$, p, $\bar{h}$, h_0, and W_0. Hence two additional relationships are necessary before the time dependence of the depressurization transient can be determined.

The stagnation enthalpy h_0 immediately inside the discharge nozzle must be expressed in terms of the average enthalpy, density, and other system properties. If the blowdown is so rapid that the separation between the liquid and vapor phases can be neglected, h_0 can be set equal to $\bar{h}$. During the two-phase blowdown, however, separation effects tend to be important, and a phase separation model such as one of those described below is called for. The discharge flow rate W_0 must also be expressed in terms of the pressure and enthalpy inside the vessel and the containment pressure. For this purpose, critical flow models, similar to that discussed in the preceding section, must be generalized to encompass two-phase flow situations.

Phase Separation. During subcooled blowdown only a single phase is present, and hence

$$h_0 = \bar{h}. \tag{8-49}$$

When the vessel pressure is decreased to the saturation level, steam bubbles will form whose buoyancy will cause them to drift upward relative to the surrounding liquid mass. If the time span over which the two-phase blowdown takes place is small compared to the time required for the bubble to rise to the top of the vessel, these buoyancy effects may be neglected, yielding a homogeneous mixture such as that pictured schematically in Fig. 8-17a. In this circumstance Eq. (8-49) is still valid.

At the other extreme, if the blowdown is due to such a small leak that it occurs over a time much longer than that required for the steam bubbles to rise to the top of the vessel, near-complete phase separation will occur. The vessel can then be modeled as shown in Fig. 8-17b with the steam phase above the liquid phase. The height z of the interface between liquid and vapor phases can then be calculated as follows. The void fraction α, or ratio of the steam to total (steam plus liquid) volume, can easily be shown to be related to the average coolant density by (Tong & Weisman, 1970)

$$\bar{m} = (1 - \alpha)m_l(p) + \alpha m_v(p), \tag{8-50}$$

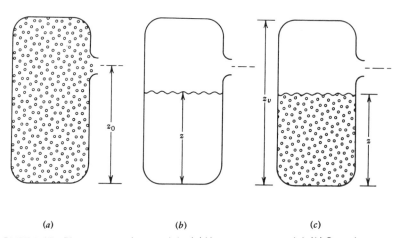

(a)   (b)   (c)

FIGURE 8-17   Phase separation models. (a) Homogenous model. (b) Complete separation model. (c) Bubble rise model.

or

$$\alpha = \frac{[m_l(p) - \bar{m}]}{[m_l(p) - m_v(p)]}, \tag{8-51}$$

where $m_l(p)$ and $m_v(p)$ are the saturated liquid and vapor densities at the pressure p. For a right cylindrical vessel, the liquid–vapor interface must be located at a height

$$z = (1 - \alpha)z_v, \tag{8-52}$$

where z_v is the vessel height. Thus z can be written as a function of $\bar{m}$ and p,

$$z = \frac{[\bar{m} - m_v(p)]}{[m_l(p) - m_v(p)]}z_v, \tag{8-53}$$

and utilized along with mass and energy balances and the equation of state in determining the depressurization transient. With z known as a function of time, the value of h_0 for a break at height z_0 is determined from

$$h_0 = \begin{cases} h_v(p), & z < z_0 \\ h_l(p), & z > z_0, \end{cases} \tag{8-54}$$

where $h_v(p)$ and $h_l(p)$ are the enthalpies, respectively, of the saturated vapor and liquid at pressure p.

Bubble rise speeds have been measured over a variety of conditions, and they are typically found to be of the order of 2 or 3 ft/sec (Tong and Weisman, 1970). This leads to rise times that often are of the same order of magnitude as the blowdown times encountered in loss-of-coolant accidents. As a result, the phase separation tends to look like Fig. 8-17c and it is necessary to take the bubble buoyancy into account. This behavior may be approximated by the following constant bubble rise velocity model.

Suppose we consider again a vessel of cross-section area A_v, height z_v, and with the break located at z_0. We assume that the coolant flashing takes place so as to result in a uniform distribution of bubbles in the liquid. The bubbles and liquid constitute a lower phase that extends up to the froth level $z(t)$. The buoyancy of the bubbles, however, causes them to drift upward with a uniform velocity v_b and form a second upper phase of saturated steam. We refer to this upper phase as the steam dome.

Now, define M_l, M_{gb}, and M_{sd} as the liquid mass, the gas bubble mass, and the steam dome mass. Clearly, the total mass of steam is

$$M_s = M_{gb} + M_{sd}, \tag{8-55}$$

and the total mass of coolant in the vessel is

$$M = M_l + M_s. \tag{8-56}$$

We now may write a mass balance by noting that the steam dome can only lose mass through the break at rate W_0, and then only if the break is located above the froth level. It can only gain mass as a result of gas bubbles rising through the froth level. Since we have assumed that the gas bubbles are distributed uniformly in the lower phase, the mass flow rate of bubbles into the steam dome will be $M_{gb}v_b/z$. Hence the steam dome mass balance is

$$\frac{dM_{sd}}{dt} = -\mathcal{H}(z_0 - z)W_0 + M_{gb}\frac{v_b}{z}, \tag{8-57}$$

where $\mathcal{H}$ is the Heavyside function defined by

$$\mathcal{H}(z_0 - z) = \begin{cases} 1, & z_0 > z \\ 0, & z_0 < z \end{cases}. \tag{8-58}$$

The masses M_l and M_{sd} may be eliminated from this mass balance by noting that

$$M_s = \alpha m_v(p)z_v A_v, \tag{8-59}$$

$$M_l = (1 - \alpha)m_l(p)z_v A_v, \tag{8-60}$$

and

$$M_{sd} = m_v(p)(z_v - z)A_v, \tag{8-61}$$

where α, the void fraction, is defined in terms of the densities in Eq. (8-51). Now Eq. (8-55) may be combined with these relationships to yield

$$M_{gb} = [z - (1 - \alpha)z_v]m_v(p)A_v, \tag{8-62}$$

and we may reduce Eq. (8-57) to

$$\frac{dz}{dt} = -(z - z_v)\frac{d}{dt}\ln m_v(p) + \frac{\mathcal{H}(z_0 - z)W_0}{m_v(p)A_v} - v_b\left[1 - (1 - \alpha)\frac{z_v}{z}\right]. \tag{8-63}$$

If the break is above the froth level, saturated steam will be discharged from the vessel. If it is below the froth level and discharge enthalpy will be given by

$$h_0 = \frac{M_{gb}h_v(p) + M_l h_l(p)}{M_{gb} + M_l}. \tag{8-64}$$

Thus utilizing Eqs. (8-60) and (8-62), we obtain

$$h_0 = \begin{cases} h_v(p), & z_0 > z, \\ \dfrac{(1 - \alpha)z_v m_l(p)h_l(p) + [z - (1 - \alpha)z_v]m_v(p)h_v(p)}{(1 - \alpha)z_v m_l(p) + [z - (1 - \alpha)z_v]m_v(p)}, & z_0 < z. \end{cases} \tag{8-65}$$

The foregoing phase-separation models allow the break enthalpy h_0 to be expressed in terms of the coolant mean density, enthalpy and pressure, and the known thermodynamic properties of saturated water and steam. If one of the extreme cases in which the assumption of no phase separation or of complete phase separation is applicable, only simple algebraic relations are required, whereas the use of the more realistic bubble rise model requires that an additional differential equation be solved. Phase-separation models have been improved further to take into account nonuniform distributions of bubbles in the lower phase (Moore and Rettig, 1973).

Critical Flow.   Before the blowdown calculation can be carried out, a means must be prescribed to determine the break flow rate in terms of the stagnation enthalpy and pressure in the vessel:

$$W_0 = W_0(h_0, p).$$ (8-66)

Two separate conditions must be considered: the discharge of liquid during the subcooled blowdown and the discharge of the two-phase mixture after bubbles form in the vessel.

When the subcooled water in a pressurized vessel is discharged to the atmosphere, it flashes to steam if its initial temperature is higher than the atmospheric saturation temperature. If such a discharge is through a nozzle or orifice, a superheated metastable state is thought to be caused by retardation in the formation of bubbles. The mass flow rate may be predicted from semiempirical correlations (Burnell, 1947):

$$W_0 = A_0\sqrt{2m_l(p)(p_{sat} - C_p)}$$ (8.67)

where p_{sat} is the saturation pressure corresponding to the vessel temperature, and C_p is an empirical coefficient, plotted in Fig. 8-18, that is related to the bubble delay time.

When a two-phase mixture is discharged through a nozzle, the presence of the vapor phase causes the sonic velocity to decrease markedly, and an annular flow pattern forms with a vapor core and liquid annulus. Based on thermodynamic arguments, and the assumption of an annular flow pattern in which the vapor travels faster than the liquid, models have been developed for the critical mass flow rate in two-phase mixtures (Fauske, 1962; Moody, 1965). At present, the Moody model is most widely used in depressurization calculations. Figure 8-19 shows Moody's results for the mass flow rate per unit area in terms of the vessel pressure and stagnation enthalpy h_0 just inside the nozzle.

When saturated steam is discharged a homogeneous critical flow pattern again exists, and the flow rate can be determined from the saturated vapor line in Fig. 8-17.

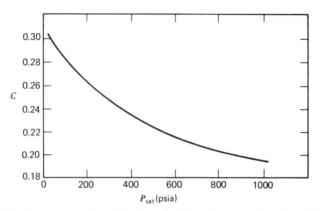

FIGURE 8-18 Pressure coefficient for Burnell critical-flow equation. From J. C. Burnell, "Flow of Boiling Water Through Nozzles, Orifices, and Pipes," *Engineering*, **164**, 572 (1947). Used by permission.

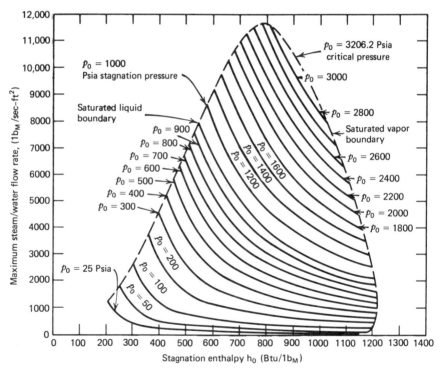

FIGURE 8-19 Maximum steam–water mass flow and local stagnation conditions. From F. J. Moody, "Maximum Flow Rate of a Single Component Two-Phase Mixture," *J. Heat Trans.*, **87**, 134 (1965). Used by permission.

Time-Dependent Solutions. The foregoing critical flow and phase separation models stipulate W_0 in terms of h_0 and p; and h_0 in terms of $\bar{h}$, $\bar{m}$, and p, respectively. Thus it is possible to solve equations (8-46), (8-47), and (8-48) numerically for $\bar{h}$, $\bar{m}$, and p as a function of time and thereby describe vessel depressurization transients. The calculations are carried out most directly if the equation of state, Eq. (8-48), can be expressed as

$$p = p(\bar{h}, \bar{m}). \tag{8-68}$$

Unfortunately, steam tables are not presently in this form, and therefore it is customary to solve Eq. (8-68) iteratively for p in terms of $\bar{m}$ and $\bar{h}$ at each time step. Alternately, instead of applying the equation of state directly, differential thermodynamic relationships may be used (Navahandi, 1969).

A comparison of the results of a numerical blowdown model to an experimental depressurization curve is shown in Fig. 8-20. The initial conditions and the proportions of the vessel are chosen to simulate a loss-of-coolant accident in a pressurized-water reactor. The subcooled blowdown is characterized by an extremely rapid depressurization that is associated with the nearly incompressible behavior of the liquid within the

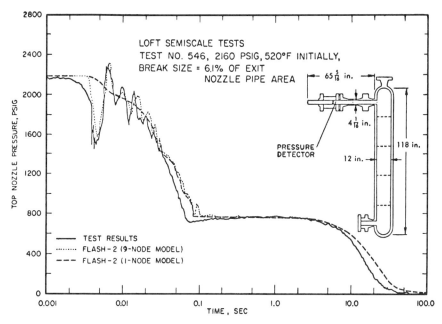

FIGURE 8-20 Comparison of FLASH-2 computer results with LOFT test. From J. A. Redfield and J. H. Murphy, "The FLASH-2 Method for Loss-of-Coolant Analysis," *Nucl. Appl.*, **6**, 127–136 (1966). Used by permission.

vessel. The saturation pressure is reached in less than 0.1 sec. The oscillations in the experimental curve during subcooled blowdown are caused by the propagation of sonic decompression waves that emanate from the break and reflect from the vessel walls. These oscillations are not present in the numerical results for the one-node model because the entire vessel is assumed to be at a single pressure. With the initiation of boiling at the end of the subcooled blowdown, the break flow rate decreases and the depressurization proceeds, with the pressure remaining nearly constant over the first several seconds of the two-phase blowdown.

This simple vessel blowdown illustrates some of the more important phenomena occurring during a primary system depressurization. The simulation of postulated reactor accidents with sufficient accuracy to determine a design bases for core structures, emergency core-cooling systems, and other safety-related equipment, however, requires that a great deal more detail be included in the numerical modeling of the accident. To this end, systems of large, interacting computer codes must be utilized (Ybarranto et al., 1972).

To simulate the hydraulic behavior of a primary system during a loss-of-coolant accident, the single-node vessel model is generalized to a network treatment of nodes and interconnecting flowpaths. Each node is represented by mass and energy balance along with a phase separation model. The equations governing each node's behavior are the same as Eqs. (8-46), (8-47), and (8-63) with the exception that the break flow is replaced by the sum of flow from the node. Thus Eqs. (8-46), (8-47), and (8-63) must be modified by letting

$$W_0 \to \sum_i W_i, \tag{8-69}$$

$$h_0 W_0 \to \sum_i h_i W_i, \tag{8-70}$$

and

$$\mathcal{H}(z_0 - z) W_0 \to \sum_i \mathcal{H}(z_i - z) W_i, \tag{8-71}$$

respectively. The critical flow through the break is treated as before, but each flow path within the primary system is represented by an incompressible flow model such as Eq. (7-30).

A typical representation of a PWR primary system by such a network of nodes and flow paths is shown in Fig. 8-21. Increasingly, compressible, two-phase, and other effects are being incorporated into the flow path models, and more sophisticated representations of phase separation, pump, valve, and other hydraulic behavior are being incorporated into the models (Moore and Rettig, 1973). In particular, a great deal of detail is needed if the

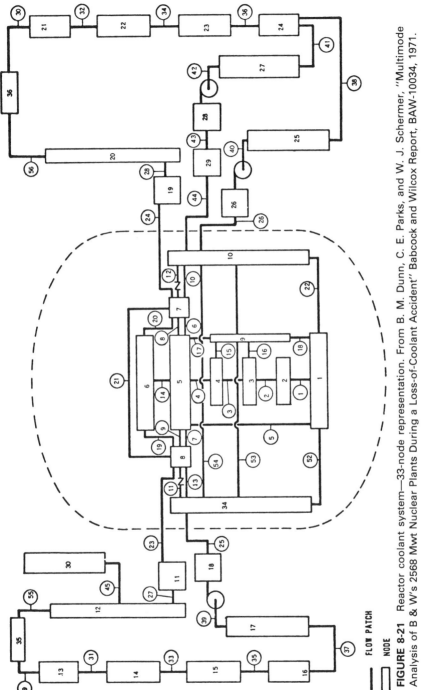

FIGURE 8-21 Reactor coolant system—33-node representation. From B. M. Dunn, C. E. Parks, and W. J. Schermer, "Multimode Analysis of B & W's 2568 Mwt Nuclear Plants During a Loss-of-Coolant Accident" Babcock and Wilcox Report, BAW-10034, 1971.

sonic decompression waves that appear during subcooled blowdown are to be described accurately.

The hydraulic modeling of the depressurization transient usually includes a lumped parameter thermal model of the reactor core in order to take into account the heat entering the coolant during the transient. This thermal model, however, may not be sufficient for determining maximum fuel and cladding temperatures. Therefore it is usually necessary to utilize a more detailed core thermal-hydraulic model that incorporates the reactor kinetics equation and utilizes the core mass flow rate and inlet conditions determined from the hydraulic model of the primary system to carry out a detailed analysis of the core. Additional computer codes may be required for the modeling of the core heat-up and/or operation of emergency core-cooling systems, and for the treatment of small breaks or other distinctly different accident transients (Ybarranto, 1972).

Depressurization Transients

No significant loss-of-coolant accident has occurred in the nuclear industry to date. Furthermore, the likelihood that one will occur in the foreseeable future is very small. Nevertheless, the accident has been chosen as a design basis for many of the safety systems in water-cooled reactors. As a result, extensive experimentation has been carried out to understand the thermal and hydraulic phenomena that may affect the course of such an accident. Likewise, the computational methods used to predict numerically the course of loss-of-coolant accidents have been continuously refined. From experimental and numerical studies, it is possible to piece together the likely sequence of events in either pressurized- or boiling-water reactors (Brockett et al., 1973; Ybarrondo et al., 1972).

Pressurized-Water Reactors. Figure 8-22 shows two loops of a typical multiloop pressurized-water reactor. The coolant loops, and particularly the inlet and outlet pipes to the primary vessel, are located above the core to ensure that the core can be recovered with water following a break anywhere in the coolant loops. Emergency core-cooling systems are provided for both large and small ruptures. Passive accumulators and low-pressure injection systems (LPIS) deliver large quantities of water to reflood the vessel in the event of a large break. The coolant inventory is maintained following small breaks through the use of the coolant charging pumps as well as high-pressure injection systems (HPIS).

In a pressurized-water reactor the two most serious accidents are the double-ended rupture of an inlet or outlet coolant pipe near the reactor

- - - - ▶ LOCA Flow
───────▶ Normal Operation Flow

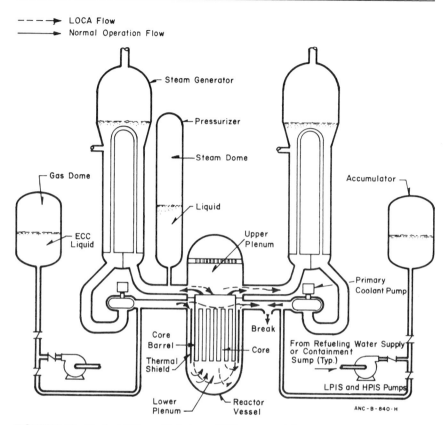

FIGURE 8-22   Typical U.S. multiple loop PWR system. Adapted from G. F. Brockett, R. W. Shumway, J. O. Zane and R. W. Griebe "Loss of Coolant, Control of Consequences by Emergency Core Cooling," *Proc. 1972 Conf. on Nuclear Solutions to World Energy Problems,* Washington, D.C., November 1972. Used by permission.

vessel. In Fig. 8-22, the inlet pipe is shown broken. Figure 8-23 illustrates the general characteristics of the reactor system during both accidents. The initial phase of the accidents is subcooled blowdown during which time the primary system pressure drops from greater than 2000 psi to the saturation pressure of about 1000 psi. For a large break the subcooled blowdown lasts less than 0.1 sec and is dominated by sonic decompression waves (not shown) that propagate through the primary system. These waves can create large transient pressure differentials across the fuel elements, core barrel and other structural components. The resulting dynamic loadings are typically greater than those anticipated from the design basis earthquake. Thus

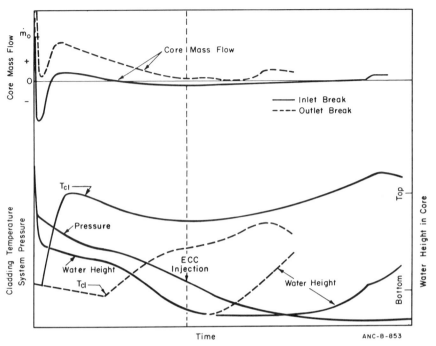

FIGURE 8-23 Generalized loss-of-coolant behavior for large pipe breaks in a PWR. Adapted from G. F. Brockett, R. W. Shumway, J. O. Zane and R. W. Griebe, "Loss of Coolant, Control of Consequences by Emergency Core Cooling," *Proc. 1972 Conf. on Nuclear Solutions to World Energy Problems,* Washington, D.C., November 1972. Used by permission.

extreme care is required in the structural design of the core and its supports to ensure that core barrel buckling, control rod levitation, or even structural collapse is not possible.

Following the subcooled blowdown, the primary consideration shifts from the structural integrity of the core to that of providing cooling to the fuel elements. We first consider the effects of an inlet break, represented by the solid lines in Fig. 8-23. The depressurization appears first at the core inlet, causing a rapid flow reversal. This is followed by a sharply deteriorated flow rate, which is associated with the coolant voiding in the core during the two-phase blowdown. Even in the absence of a trip, the rapidly decreasing coolant density in the core causes the power level to decrease to the level of decay heat within a few tenths of a second. A boiling crisis nevertheless takes place quite rapidly and results in nearly complete insulation of the fuel from the coolant. The sharp cladding temperature rise is caused by the redistributing of stored heat from fuel to cladding, resulting in transient behavior being similar to that described in Fig. 6-8.

After a few seconds the heat is redistributed between fuel and cladding, and the cladding temperature increase terminates. The cladding temperature is thereafter determined by the competition of decay heat production and heat transfer from the limited flow through the core. The flow, however, is quite small because of the nearly equal hydraulic resistances between the core inlet and outlet and the break location. As the depressurization transient is completed, typically after about 10 to 20 sec, the core becomes more completely vapor bound and the cladding temperature begins slowly to rise once more as the core heats up adiabatically from decay heat production. It is also at about this point that the emergency core-cooling injection begins.

The break of the outlet that results in the same depressurization curve will induce a somewhat milder cladding transient. Because the depressurization occurs at the core outlet, there is no core flow reversal. Moreover, a higher core flow rate is maintained during the two-phase blowdown because the hydraulic resistance from the core outlet to the break is much less than that from the core inlet to the break. Consequently the boiling crisis does not occur until after the decreased coolant density has caused the reactor to be shut down and a sufficient period of time has elapsed for the sensible heat stored in the fuel to be dumped into the coolant. As a result, the cladding temperature rise initiated by the boiling crisis is milder, being driven only by the production of decay heat. When emergency core-cooling injection is initiated, the cladding is at a substantially lower temperature than that for the inlet break.

Figure 8-23 clearly indicates that for either inlet or outlet break the emergency core-cooling injection is not immediately effective in stemming the cladding temperature rise. Since similar thermal hydraulic phenomena delay the effectiveness of the emergency cooling for pressurized- and boiling-water systems, we defer our discussion of them until after we have outlined the depressurization transient in a boiling-water reactor.

Boiling-Water Reactors.   Figure 8-24 shows a schematic representation of a typical direct-cycle boiling-water reactor. The reactor core is enclosed in a barrel and jet pump shroud arrangement in such a way that even if one of the pipes used to recirculate coolant through the core were broken at its lowest elevation, the core could be reflooded. In one current boiling-water reactor design the recirculation lines are eliminated entirely by placing the recirculation pumps inside the vessel. Consideration of recirculation line breaks then

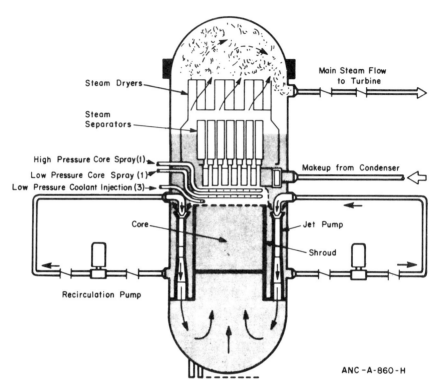

Steam Dryers

Steam Separators

High Pressure Core Spray(1)

Low Pressure Core Spray (1)

Low Pressure Coolant Injection (3)

Core

Recirculation Pump

Main Steam Flow to Turbine

Makeup from Condenser

Jet Pump

Shroud

ANC - A - 860 - H

FIGURE 8-24 Typical U.S. BWR system. Adapted from G. F. Brockett, R. W. Shumway, J. O. Zane and R. W. Griebe, "Loss of Coolant, Control of Consequences by Emergency Core Cooling," *Proc. 1972 Conf. on Nuclear Solutions to World Energy Problems,* Washington, D.C., November 1972. Used by permission.

becomes unnecessary. Emergency core cooling in boiling-water reactors is provided by high-and low-pressure systems for spraying water onto the top of the core and by injecting water at low pressure to fill the vessel.

Representative transients following steam and recirculation line breaks are shown in Fig. 8-25. Since the reactor outlet is operated at saturation pressure, the rapid initial depressurization characterizing the subcooled blowdown is absent. Moreover, the two-phase depressurizations are somewhat slower than those for a pressurized-water reactor. This is due in the case of the steam line break to the lower mass flow rates associated with the discharge of saturated steam, and in the case of the recirculation line break to the relatively small diameter of the ruptured line. Of the two accidents, the recirculation line break is the more severe. We discuss it first.

The immediate effect of the recirculation line break is to lose a substantial part of the forced convection and to cause the core flow rate to drop rapidly to about one-half of its initial value. Water continues to enter the vessel through feedwater lines and to exit through both the steam lines and the break. The isolation valves in the steam lines close a few seconds after the accident initiation, causing all the coolant to exit through the break. For the jet pump design shown, the recirculation line break causes the core outlet to depressurize faster than the inlet because of the smaller hydraulic resistance between the core outlet and the break.

Forced recirculation, albeit at a reduced level, continues to cool the cladding until the froth level falls to a point at which the jet pumps become uncovered and cavitate. Forced recirculation then is lost and the core flow rate drops to zero, causing a boiling crisis and vapor binding in the coolant channels. With the boiling crisis, the cladding temperature increases sharply because of the redistribution of sensible heat from fuel to cladding. The cladding temperature increase would continue in a manner analogous to that depicted in Fig. 6-8 were it not for the flashing of water in the lower plenum. As the water in the lower plenum, which is initially slightly subcooled, reaches saturation, it begins to flash violently to steam. This causes the froth level to surge into the core to the extent that the fuel element cooling may be sufficient to bring the cladding back down to the steam saturation temperature as indicated in Fig. 8-25. As the coolant inventory in the lower plenum is exhausted, the system pressure continues to decrease, and core flow starvation again causes a boiling crisis to occur. The cladding temperature therefore starts to rise rapidly once again. However, the temperature transient is not as severe as it would have been if the lower plenum flashing had not taken place, since more of the stored heat in the fuel has been removed to the coolant. At the time core spray is initiated, the temperature redistribution in the fuel and cladding is nearly completed, and further heat-up is driven primarily by decay heat production.

The steam line break in a boiling-water reactor results in a somewhat faster vessel depressurization. Since the steam expulsion takes place from the highest part of the vessel, however, the liquid fraction remains high at core level, and all the recirculation loops continue to function. The core flow rate remains quite high until system depressurization finally causes the recirculation lines to cavitate. Even after cavitation, the boiling of the fluid in the core and lower plenum results in sufficient flow upward through the core to delay the boiling crisis until almost all the sensible heat has been removed from the fuel. Thus the cladding stays near the coolant saturation temperature until after the initiation of core spray. Then, when the deteriorated flow conditions finally cause the boiling crisis to occur, the rate of cladding temperature increase is determined primarily by the decay heat production.

Emergency Core Cooling

Following a major loss-of-coolant accident the ultimate function of emergency core-cooling systems in either pressurized- or boiling-water reactors is to reflood the core so that the decay heat can be removed by stable nucleate boiling. The emergency cooling systems supply cool water continuously to the reactor vessel, to replace that which boils away in the core. The steam that is thus formed escapes first to the upper plenum of the vessel and then through the primary system breach and into the containment atmosphere.

The severity of the transient occurring between the end of the vessel blowdown and the establishment of the equilibrium heat transport path for decay heat removal determines whether the loss-of-coolant accident can be sustained without unacceptable damage to the reactor core. If core cooling is delayed too long, the cladding will reach excessive temperatures. As indicated in Section 6-4, at cladding temperatures in excess of 1800°F metal–water reaction begins to cause cladding embrittlement. If cladding temperatures exceed 2000°F for more than a few seconds the embrittlement becomes so extensive that when the core is reflooded the quenching of the embrittled cladding may cause it to shatter. The cladding debris will then obstruct coolant channels, and subsequent fuel element collapse may form a geometrical configuration with a greatly decreased heat transfer area and a greatly increased flow resistance. The emergency cooling systems then are not likely to be capable of removing decay heat, and the core will melt down.

To assure that the foregoing scenario does not take place, the emergency core-cooling system must stem the rising cladding temperature before embrittlement becomes excessive and then reduce it to near the saturation temperature of the depressurized vessel so that stable nucleate boiling can be established. Prompt action of the reactor protection systems coupled with the use of fast-starting pumps and valves allows the injection of emergency coolant to be initiated at about the end of the blowdown. As indicated in Figs. 8-23 and 8-25, however, a substantial period of time may elapse following the initiation of this injection before the cladding temperature rise can be stopped. Such delays result from the detrimental effects of two closely coupled thermal hydraulic phenomena: inability of water to wet (and hence efficiently cool) hot metal surfaces, and difficulty in achieving a sufficient pressure drop (and therefore flow rate) across the core.

The rewetting problem can be illustrated by considering the pool boiling curve shown in Fig. 6-3. Following the boiling crisis, the heated-surface temperature rises rapidly, say to point B, and the resulting heat transfer coefficient is quite small. If water then is brought into contact with the surface, it will bead up and be repelled, since no wetting can take place above

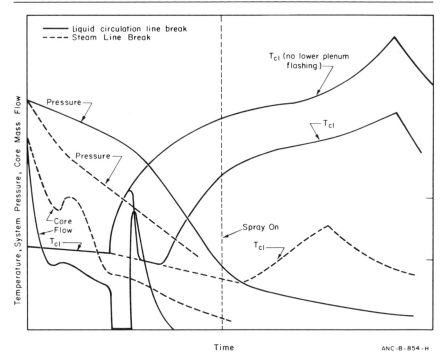

Time   ANC-B-854-H

FIGURE 8-25   Generalized loss-of-coolant behavior for large pipe breaks in a BWR. Adapted from G. F. Brockett, R. W. Shumway, J. O. Zane and R. W. Griebe, "Loss of Coolant, Control of Consequences by Emergency Core Cooling," *Proc. of the 1972 Conference on Nuclear Solutions to World Energy Problems,* Washington, D.C., November 1972. Used by permission.

the Leidenfrost temperature at point C. Consequently, the surface must be cooled by the relatively poor mechanism of forced vapor convection, radiation, and film boiling until the surface can be reduced to below the Leidenfrost temperature. The water will then wet the surface, rapidly quenching the temperature to where nucleate boiling can be reestablished. Following blowdown, the cladding temperature may be approaching 2000°F. The Leidenfrost temperature varies with pressure from about 300°F at 30 psia to 800°F at 1000 psia (Tong and Weisman, 1970). Thus in the depressurized atmosphere it may be necessary to reduce the cladding temperature by 1000°F or more before nucleate boiling can be reestablished.

Pressurized-Water Reactors.   Figure 8-23 indicates that the cladding temperature at a typical point within the pressurized-water reactor core will continue to rise, albeit at reduced rates, until the quench finally causes the cladding temperature to decrease. The time delay before cladding temperature reduction takes place has several causes. First, the lower plenum must be refilled with water. Then as water enters the core, it will not immediately wet the cladding surface, but will sputter from the wall and form a droplet flow pattern such as that shown in Fig. 8-26a. The resulting flow film boiling somewhat enchances the heat transfer and slows the rate of cladding temperature increase. After some period of time the lower ends of the fuel elements will be cooled to the Leidenfrost point and will be quenched. A quench front then develops and progresses upward through the core. The break in the cladding temperature curve in Fig. 8-23 is determined by the time at which the quench front passes through the point in the core at which the cladding temperature is being measured. The rate at which the quench front passes upward through the core depends on the rate at which the unstable film boiling at the front removes heat from the cladding, and this in turn can be correlated to the coolant mass flow rate and temperature, the initial cladding temperature, and the system pressure (Caked et al., 1971; Duffy and Porthouse, 1973).

Of the previously mentioned factors the cladding temperature and system pressure are more or less fixed by the conditions at the end of the blowdown.

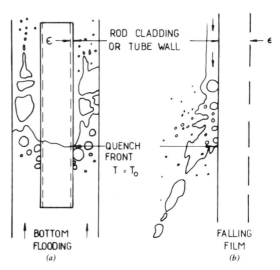

FIGURE 8-26 Physical phenomena in rewetting. From R. B. Duffy and D. T. C. Porthouse, "The Physics of Rewetting in Water Reactor Emergency Core Cooling," *Nucl. Eng. Des.* **25**, 379–394 (1973). Used by permission.

The emergency cooling water is normally injected at about room temperature. Thus the primary factor determining how effectively an emergency core-cooling system can deal with a given accident situation is the rate at which coolant can be forced through the core. The hydraulic situation that exists in the primary system following the reactor vessel blowdown, however, may place stringent limitations on the flow that can be achieved. We illustrate by considering the case of the inlet break.

The hydraulic conditions that exist during the operation of the coolant injection system may be viewed in terms of Fig. 8-22. The coolant is injected through the intact coolant loops, falls down through the downcomer annulus (between the core barrel and vessel wall), and fills the lower plenum. The water level outside the core barrel can rise no higher than the level of the break, since it will then simply spill from the break into the containment. Assuming the annulus remains filled with water, the pressure at the core inlet can be no higher than

$$p_i = p_c + m_l g \, \Delta z, \qquad (8\text{-}72)$$

where p_c, m_l, and Δz are the containment pressure, the liquid coolant density, and the elevation of the break above the core inlet.

Most of the emergency core-cooling water boils as it passes upward through the core, and the resulting steam must pass through the ruptured coolant loop before it can escape through the break and into the containment. The pressure drop between the core inlet and the containment can be written as the sum of a series of pressure drops through the core and through the ruptured coolant loop. Ignoring the hydrostatic and inertial effects of the water–steam combination in the core and coolant loop, we may determine the core mass flow rate W from

$$p_i - p_c = \frac{\tilde{\mathcal{H}}_c}{2\bar{m}} W^2 + \frac{\tilde{\mathcal{H}}_l}{2m_v} W^2, \qquad (8\text{-}73)$$

where $\tilde{\mathcal{H}}_c$ and $\tilde{\mathcal{H}}_l$ are the hydraulic resistances of the core and coolant loop, and $\bar{m}$ and m_v are the densities of the fluid in the core and the steam in the coolant loop, respectively. Combining Eqs. (8-72) and (8-73), we obtain

$$W = \left[\frac{2 m_l g \, \Delta z}{\tilde{\mathcal{H}}_c/\bar{m} + \tilde{\mathcal{H}}_l/m_v} \right]^{1/2}. \qquad (8\text{-}74)$$

From this expression it is clear that once the downcomer annulus is filled with water, the core flow rate is independent of the rate at which emergency coolant is pumped into the primary system so long as the injection system supplies water fast enough to keep the annulus filled. The flow rate is sensitive to the hydraulic resistances of the core and coolant loop. Unfortunately, these may be expected to be quite large. The boiling coolant in the core causes large acceleration losses to occur. The saturated steam in the upper plenum must pass through both the small-diameter tubing of the steam generator and the vapor-bound coolant pump before venting to the containment atmosphere. Furthermore, the secondary side of the steam generator is likely to contain large quantities of water at a much higher pressure and temperature. The saturated steam will therefore pick up additional heat as it passes through the steam generator tubing, causing it to become superheated. As a result, additional acceleration losses will be incurred in the tubing, causing the value of $\tilde{\mathcal{H}}_l$ to increase substantially, and the steam density m_v to decrease even further.

Boiling-Water Reactors. In boiling-water reactors, similar thermal-hydraulic phenomena govern the time delay before the core spray systems can effectively quench the cladding. When water is initially sprayed on top of the core, it spatters away from the hot surfaces. However, the top of the core is eventually cooled below the Leidenfrost point, and a liquid film forms at the top of the hot surfaces. A quench front then develops, as shown in Fig. 8-26b, and moves progressively downward as the unstable film boiling at the front cools the surface to the Leidenfrost point. As in the case of a pressurized-water reactor, the rate of quench front progression is correlated to the coolant mass flow rate and temperature, the initial cladding temperature, and the system pressure (Yamanouchi, 1968; Duncan and Leonard, 1971; Duffy and Porthouse, 1973).

In practice, the emergency cooling from core sprays is a two-stage process. The shrouding of boiling-water reactor fuel assemblies (see Fig. 7-8b) constitutes surfaces parallel to the fuel pins that have no internal heat source. As a result, the quench front proceeds down the shrouding at a much faster rate than down the fuel element cladding. Thus the shrouding is quenched first. There is then a substantial period of time over which the primary mechanism of heat removal is by radiation from the fuel elements to the quenched shrouding. Finally, the combined effects of radiation heat transfer, downward progression of the quench front on the cladding surfaces, and reflooding of the core cause the fuel elements also to be quenched. Hydraulic considerations play an important role in determining how fast coolant can enter the core and thereby advance the shrouding and cladding quench fronts. The boiling within the core tends to cause vapor binding of the fuel assemblies and thereby slow the entry of cooling water into the core.

Long-Term Cooling. The loss-of-coolant transient is effectively terminated once the cladding is quenched and the core reflooded. If eventual core meltdown is to be avoided, however, the emergency core-cooling system must continue to provide a heat transport path for orderly decay heat removal over a long period of time until the fuel elements can be removed from the core. In both pressurized- and boiling-water reactors it therefore is necessary to provide a continuing supply of cool water to the reactor vessel, and to ensure that the pressure in the containment vessel does not become so high that it hinders the exhaust of steam from the primary system.

Although details differ, the provisions for long-term decay heat removal typically resemble the pressurized-water reactor system shown in Fig. 8-27. Following the emptying of the accumulators, cooling water is pumped from a large, refillable reservoir, such as the refuelling water storage tank. The containment pressure is maintained at a suitably low level by condensing the steam with a spray system. The spray water also is supplied by the reservoir. After some period of time, the reservoir is depleted and large volumes of water—from both primary system boil-off and the containment spray—will collect in the containment sump. The core-cooling and containment sprays can then be switched to a recirculation mode in which water is pumped from the containment sump, through a heat exchanger and into the primary system, and into the containment spray. The decay heat thus is removed through the heat exchanger and dissipated in the environment. In all these features redundancy is provided to increase the system's reliability.

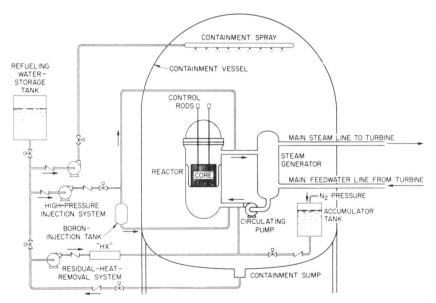

FIGURE 8-27 PWR emergency cooling system. From W. B. Cottrell, "The ECCS Rule-Making Hearing," *Nucl. Saf.* **15**, 30–55 (1974).

PROBLEMS

8-1. A steel pressure vessel has an inside diameter of 14 ft and is 8.5 in. thick. The operating coolant pressure is 2300 psia. In the vessel there is an elliptical crack 1 in. deep and 9 in. long located in the inside surface parallel to the axis of the vessel. At operating temperature the critical stress intensity factor is $K_c = 125 \times 10^3$ psi in.$^{1/2}$. At room temperature it is 40×10^3 psi in.$^{1/2}$. The vessel yield strength is 40,000 psia.

a. Calculate the stress in the vessel under operating conditions.

b. At operating conditions calculate the safety factor K_c/K for the crack as well as the critical crack depth assuming that the length-to-depth ratio remains constant.

c. If the vessel is severely overpressurized, will the crack propagate before the vessel yields plastically? If propagation takes place, at what vessel pressure will it occur? If yield takes place, at what pressure will it occur?

d. Suppose the vessel is subjected to high pressures at room temperature; repeat part (c) in this situation.

8-2. An elliptical crack of initial length a_i is present in a reactor coolant pipe. At operating stress levels the stress intensity caused by a crack of this length is K_i. During changes in power level the pipe undergoes stress cycling of magnitude $\Delta\sigma$.

a. Show that fatigue, governed by Eq. (8-10), will cause the crack to grow as

$$a = a_i \left[1 - \frac{N C_0}{a_i} K_i^4 \left(\frac{\Delta\sigma}{\sigma} \right)^4 \right]^{-1},$$

where N is the number of cycles.

b. Suppose that $C_0 = 4 \times 10^{-24}$ (1/psi^4-in-cycle), $a_i = 2$ in, and K_i is one-half the critical stress intensity factor of $K_c = 1.0 \times 10^5$ psi in$^{1/2}$. Determine how many stress cycles will occur before failure when

$$\frac{\Delta\sigma}{\sigma} = 0.25, \qquad \text{When} \quad \frac{\Delta\sigma}{\sigma} = 0.5.$$

8-3. A design flaw in an LMFBR results in a 1 inch diameter sodium purification line being located so that its lowest point is 15 ft below the sodium level in the reactor vessel and the length of the line between the vessel and this low point is 30 ft. There are four 90° bends in the line within this 30 ft segment. If the line ruptures at its lowest point, how fast will sodium initially be discharged from the vessel? Assume that the sodium density is $54 \, \text{lb}_m/\text{ft}^3$, the thermal expansivity is $1.6 \times 10^{-4}/°\text{F}$, and the cover gas in the vessel is at 3 atm pressure.

8-4. Verify Eqs. (8-26) through (8-28).

8-5. The primary system of a helium-cooled reactor is at 1200 psia and 800°F. The coolant volume is $35,000 \, \text{ft}^3$. Estimate the isothermal depressurization constant τ^* as a function of rupture areas between 0.1 and 100 in.^{2}

8-6. Suppose that a vessel filled with a perfect gas depressurizes adiabatically. Then $(p/p_0) = (m/m_0)^\gamma$.

  a. Show that the depressurization transit is now given by

$$p = p_0 \left[1 + \frac{(\gamma - 1)t}{2\tau^*} \right]^{[-2\gamma/(\gamma - 1)]}$$

  where τ^* is the isothermal blowdown constant given by Eq. (8-33).
  b. For a helium reactor with $\gamma = 1.33$, plot p/p_0 versus t/τ^* for both isothermal and adiabatic blowdown.
  c. Assume that the primary system is initially at 800°F and 1000 psi, the volume is $50 \times 10^3 \, \text{ft}^3$, and the break area is 25 in.^{2} According to the preceding two blowdown models, how much time will be required before twice atmospheric pressure is reached?
  d. For the transient described in part (c), after what time and pressure will the flow no longer be choked according to each of the models?
  e. Do the two models under- or overpredict the rate of blowdown following the termination of choked flow?

8-7. Verify Eqs. (8-41) and (8-42).

8-8. Verify Eqs. (8-62) through (8-65).

8-9. A 3000 MW(t) PWR has undergone a loss-of-coolant accident and depressurized to 20 psia. One hour after the accident, at what rate (in gal/min) must 100°F water be furnished to the core by the emergency core-cooling system to remove decay heat and thereby prevent the core temperature from increasing?

ACCIDENT CONTAINMENT

Chapter 9

9-1 INTRODUCTION

Nuclear power plants and their associated protection and safety systems are designed to operate so that even in the event of a major (albeit very improbable) accident, the reactor core will not melt. This emphasis on reactor safety is justified, since so long as the reactor core remains essentially intact, even potentially severe accidents will result only in minimal releases of radioactive materials to the environment. To reduce further the risk associated with the large inventory of radioactive material residing in the reactor core, it has become nearly universal practice to require that the primary system envelope be enclosed in one or multiple containment structures. Normally, such structures play no part in the routine operation of the nuclear plant. They are, instead, essentially leaktight barriers designed to trap those radioactive materials that may escape from the fuel and penetrate the primary system envelope and prevent them from entering the environment.

If the containment is to fulfill this function, careful account must be taken of the stresses to which it may be subjected under the severest accident conditions. Before a determination can be made of the pressures, temperatures, and impact and other loadings to which the containment may be subjected, the strength of the energy sources and the modes of the load transmission that appear in the design basis accidents must be ascertained. For this reason we place emphasis in this chapter on the nature of these limiting accidents that place the severest stresses on the containment structure and therefore on the design of the containment system.

The accident loadings for which containment systems must be designed have their origins in five sources:

1. Stored energy.
2. Nuclear transient energy.
3. Decay heat.
4. Chemical reaction energy.
5. Site-related energy sources.

The stored energy consists of sensible and latent heat stored in the fuel, coolant, and other primary system components at the time the accident is initiated. Nuclear transient energy is the heat produced promptly by fission, should an accident result in a nuclear excursion. The decay heat is produced continuously from the fission products and actinides present in the core at the initiation of the accident. Exothermic chemical reactions are a potential cause of accidents as well as sometimes being by-products of them. Finally, site-related energy sources refer to both man-made accidents and natural forces that originate from outside the nuclear plant. Although more than one of these energy sources are likely to contribute significantly to the damage incurred from a major accident, it is useful to preview the characteristics and situations in which each of these sources is likely to make a predominant contribution.

In the loss-of-coolant accidents, discussed in Chapter 8, the breach of the primary system envelope brings the reactor coolant into contact with the containment structures. The energy stored in high-pressure and high-temperature gas or water coolants is the predominant contribution to the increase in containment pressure and normally determines the combination of containment strength and volume required to ensure that unacceptable leakage rates to the environment do not occur. An upper limit on the amount of stored energy available for containment pressurization can be determined quite accurately from the reactor characteristics, primary system volume, coolant temperature and pressure, and so on. Thus upper limits can be set on the pressure and temperature conditions that result from the expulsion of coolant.

In superprompt-critical excursions the fission energy generated is not likely to be more than that caused by a few seconds of full-power operation of the reactor. However, this energy is generated in a rapid burst as discussed in Chapter 5, and power levels of many times the rated reactor power value may be reached over a time span that is too short for heat removal to take place by the normal mechanisms. In this case the same fuel may vaporize, causing high pressure that creates blast effects similar to a rather inefficient chemical explosive. Even if the fuel vapor pressure is not significant in itself, the fuel fragmentation that accompanies vaporization may cause very rapid transfer of heat from the molten fuel to the coolant. Under these circumstances vapor explosion may result from the high-pressure vaporization of the more volatile coolant.

Whether caused by rapid vaporization of fuel or of coolant, the resulting blast effects may cause impact loading to the pressure vessel walls and, in the case of liquid-cooled systems, create water hammers that endanger the vessel head hold-down mechanism. In either case the propagation of the damage through the primary system envelope would then cause loading on the containment structures. One of the most difficult tasks in assessing the destructive force of nuclear excursions is the determination of the magnitude of the energy release. For in contrast to stored energy, which is readily calculable from the reactor design parameters, nuclear transient energy depends on several of the least-well-established characteristics of the postulated accident: the reactivity insertion rate, the inertial delay in fuel motion, the degree of fuel fragmentation, and the behavior of molten and/or vaporized fuel and/or coolant.

Like stored coolant energy, the energy available from decay heat is readily predictable. Moreover, it has little effect during the first seconds of most accidents, since it is released at a rate of only a few percent of full reactor power. Decay heat, however, is the most persistent of the energy sources in that its production continues for extremely long times independent of the conditions following the accident. For this reason no matter how innocuous the initial stages of an accident may appear, adequate cooling of the core must be maintained for an indefinitely long period of time. Otherwise, decay heat will eventually cause fuel melting and lead to core meltdown, in which the molten fuel mass causes the core structure to collapse, the reactor vessel to melt through, and the containment structure to be subjected to severe thermal loadings.

Energy from chemical reactions may be the prominent determinant of the course of an accident, or it may exacerbate accidents arising from other causes. For example, the breach of a low-pressure sodium-cooled system causes no significant release of stored energy. But if the containment atmosphere is not inert, it will be pressurized from the ensuing sodium fire. In contrast, metal–water reactions may contribute to the meltdown of water-cooled reactors following a loss-of-coolant accident, but their effects become significant only if the inadequacy of the emergency core-cooling system causes the cladding to reach excessive temperatures following the depressurization transient. As with nuclear transients, an upper limit on the energy release from chemical reactions is difficult to predict. Exceedingly large—and unacceptable—releases would result, for example, in a sodium-cooled system if all the coolant inventory were to burn, of if all zircaloy cladding were to undergo a metal–water reaction. At the same time it is difficult to perceive how situations could arise in which sufficiently complete mixing could occur under the required thermodynamic conditions to cause more than a small fraction of the materials to react. However, it is often difficult to understand the accident conditions well enough to predict a reasonable upper bound on the amount of material that will interact.

Under the heading of site-related energy sources we lump both man-made and natural forces originating outside the nuclear power plant. Site-induced accidents of the man-made variety include aircraft impact, industrial explosions, and so on. Upper limits on the magnitude of such accidents can normally be determined, and probabilistic arguments must then be used to decide the damage level that is acceptable. Likewise, containment structures must be designed to withstand damaging tornadoes, high winds, flooding, high waves, and other natural events. Although some naturally occurring events tend to have approximate limits in magnitude, such as the wind speeds from hurricanes or tornadoes, others appear to have none. Of the latter variety, earthquakes perhaps are the most important from the viewpoint of reactor safety. For although the frequency of earthquakes is a decreasing function of severity, it is difficult to place an upper bound on seismic activity. Moreover, the ground shaking affects not only the containment structure but the entire plant.

In the following sections we examine the severe accidents that may arise from the previously stated energy sources and examine their effects on containment in more detail. In Section 9-2 containment pressurization caused by primary system breaches is discussed. Section 9-3 is devoted to the explosive core disassembly accidents that frequently serve as a design basis for fast reactor containments. In Section 9-4 the characteristics of core meltdown accidents are discussed. We conclude, in Section 9-5, with a discussion of site-induced accidents, placing primary emphasis on severe earthquakes.

9-2 CONTAINMENT PRESSURIZATION

The primary function of the containment is to serve as an essentially leaktight barrier that will trap any radioactive materials that may leak from the primary system and prevent their release to the atmosphere. To accomplish this task, the containment must be capable of withstanding the maximum pressure that may be generated in conjunction with the design basis accidents. Such pressures are brought about in many instances as a result of the breach of the primary system, with a subsequent expulsion of the reactor coolant into the containment atmosphere. In high-pressure gas- or water-cooled systems, the stored heat of the coolant accounts for the pressure rise. In sodium-cooled systems the exothermic sodium–oxygen reactions may cause the pressure to rise if the containment atmosphere is not inert.

Even though the design pressure of the containment is sufficiently high to withstand the pressurization of a hypothesized accident, it is impossible to achieve a zero leak rate. The leak rate may be specified to be quite small, typically of the order of 0.1 volume percent per day at design pressure. This rate, however, is an increasing function of the containment pressure, and hence it is desirable to reduce the containment pressure rapidly following the explusion of the coolant to minimize leakage of radioactive materials to the environment. For this reason heat exchangers, spray systems, various means of vapor suppression, or other active safety systems normally are provided to cool the containment atmosphere.

In this section we first examine the expulsion of a high-pressure water coolant into a dry containment volume and then examine the means by which the associated heat is removed from the containment atmosphere. We next examine the techniques that are used to reduce the pressure in water-cooled reactor containments. We then examine the problem of the containment pressures generated by fires in sodium-cooled systems, and we conclude with a discussion of some of the difficulties encountered in designing containment shells to have very small leak rates.

Dry Containment

Consider the situation in which a primary system breach in a water-cooled reactor causes the coolant to be expelled into a containment shell. To estimate the pressures reached within the shell, we apply macroscopic mass and energy balances to the containment volume in a manner reminiscent of their application to primary system depressurization in Section 8-5. Once again the effects of potential and kinetic energy are neglected, and the constituents of the containment atmosphere are assumed to be in thermal equilibrium, with no pressure or temperature gradients present. Now, however, we must take into account the nitrogen, oxygen, and other gases that make up the initial containment atmosphere, as well as the coolant expelled from the primary system.

We modify the mass and energy balances, Eqs. (8-43) and (8-44), to be applicable to the air–water mixture by assuming that the air can be represented as a perfect gas. We have

$$V_{ct}\frac{dm}{dt} = W_0 \qquad (9\text{-}1)$$

and

$$V_{ct}\frac{d}{dt}[mu + m_a c_a T_{ct}] = h_0 W_0 - Q'_{ct}, \qquad (9\text{-}2)$$

where

V_{ct} = containment volume,

m = H_2O density in containment,

u = H_2O internal energy,

m_a = initial air density,

c_a = air specific heat per unit mass at constant volume,

W_0 = mass flow rate from the primary system,

h_0 = enthalpy of coolant discharged from primary system,

Q'_{ct} = rate of heat loss from containment atmosphere,

T_{ct} = absolute containment atmosphere temperature.

In this we neglect the small leakages to the environment and conservatively assume that the coolant discharge to the containment is an isenthalpic process.

In addition to the mass and energy balances we use the equation of state for water. For present purposes we write the equation of state as

$$u = u(m, T_{ct}). \tag{9-3}$$

Almost immediately following a large breach in the primary coolant system, the containment atmosphere will become saturated. Therefore the equation of state is more conveniently represented in terms of saturated liquid and vapor properties:

$$mu = (1 - \alpha)m_l u_l + \alpha m_v u_v. \tag{9-4}$$

The quantities m_l, u_l, m_v, and u_v may be written as a function only of T_{ct} and the void fraction

$$\alpha = \frac{m_l - m}{m_l - m_v}, \tag{9-5}$$

which is a function of both m and T_{ct}.

The initial conditions, containment volume, primary system discharge rate, and enthalpy are given. Hence Eqs. (9-1), (9-2), and (9-4) can be solved for u, m, and T_{ct} as a function of time provided $Q'_{ct}(t)$ can also be estimated. The law of partial pressures may then be used to determine the containment pressure transient:

$$p_{ct} = p_w + p_a. \tag{9-6}$$

Under saturated conditions the water partial pressure p_w is a function only of the temperature T_{ct} of the containment atmosphere. The partial pressure p_a caused by the air can be determined by applying the perfect gas law to that part of the containment volume occupied by the steam–air mixture:

$$p_{ct} = p_w(T_{ct}) + \frac{m_a}{\alpha} R_a T_{ct}, \tag{9-7}$$

where R_a is the gas constant per unit mass of air.

To determine the time history of the containment pressure transient, the explicit effects of the containment heat removal mechanisms must be incorporated into Eqs. (9-1), (9-2), and (9-4), and these equations must be solved as a function of time. A pessimistic estimate of the peak pressure may be made, however, by initially neglecting the heat removal mechanisms and integrating Eq. (9-2) from the accident initiation to the end of the primary system blowdown. We have

$$(mu + m_a c_a T_{ct})|_{t=t_b} - (mu + m_a c_a T_{ct})|_{t=0} = \frac{1}{V_{ct}} \int_0^{t_b} W_0 h_0 \, dt - \frac{1}{V_{ct}} \int_0^{t_b} Q'_{ct} \, dt$$

(9-8)

where $t = 0$ and t_b correspond to the initiation and the end of the depressurization, respectively. We assume that initially there is no water vapor in the containment, and therefore $m|_{t=0} = 0$. Moreover, at the end of the depressurization transient a saturated mixture of liquid and vapor will be present, and therefore Eq. (9-4) is applicable. Making these substitutions in Eq. (9-8), we have

$$[1 - \alpha(m, T_{ct})]m_l(T_{ct})u_l(T_{ct}) + \alpha(m, T_{ct})m_v(T_{ct})u_v(T_{ct}) + m_a c_a (T_{ct} - T_0)$$

$$= \frac{1}{V_{ct}} \int_0^{t_b} W_0 h_0 \, dt - \frac{1}{V_{ct}} \int_0^{t_b} Q'_{ct} \, dt,$$

(9-9)

where for brevity we have replaced $T_{ct}|_{t=t_b}$ and $T_{ct}|_{t=0}$ by T_{ct} and T_0, respectively.

The coolant density and discharge enthalpy in this expression is determined by assuming that the primary system volume is initially filled with saturated liquid at temperature T_{pr} and that at the end of blowdown all the coolant has been expelled to the containment. We have

$$m = \frac{m_l(T_{pr}) V_{pr}}{V_{ct}}$$

(9-10)

and

$$\frac{1}{V_{ct}} \int_0^{t_b} W_0 h_0 \, dt = \frac{m_l(T_{pr}) u_l(T_{pr}) V_{pr}}{V_{ct}}.$$

(9-11)

With known primary system properties these expressions can be numerically evaluated. Then, if the blowdown is fast enough that the heat transfer out of the containment, represented by the last term to the right of Eq. (9-9), can be neglected, steam table data can be used to solve Eq. (9-9) numerically for the containment temperature, and therefore for the pressure. This evaluation has been carried out elsewhere using standard atmospheric conditions for the initial containment atmosphere (Slaughterbeck, 1971). In Fig. 9-1 are

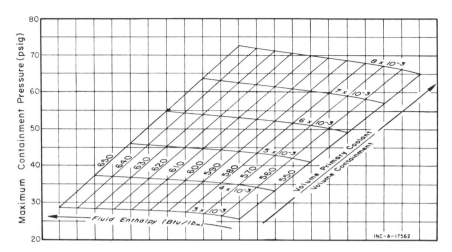

FIGURE 9-1 Relation among maximum containment pressure, fluid enthalpy, and ratio of primary coolant volume to containment volume. From D. C. Slaughterbeck, "Correlations to Predict the Maximum Containment Pressure Following a Loss-of-Coolant Accident in Large Pressurized Water Reactors with Dry Containments," Idaho Nuclear Corporation Report, IN-1468, May 1971.

plotted the results for the containment pressure as a function of V_{pr}/V_{ct}, the primary system-to-containment volume ratio, and the initial enthalpy of the primary system.

The foregoing analysis neglects several energy contributions that may significantly affect the containment pressurization. During the course of the blowdown, for example, heat may be transferred from the pressure vessel and core internals to the coolant. This may add a few percent to the enthalpy of the fluid discharged to the containment. In a pressurized-water reactor with a double-ended pipe rupture, much of the coolant passes through a steam generator, and as much as 10 percent of the coolant energy may be dissipated before it enters the containment. Finally, heat is transferred to the walls of the containment vessel by natural convection. This process may be represented roughly by Newton's law of cooling:

$$Q_{ct} = A_{ct}\tilde{h}_{ct}(T_{ct} - T_{wl}),\qquad(9\text{-}12)$$

where A_{ct} is the effective wall area of the containment structure, T_{wl} is the average wall temperature, nd $\tilde{h}_{ct}$ is the heat transfer coefficient. For a typical steel-lined containment, $\tilde{h}_{ct}$ is typically of the order of 150 Btu/(ft^2)(hr)(°F) (U.S. Atómic Energy Commision, 1974b).

Using representative numbers, rough correlations have been made to take into account the decreases in peak containment pressure caused by the heat transfer to the steam generators and to the containment walls (Slaughterbeck, 1971). These are given in Figs. 9-2 and 9-3, respectively. To obtain the corrected maximum containment pressure p'_{ct} we take

$$p'_{ct} = p_{ct} - \Delta p_1 - \left[200 \frac{V_{pr}}{V_{ct}} - 0.1 \right] \Delta p_2, \qquad (9\text{-}13)$$

where p_{ct} is read from Fig. 9-1, and $-\Delta p_1$ and $-\Delta p_2$ are taken from Figs. 9-2 and 9-3, respectively.

Although this graphical technique provides a first estimate of the peak containment pressure, several additional factors are taken into account in computational methods that are aimed at providing a more comprehensive picture of containment transients (Richardson et al., 1967). The assumption of thermodynamic equilibrium may introduce substantial errors at larger times when a pool is formed in the sump of the containment vessel because of water condensing from the containment atmosphere or pouring from the reactor vessel following the initiation of emergency core cooling. The thermal contact between the pool and the containment atmosphere may be quite poor, requiring a nonequilibrium phase separation model to account for the effect of its formation on the pressure. In addition, walls or other

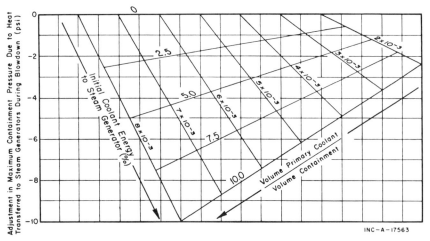

FIGURE 9-2 Effect of heat transferred to steam generators during blowdown on the maximum containment pressure as a function of the ratio of primary coolant volume to containment volume. From D. C. Slaughterbeck, "Correlations to Predict the Maximum Containment Pressure Following a Loss-of-Coolant Accident in Large Pressurized Water Reactors with Dry Containments," Idaho Nuclear Corporation Report, IN-1468, May 1971.

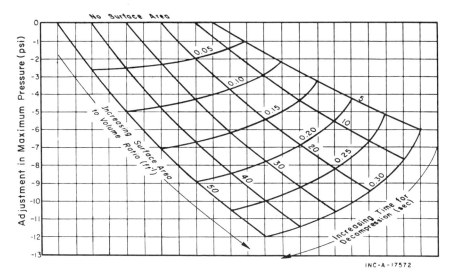

INC-A-17572

FIGURE 9-3 Reduction in maximum containment pressure as a result of energy absorption in containment structures for various ratios of surface area to volume and decomposition time. D. C. Slaughterbeck, "Correlations to Predict the Maximum Containment Pressure Following a Loss-of-Coolant Accident with Dry Containments," Idaho Nuclear Corporation Report, IN-1468, May 1971.

structures may introduce flow paths with relatively high hydraulic resistances between compartments within the containment shell. A more elaborate analysis with nodes and flow paths, such as described for the primary system in Section 8-5, may then be required both to calculate the pressures in the containment shell and to determine pressure differences across internal structures.

The peak pressure in a simple containment shell is expected to occur roughly at the end of the primary system blowdown. Following blowdown, however, steam and hot water will continue to be discharged from the primary system to the containment as a result of the removal of decay heat from the core by emergency core-cooling systems. If adequate heat transport paths away from the containment are not provided, the pressure will continue to rise, leading to eventual failure. Some heat removal capability is provided for naturally by the transport of heat through the containment walls. Particularly for the steel-lined, prestressed concrete containment structures that are in wide use, however, the thermal resistance of the wall severely limits the rate at which heat can be removed by this mechanism. As a result, active safety systems must be provided for heat removal.

Pressure Reduction Systems

From the foregoing discussion, the need for active systems to remove heat from water reactor containments is readily seen. These systems are used not only to prevent primary system boil-off due to decay heat from causing a continually increasing pressure, but also to reduce the pressure rapidly and thereby reduce the leak rate following an accident. In addition, safety systems often are incorporated into the containment to condense some of the primary system discharge before the blowdown is completed, and thereby reduce the peak pressure. This in turn allows a smaller containment volume and/or strength to be utilized. As indicated for a pressurized-water reactor in Fig. 8-27, these containment pressure reduction devices are incorporated into the overall scheme for the transport of decay heat away from the plant following a loss-of-coolant accident.

In dry containments, recirculation systems are often used to cool the containment atmosphere as well as to remove fission products from it. Fans are utilized to force the containment atmosphere through a filter, for fission product removal, and a heat exchanger, for containment cooling. These systems are easily tested and in some cases they may be run continually, obviating the start-up of pumps, fans, or other equipment upon detection of an accident. The effect of the heat exchanger may be approximately accounted for by modifying the heat removal term in the foregoing equations to

$$Q'_{ct} \rightarrow Q'_{ct} + c_{ct} W_{hx} (\Delta T)_m, \qquad (9\text{-}14)$$

where W_{hx} is the mass flow rate through the heat exchanger, $(\Delta T)_m$ is a logarithmic temperature drop determined by the heat exchanger design, and c_{ct} is the specific heat of the containment atmosphere.

Spray systems, such as that indicated in Fig. 8-27, have become a near universal feature of water reactors for more rapid reduction of the containment pressure and, as discussed in Section 10-3, for the reduction of the concentration of fission products in the containment atmosphere. The effect of a spray system on heat removal from the containment atmosphere can be approximated by letting

$$Q'_{ct}(t) \rightarrow Q_{ct}(t)' + c_p W_{sp} \epsilon_{sp} (T_{ct} - T_{sp}), \qquad (9\text{-}15)$$

where W_{sp}, c_p, and T_{sp} are the mass flow rate, specific heat, and temperature of the water delivered to the spray nozzles. If the spray water comes to thermal equilibrium with the containment atmosphere before falling to the sump, the efficiency ϵ_{sp} is set equal to one. Although perfect efficiency is not possible, the efficiency is maximized by producing spray droplets as small as practicable and by maximizing the distance between the spray nozzles and the sump. As indicated in Fig. 8-27, once the water reaches the sump it is circulated through a heat exchanger and is fed back to the spray and/or emergency core-cooling system.

Figure 9-4 shows some typical pressure traces in a dry containment system following a loss-of-coolant accident. If no safety systems, including the emergency core cooling (here called safety injection), function, the combined effects of decay heat and metal–water reaction will deliver enough energy to cause a continuous rise in containment pressure. With the emergency core-cooling system operating along with various of the containment heat removal systems, the increase is stemmed, and the pressure eventually is reduced to atmospheric level, the decrease being particularly sharp with the operation of the spray system.

In recent years the increasingly large capacities of water-cooled reactors have caused dry containments to become less attractive because of the large volumes required. As a result, passive vapor suppression concepts have been

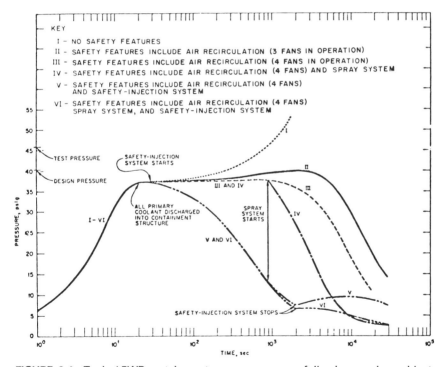

FIGURE 9-4 Typical PWR containment-pressure response following a major accident. From Department of Water and Power of the City of Los Angeles, "Preliminary Hazards Summary Report, Malibu Nuclear Plant No. 1," U.S. Atomic Energy Commission Docket No. 50-214, 1963.

introduced to reduce significantly the peak pressure and therefore the required size of the containment structure. Both water and ice have been utilized for vapor suppression. We consider first a boiling-water reactor system utilizing water as a medium for vapor suppression.

Although several more recent design variants have been utilized, the original boiling-water reactor system shown in Fig. 9-5 illustrates the primary features of the vapor suppression systems (Wade, 1974). The system consists of a dry well, which houses the reactor vessel and coolant recirculation system; a wet well (or pressure suppression chamber), which stores a large volume of cool water; a set of vertical connecting vents from the dry well to the wet well; and isolation valves in the steam lines. A pipe rupture in the primary system causes dry-well and wet-well pressure transients such as those shown in Fig. 9-6. The flashing of the primary coolant causes the dry-well pressure to increase. The increasing dry-well pressure, however, forces the water to clear from the vents into the wet well, causing steam and air from the dry well to vent into the wet well. The steam rapidly condenses in the cool water, causing the dry-well pressure to sharply drop as it comes into equilibrium with the wet-well atmosphere. Early in the transient, the steam line isolation valves (not shown) are closed to complete the closure of the containment shell.

Following the blowdown transient a decay heat removal path out of the vapor suppression system must be established. If none of the safety systems used for this purpose functions, decay heat and heat generated from the metal–water reactions lead to overpressurization of the containment, as

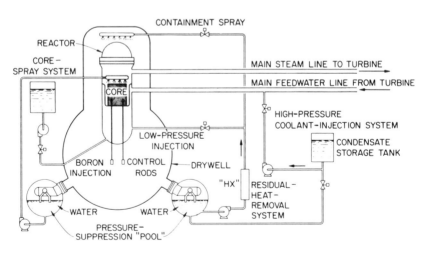

FIGURE 9-5 BWR vapor suppression containment system. From W. B. Cottrell, "The ECCS Rule-Making Hearing," *Nucl. Saf.* **15**, 30–55 (1974).

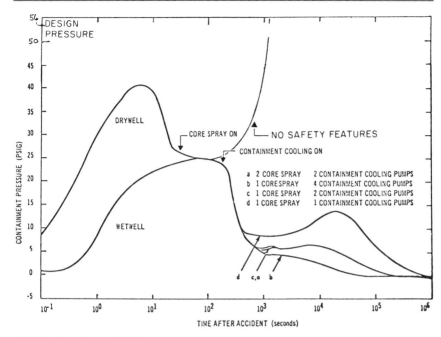

FIGURE 9-6 Typical BWR containment-pressure response following a major accident. From Tennessee Valley Authority, "Brown's Ferry Nuclear Power Station Design and Analysis Report," U.S. Atomic Energy Commission Docket No. 50-259, 1963.

indicated in Fig. 9-6. With the operation of the emergency core-cooling and containment-cooling systems, however, the pressure transients are quenched, causing the dry well and wet well to return to near atmospheric pressure. Containment cooling is provided by a dry-well spray and heat exchanger system similar in arrangement to those used in dry containment systems.

In some pressurized-water reactors, beds of ice are utilized in vapor suppression systems that in principle are quite similar to the water system discussed here (Weems et al., 1970). A bed of ice cubes is housed in a cold-storage compartment surrounding the primary system and is kept frozen by conventional refrigeration equipment. The containment is divided into an upper and a lower chamber separated by a compartment structure enclosing the bed with the primary system located in the lower chamber. Pressurization of the lower chamber opens access panels located at the inlet of the ice storage compartment permitting air and steam to flow through the ice bed. The steam is condensed, and the air passes into the upper chambers via the outlet access panels.

The calculation of the pressure transients in vapor suppression systems utilizing either water or ice is more involved than for the dry containment. Each requires that mass and energy balances be carried out in at least two volumes. More important, careful modeling is required to predict the vent clearance and flow to the wet well in the water system (Carmichael and Marko, 1969) and to simulate the opening of the access panels and the flow through the bed of the ice condenser (Salvatori, 1974).

Sodium Fires

Chemical reactions can contribute to containment pressurization in most reactor types. The foregoing discussion of water reactors indicates that the initial containment pressure peak is due entirely to the flashing of the coolant, but the energy released from metal–water reactions in the reactor core, when combined with decay heat can lead to overpressurization of the containment if an adequate containment-cooling system is not provided. Similarly, the primary containment pressure peak in a gas-cooled graphite-moderated reactor is expected to result from the expulsion of the high-pressure coolant following the breach of the primary system. But if water or oxygen then comes into contact with the overheated core, additional energy from graphite combustion may reach the containment. In contrast to these systems, the coolant vapor pressure in sodium-cooled reactors has a negligible effect on the pressurization of the containment shell. Instead, fires are the dominant mechanism by which pressurization of the outer containment shell may be postulated (Graham, 1971; Baurmash et al., 1974).

Sodium may undergo chemical reactions with both water and oxygen:

$$Na + H_2O \rightarrow NaOH + \tfrac{1}{2}H_2,$$
$$3Na + H_2O + O_2 \rightarrow NaO + 2NaOH, \qquad\qquad (9\text{-}16)$$
$$4Na + O_2 \rightarrow 2Na_2O,$$
$$2Na + O_2 \rightarrow 2NaO,$$
$$4Na + 2O_2 \rightarrow 2Na_2O_2.$$

The complete combustion of the sodium inventory of a reactor would result in an extremely large release of energy. However, it is incredible that this should happen since a great deal of care is taken to ensure that large amounts of the reactants cannot come into contact. Moreover, if they did, the incomplete mixing of the reactants and the thermodynamic limitations would permit no more than a small fraction of the sodium to burn. One of the foremost difficulties in performing containment pressurization calculations for sodium-cooled systems lies in placing reasonable upper bounds on the fraction of the sodium inventory that is likely to undergo combustion.

The layout of sodium-cooled fast reactors provides a major barrier to the combustion of sodium from the primary system, as may be seen by examining Figs. 1-15c and 9-7. The primary potential for sodium–water reactions lies in a breach of the steam generator. With the intermediate loop configuration such as shown in Fig. 1-15c, however, such a reaction would not directly involve the primary system, and moreover the fire would take place outside the containment shell. The likelihood of sodium–oxygen reactions is greatly reduced by placing the primary system in a vault structure, such as is shown in Fig. 9-7, that has an inert atmosphere, with an outer containment shell placed over the vault head. Most potential leak paths from the primary system would vent to the inert atmosphere of the vault. Only if there was also a failure in the partition between the vault and the outer containment shell would significant amount of oxygen and sodium come together. Thus, unless there is a lifting of the reactor vessel head or one of its penetrations, there is no single barrier failure that will bring sodium and oxygen into direct contact.

These precautions notwithstanding, it is instructive to examine the properties of sodium–oxygen fires. For purposes of analysis three types of sodium oxidation are normally referred to (Graham, 1971): pool fires, spray fires, and explosive sodium ejection. The most likely of these to occur, but the potentially least destructive, is the pool fire. Since sodium systems are at low pressure, leakage is likely to be slow. Even if a major pipe rupture occurs,

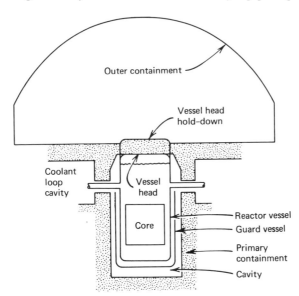

FIGURE 9-7 Typical LMFBR containment features.

normal operating pressures in sodium systems are low enough that the sodium will remain a liquid, pouring forth as a solid stream and forming a pool on the surrounding surface. No flashing will occur since even at atmospheric pressure the operating temperatures of sodium systems are well below the boiling point. If the pool is in an atmosphere with greater than 4% oxygen, the pool will ignite at temperatures somewhere between 400 and 600°F, depending on whether the surface is disturbed. The interaction is characterized by a low flame and produces a dense white smoke. Experiments indicate that such fires tend to burn at a rate of about 5 lb of sodium per hour per square foot of surface area, with the combustion liberating between 4100 and 4850 Btu/lb of sodium depending on which of the reactions given in Eqs. (9-16) are most prominent.

The containment pressurization resulting from a pool fire is determined by the heat transfer considerations as well as the reaction rate. A substantial part of the heat produced is conducted downward into the sodium pool because of its high conductivity. This competes with the heat transferred into the atmosphere above the pool by convection and radiation. There is a further heat loss through the walls of the containment enclosure. In addition, a recirculation system may be located in the enclosure to further aid heat removal and thus limit the pressure buildup. As the fire progresses the oxygen content in the atmosphere decreases, slowing the combustion rate. With about 75% of the sodium consumed, a further decrease in burning rate is likely to result from the buildup of an oxide crust from the pool base through the surface (Shire, 1970).

Models have been developed that utilize empirical heat transfer coefficients in conjunction with the system geometry and heat conduction properties to predict containment transients from pool fires. Figure 9-8 illustrates the general properties of the results from such a model. In this example it is postulated that 270 tons of 1050°F sodium is spilled into the containment enclosure of the Fast Flux Test Facility. The peak pressure in the steel containment sphere is reached after about one hour. Further increase in pressure is prevented by the heat transfer through the containment shell and the decreasing reaction rate caused by oxygen depletion. The fire terminates after 37 hours because of oxygen depletion. In this example no active containment-cooling system is utilized. The use of active heat removal systems would be expected to reduce the pressure transient substantially.

The second reaction category is the spray fire. If a high-pressure liquid sodium spray occurs, the sodium–oxygen reaction takes place primarily while the sodium is in flight. For given sodium and atmospheric conditions, the reaction rate is proportional to the sodium surface exposed, which in turn increases with increased mass flow rate and with decreased droplet size. The higher pressure required to initiate a spray fire would most likely occur

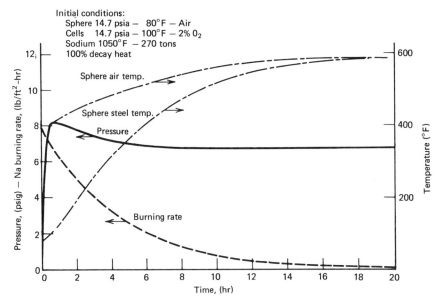

FIGURE 9-8 Sodium fire containment transient. Adapted from *Lecture Series in Fast Reactor Safety Technology and Practices,* Vol. 2, P. L. Hoffman (Ed.), Battelle Northwest Laboratory Report, BNWL-SA-3093.

under accident conditions. For example, a nuclear excursion causing an overpressure in the primary system might cause the pressure vessel lid to lift slightly, allowing sodium to be ejected, or it might cause the rupture of small-diameter piping causing a jet to occur. A secondary mechanism for causing the spray fire might be the impingement of a high-velocity stream of sodium on a piece of piping or other equipment, causing it to break up into droplets.

The high-pressure conditions leading to the generation of sprays are less likely to occur than conditions leading to pool fires. In addition, it is difficult to envision situations in which the amounts of sodium dispersed as a spray could be as large as that appearing in pool fires. Should a given amount of sodium interact in a spray fire, however, the reaction would tend to be more violent because the total surface of the sodium droplets would be large, leading to a rapid reaction rate, and the heat of reaction would be transferred directly to the gas atmosphere, with little being dissipated as sensible heat in the sodium, or to the containment walls or other structures.

Efforts to predict the pressurization caused by spray fires have for the most part been based on chemical equilibrium calculations. Because such reactions are observed to occur rapidly compared to the time required for significant heat transfer to occur, the adiabatic approximation may be used to compute temperature rises. If the assumption that complete reaction takes place is made, thermodynamic arguments can be used to predict maximum temperature and pressure once the initial conditions are known. Results of such calculations are shown in Fig. 9-9. The peak pressure is sensitive to the ratio of sodium to oxygen in the spray atmosphere. As may be seen the maximum occurs at approximately $7:1$ Na/O_2 molar ratio. In air a $1:1$ ratio corresponds to 12.3 lb of Na dispensed in 1000 ft^3 of air. The curves are also dependent on initial temperatures of the sodium and atmosphere and on the oxygen content of the air. At $7:1$ Na/O_2 ratio the pressure increases only about 2% per 100°F of sodium temperature. Interestingly, the peak pressure is inversely proportional to the initial atmosphere temperature. When the oxygen concentration is reduced in the atmosphere, so are the peak temperatures and pressure. However, only a one-third pressure reduction occurs when the oxygen concentration is reduced by a factor of 2 (Shire, 1970). From a practical viewpoint the degree to which perfect mixing is achieved is quite important. In Fig. 9-9 are shown results in which varying amounts of sodium are mixed with efficiencies of 10, 50, and 100%.

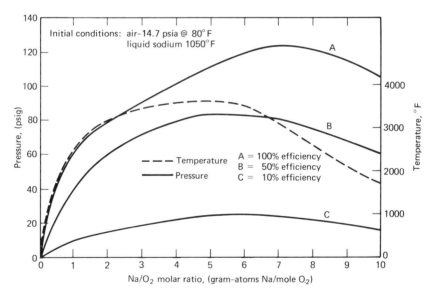

FIGURE 9-9 Adiabatic pressures and temperatures for sodium-air reaction. Adapted from *Lecture Series in Fast Reactor Safety Technology and Practices,* Vol. 2, P. L. Hoffman (Ed.), Battelle Northwest Laboratory Report, BNWL-SA-3093.

Explosive ejections are the least likely form that a sodium fire may take. Conceivably one could take place if an explosive nuclear excursion were powerful enough to eject the vessel head into the containment shell. If such an event were to occur, however, blast and missile damage to the containment shell would be a more immediate concern.

Containment Leakage

Once the pressure transients associated with design basis accidents have been determined, containment strength requirements can be set so that structural failure will not occur and cause the release of the containment atmosphere to the environment. In addition to the specification of a design pressure that will not be exceeded, a maximum leak rate at pressures below this maximum must also be specified since an exactly zero leak rate is unobtainable. Containment shells for modern power reactors have very low leak rates required, 0.1% of the containment atmosphere per day at design pressure being typical. In these systems the leak rate is so low as to have essentially no effect on the pressure transient within the shell. This situation is in contrast to some research testing and small power reactors that in the past have been protected by so-called confinement systems (Thompson and McCullogh, 1973). Although the designs of confinement systems varied greatly, they usually involved the holdup or controlled release of the confined atmosphere to the environment. This release usually was preceded by scrubbing, filtration, or some other means of reducing the concentration of radioactive materials in the effluent.

Containment Penetrations. It is possible to design the walls of the containment shell with extremely small leak rates. Small leak rates through the containment structures are made difficult to achieve, however, by the large number of penetrations required in the shell wall. Mass and energy must pass through the containment barrier at high rates in the form of feedwater from the condenser and steam to the turbine. Furthermore, there are many electrical penetrations for instrumentation and control, piping penetrations for service water, auxiliary cooling, and other purposes, and larger access interlocks for maintenance and inspection personnel and replacement equipment. In addition to the many containment penetrations required for normal plant operation, essential safety systems required during accidents require additional penetrations through the containment envelope. These often include coolant lines for the emergency core- and containment-cooling systems, emergency power supplies, protection system cables, and so on. Finally, provisions must be made for moving the highly radioactive spent fuel through the containment envelope. Designs for this function

range from interlock systems that permit fuel transfer during power operation to systems in which the containment envelope must be removed during
refueling shutdowns.

The many containment penetrations required for operational, safety, and
refueling systems result in low leak rate requirements being incompatible
with normal plant operations. Therefore the containment cannot be completely passive. Rather, an active isolation system must be incorporated to
close off all penetrations that are not required under accident conditions.
Great care is required in coordinating the containment isolation with the
functioning of the safety systems. Since the containment isolation may
interrupt the normal flow of mass and energy to and from the turbine-
condenser, it is essential that the reactor be shut down and that the auxiliary
core-cooling system function properly. Penetrations relating to emergency
power, safety instrumentation, and other systems essential to the safe
shutdown and removal of decay heat must also remain intact without
resulting in excessive leakage under accident conditions. At the same time,
spurious operation of the isolation system must be avoided, since the result is
a sudden plant shutdown with the associated thermal shock and economic
loss.

The variety of penetration types requires that a number of different
strategies be used within the isolation system. Most often electrical penetrations are made sufficiently leaktight that no additional isolation is required.
Interlocks for personnel are locked in place upon isolation except for
specially designed systems for rescue of trapped personnel or other necessary safety functions. The substantial number of piping systems passing
through the containment must be isolated in a number of different ways
depending on their function. All sections of such systems designed for
pressure and temperature conditions that are less severe than those in the
containment must be isolated by quick-acting valves or other means. Four
methods of system isolation are shown schematically in Fig. 9-10. System A
is used for those systems that must remain operational while the containment is isolated, but that cannot be placed entirely within the containment.
Emergency core- and containment-cooling loops are examples of such
systems. Configuration B or C is applicable to systems that are not required
to be functional during isolation. The service water system might use system
B, whereas the working fluid or intermediate coolant loops in indirect-cycle
systems might appear as in the C configuration. In system D radioactive
material is allowed to pass into a compartment outside the containment
shell; it is then allowed to vent to the atmosphere only if the level of
radioactivity is sufficiently low.

Even with the inclusion of a containment isolation system the requirements for extremely low leak rates through the containment shell under

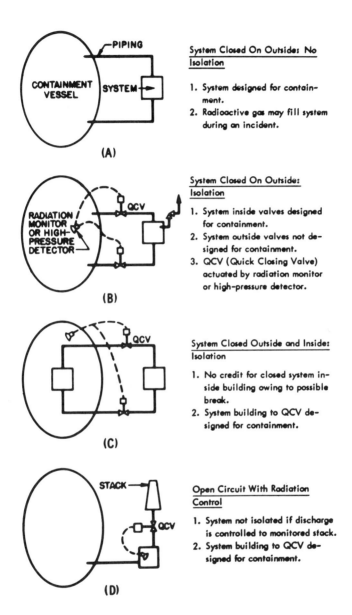

System Closed On Outside: No Isolation

1. System designed for containment.
2. Radioactive gas may fill system during an incident.

(A)

System Closed On Outside: Isolation

1. System inside valves designed for containment.
2. System outside valves not designed for containment.
3. QCV (Quick Closing Valve) actuated by radiation monitor or high-pressure detector.

(B)

System Closed Outside and Inside: Isolation

1. No credit for closed system inside building owing to possible break.
2. System building to QCV designed for containment.

(C)

Open Circuit With Radiation Control

1. System not isolated if discharge is controlled to monitored stack.
2. System building to QCV designed for containment.

(D)

FIGURE 9-10 Containment penetration systems. Reprinted from "Plant Structures, Containment, Design and Site Criteria" by N. J. Peters and J. G. Yevick in *Fast Reactor Technology and Plant Design,* J. C. Yevick (Ed.), 1966, by permission of the M.I.T. Press, Cambridge, Mass.

accident conditions raise some challenging operational problems. As barriers to radioactive release, containment systems differ from the fuel elements and primary system envelope in that they are not a functional part of the power plant under normal operating circumstances. They are subjected to loading only under the severest and most improbable of accident conditions. This fact complicates acceptance testing, and makes it more difficult to formulate and implement surveillance programs that ensure that unacceptable deterioration is not taking place during plant operation. It is disruptive to normal plant operation, and in some cases impossible, actively to raise the containment atmosphere to design pressure and temperature for purposes of leak testing. Therefore elaborate measures must be developed by which design behavior is extrapolated from visual inspection, low-pressure-leak testing, or other means that can be carried out routinely (Zapp, 1968).

Multiple Containment.   One method for further reducing the leak rate to the environment is to place a secondary containment shell around the primary or pressure-resistant containment. The secondary shell is not expected to withstand large internal pressures. Rather its volume is kept slightly below atmospheric pressure so that there can be no leakage to the environment. Then if an accident causes high pressure inside the primary containment, any effluent leaking from the primary containment will be trapped within the secondary shell. The subatmospheric pressure is maintained within the secondary containment by.extracting part of the atmosphere, filtering or scrubbing it to reduce the concentration of radioactive material, and then exhausting it to one of two places. It may be pumped back into the primary containment if under accident conditions it is too radioactive to be discharged to the environment, or, if the air-cleaning procedure reduces the concentration of radioactive material to acceptable levels, the effluent from the secondary system may be discharged to the environment.

The use of two concentric and independent containment shells in theory provides a leak rate that is much smaller than the individual leak rates. Moreover, with a subatmospheric secondary containment, an essentially zero leak rate should be obtainable, provided that the accident leak rate from the primary containment does not become so large as to overpower the pressure control apparatus for the secondary containment. Again, the effects of penetrations through the containment structures must be considered. Pipes, instrument cables, trays, or other items that pass through both shells may cause leakage directly from the primary containment atmosphere to the environment. Thus a great deal of care must be taken in blocking these leak paths. Otherwise these penetrations are likely to dominate the leakage behavior of multiple containment shells.

Multiple shells may take a variety of forms. They may, for example, be in the form of two concentric shells that are separated only by a thin annulus of air or porous material. In this situation most penetrations must go through both shells and therefore present a particular problem. If the primary containment structure is sufficiently small, a more popular configuration is to surround it with a substantially larger leaktight reactor building. Since the building is larger, a substantial amount of emergency cooling and other equipment may be located within it. This serves to reduce the number of common penetrations through primary and secondary containment shells. This building also can serve to enclose the spent-fuel storage pool, since this is also required to be within a leaktight structure. Moreover, the secondary containment is necessary anyway in plants where the primary containment must be opened for refueling operations. At present, secondary containments are used in conjunction with boiling-water reactor vapor suppression systems, since the dry-well head must be removed for refueling. In sodium-cooled reactors, the inerted reactor vault serves as the primary shell and is capable of withstanding substantial blast pressure, as is discussed in the following section. The outer containment shell then is designed to withstand only the more nominal pressures that are likely to be produced by sodium fires.

Failure Modes.   Our attention thus far has been directed toward maintaining the containment leak rates within the required limits when the containment is subjected to its design pressure. In the event that an accident should create pressure in excess of that permitted by the containment design criteria, the containment leakage will grow and may become unacceptably large. The leakage rate may increase continuously with pressure above the design limit until some ultimate pressure is reached at which the containment envelope fails structurally. At this point the containment may be expected to depressurize rapidly, emptying its contents to the environment. Thus with structural failure of the containment there is likely to appear a large increase in radiological consequences of an accident. For this reason it is useful to examine briefly the mechanisms that may lead to such failure.

The failure mechanisms of water-reactor containment constructed either of ductile steel or of steel-lined reinforced concrete have been reviewed in conjunction with the U.S. Atomic Energy Commission (1974b) Reactor Safety Study. Steel shells are constructed in compliance with Section III of the ASME Pressure Vessel Code. As such, the design stress levels are well below the yield point both in the base metal and around the reinforced penetrations and other stress risers. Because the metals are ductile, plastic deformation will tend to prevent rupture well beyond the yield point. One could argue for the ultimate strength of the steel as the failure point. This, however, would assume that all the welds have a strength and ductility equal to the base metal and that the structure would be free to deform as required to accommodate load bending or other local secondary stresses. Prudently assuming that these conditions cannot be met all the way to the ultimate strength, it has been suggested that a value midway between yield and ultimate strength be taken as the failure point. For a boiling-water reactor vapor suppression system with a design pressure of 56 psia, the failure pressure is estimated to be 175 psia (U.S. Atomic Energy Commission, 1974b).

For concrete containment structures, the interaction between reinforced-concrete and the ductile steel liner must be considered in order to predict the failure pressure. The leaktightness of the system is maintained by the ductile steel liner; the structural strength is provided by the reinforcing steel in the concrete. The concrete transmits the internal pressure load from the liner to the reinforcing steel. The steel liner strength is small, but if the structure as a whole is to function, its integrity must be maintained. If liner integrity is lost, internal pressures cannot be maintained because of the porosity and cracking of the concrete, even though the reinforced concrete is structurally intact. Because the reinforcing steel is the main load-bearing element in the structure, its failure is synonymous with the failure of the structure.

Reinforced-concrete vessels are designed in accordance with the ACI for reinforced-concrete structures. As such there is a safety factor of 1.67 on the yield strength of the reinforcing steel. It has been suggested that the ultimate strength of such vessels is reached when the yield point of the reinforcing is reached, since the large strain associated with stress levels beyond yield result in separation of the steel from the concrete and the loss of strength as a result of concrete cracking or crumbling. The concrete no longer transmits the internal pressure to the reinforcing steel and the liner ruptures. This mechanism assumes that failure will not occur at lower internal pressures because of local stresses around inadequately reinforced penetrations or other stress risers. It also assumes that the ductile liner is free to expand with the concrete–reinforced-steel aggregate, a condition requiring that the improper anchoring of the liner to the concrete or penetration does not cause local stress risers during expansion. These matters attended to, it is found that a vessel whose design pressure is 60 psia may be expected to fail structurally at a pressure of 100 psia (U.S. Atomic Energy Commission, 1974b).

9-3 EXPLOSIVE ENERGY RELEASES

In this and the following section we examine the severest of reactor accidents: those in which the reactor core is partially or wholly destroyed. In sound reactor design, a great many precautions are taken to prevent initiating events that could lead to such accidents and to incorporate safety systems into the reactor that will mitigate accident consequences, thus preventing the point of core destruction from being reached. As a result, the probability of a core-destructive accident is made so small that for some reactors such accidents are not included within the design basis for the containment structures. Even though the likelihood of core destruction is extremely small, however, it is often deemed necessary to ensure that a direct path for leakage through the primary system envelope and containment structures to the atmosphere does not result. For with the destruction of the fuel elements a large inventory of fission products would be available within the primary system to escape to the environment.

For purposes of analysis, the spectrum of accidents that are capable of causing core destruction and a threat to the other fission product barriers may be divided roughly into two categories. In explosive nuclear excursions (taken up in this section) nuclear transient energy is released at such a rapid rate as to cause fuel and/or coolant vaporization, the generation of high pressures, and subsequent blast or other mechanical loading on the primary system envelope and/or containment structure. In core meltdowns (taken up in the following section) energy is generated at a more moderate rate as the result of milder transients and/or decay heat following reactor shutdown. The failure to remove this heat in an orderly fashion, however, will eventually cause the core to melt and create thermal loadings on the core support structures, reactor vessel, and eventually the foundations of the containment structure as the molten fuel debris sinks downward in what in the United States has come to be known as the China syndrome.

Initiating Mechanisms

Mechanical damage caused by pressure generation during nuclear excursions has been observed in the destructive tests of the experimental BORAX and SPERT reactors, and in experiments conducted within the TREAT reactor (Thompson, 1964). Probably the most impressive example of such damage, however, occurred in the SL-1 accident, in which a small metal-fueled, water-cooled reactor was destroyed by a violent nuclear excursion. As a result of the pressures generated in this accident, the control rods and their drive mechanisms were ejected out of the reactor vessel, and the entire vessel was lifted approximately 11 ft, shearing the coolant loops in the process. These experiments and accidents all took place on relatively small, thermal reactors for which the fission product inventories were minimal. They were all driven by reactivity ramps caused by the ejection of a high-worth control rod.

In the thermal reactors presently used for commercial power production, this mechanism has been eliminated as a potential cause of destructive nuclear excursions. Many independent, low-worth rods are used for reactor control, and mechanical restraints are frequently incorporated into control rod designs to restrict the ejection rate in the event of an accident. As a result, the ejection of a control rod whose worth is abnormally large may result in some fuel melting, but in current reactor designs the destructive work energy is negligible. Moreover, other mechanisms are not present in thermal reactor cores that can lead to the sustained reactivity insertion rates of tens of dollars per second required to cause explosive energy releases. Generally these low-enrichment, undermoderated systems are in their most reactive state, and loss of coolant, fuel overheating, or fuel slumping all lead to decreases in reactivity. Even where coolant bubble collapse or other rapid reactivity insertion mechanisms are possible, the resulting reactivity ramps tend not to be sufficiently sustained to cause fuel vaporization. As a result, pressure generation and mechanical work do not play a significant role in the evaluation of present-day thermal reactors used for commercial power production.

As discussed in Chapters 4 and 5, the situation is quite different in fast reactor cores. There more highly enriched systems are not in their most reactive configuration, and, moreover, in the absence of coolant and cladding material there is enough fuel present to form several critical masses. The boiling away of the liquid coolant, the melting away of cladding material, or the compaction of the fuel mass each may produce substantial increases in reactivity. Thus the potential for destructive nuclear excursions cannot be categorically eliminated. As a result, a great deal of effort is being expended to ascertain the amount of destructive work that could be generated by fast reactor excursions, and to evaluate the resulting mechanical loading on the core support structures, primary system envelope, and safety systems.

In Chapter 5 the effects of reactor size, reactivity insertion rate, and Doppler coefficient on explosive energy release are discussed. From this discussion it is clear that the results obtained for destructive work energy have decreased with time because of both the evolution of fast reactor core designs and the refinement of the analysis. With increased reactor core size and the associated larger value of the sodium void reactivity coefficients, one might expect an increase in the work energy. However, this has been more than compensated by the substantial negative Doppler coefficient that appears in the more recent designs, and by the refinement of the analytical tools for placing upper bounds on the rates at which reactivity can mechanistically be added to particular reactors. Disassembly calculations for early, metal-fueled fast reactors resulted in destructive energy releases of the order of 1000 MW-sec (Koch et al., 1957; Power Reactor Development Company 1961), whereas the more refined analysis of later oxide-fueled cores tend to yield smaller work energies even though the reactors have higher power ratings (Kintner et al., 1972).

Since the initial Bethe–Tait calculations, in which it was assumed that the upper half of the reactor core simply fell with gravitational acceleration into the lower half, a good deal of insight has been gained into the mechanisms and magnitudes of reactivity insertion that are possible in fast reactor cores (Jackson et al., 1974; Boudreau and Jackson, 1974). For current sodium-cooled core designs, two sequences of events can be identified that might permit the point of a superprompt-critical power burst to be approached. But in each case two highly improbable failures of normally independent systems must occur. In the transient overpower accident an uncontrolled rod withdrawal must be coupled with a failure of the shutdown system for the possibility of a power burst to arise. In the loss-of-flow accident, the loss of power to all the coolant pumps must be coupled with a failure of the shutdown system. In cores of the size of 300 MW(e) or less, these accidents do not necessarily lead to superprompt-critical bursts, but there is still enough uncertainty in the analysis of the core behavior that they cannot be precluded. The behavior of sodium-cooled cores that could lead to a destructive power burst differs substantially for the two accidents, and therefore we discuss them separately.

Transient Overpower Accident.   The transient overpower accident with failure to trip may not immediately lead to prompt criticality. Nevertheless, the fuel temperature will increase and lead to fuel melting—and possibily partial vaporization. Fuel failure will occur before either cladding melting or boiling of the sodium coolant takes place. Observe from Fig. 6-12 that the fuel enthalpy required to cause failure decreases with increased initial cladding temperature. Therefore failures are most likely to occur above the core midplane where the cladding is hottest. On fuel failure, liquid and/or vaporized fuel and fission product gases are ejected into the coolant channels, causing sodium voiding. At this point there are substantial reactivity effects caused both by the sodium voiding and by the axial fuel relocation. The reactivity effect of the sodium void will be positive unless the fuel failure is so close to the core outlet that leakage effects cause the local void coefficient to become negative, as indicated, for example, in Fig. 4-1. The reactivity effect of the fuel ejection may be positive or negative depending on whether the net motion of fuel mass is toward or away from the core midplane.

Analytical modeling done in conjunction with experimental investigations tends to indicate that the net effect of such fuel failures should be a loss of reactivity leading to shutdown of the reactor without the occurrence of a superprompt-critical power burst (Project Management Corp., 1975; Wider et al., 1974). At the time of fuel failure, the sodium is still several hundred degrees subcooled, causing the stream of ejected fuel to be quenched and shattered into small aerosol-like particles. Because the sodium is traveling at a high velocity upward through the core, it will tend to sweep fuel debris and fission product gases upward and out of the core. This removal of fuel from the core will more than compensate for any reactivity addition caused by the coolant voiding, and hence the nuclear transient will be terminated. Only in the unlikely event that the initial fuel failure takes place in the region of cooler cladding around or below the core midplane would the excursion be aggravated. In the case of midplane failure, movement of molten fuel within the pins toward the failure location can temporarily add more reactivity than is lost by fuel sweep out from the coolant channels. In the case of failure below the midplane, the coolant will sweep the ejected fuel through a region of higher reactivity worth, and the coolant voiding will also be more extensive. Even then the energy release from the resulting power burst is thought to be relatively small compared to releases that might result from the loss-of-flow accident for current core designs.

Loss-of-Flow Accident. The sequence of events that may lead to a superprompt-critical burst in the event of a loss of flow coupled with a shutdown system failure is quite different (Project Management Corp., 1975; Jackson and Weber, 1975). During the coolant pump coastdown the coolant outlet temperature will increase as indicated by the analysis of Section 7-4. There also will be an increasing level of power production caused by the positive temperature coefficient of the coolant. As the core outlet reaches the sodium saturation temperature, coolant voiding will be initiated at the outlet and will spread downward into the core. In present-sized cores [of less than 300 MW(e)] this voiding is insufficient of itself to cause a destructive power burst. The lack of cooling and increasing power level cause the cladding near the core outlet to melt, however, and as the cladding melting subsequently spreads downward into the core, it has two effects; if the neutron-absorbing cladding is swept out of the core by the coolant vapor or runs downward to the core inlet, it causes an additional reactivity increase. The loss of the cladding also destroys the structural integrity of the fuel element. The progression of the accident thereafter depends strongly on the fuel motion that takes place following the cladding melting. It is thought most likely that the molten fuel and entrained fission products will erupt into the now voided coolant channels and be ejected from the core, coming to rest in the outlet blanket. Scenarios can be envisioned, however, in which the fuel melting at the midplane causes the upper fuel element segments to fall downward toward the core midplane, causing a large reactivity increase. In an alternate scenario fuel debris that has frozen to the upper blanket may remelt and then at a later time drop back into the core to cause the formation of a supercritical mass.

Future Fast Reactors. The foregoing sequences of events have the potential for bringing about a superprompt-critical power burst strong enough to cause shutdown by core disassembly. Consequently, they are used as mechanisms for predicting the reactivity insertion rate and initial conditions for the Bethe–Tait analysis of the nuclear energy release and for an evaluation of the destructive work potential. In the future, as larger reactor cores are designed, the nature of the transients that lead to superprompt-critical power bursts may change significantly. For example, there will be a substantially larger positive void coefficient in a sodium-cooled system of, say, 1000 MW(e). Then the voiding initiated at the coolant outlet in the event of a loss-of-flow (without trip) accident may be sufficient to cause a superprompt-critical reactivity ramp even before cladding or fuel melting begins.

Work is also underway in the evaluation of potentially destructive accidents in gas-cooled fast reactors (U.S. Atomic Energy Commission, 1974c). In these systems the coolant density has a much smaller effect on reactivity. However, cooling failures or power transients coupled with the failure of the shutdown could lead to cladding melting. In such an event, the relative axial motion of molten cladding and coolant within the core once again play an important role in determining whether a superprompt-critical configuration will form. In gas-cooled systems the flow is downward through the core. Therefore in either an overpower or cooling-failure transient, fuel failure and cladding melting would most likely occur below the core midplane, where the cladding is hotter. The coolant would tend to sweep the fuel and cladding debris downward and away from the core midplane. The reactivity then would go up or down, depending on whether the loss of the cladding or of the fuel had the dominant effect. A mjaor question centers about whether the molten material would collect in the outlet blanket blocking further flow and causing fuel slumping from the core into a more compact and therefore more reactive configuration at the core outlet. Likewise, any mechanism that could cause fuel melting and failure above the midplane must be scrutinized, since then molten fuel would slump or be swept by the coolant toward a region of higher reactivity worth.

Fuel Vapor Expansion

Once sufficient analysis has been carried out to establish a rate of reactivity increase and the configuration of the reactor core at the onset of the superprompt-critical burst, a coupled neutronic-hydrodynamic disassembly calculation, such as is discussed in Chapter 5, can be carried out to estimate the energy release from the transient. Not all of this energy, however, is available to do destructive work. The energy required to raise the fuel temperature to its melting point and that required to overcome the latent heat of fusion, for example, offer no direct threat to the structures surrounding the reactor core. If fuel is vaporized in a volume whose expansion is limited by the inertia of the surrounding medium, however, high-pressure fuel vapor may be produced. The time scale over which nuclear excursions take place are of the order of milliseconds, which is too slow for significant shock waves to be produced, for the shock wave production would require a detonation at a microsecond time scale as occurs in TNT or other high explosives. Blast caused by the expansion of the resulting vapor bubble thus will be the dominant mechanism for the fuel vapor pressure to be converted directly to damaging work.

In reality, the hot vapor is likely to lose a significant part of its energy to the surrounding medium by radiation and other heat transfer mechanisms. A bound, however, can be set on the destructive work potential by assuming that the bubble expands adiabatically. Moreover, the $p-v$ curve that thus can be generated characterizes the variation of work production with pressure and volume, providing useful information for the evaluation of structural damage.

To estimate the work done by the expanding fuel vapor, the model contained in the following paragraph can be used (Cho et al., 1974; Cho and Epstein, 1974). It is assumed that the coolant has been expelled from the reactor core. By the time that the reactor has become subcritical from the disassembly reactivity, the fuel has become molten and dispersed, creating a two-phase fuel mixture of droplets and vapor that are in thermal equilibrium with one another. The destructive work is then produced by the adiabatic expansion of this two-phase mixture.

An approximate calculation can be made by assuming that the entire core is at the core-average temperature, say T_I, which is above the solidus temperature of the oxide fuel. The quality χ, or fraction of the fuel mass in the vapor state, can be expressed in terms of specific volumes

$$\chi = \frac{\mathcal{V} - \mathcal{V}_l}{\mathcal{V}_v - \mathcal{V}_l}, \tag{9-17}$$

where $\mathcal{V}_v$ and $\mathcal{V}_l$ are the saturated vapor and liquid specific volume of the fuel and $\mathcal{V}$ is the specific volume of the two-phase mixtures. The initial value of $\mathcal{V}$ can be estimated by assuming that the total fuel mass M_f is uniformly dispersed over the core volume V_c:

$$\mathcal{V}_I = \frac{V_c}{M_f}. \tag{9-18}$$

If coolant or structural material remain in the core, the volume of V_c is reduced to include only the fuel volume plus any free volume within the core not occupied by the other materials.

Before the inital pressure can be estimated and the expansion characteristics evaluated, an approximate fuel equation of state must be formulated. We utilize the following simplified model. The liquid phase is assumed to be incompressible, yielding a value of $\mathcal{V}_l$ that is independent of pressure. The vapor phase is assumed to be an ideal gas. Thus

$$p\mathcal{V}_v = R_f T, \qquad (9\text{-}19)$$

or correspondingly,

$$p\mathcal{V} = \chi R_f T, \qquad (9\text{-}20)$$

where R_f is the gas constant per unit mass of fuel. Finally, the vapor pressure is represented by the Clausius–Clapeyron equation:

$$\frac{dp}{dT} = \frac{JL}{T(\mathcal{V}_v - \mathcal{V}_l)}, \qquad (9\text{-}21)$$

where

$$L = h_v - h_l \qquad (9\text{-}22)$$

is the latent heat of vaporization, h_l and h_v are the liquid and vapor enthalpies, respectively, and J is the thermal-to-mechanical energy conversion constant.

We make the further assumptions that, over the range of temperatures under consideration, L can be assumed to remain constant and that $\mathcal{V}_l \ll \mathcal{V}_v$. Therefore we can write

$$\frac{dp}{dT} = \frac{JLp}{R_f T^2}, \qquad (9\text{-}23)$$

which can be integrated to yield the vapor pressure curve

$$p = p_0 \exp\left[\frac{JL}{R_f}\left(\frac{1}{T_0} - \frac{1}{T}\right)\right], \qquad (9\text{-}24)$$

where p_0 and T_0 are reference conditions. For uranium dioxide, the following data have been used (Sha and Hughes, 1970):

$$p = 2.1925 \times 10^5 \exp\left(-\frac{43,957}{T}\right), \qquad (9\text{-}25)$$

where p is in atmospheres and T in degrees Kelvin. More general agreement with experimental data can be obtained by the slightly more general form

(Sha and Hughes, 1970)

$$p = 10^{-6} \exp\left\{-4.34 \ln T - \frac{76,800}{T} + 69.979\right\}, \qquad (9\text{-}26)$$

where the units are again atmospheres and degrees Kelvin.

With the vapor pressure curve specified, we can determine the initial core pressure p_I from T_I. Since we also know $\mathcal{V}_I$ (or the corresponding χ_I), we have specified the initial state from which the fuel is to begin its adiabatic expansion. For an adiabatic expansion, we have

$$\frac{1}{J} p \, d\mathcal{V} = -du, \qquad (9\text{-}27)$$

where $p \, d\mathcal{V}$ is the incremental work done per mass of fuel and u is the fuel internal energy. Writing the internal energy in terms of the fuel enthalpy,

$$u = h - \frac{p\mathcal{V}}{J}, \qquad (9\text{-}28)$$

we can express the incremental expansion work as

$$\frac{1}{J} p \, d\mathcal{V} = -dh + \frac{1}{J} d(p\mathcal{V}). \qquad (9\text{-}29)$$

The enthalpy of the two-phase mixture is given by

$$h = h_l + \chi(h_v - h_l), \qquad (9\text{-}30)$$

or

$$h = c_p(T - T_0) + \chi L. \qquad (9\text{-}31)$$

In this we assume that the specific heat c_p and latent heat of vaporization L are constants. Hence

$$dh = c_p \, dT + L \, d\chi \qquad (9\text{-}32)$$

Combining this expression with Eqs. (9-20) and (9-29) yields

$$\frac{1}{J} p \, d\mathcal{V} = -c_p \, dT - L \, d\chi + \frac{1}{J} d(\chi R_f T), \qquad (9\text{-}33)$$

or

$$\frac{1}{J} \int_{\mathcal{V}_I} p \, d\mathcal{V} = c_p(T_I - T) - L(\chi - \chi_I) + \frac{R_f}{L}(\chi T - \chi_I T_I). \qquad (9\text{-}34)$$

The work energy is obtained by multiplying by the fuel mass:

$$W(p) = M_f \left[c_p(T - T_I) - L(\chi - \chi_I) + \frac{R_f}{L}(\chi T - \chi_I T_I) \right]. \qquad (9\text{-}35)$$

Before the work can be evaluated, an auxiliary relationship must be derived to express χ in terms of T, p, and/or $\mathcal{V}$. To do this we cancel the $pd\mathcal{V}$ term from both sides of Eq. (9-29) and replace dh by Eq. (9-32):

$$c_p \, dT + L \, d\chi = \frac{\mathcal{V} \, dp}{J}. \qquad (9\text{-}36)$$

Now by taking $\mathcal{V}_l \ll \mathcal{V}_v$ and noting from Eqs. (9-19) and (9-20) that $\mathcal{V}_v = \mathcal{V}/\chi$, Eq. (9-21) may be rewritten as

$$dp = \frac{J\chi L}{\mathcal{V}} \frac{dT}{T}. \qquad (9\text{-}37)$$

Now substitute this expression into Eq. (9-33) and rearrange terms to yield

$$d\chi - \chi \frac{dT}{T} = -c_p \frac{dT}{L}. \qquad (9\text{-}38)$$

Divide this expression by T and collect terms on the left:

$$d\left(\frac{\chi}{T}\right) = \frac{-c_p}{L} \frac{dT}{T}. \qquad (9\text{-}39)$$

Integration then yields the desired expression for χ:

$$\chi = T\left[\frac{\chi_I}{T_I} + \frac{c_p}{L} \ln\left(\frac{T_I}{T}\right) \right]. \qquad (9\text{-}40)$$

We are now in a position to place an upper limit on the destructive work potential. Suppose the vapor bubble expansion continues until the bubble pressure drops to 1 atm. The equation of state—Eq. (9-24), (9-25), or (9-26)—can be inverted to determine the final fuel temperature. With this temperature T known, Eq. (9-40) can be used to determine the final fuel quality χ. With χ and T known, Eq. (9-35) can be used to determine the total work. It also may be desirable to estimate the size of the bubble at the end of the expansion. For this purpose we combine Eqs. (9-17) and (9-19) to yield the specific volume as

$$\mathcal{V} = \mathcal{V}_I + \left(\frac{R_f T}{p} - \mathcal{V}_I\right)\chi. \qquad (9\text{-}41)$$

Therefore once p, T, and χ are known, the bubble volume $M_f\mathcal{V}$ can be evaluated from this expression.

Expanding the fuel vapor bubble all the way to atmospheric pressure may place an upper bound on the work energy. For the analysis of structures, however, it is more relevant to have the work energy calculated as a function of bubble pressure and/or size. This may be accomplished with a series of calculations using decreasing values of T, calculating the corresponding values of p and χ from Eqs. (9-25) and (9-40), and then estimating the work energy and volume corresponding to each value of p from Eqs. (9-35) and (9-41), respectively.

Results of the foregoing model have been obtained for the work energy as a function of pressure for the Fast Flux Test Facility corresponding to a number of initial core temperatures. It is seen in Fig. 9-11 that a substantial part of the work does not appear until relatively low pressures are reached. For reasons taken up in a subsequent subsection, the low-pressure part of the bubble expansion is not very effective in causing damage to the primary system structures. Similar curves can be derived for the work energy as a function of the fuel bubble volume, and as will be seen in subsequent

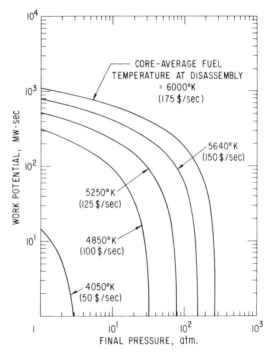

FIGURE 9-11 Thermodynamic work potential of fuel expansion for 3000 Kg of fuel. From D. H. Cho and M. Epstein, private communications, Argonne National Laboratory, 1974.

discussions, this characterization is very important for predicting the damage to the pressure vessel head and closure devices that may occur if the sodium above the core is impacted against the vessel head.

Sodium Vapor Expansion

Inasmuch as the preponderance of energy generated during a nuclear excursion is deposited in the fuel, expansion of the resulting fuel vapor bubble is the most immediate mechanism for producing destructive work. If, however, molten and/or vaporized fuel should make intimate thermal contact with the liquid coolant, causing large amounts of energy to be transferred very rapidly to the coolant, larger work energies may result, since the higher vapor pressure of the coolant causes it to be a more effective working fluid than the fuel. The concern over such vapor explosion phenomena in the design of sodium-cooled fast reactors was raised first by the calculations of Hicks and Menzies (1965) and has since been investigated extensively. A necessary prerequisite for the rapid transfer of heat from fuel to sodium is that the fuel be shattered into very fine particles or droplets. The fine particles must then mix intimately with a mass of sodium to produce significant work energy. The determination of accident situations in which sufficient fuel–coolant thermal contact would take place to cause such an emergency release is problematical. If perfect thermal contact is postulated, however, thermodynamic arguments can be used to place upper bounds on the work done by the expanding sodium vapor.

Two thermodynamic models have been proposed. In the Hicks–Menzies model, instantaneous thermal equilibrium is assumed to occur on contact of molten fuel and sodium. The fuel and sodium then remain in thermal equilibrium as the sodium expands until atmospheric pressure is reached. In the modified Hicks–Menzies model instantaneous thermal equilibrium between fuel and sodium is again assumed to take place. Now, however, the sodium is assumed to expand adiabatically to produce the work energy. The original calculations using these models used extremely simple equations of state and assumptions regarding fuel and sodium properties. Later elaborations use refined representations of the fuel and sodium properties (Padilla, 1971).

The thermodynamic models are conveniently visualized on a temperature-entropy diagram shown in Fig. 9-12. Sodium coolant is assumed to come into rapid contact with the fuel and to be heated along the saturation line from its initial temperature T_1 until it comes into equilibrium at temperature T_2. The expansion process may then take place between two extremes. If an infinite heat transfer rate is assumed, the sodium can expand from T_2 to the terminal isocore p_0 in thermal equilibrium with the fuel along

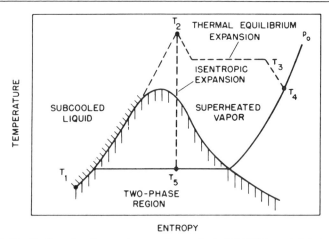

ENTROPY

FIGURE 9-12 Sodium heating and expansion paths in SOCOOL model. Adapted from A. Padilla, Jr., "Analysis of Mechanical Work Energy for LMFBR Maximum Accident," *Nucl. Tech.,* **12**, 348–355 (1971). Used by permission.

the path $T_2 - T_3 - T_4$. The sodium first expands as the molten fuel cools to its melting point T_3. It then undergoes isothermal expansion as the fuel solidifies. Finally, it expands to T_4 in thermal equilibrium with the solid fuel. On the other extreme, if it is assumed that there is no heat transfer during the sodium expansion, as in the modified Hicks–Menzies model, the sodium then expands adiabatically and reversibly (i.e., isentropically) to temperature T_5.

The path in the temperature-entropy diagram is sensitive to the ratio of fuel to sodium in the mixture as well as to the initial fuel temperature. When the initial fuel temperature is very large or the sodium-to-fuel mass ratio is small, the equilibrium point T_2 is above the critical point of the sodium, as shown in Fig. 9-12. For higher sodium–fuel mass ratios, the equilibrium temperature would be smaller, and the expansion would then take place in the two-phase region. In Fig. 9-13 the mechanical work potential is compared for the isentropic and thermal equilibrium expansion models over a range of coolant-to-fuel mass ratios. Note that if no heat transfer occurs during the expansion phase, there is a reduction in work energy potential by about a factor of 3 over the situation when the fuel and coolant remain in intimate thermal contact.

Figure 9-13 indicates a general increase in the work energy per unit of fuel as the coolant–fuel ratio increases, with the exception of a slight break in each curve, which corresponds to the points where complete vaporization of the sodium has taken place. However, as the fraction of sodium becomes large, one would expect the thermal equilibrium model to become less

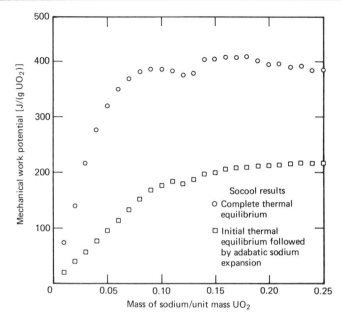

FIGURE 9-13 Comparison of thermodynamic models for UO_2 sodium reactions. Adapted from A. Padilla, Jr., "Analysis of Mechanical Work Energy for LMFBR Maximum Accident," *Nucl. Tech.*, **12**, 348–355 (1971). Used by permission.

viable, tending to make the lower of the two curves more realistic. Using such thermodynamic models, thermal to mechanical conversion efficiencies of from 10 to 30% are obtained when thermal equilibrium is assumed throughout the sodium expansion. These efficiencies are reduced to 3 to 7% if the sodium is assumed to expand adiabatically, and to 2 to 5% if the effects of cladding heat capacity are included (Kintner et al., 1972).

Even with the improvements in thermodynamic fuel–coolant interaction models, the conversion efficiencies are much larger than those found in experimental reactor transients. Several tests run with the TREAT reactor to simulate accident conditions yielded conversion efficiencies as low as 0.1% (Kintner et al., 1972). Such results have focused attention on the finite fuel–sodium heat transfer rates that are actually obtainable under accident conditions and the extent to which the fuel vapor bubble would actually mix with the surrounding sodium.

Recent modeling efforts have taken these effects into account. The analysis of the Fast Flux Test Facility (Cho and Epstein, 1974) assumes that no sodium is present in the core at the time of core disassembly. The two-phase mixture of vapor and droplets of molten fuel is ejected from the core and expands in a single bubble constrained by the inertia of the sodium pool directly above the core. Some of the liquid sodium from the surrounding pool is assumed to be entrained in the expanding bubble. Finally, heat transfer rates are estimated in determining how fast the sodium droplets will come to thermal equilibrium with the fuel. The heat transfer mechanism is postulated as follows: The sodium droplets are heated to their saturation temperature at bubble pressure and then vaporize as heat is transferred from the surrounding fuel vapor, condensing some of the fuel vapor in the vicinity of the droplet. The saturated droplets then vaporize by radiation and conduction through sodium vapor layers that form around them. On formation, the sodium vapor is rapidly superheated to the fuel temperature by mixing of sodium and fuel vapors. Thus there is never any direct contact between sodium droplet and the liquid fuel droplets.

Using rate models such as this, the entrainment of sodium in the expanding fuel debris is found to increase or decrease the work energy, depending on the sodium–fuel ratio. For example, in Fig. 9-14 is plotted the work energy generated up to the time of sodium impact with the head of the Fast

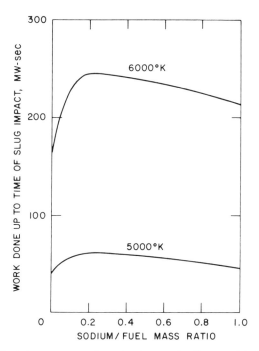

FIGURE 9-14 Work done up to sodium slug impact for disassembly fuel temperatures of 5000 and 6000°K; effect of mass ratio between entrained sodium and fuel (fuel ejection area = 11,700 cm²). From D. H. Cho and M. Epstein, private communications, Argonne National Laboratory, 1974.

Flux Test Facility vessel. A zero sodium–fuel mass ratio corresponds to the expansion of a bubble of pure fuel vapor discussed in the preceding subsection. There is an amplification of the work energy with the entrainment of small amounts of sodium, with the optimum value occurring at a sodium–fuel mass ratio of 0.2 or less. When large amounts of sodium are entrained, they have a quenching effect on the bubble, causing the work energy to drop below that of the pure fuel expansion model. Even for optional mixing, however, the work energy is never more than twice that caused by the fuel vapor expansion with no entrainment. This analysis yields much smaller work energies than predicted by the thermodynamic models in which instantaneous mixing and infinite heat transfer are assumed. For example, with an initial temperature of 5000°K work energies of up to 1000 MW-sec to time of impact are estimated using thermodynamic models. These estimates are more than an order of magnitude greater than the rate-limited results of Fig. 9-14 (Cho and Epstein, 1974).

The foregoing vapor bubble model assumed that after the high-pressure two-phase fuel is ejected into the sodium pool, liquid fuel never makes direct contact with liquid sodium. The sodium droplets contact fuel vapor, which then begins to condense, but liquid–liquid contact is then prevented by the sodium vapor generated at the droplet surface. Without liquid–liquid contact there can be no vapor explosion (Fauske, 1974), thereby explaining the very nominal increases in work energy by the presence of the entrained sodium. Although the rate-limited model is still a highly idealized approximation to the situation that is likely to exist in the event of a reactor accident, experimental evidence does seem to support the assumption. TREAT reactor tests (Epstein and Cho, 1974) and out-of-pile tests (Henry et al., 1974; Johnson et al., 1974) indicate that when a high-temperature two-phase fuel is injected into a pool of sodium, the sodium is pushed back by the expanding bubble with little entrainment. Analysis of such experiments indicates that no explosive interaction takes place between the injected UO_2 and the sodium pool and that the fuel bubble is simply quenched on contact with the cold sodium, leaving a debris of fine particles of solidified fuel. Because the accumulated experimental evidence indicates that an explosive fuel–coolant interaction is not likely to occur in a fast reactor accident, the expansion of the fuel vapor bubble with no additional destructive effect produced by the fuel–coolant interaction is being used to evaluate work energy potentials (Project Management Corp., 1975).

Structural Response

The work energy resulting from the high pressures generated in severe nuclear excursions may cause structural damage of a variety of forms.

Deformation or crushing of instrumentation, control rod guides, thermal shields, and other internals to the reactor vessel are most likely to appear. If progressively more explosive accidents are considered, coolant loops and/or auxiliary piping may be breached and control drives, access plugs, or other pressure envelope fittings may be ejected from the primary system. A truly catastrophic event may lead to the lifting of the reactor vessel head or rupture of the vessel walls. To preclude such unacceptable consequences, a conservative overestimate of the work energy from the design basis nuclear excursion is made, for example, by the methods discussed in the preceding subsections. This work energy then serves as a design basis for evaluating the structural response of the primary system to explosive power bursts. In performing the design basis evaluation it must be established that the reactor vessel will not rupture and that missiles or sodium expulsion from the primary system will not expose the containment structures to excessive blast or impact loadings. It is also necessary to establish that a means of cooling the core—or its debris—following the excursion will remain intact, since if this is not the case, a core meltdown will follow.

Blast Damage.   Both experimental and analytical methods have evolved for the evaluation of structural damage from severe nuclear excursions. Experimental work first centered about equating the estimated work energy to an equivalent charge of TNT,

$$1 \text{ lb TNT} = 2 \text{ MW-sec},$$

and then examining the destructive effects of charges of TNT on scale model structures. In this way Wise and Proctor (1965) developed expressions for the weight of TNT that could be withstood without rupture for cylindrical steel vessels as a function of vessel diameter and thickness. Scale models of vessels provided insight not only into vessel rupture thresholds but also into water hammer effects capable of lifting vessel heads. Perhaps the most dramatic example of the latter was the modeling of the FERMI-I reactor, which indicated that the design basis of 1000 lb of TNT could cause a vessel head to lift by as much as 100 ft if no hold-down mechanism were incorporated into the design (Wise et al., 1962).

Experiments with TNT have provided many useful relationships for bounding the damage from nuclear excursions. Their usefulness for the detailed modeling of the pressure histories likely to result from reactor accidents, however, is limited because of the differences in the energy release mechanisms between high explosives and hypothesized nuclear excursions. A TNT explosion consists of a detonation lasting for only microseconds but creating local pressures of the order of 50,000 atmospheres. Hypothesized energy releases from nuclear excursions resemble the conflagrations of propellant types of explosions more closely, generating pressures of only a few hundred atmospheres but lasting over a time scale of milliseconds. The forms of the energy release and of the resulting structural damage differ significantly between a high explosive detonation and a propellant conflagration. In TNT, for example, about half of the energy is carried away by the shock wave generated by the detonation, whereas the remainder of the energy is in the form of blast pressure from the expansion on a millisecond time scale of the high-temperature vapor bubble of detonation products. Nuclear excursions, like deflagrations, take place too slowly to spawn a significant shock wave; therefore the damage is due predominantly to the blast from high-temperature vapor expansion (Hicks and Menzies, 1965).

The type of damage caused by the shock wave and the blast pressure tends to be different. The intense pressure of very short duration occurring in the shock wave tends to cause severe local deformations in the surrounding internal structures and primary system envelope. The impulse is sufficiently small, however, so that the shock is not effective for moving massive objects such as the reactor vessel head. In contrast, the lower pressure resulting from a blast of longer duration tends to produce less severe local deformations, but such a blast is more effective in moving massive objects. As a result, TNT simulation tends to yield conservatively pessimistic criteria of damage to vessel internals and for the rupture of the reactor vessel walls. On the other hand, TNT would be expected to be less effective than a slow conflagration of the same total energy release in accelerating the large mass of liquid coolant above the core upward to impact the vessel head.

More refined simulation of work energy from nuclear excursions has required that explosive charges be developed that more closely resemble the $p\text{-}v$ characteristics discussed in the previous subsections (Florence and Abrahamson, 1973). The use of such charges in conjunction with the refined modeling of reactor primary systems provided an increasing data base against which analytical damage estimation techniques can be checked.

Response Modeling. In addition to damage prediction methods based on gross thermodynamic and energy arguments used in conjunction with empirical correlations obtained from chemical explosive tests, an extensive effort is being made to develop analytical techniques that will predict structural damage directly, using the $p\text{-}v$ expansions of the vapor bubble impact along with the system geometry, the fluid properties, and the constitutive laws governing the behavior of the materials of the primary system structures.

A method for predicting the blast loading on reactor vessels has been formulated by assuming that the coolant is incompressible, using combinations of one-dimensional Newtonian motions with axial and radial directions and assuming constitutive relationships to determine the vessel deformations (Fox, 1969). Increasingly refined analyses of structural responses are becoming commonplace with methods that combine two-dimensional compressible hydrodynamic treatments of the coolant behavior with detailed solid mechanics analysis of the pressure vessel and other primary system components (Fistedis, 1975). Such methods are capable of providing transient stresses and strains and final deformations in the reactor vessel, the vessel cover, and the cover hold-down bolts and of predicting the pressure pulses that may propagate through the primary system piping systems.

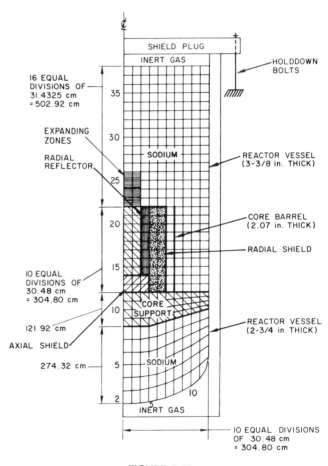

FIGURE 9-15a

Through the use of present analytical techniques, properly verified against chemical explosive model experiments, structural response from explosive nuclear excursions can be understood. In Fig. 9-15a is shown the mathematical discretization of the Fast Flux Test Facility utilized in the REXCO code. In Fig. 9-15b is shown the distortion in the mesh structure at the time that dynamic equilibrium is reached following an excursion resulting from a reactivity ramp of 125 dollars per second. Figure 9-16 shows a typical partition of energy as a function of time from a Fast Flux Test Facility excursion. The work energy produced by the expansion of the fuel-coolant vapor bubble first appears as radial and axial kinetic energy of the coolant.

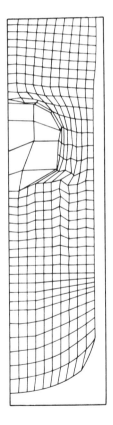

FIGURE 9-15b Mathematical model of the Fast Flux Test Facility primary system for a $100/sec nuclear excursion. From D. H. Cho and M. Epstein, private communications, Argonne National Laboratory, 1974.

This deforms the structures surrounding the core, passes downward on the core support structures, deforms the core barrel and vessel wall radially outward, and begins to accelerate the sodium pool above the core upward. The downward kinetic energy is quite small because the core support structure obstructs the coolant motion. The radial kinetic energy is transformed to strain energy by the plastic deformations of the core barrel and vessel walls in the region of the core. The latter strain energies increase progressively with time until the pressure in the primary system falls to where the stress levels are reduced to below the yield points of the material. At this time a state of dynamic equilibrium is reached.

The upward axial kinetic energy appears in the acceleration of the slug of sodium above the core until it impacts against the reactor vessel head. At the time of slug impact much of the coolant kinetic energy is transmitted to the vessel head and deforms the hold-down bolts plastically. If the bolts fracture, the head will be lifted from the vessel. If, as indicated in Fig. 9-16, they do not fracture, some of the remaining kinetic energy of the sodium slug is deflected radially outward, causing plastic deformation of the upper reaches of the reactor vessel wall. Thus, barring the failure of the plug hold-down bolts, the kinetic energy of the coolant slug is transferred into strain energy of both the hold-down bolts and the vessel wall.

The primary coolant loops and other auxiliary piping are also affected by the expansion of the fuel-coolant vapor bubble. A pressure pulse appears at the inlet and outlet nozzles and propagates through the loops. Interference

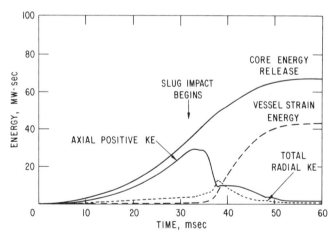

FIGURE 9-16   Fast Flux Test Facility primary system energy distribution for a $100/sec nuclear excursion. From D. H. Cho and M. Epstein, private communications, Argonne National Laboratory, 1974.

of the resulting pressure waves as well as the collapse of vapor cavities within the heat transport systems may cause severe local pressure pulses. The magnitude of the piping system pressure pulse, however, is damped by a number of energy-absorbing mechanisms: fluid-induced pipe support motions, plastic deformations of the piping and fittings, and frictional fluid flow resistance (Project Management Corp., 1975).

The partitioning of energy between the various destructive mechanisms is a function of the relative strengths of the energy release, core barrel, vessel walls and innumerable other variables. This may be illustrated by the following trade-off. If the core barrel and vessel walls are strong and rigid, they undergo very little plastic deformation. Then a gun barrel effect is created in which the preponderance of the bubble expansion work is directed upward to accelerate the sodium slug into the vessel head. In the upper limit, where the energy absorbed by strain on the core barrel and reactor vessel is insignificant, the kinetic energy with which the sodium slug impacts the vessel head may be determined from

$$M_{sl}\frac{v^2}{2} = \int_{V_I}^{V_I+V_{cg}} p(V)\,dV, \qquad (9\text{-}42)$$

where the effects of elastic energy and cover gas compression are neglected. We have defined

M_{sl} = sodium slug mass,

v = slug velocity on impact,

$p(V)$ = expansion characteristic of the vapor bubble,

V_I = initial vapor bubble volume,

V_{cg} = initial cover gas volume.

On the other hand, if the core barrel and reactor vessel are less rigid, they will undergo extensive plastic deformation and absorb significant amounts of energy before the sodium slug is accelerated significantly. In this situation the volume of the vapor bubble will increase, say by amount ΔV, because of the bulging of the reactor vessel. The kinetic energy of the sodium slug at slug impact is then given by

$$M_{sl}\frac{v^2}{2} = \int_{V_I+\Delta V}^{V_I+\Delta V+V_{cg}} p(V)\,dV, \qquad (9\text{-}43)$$

which must be smaller than Eq. (9-42), since $p(V)$ decreases with increasing V. In the extreme case where the reactor vessel ruptures, there obviously is little or no driving force to accelerate the sodium upward.

A final word is in order concerning the definition of the destructive work energy. It will be recalled that the work energy potential is defined by expanding the vapor bubble all the way down to atmospheric pressure, and it is this number from which the equivalent weight of chemical explosive is determined for purposes of experimental modeling. However, we see from the examples in Fig. 9-11 that the available work energy strongly decreases as the final pressure is increased. It follows that the maximum energy available for slug impact or vessel deformation is much smaller. Clearly, from Eq. (9-42), the energy available for slug impact is limited by the pressure when the bubble has expanded only to fill the cover gas volume. When the gas bubble pressure has further decreased to where the stresses in the reactor vessel and other primary system envelope components are in the elastic range, there is no further deformation and a state of dynamic equilibrium is reached. Below this pressure no more work energy is produced, since the bubble can no longer expand, and the remaining energy is converted to thermal energy of the coolant, core components, and primary system components.

9-4 CORE MELTDOWN

The sequence of events that we group under the heading of core meltdown may be initiated by a variety of accidents, depending on the reactor type. The meltdown may be the consequence of an explosive disassembly accident in which the core has been destroyed and dispersed by the pressures generated by the nuclear excursion or by overpower transients that are somewhat less severe but that nevertheless cause extensive fuel melting. Cooling failures ranging from the total loss of coolant to the failure of residual heat-removal systems following normal reactor shutdown may also lead to core meltdown. Whatever the initiating accident may have been, if the reactor is left in a configuration from which decay heat cannot be removed in an orderly manner, a meltdown eventually will result. Following severe nuclear excursions, the decay heat removal is most likely to be hindered by coolant channel blockages or other damage to the highly optimized heat transport geometry within the reactor core. In the case of cooling-failure accidents, the core heat transport geometry may remain intact following the initial accident transient, but all means of removing decay heat by forced or natural convection will have been rendered inoperative.

With the blockage of decay heat removal, a sequence of events such as the following is likely to occur. The fuel elements will melt and slump, causing intolerable thermal loading in the core support structure. The support structure will then collapse, dumping the molten core debris into the bottom of the reactor vessel. The heat stored in the molten debris, as well as the continued generation of decay heat, will cause the vessel to melt through, dropping the molten mass onto the containment floor. The floor will then be penetrated, and the molten mass will sink into the earth until sufficient dispersion within the surrounding ground has taken place to form a stable configuration for the transport of decay heat away from the core debris and into the surrounding earth.

The central question to be answered in the analysis of meltdown accidents is whether there will be a breach in the containment structures above ground level, creating a direct path for the gases and volatile fission products given up by the molten fuel mass to escape to the atmosphere. The dispersal of core debris into the ground beneath the containment foundation is a less serious matter because the fission products remaining with the debris would enter the environment only by the sluggish and less consequential mechanism of ground water movement.

The points at which the reactor core collapses into the reactor vessel and at which the molten core debris drops onto the containment floor are likely to be critical in determining whether the above-ground integrity of the containment shell can be maintained. At these junctures, the molten core debris may suddenly come into contact with new materials, creating the possibility of violent physical or chemical reactions that may cause overpressure or missile damage to containment structures. In particular, if water, sodium, or other liquids have collected in the bottom of the reactor vessel or on the containment floor, the sudden dropping of the molten fuel mass onto the pool of higher vapor pressure liquid under some conditions may cause violent vapor explosion and result in containment overpressure and possible shock or missile damage to surrounding structures.

In addition, contact of the molten core debris with the containment floor or other concrete structures will cause violent spalling because of the boiling of the water of hydration in the concrete (McAfee et al., 1976). The rapid decomposition of the concrete at high temperatures leads to the production of large quantities of steam and carbon dioxide. The generation of these gases will further contribute to the pressurization of the containment structure. In water-cooled systems the production of carbon dioxide may be the more serious problem because it is noncondensible and therefore unaffected by the various safety systems that are often used to condense steam and therefore reduce containment pressure. In sodium-cooled systems the steam generation may lead to the exothermic combustion of the coolant that has spilled onto the containment floor.

To examine the details of possible core meltdown accidents, it is necessary to consider the aftermaths of particular accident transients that may be hypothesized to lead to core meltdown for individual reactor types. In what follows we examine the meltdown in sodium-, water-, and gas-cooled reactors and then discuss methods for mitigating the effects of the accidents through the use of core-catching devices.

Sodium-Cooled Reactors

In sodium-cooled fast reactors energetic nuclear excursions are thought to be the most likely cause of enough damage to prevent the orderly removal of decay heat. As described in the preceding sections, these accidents are not likely to cause major breaches in the primary system envelope, and therefore following the excursion the reactor vessel may be presumed to remain filled with sodium. In such transients, however, a substantial part of the fuel may have become molten, causing a loss of structural integrity of the fuel element. Moreover, some of the fuel may have been vaporized, causing a two-phase mixture of liquid and vaporized fuel to be ejected from the core under substantial pressures.

Even with sodium remaining in the vessel, the removal of decay heat from a core that has undergone extensive damage is problematical. In the lower power density regions of the core, where extensive fuel melting has not occurred, the fuel elements remain structurally intact. Since the primary system of sodium-cooled reactors normally is designed to provide decay heat removal by natural convection, these regions of the core may be cooled in place, provided that the pressure generated in the higher power density regions of the core has not crushed the fuel assemblies to the point where the coolant flow is greatly obstructed.

In the higher power density regions of the core the molten and/or vaporized fuel must eventually come into contact with much colder sodium, either as a result of the ejection of the fuel from the core or from reentry of sodium back into the core region. It is found experimentally that when molten oxide fuel is injected into cold sodium, it shatters and freezes and forms a debris of solid particulate (Johnson et al., 1974). The more volatile fission products escape before the particulate debris is formed. However, the fission products generating about 70% of the total decay heat are retained in the particulate rubble (Project Management Corp., 1975).

The postaccident heat removal problem is centered about the cooling of the oxide fuel particulate. This particulate, along with cladding debris, tends to collect on horizontal surfaces mixed with sodium in a form resembling a fluidized bed. The decay heat generated by the particulate is removed by the combined effects of sodium boiling and natural convection within the bed (Gabor et al., 1974). For the power densities in present-generation fast reactors such beds remain stable for a maximum thickness of the order of 2 to 4 in. If the heat generation is too great, the bed will dry out, however, and particulate recoalescence can occur, forming a molten pool of fuel and steel that may then melt through the supporting surface.

Conjecture can be made regarding the disposition of the fuel debris following disassembly accidents. If forced convection of sodium is maintained upward through the damaged core, much of the particulate is expected to come to rest on horizontal surfaces in the outlet plenum. Smaller amounts may be expected to collect at elbows or recessed areas in the coolant loops, and some might eventually find its way to the bottom of the reactor vessel. Significant fractions of the fuel debris may be coolable in these locations from the formation of the aforementioned debris beds. For example, in the Clinch River breeder reactor design, it is estimated that between 28 and 46% of the fuel inventory could be cooled on the horizontal surfaces of the upper plenum, whereas 0.8 to 8.0% could be cooled in the bottom of the reactor vessel (Project Management Corp., 1975).

Compacted fuel debris may settle within the core region, and the cooling of this material can only take place to the extent that flow passages up through the rubble are available. While the fluidized bed effect may provide adequate cooling, if a large part of the core has suffered destruction, it is difficult to preclude the possibility that the debris will coalesce and cause complete flow blockages in some core regions. The particulate rubble may then form into a larger mass of molten fuel and slump downward under the force of gravity.

The size of such a molten fuel mass that may form in the core or for that matter anywhere within the system, is limited by the onset of thermal instabilities. To illustrate, consider a sphere of uranium dioxide of radius R. Decay heat is generated uniformly within the sphere at a rate q_d''' that depends on the original specific power of the reactor, the fraction of the fission products retained in the fuel debris, and the time since shutdown. If the decay heat is dissipated by conduction to the outer surface of the sphere, at temperature T_s, then the temperature at the center of the sphere is easily shown to be

$$T = T_s + \frac{q_d''' R^2}{6k_f}, \qquad (9\text{-}44)$$

where k_f is the thermal conductivity of the molten fuel. Clearly, the interior temperature rises as the sphere grows in size. At the radius for which T reaches the fuel boiling temperature, vapor pressure will be generated at the center of the sphere, causing the fuel to disperse.

Ivins and Hesson (1969) have estimated the maximum dimensions of thermally stable configurations of molten uranium dioxide in sodium. The results for spheres, infinite slabs, and cylinders are shown in Fig. 9-17 for

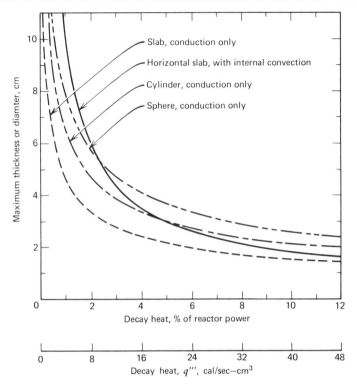

Slab, conduction only

Horizontal slab, with internal convection

Cylinder, conduction only

Sphere, conduction only

Decay heat, % of reactor power

Decay heat, q''', cal/sec–cm^3

FIGURE 9-17 Maximum dimensions of thermally stable shapes. From R. O. Ivins and J. C. Hesson, "Post Accident Heat Removal, Reactor Development Program Progress Report," Argonne National Laboratory Report, ANL-7640, 1969.

decay heat generation rates consistent with fast reactor power densities. For the slab case the effect of natural convection within the molten fuel mass in increasing the maximum stable size also is estimated.

From the dimensions given in Fig. 9-17, it is clear that the degree of coalescence of fuel debris into molten masses is limited by thermal instabilities. Thus, should meltdown of the core take place, it is not expected to be characterized by the slumping of a continually growing molten fuel mass. Rather, molten fuel masses may coalesce, grow, but then erupt as they become unstable. The fuel dispersed from the eruption will then shatter on contact with colder sodium and refreeze as particulate debris. If adequate cooling is not available the process will continue in a disorderly cyclic fashion.

The dispersal of the fuel from thermal instabilities has another important consequence. It hinders the formation of fuel masses that are large enough to become supercritical. For most configurations, whether molten fuel masses within the core region or pools of fuel in the bottom of the pressure vessel, the rate of decay heat generated in fast reactor cores is sufficient to minimize the problem of recriticality. Only if external pressures can cause the compaction of a large mass of fuel over a short enough time period that criticality can be achieved before the interior fuel temperature reaches the fuel boiling point is there the possibility of a nuclear excursion. But under the conditions expected following postulated accidents it is highly improbable that the energy release from such an event would add significantly to the energy produced by decay heat.

From the foregoing discussion it is clear that sodium circulation through particulate fuel debris may be capable of cooling the rubble of partially destroyed cores within the primary system. However, as accidents in which increasingly large fractions of the core have been melted or vaporized are considered, the prevention of core meltdown and reactor vessel melt-through becomes increasingly problematical.

Vessel melt-through is inevitable if the primary system envelope, including the guard vessel, is breached in such a way as to terminate the natural circulation of sodium through the system. The lack of a heat-removal path results in the boil-off of sodium and the eventual slumping of the vapor-bound fuel debris into the bottom of the uncooled reactor vessel. Such a breach could be the result of excessive blast damage to the reactor vessel or coolant piping. The accumulation of fuel debris in beds of excessive depth in the lower head of the reactor vessel or at a recessed area in the coolant piping also may lead to bed dry-out, subsequent meltthrough of the supporting surfaces, and the dumping of the sodium from the primary system.

Even in the absence of a primary system breach, excessive core damage may block the flow path through the molten core and cause the meltdown of the core onto the core support structure. The support structure would partially melt and then collapse from the weight of the core debris. The process is then repeated in the lower head of the reactor vessel and in the guard vessel, dumping core debris and sodium onto the containment floor. In a sodium reactor of current generation design, such a melt sequence is conservatively estimated to require between 15 and 30 minutes before the fuel would reach the containment floor (Project Management Corp., 1975).

Water-Cooled Reactors

In water-cooled reactors nuclear excursions are considered much less likely than cooling failures to cause core meltdowns. The most widely studied of such failures is the loss-of-coolant accident initiated by a rupture of a major cooling pipe. As indicated in Section 8-5, core destruction is prevented by the use of emergency cooling systems. If, following such an accident, however, there also is a total failure of the emergency core-cooling system, a core meltdown will result. The initial phases of the meltdown transient would be a strong function of the time at which the core-cooling system failed. If, during the design basis accident, there is a failure to initiate emergency core cooling following the coolant blowdown, the primary vessel could be devoid of water in as little as 10 to 20 seconds after the accident initiation. On the other hand if the emergency core-cooling system effectively quenches the core but then fails at a later time, the transient would begin with the vessel filled with relatively cool water, at near atmospheric pressure, and there would be a period of boil-off before the core would become vapor bound. The latter sequence of events may take place, for example, if the reservoirs for cooling water become exhausted and remedial steps are not taken.

In addition to the various sequences originating from loss-of-coolant accidents, a core meltdown may result following a normal reactor shutdown if all the decay heat removal systems are lost. Such would likely be the case, for example, if all on-site and off-site power were lost. In this situation the primary system would initially contain a full inventory of coolant. With the loss of cooling, however, the primary system would heat up and overpressurize. Safety valves would then blow and relieve the pressure, and there would be a boil-off of the coolant inventory through safety and relief valves until the core became vapor bound and began to heat up.

Following any of the preceding events, boil-off will eventually cause the core to become vapor bound. The fuel elements overheat and substantial energy is generated from metal–water reactions as the zirconium cladding rises above about 2000°F. At about 3400°F the zirconium cladding begins to melt and the structural integrity of the fuel elements is lost. With the loss of structural integrity, the still solid uranium dioxide fuel pellets may wedge in the cooling channels, since they are prevented from falling from the core by the close packing of the fuel elements in the reactor lattice. With the generation of additional decay heat, the fuel will melt and begin to coalesce into a growing mass of molten debris in the higher power density regions of the core.

The behavior of such molten fuel masses is subject to uncertainty. Conceivably, the molten cladding and fuel material may drip down onto the core support structure and lower head of the reactor vessel. In this case there would be a gradual buildup of core debris in the bottom of the vessel as the core melts. This sequence of events, however, is thought to be much less likely than a buildup of a molten fuel mass within the core region until 50 to 80% of the core is molten (U.S. Atomic Energy Commission, 1974). The thermal loading on the core support structure would cause it to collapse, dumping the large molten fuel and cladding mass onto the lower head of the reactor vessel.

The large mass of molten fuel is thought to remain with the core volume until well over half of the fuel has melted, because after fuel or cladding melt, they are most likely to refreeze as they slump into regions where the fuel elements are cooler. If some water remains in the bottom of the core, it rapidly quenches the downward-moving fuel. In either event a solid plug is likely to form and support the mass of molten debris. Some boiling within the mass would take place because of the thermal instability effects, discussed in conjunction with sodium reactors in the preceding section, and this would enhance convection within the mass. The lower specific powers in thermal reactors permit the dimensions of the fuel mass to become much larger than in fast reactors before internal eruption commences. There is no possibility of recriticality, however, owing to the low enrichment of thermal reactor fuels. The dissipation of decay heat from the molten mass may be aided by a number of mechanisms. Downward conduction into the solid masses of fuel would dissipate heat by creating a melt front moving downward through the core. The internal convection of the molten mass would also create heat losses from the top of the pool by radiation and to a lesser extent by boil-off of fuel vapor. This would create further melting of fuel above the pool and cause molten debris to drop into the pool (U.S. Atomic Energy Commission, 1974b).

The rate of heat-up, melting, and collapse of the core can be strongly affected by the presence of water or steam in the core region. If the core is initially filled with water when failure of the cooling system occurs, the boil-off will delay the initiation of core heat-up. Later, even as melting is occurring above the core midplane, water still present at the fuel inlet will tend to quench the downward movement of molten debris. Although the presence of water or steam in the core can slow the early stage of the meltdown processes, at later times its presence may become detrimental. After the cladding has reached temperatures at which metal–water reactions are prominent, the small cooling effects of the passage of steam up through the core are overwhelmed by the heat produced by the more rapid rate of metal–water reactions induced by the adequate steam supply. For example, if a partially disabled core-cooling system continues to supply some water to the lower head of the reactor vessel, this water will boil on contact with the high temperature of the vessel wall and supply a continuous flow of steam into the core region, thereby aggravating the metal–water reaction.

The preceding effects make it difficult to predict accurately the time scale on which the core meltdown sequence will take place. Estimates of one hour have been made for the time required for core collapse to take place following the initiation of meltdown in the absence of water in the vessel.

Following collapse of the core another half hour or more is required for the molten mass to heat the lower vessel head to a point where structural failure occurs, dropping the molten debris onto the containment floor. This time is lengthened if the lower vessel head is partially filled with water at the time of core collapse. About an additional half hour may be required for boil-off before it melts through. At the same time the sudden dropping of a large molten fuel mass into a water pool may produce an energetic vapor explosion. It is estimated that if 20% or more of the fuel mass of a large reactor core reacts with the water, enough pressure will be generated to cause breach of the containment structures (U.S. Atomic Energy Commission, 1974b).

Following reactor vessel melt-through, the molten mass of fuel, cladding, and structural steel is dumped on the containment floor. If there is water in the containment sump the possibility for a vapor explosion causing further pressurization of the containment again exists. And in any event, the molten mass will attack the concrete foundation of the containment.

Gas-Cooled Reactors

The designs of fast and thermal gas-cooled reactors are sufficiently different that the discussion of meltdown of the two core types is considered individually. In proposed gas-cooled fast reactors, the cores consist of oxide fuel pins in stainless steel cladding. The fuel assemblies are supported from above, and the helium flow is downward through the core.

In the event of a loss of forced circulation through the core following shutdown, the core would heat up and melt in a manner qualitatively similar to a water reactor core in the absence of steam. The cladding would first melt and tend to flow downward, and this would be followed by fuel melting. Whether the molten cladding and fuel would fall free of the core or refreeze and form a plug in the lower regions of the core or axial blanket has not been totally resolved. In the event of plugging, however, a mass of fuel and cladding would coalesce within the core or lower blanket. Thermal instabilities would then be expected to develop, causing eruption, dispersion, and rearrangement of such a mass.

If the loss of forced circulation has occurred in conjunction with a trip failure, examination of clad slumping, fuel bowing, and related phenomena is necessary to determine whether a reactivity excursion is likely. In particular, careful consideration would be necessary to ensure that the thermal instabilities would be sufficient to prevent the core from forming into supercritical configuration. At some point the meltdown will progress to where excess thermal loading on the support structure will cause the collapse of the core. It will then fall and melt through the thermal shielding and be dumped onto the floor of the reactor vessel. Since gas-cooled fast reactors are proposed to be built in a prestressed concrete vessel, the further behavior of the molten core will resemble that of sodium- or water-cooled systems when molten fuel debris comes into contact with the concrete containment floor.

Should the decay heat removal capability be lost for a graphite-moderated reactor, a meltdown might eventually take place. In these reactors, however, the rate at which such an accident could progress would be much slower than that for any of the other reactor types. The low power densities in association with the heat capacity of the large mass of graphite moderator result in adiabatic heat-up rates from decay heat that are only a few degrees per second. This, coupled with the very high temperature that must be reached before graphite loses its structural support capability, seems to indicate that the core of such a graphite-moderated system would be very slow in slumping to the bottom of the concrete reactor vessel.

Containment Penetration

The last stage of a reactor meltdown accident commences when the molten core debris falls to the floor of the containment structures. The debris then attacks and penetrates the foundation and sinks into the earth. Some details may differ considerably, depending on whether the containment floor is in a sodium or a water pool, a steam–air atmosphere, or an inert gas atmosphere. The general features of the molten fuel behavior may be illustrated, however, by considering the containment penetration in water-cooled systems (U.S. Atomic Energy Commission, 1974b). The molten mass of fuel, zirconium, zirconium oxide, steel, iron oxide, and other core constituents interacts with concrete in a complex and not fully understood manner. As the molten mass falls onto the concrete, the free water below the surface vaporizes and causes spalling. The initial spalling causes quite rapid penetration with an initial rate of from 15 to 30 ft/hr. As the temperature of the concrete rises further, decomposition takes place: At about 700°F the water of hydration is given up, and between 1400 and 1600°F the limestone will decompose, emitting carbon dioxide gas. The resulting steam and carbon dioxide are released, adding to the pressure of the containment atmosphere, and the remaining concrete constituents—primary calcium or silicone oxides—are absorbed or dissolved in the molten core mass.

As the concrete decomposition products mix with the molten core material, the temperature decreases until some mixture constituents begin to precipitate. The uranium dioxide would begin to precipitate from the molten mixture after the penetration of about 1.5 ft of concrete. Driven by decay heat, the viscous mixture of core and concrete debris continues to progress more slowly downward through the concrete. Steam, carbon dioxide, silica, and calcia pass upward through the molten mass, because of their lower density. The resulting agitation in conjunction with natural convection currents prevents excessive temperatures from being reached at the center of the molten mass. The molten mass is likely to become covered with a solid crust. The length of time required to penetrate a containment foundation may vary substantially, depending on how much horizontal spreading takes place as it melts downward. For a typical water reactor core, it is estimated that the penetration of a 10-ft thick concrete containment vessel would require on the order of 18 hours (U.S. Atomic Energy Commission, 1974b).

After the concrete containment foundation is penetrated, the molten core and concrete debris sink into the underlying gravel and then into the earth. The dimensions of the molten debris mass will continue to grow, engulfing increased volumes of the surrounding earth. Estimates for the size of the molten debris mass in limestone and in dry sand are shown in Fig. 9–18. It is seen that the heat carried into the surrounding limestone sand or other media is not sufficient to limit the growth of debris mass until more than two years after the accident. It is thought, however, that it is not likely that the debris will proceed more than 10 to 50 ft beneath the bottom of the containment foundations unless the core materials are able to penetrate the underlying media without mixing and being diluted by the decomposition products of the earth (Fontana, 1968).

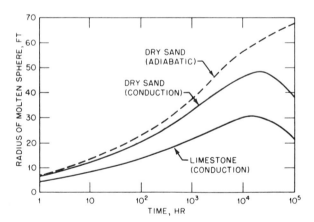

FIGURE 9.18   Conduction-controlled expansion of a molten sphere containing 1-hr-aged fission products from a 3200-MW(t) reactor (20,000 MWd/metric ton) in an infinite medium of dry sand or limestone. From M. H. Fontana, "Core Melt Through in LMFBR's—A Condensed Review," Oak Ridge National Laboratory Report, ORNL-TM-3504, 1971.

Core-Catching Devices

The overwhelming emphasis in reactor safety is on the prevention of core meltdown. Moreover, although there have been several accidents that have resulted in some fuel melting, to date there have been no accidents severe enough to cause the syndrome of core collapse, reactor vessel melt-through, containment penetration, and dispersal into the ground. Nevertheless, a number of proposals have been made for the design of core catcher systems to control or stop the motion of the molten core mass should such an accident take place. Core catchers may differ in both their location within the reactor system and in the mechanism that is used to cool and control the motion of the core debris.

As indicated in Fig. 9-19, a core catcher can be located in one of three locations: within the reactor vessel, as an integral part of the reactor vessel, or in a sump below the reactor vessel. The core catcher generally may be classified as catcher trays, cooled crucibles, or sacrificed beds. The choice of one of these core catcher concepts depends strongly on its location within the system, the reactor coolant, and the class of postulated accidents against which it is supposed to protect.

Core catchers may consist of one or an array of trays into which the core can fall and be cooled by the surrounding fluid. The tray must have sufficient area that the molten part of the core will spread out over a large area and not increase to a sufficient thickness that thermal instabilities violent enough to damage the tray or their supports will result. As can be surmised from the slab results in Fig. 9-17, such instabilities can result for quite small thicknesses, and therefore large areas may be required. A conceptual design for such a tray array in a sodium-cooled reactor is shown in Fig. 9-20. Other geometries are possible; the sodium-cooled Fast Flux Test Facility has a domed structure beneath the core that serves a similar purpose in being able

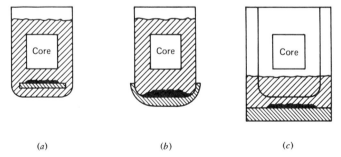

(a)   (b)   (c)

FIGURE 9.19 Schematic core catcher configurations. (a) Within reactor vessel. (b) Reinforced reactor vessel. (c) Below reactor vessel.

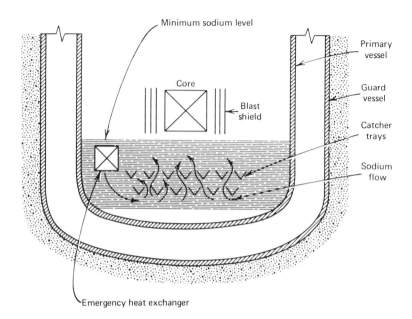

FIGURE 9.20 Catcher trays in sodium pool. From M. H. Fontana, "Core Melt Through in LMFBR's—A Condensed Review," Oak Ridge National Laboratory Report, ORNL-TM-3504, 1971.

to catch and hold a substantial amount of core rubble that can then be cooled by the surrounding sodium coolant (Hanford Engineering Development Laboratory, 1972). If supplemental cooling were to be added to the underside of the reactor vessel, conceivably a reinforced bottom head of the vessel might also serve as a core catcher.

The core-catching trays within the reactor vessels are proposed for use in sodium-cooled systems, where it is expected that even if core meltdown should take place, the sodium inventory would remain within the vessel, allowing heat removal by natural convection. Such systems would not be applicable to gas-cooled reactors, since the convective heat transfer would be completely inadequate for decay heat removal. Neither is the within-vessel system very promising for water-cooled reactors, since the design basis accident stems from a loss of coolant and therefore the vessel would not be expected to remain filled with water. Even in sodium systems where within-vessel core catchers appear to be most reasonable, structural difficulty may arise from the blast loading of the nuclear excursion or from thermal shock as the hot core debris falls suddenly into the trays.

Another possible core catcher configuration consists of a low thermal conductivity crucible capable of withstanding very high temperatures. Depleted uranium dioxide and graphite are the materials most often considered. The crucible catches the core and heat is removed by a variety of mechanisms, depending on the system. Graphite or other materials have been suggested to line the vessel of sodium-cooled reactors in order to decrease thermal loading on the vessel walls (Fontana, 1971). Crucible core catchers have also been proposed for gas-cooled reactors (Holcomb, 1971). Here the crucible consists of a heavy lining to protect the prestressed concrete vessel from direct contact with the molten fuel. The heat is removed from the molten fuel debris by downward heat conduction through the crucible to a forced convection cooling system. Heat is also transferred upward by radiation to the steel liner of the vessel, where it is removed by the linear cooling system.

Crucible core catchers have been proposed for sodium- and water-cooled systems located beneath the reactor vessel under a pool of coolant in the reactor sump. In these systems the molten core debris would melt through the vessel and fall into the crucible, where the heat would be removed by the overlying coolant pool to a system of heat exchangers.

A third method for the control of molten core material is to place a bed of sacrificial material under the reactor vessel. The material properties are chosen so that when the molten core debris drops onto the material, it will diffuse into the bed, causing its temperature to decrease first as the molten fuel becomes diluted by the bed and then as the decay heat rate decreases with time. If the debris can be sufficiently diluted it may be possible to achieve a stable configuration from which heat is removed to the surrounding media. More often active means of heat removal are incorporated into the design to ensure that the molten mass will not penetrate the outer confines of the containment structure. For example, the magnesia sacrificial bed core catcher proposed to be located in the vault below the Clinch River breeder reactor will include provisions for forced convection heat removal from the sides and bottom of the catcher. In addition, the catcher will be submerged in a pool of sodium, allowing for heat transfer out through the upper surface of the bed (Project Management Corp., 1974).

9-5 SITE-INDUCED ACCIDENTS

The accidents studied in reactor safety analysis to a great extent are assumed to be initiated by malfunctions or failures arising within the nuclear plant. These faults in turn may be due to faulty design, equipment failures, operator error, or a number of other interrelated causes. In addition, both natural and man-made events exterior to the plant can have the potential for inducing serious accidents. Hence they must be taken into consideration in setting design bases and operating procedures. Particular scrutiny of the various site-induced accidents is necessary because they often have the potential for introducing common-mode failures of normally independent plant components if protective measures are not designed properly to accommodate their effects.

Protection from Natural and Man-made Catastrophes

Although much remains to be done, a great deal of effort has been expended both in estimating the likelihood of major site-induced accidents and in assessing the damage that they can cause (Cottrell, 1975). Of the myriad of natural and man-made disasters that may occur outside the nuclear plant, those likely to have a sufficiently high probability and intensity to bear significantly on the safety of the plant can be categorized broadly as tornadoes and high winds, floods and high waves, earthquakes and other ground motion, crashes of aircraft or other vehicles, sabotage, and acts of war.

In general, plants that can withstand severe tornadoes automatically meet the wind-loading criteria imposed by hurricanes or other storms, since the whirlwind effects of a tornado produce wind speeds and pressure differentials that greatly exceed those of other storm types. The extensive study of the characteristics and destructive capability of tornadoes has led to the specification of a design basis tornado that must be withstood by the nuclear power plânt (Doan, 1970).

The funnel cloud is characterized by a tangential wind speed about its center as well as a translational speed that describes the net movement of the cloud. The pressure in the interior of the cloud is substantially less than that of the surrounding atmosphere. Thus a maximum pressure decrease and a maximum rate of pressure drop are also specified. In the United States, nuclear plants east of the Rocky Mountains are required to withstand the worst (class I) tornadoes (McDonald et al., 1974):

Tangential speed	290 mph
Translational speed	5–70 mph
Pressure drop	3 psi
Pressure drop rate	2 psi/sec
Maximum diameter	150 ft

Somewhat more moderate (class II and III) design basis tornadoes are stipulated west of the Rocky Mountains, where the frequency and severity of tornadoes is much smaller.

The structures essential to reactor safety must be capable of withstanding the pressure transients from the air flow over and around structures. In addition to the wind speeds of up to 350 mph, the vital plant structures must withstand the sudden drop and then increase in pressure as the tornado vortex passes over. The pressure drop causes differential pressures across structures. In addition, the sudden decrease in exterior pressure tends to explode low-strength enclosures that are nearly airtight. Finally, structures must be designed against missiles generated by tornado winds. For purposes of analysis, design basis missiles are also stipulated. These include various steel pipes and rods, a utility pole, and an automobile (McDonald et al., 1974). It is noteworthy that the number of missiles generated is likely to increase substantially if there is construction on the reactor site, since large quantities of unsecured materials would then normally be present.

In addition to wind and other atmospheric phenomena, attention must be given to the possibility of flooding or damaging waves. Flooding of river basins or coastal lowlands because of excessive rainfall or hurricanes does not occur without warning (Wall, 1974a). Thus if water levels are predicted that exceed that for which the reactor can safely remain in operation, there is time to shut and cool down the reactor in an orderly fashion. Provisions must then be made for the orderly removal of decay heat, even though the water rises to the level of a highly improbably design basis flood and remains there for an extended period of time.

Careful analysis is also required to prevent the more sudden inundation of a nuclear plant that conceivably could be caused by tsunamis, seiches, dam failures, and the like. Protection is provided by adequate elevation where possible, coupled with seismically resistant sea walls for the protection of water intakes and other low-lying equipment. The risk of flooding as a result of dam failures is minimized either by avoiding plant sites where the possibility exists or by ensuring that the probability of the dam failing is so low as to not contribute significantly to the risk.

Movement of the ground as well as that of air or water may endanger the integrity of a nuclear plant. The dangers of landslides, avalanches, and even volcanic activity are minimized through careful site selection. Of all the site-induced accidents, however, earthquakes receive the most attention (Lomenick et al., 1970). There are a number of reasons for this (Cottrell, 1975): Although particular regions—such as the west coast in the United States—are particularly susceptible, earthquakes may occur over wide areas, including the entire United States. Earthquakes have a wide range of potential severity. The frequency decreases with increasing earthquake intensity, but it is difficult to extrapolate from historical experience to place an upper limit on the intensity of seismic events that may be expected at a particular reactor site. Furthermore, methods for predicting the time and location of earthquakes are in their infancy. Finally, plant protection against the destructiveness of the interrelated phenomena associated with earthquakes presents difficult design problems.

The effects of earthquakes and the means by which a nuclear plant is protected from them are varied. The large ground displacements sometimes associated with earthquakes are avoided by not siting nuclear plants over or near faults that are active. Extensive analysis of the geology of the plant site also is necessary to ensure that liquifaction or other soil-related problems do not cause failures of the foundation or other essential structures following an earthquake. Precautions also are taken not to locate the plant where it is susceptible to seismically induced landslides, dam failures, or other related mechanisms, and a sea wall must be constructed if the location is subject to damage from tsunamis. Perhaps the most pervasive mechanism for direct damage to the plant, however, and indeed the cause of the aforementioned effects, is seismic ground shaking, for every plant component is simultaneously subjected to transient stress from it. As a result, a large body of literature has developed around the characterization of seismic ground shaking and the response of plant structures and components to it. We defer discussion of ground shaking and its consequences to the following subsections.

In addition to the preceding natural phenomena, nuclear power plants may be subject to damage from a variety of accidents that originate off site: explosions of railroad tank cars, barge collisions with water intakes, and the like. Moreover, unique accidents may become important for novel siting conditions. For example, nuclear plants placed on small artificial islands can be damaged from collisions of tankers or other large ships. The site-induced accident most considered is the crash of an aircraft into the power plant.

In considering the accident risk incurred as a result of aircraft crashes, an analysis must again be made of probability versus consequences (Wall, 1974). In general, only the effects of large commercial or military aircraft need be considered, since the plant structures invariably are of sufficient strength to withstand the impact of smaller, general aviation aircraft without incurring significant damage. Even for a large aircraft the response of the reactor building as a whole would be milder than that for a design basis earthquake, as may be verified from momentum conservation arguments. Localized damage to specific structural components, however, may be significant. The concentrated impact of the aircraft engine may cause structural perforation, for example. Lightweight structures may also collapse on impact, causing the aircraft and/or structural debris to destroy the enclosed equipment. Finally, since fire results from about 50% of commercial aircraft crashes, its effects on impacted structures require consideration. Generally it is found that for a site more than about five miles from an airport and not within a busy air corridor, the probability of a crash into the plant is small enough that no safety measures beyond those already taken are required. To minimize the risk, plants at sites more closely located to aircraft traffic may require some additional hardening (Wall, 1974).

With the exception of ground shaking, which may affect all plant components, the damage resulting from the variety of site-induced accidents discussed previously tends to be of a different nature from that from accidents that originate within the primary system. In the preponderance of cases the reactor can be shut down, and although the containment vessel may be damaged, it serves to protect the primary system from direct damage. A scenario for the release of radioactive materials is more likely to stem from the incapacitation of the residual decay heat removal system following reactor shutdown. Decay heat removal requires both power and service water, and water intakes and the supply of off-site power are likely to be the most vulnerable to site-induced accidents. Thus it is extremely important that the on-site emergency power supply system and the supply of essential service water be protected from external events, particularly when they are located outside of the containment building.

Precautions must also be taken to protect fuel storage pools from external events. Although fuel located in these storage facilities has had sufficient decay time for all but the longer-lived isotopes to die away, it still constitutes a sizable radioactive inventory. Thus, as is often the case, the pool is located outside the containment shell in an airtight but lightweight structure, a close examination must be made to ensure that wall penetration, structural collapse, fire, or other mechanisms will neither cause undue mechanical damage to the stored fuel nor drain the pool of cooling water.

Thus far we have considered only damage resulting from accident events. Although there is no historical precedent, questions of sabotage also deserve examination. From the study of present-day nuclear power plants several conclusions can be drawn (Turner et al., 1970). To be successful, the saboteur would require extensive technical capability and knowledge of the plant layout in order to select and damage vital components. Moreover, he must want not simply to disable the plant but to endanger the public. A single sabotage-induced failure may disable the plant but would be well within the envelope of postulated accidents for which safety systems prevent the release of radioactive materials to the environment. In order to endanger the public the saboteur must be capable of executing multiple and coordinate failures of independent systems. The difficulty of carrying out such tasks makes it improbable that a saboteur with the intent of causing a public hazard would choose a nuclear power plant when simpler and more vulnerable targets are available. The risk of sabotage is further reduced through security measures that greatly decrease the likelihood of success. During times of civil or labor unrest, when sabotage is thought to be most likely, added security forces can further decrease the likelihood of successful sabotage.

Studies also have been made to ascertain whether acts of war directed against nuclear power plants could result in increased numbers of casualties. The only situation thought to lead to significant civil defense implications is the targeting of a nuclear power plant for destruction by thermonuclear weapons. The number of casualties likely to result from such an event has been examined for both light-water cooled reactors (Chester and Chester, 1970) and liquid-metal-cooled fast reactors (Chester and Chester, 1974). It is found in both cases that a significant increase to casualties induced by the thermonuclear detonation could occur only if the reactor were subject to essentially a direct hit (i.e., the reactor were within the crater). Given the accuracy of intercontinental missiles, this would probably be an expensive tactic, since to ensure a reasonable probability of a direct hit several missiles would have to be assigned to the same target.

Earthquake Characterization

Before measures can be taken to protect nuclear plants from seismic ground shaking, an understanding of the nature of earthquakes must be translated into quantitative description of the seismic motion (Lomenick, 1970). Earthquakes may be defined as the elastic waves that travel through the earth. The waves are transient in nature, with the strongest shocks being felt for no more than a few minutes. It is generally agreed that most damaging earthquakes are likely to result from faults in tetonic rock. Elastic strain energy gradually builds up in the vicinity of a fault, until the stress level overcomes the static friction along the faces of the fault. The energy is then released in the form of elastic shocks as displacement takes place along the fault to release the strain energy.

The nature as well as the frequency of earthquakes tends to differ with geographical location. In the United States, for example, there are substantial differences between the quakes that occur on the west coast and those in the remainder of the country. In the west, where they appear most frequently, damaging earthquakes can generally be associated with movement along well-defined faults that extend to the earth's surface. For example, the San Francisco earthquake of 1906 resulted from movement along 270 miles of the San Andreas fault, with slippages of as much as 22 ft between fault faces. In the eastern United States, damaging earthquakes are much less frequent, and those that do occur have not been associated with surface faults. Instead they originate from deeper in the earth and cannot be clearly related to faulting or other geological structures. The great historical earthquakes at Charleston, South Carolina, in 1886—generally considered to be the largest in United States history—and at New Madrid, Missouri, in 1811–1812 were of this nature.

To assess the threat of an earthquake it is necessary to have some quantitative means of describing the size of an earthquake and the damage caused by it. Through seismographic observation, the center or focus of an earthquake can be determined. The epicenter is then defined as the location on the earth's surface directly above the focus. The earthquake's magnitude refers to the energy released at the focus. It may be determined on a relative scale from seismographic measurements. The most commonly used measure at present is the Richter logarithmic scale. The earthquake intensity is determined by the amount of damage or felt motion by the disturbance at a ground level location. The Modified Mercalli (MM) scale shown in Fig. 9-21 is the most commonly used intensity scale in the United States. A more quantitative measure of earthquake intensity is the maximum acceleration occurring during the ground shaking. Estimates have been made relating the qualitative classification of the Modified Mercalli scale to the maximum ground acceleration, and to the ground acceleration at the epicenter for an earthquake of a given Richter magnitude or energy release. Some estimated equivalences are shown in Fig. 9-21. Empirical earthquake magnitude, distance, and intensity relationships have been formulated (Hudson, 1974). Thus if the focus and magnitude of an earthquake are known, the intensity at a site at a given distance from the epicenter is estimated in terms of the Modified Mercalli scale or maximum ground acceleration. In such estimates local geological conditions must be taken into account, since if the site is underlain by sediments or soils, the ground motion may be amplified by the passage of the elastic waves from the bedrock to the soils; amplification by a factor of 2 or more is not uncommon.

An estimate of the maximum acceleration is not sufficient for determining damage to structures. A more detailed description of the shaking motion is needed (Lomenick et al., 1970). One procedure is to normalize the maximum acceleration to one of the traces of acceleration versus time that has been obtained for large earthquakes. Traces of acceleration, velocity, and displacement, as shown, for example, in Fig. 9-22, yield a great deal of information in addition to the maximum acceleration. In particular, they indicate the spectrum of frequencies at which the ground shaking occurs. They can be used as an input-forcing function for computational models of structures to determine their response to the earthquakes, as illustrated below. A single or even a few such traces, however, may not be sufficient to guarantee the safety of the structure, since the acceleration trace for each earthquake has an individual character in that certain frequency vibrations tend to dominate the spectrum. If these frequencies correspond to any of the natural frequencies in the structure, the damage is maximized. On the other

Modified Mercalli Intensity Scale

Modified Mercalli Intensity Scale	Description of Effects (Masonry A, B, C, and D Are Defined Below;a From Ref. 27)	Maximum Acceleration (g)	Richter Magnitude	Energy Release (ergs)
I	Not felt; marginal and long-period effects of large earthquakes evident		M2	10^{14}
II	Felt by persons at rest, on upper floors, or favorably placed			10^{15}
III	Felt indoors; hanging objects swing; vibration like passing of light trucks occurs; duration estimated; might not be recognized as an earthquake	0.003 to 0.007	M3	10^{16}
IV	Hanging objects swing; vibration occurs that is like passing of heavy trucks, or there is a sensation of a jolt like a heavy ball striking the walls; standing motor cars rock; windows, dishes, and doors rattle; glasses clink; crockery clashes; in the upper range of IV, wooden walls and frame creak	0.007 to 0.015		10^{17}
V	Felt outdoors; duration estimated; sleepers waken; liquids become disturbed, some spill; small unstable objects are displaced or upset; doors swing, close, and open; shutters and pictures move; pendulum clocks stop, start, and change rate	0.015 to 0.03	M4	10^{18}
VI	Felt by all; many are frightened and run outdoors; persons walk unsteadily; windows, dishes, glassware break; knickknacks, books, etc., fall off shelves; pictures fall off walls; furniture moves or overturns; weak plaster and masonry D crack; small bells ring (church, school); trees, bushes shake	0.03 to 0.09	M5	10^{19}

Intensity	Description	Acceleration (g)
VII	Difficult to stand; noticed by drivers of motor cars; hanging objects quiver; furniture breaks; damage occurs to masonry D, including cracks; weak chimneys break at roof line; plaster, loose bricks, stones, tiles, cornices fall; some cracks appear in masonry C; waves appear on ponds, water turbid with mud; small slides and caveins occur along sand or gravel banks; large bells ring	0.07 to 0.22
VIII	Steering of motor cars affected; damage occurs to masonry C, with partial collapse; some damage occurs to masonry B, but none to masonry A; stucco and some masonry walls fall; twisting, fall of chimneys, factory stacks, monuments, towers, and elevated tanks occur; frame houses move on foundations if not bolted down; loose panel walls are thrown out; changes occur in flow or temperature of springs and wells; cracks appear in wet ground and on steep slopes	0.15 to 0.3
IX	General panic; masonry D is destroyed; masonry C is heavily damaged, sometimes with complete collapse; masonry B is seriously damaged; general damage occurs to foundations; frame structures shift off foundations, if not bolted; frames crack; serious damage occurs to reservoirs; underground pipes break; conspicuous cracks appear in ground; sand an mud ejected in alluviated areas; earthquake fountains and sand craters occur	0.3 to 0.7
X	Most masonry and frame structures are destroyed, with their foundations; some well-built wooden structures and bridges are destroyed; serious damage occurs to dams, dikes, and embankments; large landslides occur; water is thrown on banks of canals, rivers, lakes, etc.; sand and mud shift horizontally on beaches and flat land; rails are bent slightly	0.45 to 1.5
XI	Rails are bent greatly; underground pipelines are completely out of service	0.5 to 3
XII	Damage nearly total; large rock masses are displaced; lines of sight and level are distorted; objects are thrown into air	0.5 to 7

[a]*Masonry A.* Good workmanship, mortar, and design; reinforced, especially laterally, and bound together by using steel, concrete, etc.; designed to resist lateral forces.

Masonry B. Good workmanship and mortar; reinforced, but not designed in detail to resist lateral forces.

Masonry C. Ordinary workmanship and mortar; no extreme weaknesses like failing to tie in at corners, but neither reinforced nor designed against horizontal forces.

Masonry D. Weak materials, such as adobe; poor mortar; low standards of workmanship; weak horizontally.

FIGURE 9.21 Approximate relationships between intensity, acceleration, magnitude, and energy release. From T. F. Lomenick and NSIC staff, "Earthquakes and Nuclear Power Plant Design," Oak Ridge National Laboratory Report, ORNL-NSIC-28, 1970.

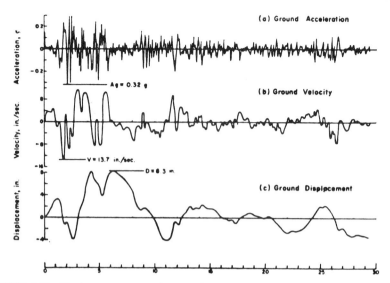

FIGURE 9.22 Motions computed from N-S component of earthquake at E1 Centro, May, 18 1940. From N. M. Newmark, "Effects of Earthquakes on Dams and Embankments," *Géotechnique* **15**, 139–160 (1965). Used by permission.

hand, if the particular earthquake chosen does not have significant contributions at the structure's natural frequencies, a misleading underestimate of damage may be obtained. For this reason, if the time history approach is used in the dynamic analysis of the nuclear plant, enough different time histories must be used to ensure that all frequencies likely to occur on the hypothesized earthquake are adequately represented.

Although the time history approach to earthquake characterization is coming into increased use, particularly in research-oriented studies, a second method is most often used in design work. The response spectrum permits one to use the time history to establish an envelope within which all earthquakes with less than a given maximum acceleration should fall. The method then utilizes the fact that the dynamic behavior of a structure can usually be reduced to combinations of simple harmonic oscillations. Thus the time history trace is codified by its effect on a simple harmonic oscillator as a function of natural frequency and damping.

Suppose that $y(t)$ is the displacement of a harmonic oscillator relative to the ground. If the ground displacement is $z(t)$, then

$$\frac{d^2y}{dt^2} + 2\beta\omega\frac{dy}{dt} + \omega^2 y = -\frac{d^2z}{dt^2}, \qquad (9\text{-}45)$$

where ω is the natural frequency of the undamped oscillator and β is the fraction of critical damping. If the displacement and velocity of the oscillator are assumed to be zero at zero time and the ground acceleration is known from a time history of an earthquake, then this equation may be solved to yield

$$y(t) = \frac{1}{\omega\sqrt{1-\beta^2}} \int_0^t \frac{d^2z(t')}{dt'^2} e^{-(t-t')\beta\omega} \sin\sqrt{1-\beta^2}\,\omega(t'-t)\,dt'. \quad (9\text{-}46)$$

Three quantities are normally deduced from this result. The spectral displacement S_d is the largest displacement occurring during the course of the time history:

$$S_d = \max |y(t)|. \quad (9\text{-}47)$$

The spectral velocity and acceleration S_v and S_a are then defined as the maximum oscillator velocity and acceleration resulting from a sinusoidal displacement wave of amplitude S_d:

$$S_v = \omega\sqrt{1-\beta^2}\,S_d, \quad (9\text{-}48)$$

$$S_a = \omega^2(1-\beta^2)S_d, \quad (9\text{-}49)$$

where $\sqrt{1-\beta^2}\,\omega$ is the damped natural frequency.

Using tripartite graph paper, the spectral displacement, velocity, and acceleration S_d, S_v, and S_a may be plotted versus the oscillator frequency for a number of damping values, as shown in Fig. 9-23. This figure illustrates several interesting points. An oscillator with a low natural frequency (long period) will experience the largest relative displacements, whereas one with a high natural frequency (short period) will experience maximum acceleration. Oscillators with natural frequencies in the intermediate range will experience maximum velocity, which is an indication of the amount of energy transferred to the oscillator by the ground shaking. Damping has the uniform effect of decreasing the response of the oscillator to the ground motion.

The jagged behavior of the response spectrum in Fig. 9-23 is due to the individual nature of the earthquake, with some frequencies having larger components in the ground motion than others. To remove the dependence on the peculiarities of a particular shock, response spectra from several different earthquakes are often calculated and normalized to the same maximum ground acceleration. A smoothed spectrum is then drawn that is an envelope to the spectra for all the earthquakes. One such envelope, used for design work, is pictured in Fig. 9-24.

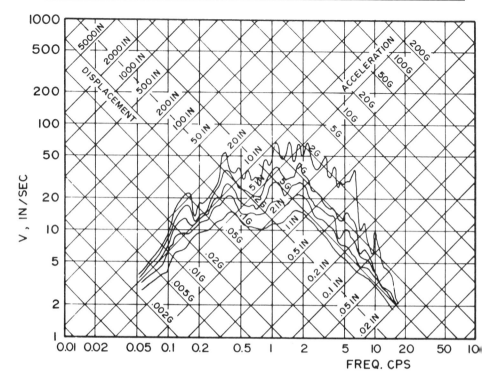

FIGURE 9.23 Response spectra for El Centra earthquake (May, 18 1940, N-S component) for 0, 2, 5, 10, 20% damping. From N. M. Newmark, "Design Criteria for Nuclear Reactors Subjected to Earthquake Hazards," *Proc. IAEA Panel on Aseismic Design and Testing of Nuclear Facilities,* Tokyo, Japan Earthquake Engineering Society, 1967.

Before a nuclear power plant can be built, the seismic characteristics of the site must be determined so that seismic design bases for vital components and structures can be set (U.S. Atomic Energy Commission, 1973c). In the United States a maximum ground acceleration is stipulated along with a response spectrum envelope such as that shown in Fig. 9-24. Earthquakes corresponding to two maximum accelerations are specified: the design basis earthquake and the operating earthquake. The maximum acceleration for the operating basis must be at least one-half of that for the design basis earthquake. It then must be demonstrated that the structures and systems vital to plant safety are designed to function properly in the environment of the design basis earthquake. Safe shutdown and orderly removal of decay heat is thereby assured. The plant must be designed to be capable of continued operation through the lesser, operating basis earthquake.

No fixed formula is used in setting the maximum acceleration for the design basis earthquake. Rather, the judgment of seismologists must be

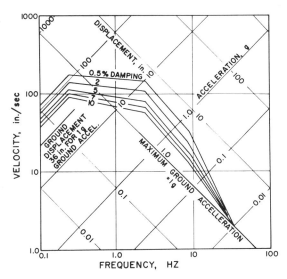

FIGURE 9.24 AEC horizontal design response spectra scaled to 1 g horizontal ground acceleration. Adapted from U.S. Atomic Energy Commission, "Design Response Spectra for Design of Nuclear Power Plants," Regulatory Guide 1.6, 1973.

relied on to take into account a number of factors that may affect a particular reactor site. On the west coast of the United States, where earthquakes are associated with surface faulting, correlations can be made between probable maximum values of the Richter magnitude, the fault length, and other characteristics. The previously mentioned magnitude-distance-intensity correlations may then be applied to estimate the maximum acceleration likely to occur in the bedrock at the site from quakes originating in the various faults in the surrounding area. In the eastern United States, where large earthquakes are less frequent and historically have not been associated with known faults, the estimation of a maximum acceleration is more problematical. The estimate is usually based on the maximum acceleration thought to be associated with the Charleston or New Madrid earthquakes. Whichever method is used to estimate acceleration, local geological conditions must also be taken into account to assure that the amplification of seismic shocks as they pass through surface soils is conservatively predicted.

Seismic Response of Nuclear Plants

Extensive analysis must be carried out to ensure that a nuclear power plant is designed to meet the criteria imposed by the operating basis and design basis earthquakes. The detail with which analysis is made is to a large extent determined by the function that each plant structure serves in preventing the release of radioactive material. To this end, the Nuclear Regulatory Commission classifies the facilities and structures making up a nuclear power plant into three classes:

1. *Class I*—those structures or equipment whose failure clearly might cause or contribute to a nuclear incident; or alternately, those structures or equipment that are believed necessary for containment of fission products and safe shutdown of the plant.
2. *Class II*—structures or equipment essential to the power operation but whose individual failure it is believed would not cause or contribute to a nuclear incident and would not impair the ability for safe shutdown or containment.
3. *Class III*—structures or equipment convenient to the operation of the facility but not essential to safety.

The Class I systems that deserve careful scrutiny and analysis include (1) the reactor core, including the structural support of fuel and control elements; (2) the primary coolant envelope, including pressure vessels, piping, pumps, valves, and heat exchangers; (3) normal and emergency coolant supply systems, including coolant storage and ultimate sources of coolant at the site; (4) control systems for matching the rate of energy release in the reactor core with the prevailing capacity of heat removal systems under normal and emergency modes of operation; (5) secondary containment and other effluent control measures; (6) radiation shielding; (7) fuel storage facilities; and (8) normal and emergency power supplies.

The treatment of the three classes of structures is quite different. For Class I structures a thorough dynamic analysis is made using the most sophisticated methods available, with whatever uncertainty occurs being covered by pessimistic assumptions. On the other hand, Class II and particularly Class III structures would more typically be designed using whatever building, piping, and other codes are applicable. Thus whereas Class I structures are designed using mathematical models that pertain to the particular situations under consideration, the nonnuclear codes used in the design of Class II and III structures provide relatively simple and therefore necessarily inexact rules for the design of large classes of structures. The details of dynamic analysis methods vary a good deal, depending on whether foundations, containment structures, primary system piping systems, or other facilities are under consideration. Moreover, rapid advances are being made in both the use of response spectrum and time history analyses for the various power plant components. Nevertheless, it is useful to sketch some of the principal features that these methods have in common (Briggs, 1970; Pahl, 1970).

Dynamic Analysis. For dynamic analysis the structure under consideration is approximated by a number of characteristic points or nodes where the
mass is assumed to be concentrated. For example, the containment structure
shown in Fig. 9-25a has been discretized as shown in Fig. 9-25b. In general,
each nodal displacement may have 6 degrees of freedom—three translational and three rotational—although symmetry arguments on the geometry
of the structure and/or the components of the earth shaking can usually be
used to reduce significantly the number of degrees of freedom of each node.

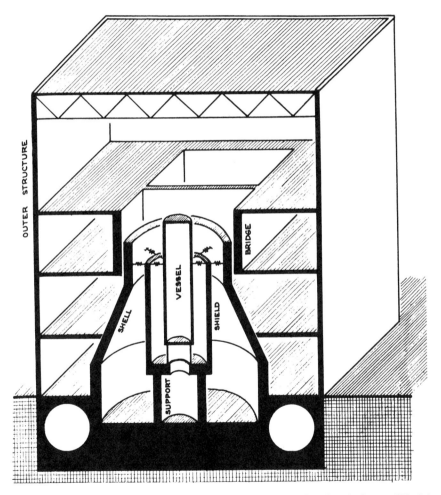

FIGURE 9.25a Perspective of containment structure. Reprinted from "Model
Responses of Containment Structures," by P. J. Pahl in *Seismic Design of Nuclear Power
Plants*, R. H. Hansen, (Ed.), 1970, by permission of the M.I.T. Press, Cambridge, Mass.

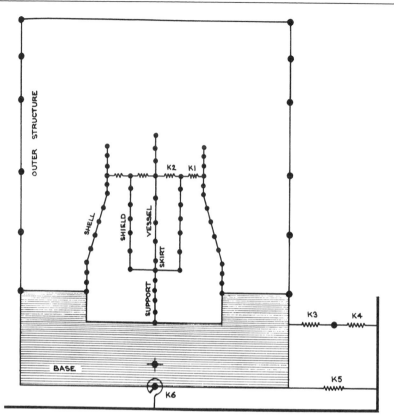

FIGURE 9.25b Model of containment structure. Reprinted from "Model Responses of Containment Structures," By P. J. Pahl in *Seismic Design of Nuclear Power Plants,* R. H. Hansen, (Ed.), 1970, by permission of the M.I.T. Press, Cambridge, Mass.

Each degree of freedom associated with a node is specified as a generalized displacement. For simplicity in illustration we assume only one degree of freedom for each of a total of N nodes.

Each node is joined to adjacent nodes in the structure by the equivalent to a spring, which obeys Hook's law, and a viscous damping force, which is proportional to the relative velocity. The structure then is modeled as a coupled set of differential equations.

$$M_i \frac{d^2 x_i}{dt^2} + \sum_j \tilde{C}_{ij} \frac{d\tilde{x}_j}{dt} + \sum_j \tilde{K}_{ij} \tilde{x}_j = 0, \qquad i = 1, 2, \ldots, N, \qquad (9\text{-}50)$$

where x_i is the absolute displacement of the ith node and $\tilde{x}_i$ is the displacement of the ith node relative to the ground. The absolute displacement is related to the ground displacement $x_g(t)$ by

$$x_i(t) = \tilde{x}_i(t) + x_g(t). \tag{9-51}$$

The M_i are the lumped masses for each node; the $\tilde{K}_{ij}$ are spring constants or stiffness coefficients that connect adjoining nodes, and the $\tilde{C}_{ij}$ are the corresponding damping coefficients. The j summations are taken over the neighboring nodes, which are connected by springs and damping (i.e., along the solid lines in Fig. 9-25b). Hence Eq. (9-50) may be written as

$$M_i \frac{d^2 \tilde{x}_i}{dt^2} + \sum_j \tilde{C}_{ij} \frac{d\tilde{x}_j}{dt} + \sum_j \tilde{K}_{ij}\tilde{x}_j = -M_i \frac{d^2 x_g}{dt^2}, \qquad i = 1, 2, \ldots, N. \tag{9-52}$$

The mass, damping, and elasticity coefficients may be determined in a number of ways, depending on the required accuracy and the resolution of the nodal discretization; they may be determined from beam or membrane theory or through the use of the finite element method. In practice, the damping or dissipation coefficients are the most difficult to obtain, for they often represent linear approximation to nonlinear dynamic processes.

It should be realized that sets of equations such as Eq. (9-52) contain idealizations, some of which can be eliminated in more sophisticated analysis. The linearity of the equations implies that the structure remains in its elastic range where damping tends to be relatively small. Should plastic deformations be considered, not only would the equations become non-linear, but the damping effects would be much larger. Moreover, only simple viscous damping is considered here; modification would have to be made to the equations to treat sliding friction, rigid constraints on motion, and so on, properly. Nevertheless, the simple coupled harmonic oscillation approxima-tion is often adequate, and even where more detailed nonlinear analysis is called for, the linear model can provide a good deal of insight into the behavior of the structure.

Having settled on a set of equations, we would like to determine the maximum displacements, velocities, and accelerations for each node during the course of the earthquake. As stated previously, there are two approaches for doing this. We can directly substitute the ground acceleration time history into the right-hand sides of Eq. (9-52) and numerically solve for the maximum values of $\tilde{x}_i(t)$ and its derivatives. The difficulty with this is that the numerical solution of the equations can be very expensive and often fraught with difficulties resulting from numerical instabilities. Moreover, for relia-bility the calculations must be repeated using several different time histories of the ground shaking.

The second, more widely used method of modal analysis makes direct use of the response spectra envelope established earlier. The first step in this procedure is to establish a set of normal displacement modes and their characteristic frequencies. To do this the following procedure is used. If the damping effects are neglected, which for this purpose is normally a reasonable approximation, then normal modes of oscillation and their natural frequencies are found by setting the source (i.e., ground acceleration) and damping terms equal to zero. Equation (9-53) then reduces to

$$M_i \frac{d^2 \tilde{x}_i}{dt^2} + \sum_j \tilde{K}_{ij} \tilde{x}_j = 0, \qquad i = 1, 2, \ldots, N \qquad (9\text{-}53)$$

If we now look for sinusoidal solutions of the form

$$\tilde{x}_i(t) = X_i \sin \omega t, \qquad (9\text{-}54)$$

we obtain

$$-M_i \omega^2 X_i + \sum_j \tilde{K}_{ij} X_j = 0, \qquad i = 1, 2, \ldots, N. \qquad (9\text{-}55)$$

This will be recognized (Wilkenson, 1965) as just the problem of finding the N eigenvalues ω_n, $n = 1, \ldots, N$, and N corresponding eigenvectors $\{X_{in}, X_{2n}, \ldots, X_{Nn}\}$, $n = 1, 2, \ldots, N$, that satisfy this set of homogeneous linear algebraic equations:

$$-M_i \omega_n^2 X_{in} + \sum_j \tilde{K}_{ij} X_{in} = 0, \qquad n = 1, 2, \ldots, N. \qquad (9\text{-}56)$$

For given values of M_i and $\tilde{K}_{ij}$ these ω_n and X_{in} can be determined by any number of standard numerical techniques. Since the stiffness coefficients are symmetric, $\tilde{K}_{ij} = \tilde{K}_{ji}$, and it may be shown that

$$(\omega_n^2 - \omega_{n'}^2) \sum_i M_i X_{in} X_{in'} = 0. \qquad (9\text{-}57)$$

Therefore if $\omega_n \neq \omega_{n'}$ we have an orthogonality relationship

$$\sum_i M_i X_{in} X_{in'} = 0, \qquad n \neq n'. \qquad (9\text{-}58)$$

Since the X_{in} are solutions to a set of homogeneous equations, we are free to normalize each of the N eigenvectors as we see fit. We normalize them by setting

$$\sum_i M_i X_{in} X_{in} = 1. \qquad (9\text{-}59)$$

We have now determined the characteristic frequencies and the displacement modes (i.e., the eigenvalues ω_n and the eigenvectors $\{X_{in} \ldots X_{Nn}\}$) of the structure. We again neglect the damping coefficients and seek a solution to Eq. (9-52) by expanding the displacements in terms of the eigenvectors

$$\tilde{x}_i(t) = \sum_{n=1}^{N} A_n(t)X_{in}, \qquad i = 1, 2, \ldots, N. \tag{9-60}$$

Substituting this expression into Eq. (9-52) with $C_{ij} = 0$, and utilizing Eq. (9-56) we obtain

$$\sum_{n=1}^{N} M_i \left[\frac{d^2 A_n(t)}{dt^2} + \omega_n^2 A_n(t) \right] X_{in} = -M_i \frac{d^2 \tilde{x}_g(t)}{dt^2} \qquad i = 1, 2, \ldots, N. \tag{9-61}$$

Finally, we multiply this expression by $X_{in'}$ and sum over i to obtain

$$\sum_{n=1}^{N} \left[\frac{d^2 A_n(t)}{dt^2} + \omega_n^2 A_n(t) \right] \sum_{i=1}^{} M_i X_{in} X_{in'} = - \sum_{i=1}^{} M_i X_{in'} \frac{d^2 x_g(t)}{dt^2}. \tag{9-62}$$

Using the orthogonality and normalization conditions of Eqs. (9-58) and (9-59), we then have

$$\frac{d^2 A_n(t)}{dt^2} + \omega_n^2 A_n(t) = -\Gamma_n \frac{d^2 \tilde{x}_g(t)}{dt^2}, \qquad n = 1, \ldots, N, \tag{9-63}$$

where

$$\Gamma_n = \sum_{i=1}^{N} M_i X_{in}. \tag{9-64}$$

The form of these equations indicates that by neglecting damping effects we are able to reduce the structural motion to a superposition of displacement modes each of which has the behavior of a simple undamped harmonic oscillator. To treat damping effects in an approximate manner, we reintroduce model damping factors into the harmonic oscillator equations. Thus Eq. (9-63) becomes

$$\frac{d^2 A_n(t)}{dt^2} + 2\beta_n \omega_n \frac{dA_n(t)}{dt} + \omega_n^2 A_n(t) = -\Gamma_n \frac{d^2 \tilde{x}_g(t)}{dt^2}, \tag{9-65}$$

where β_n is the fraction of critical damping for the nth mode. It is difficult to estimate the modal damping factors accurately. However, approximate values, such as those in Table 9-1, can often be used to approximate the general behavior of various structural types.

We are now in a position to utilize the response spectrum envelope from the preceding subsection, since Eqs. (9-45) and (9-65) are identical in form. With known values of β_n, ω_n, and Γ_n, we can determine the maximum modal displacements, velocities, and accelerations once the ground acceleration is

TABLE 9-1 Damping Values for Seismic Design (Percent of Critical Damping)

Structure or Component	Operating Basis Earthquake or $\frac{1}{2}$ Safe Shutdown Earthquake	Safe Shutdown Earthquake
Equipment and large-diameter piping systems,[1] pipe diameter greater than 12 in.	2	3
Small-diameter piping systems, diameter equal to or less than 12 in.	1	2
Welded steel structures	2	4
Bolted steel structures	4	7
Prestressed concrete structures	2	5
Reinforced concrete structures	4	7

From U.S. Atomic Energy Commission, 1973e.
[1] Includes both material and structural damping. If the piping system consists of only one or two spans with little structural damping, use values for small-diameter piping.

known. Given a maximum ground acceleration, we multiply by Γ_n to obtain the maximum value of the right-hand side of Eq. (9-65). Then for the appropriate frequency and damping, we can read the values of the maximum displacement and velocity from Fig. 9-24 for each of the n modes.

There remains the problem of how to combine the modal displacements to obtain total displacements, since it is the value of $\tilde{x}_i(t)$ and its derivatives, rather than the $A_n(t)$, from which stress levels must be estimated. Only the maximum values of the $A_n(t)$ and their derivatives are estimated in the response spectrum method. Hence Eq. (9-60) cannot be used to evaluate the maximum values of $\tilde{x}_i(t)$ exactly, because the maxima in the various $A_n(t)$ are not expected to occur simultaneously or with the same sign. An upper limit on $\tilde{x}_i(t)$ can be set by noting that

$$\max_{\text{all } t} |\tilde{x}_i(t)| \leq \sum_{n=1}^{N} |X_{in}| \max_{\text{all } t} |A_n(t)|.$$

This, however, leads to gross overestimates of the displacements. A procedure that is more often used entails the assumption that the behavior of the N modes is independent of one another so that the least squares estimate

$$\max_{\text{all } t} |\tilde{x}_i(t)| \approx \left[\sum_{n=1}^{N} X_{in}^2 \max_{\text{all } t} A_n(t)^2 \right]^{1/2}$$

is more appropriate.

Damage Characterization. From the use of response spectra and direct time history methods, a picture can be constructed as to how a nuclear power plant is likely to behave under earthquake conditions. Typically, it is found that quite different natural frequencies and levels of damping are encountered when soils, containment structures, and primary system envelopes are examined. The dynamic effects of the soil as manifested in the rocking and/or translation of the plant foundation are characterized by low natural frequencies and a relatively high degree of damping. The natural frequency of the containment shell tends to be roughly in an intermediate frequency range with moderate structural damping. In contrast, if the primary systems consist of a conventional steel vessel and piping layout, it is likely to have a higher natural frequency with little damping. Thus the effect of ground motion on the structures may originate in different portions of the response spectrum, with differing importance associated with the induced displacement, velocity, and acceleration. For example, systems with high natural frequencies may well experience accelerations in excess of that of the ground motion, whereas low-frequency systems are more prone to larger displacements if insufficient damping is available. One thus might picture the effect of an earthquake on a nuclear plant as a fairly rapid jiggling of the steel primary system superimposed on an intermediate frequency bending of the containment structure, superimposed on a slower rocking of the entire system in the soil.

Obviously, in utilizing the results of dynamic simulation, failure criteria must be set, and calculations must be carried out in sufficient detail for each system to ensure that displacements, velocities, and accelerations do not create stress levels that exceed the failure criteria. In doing this it is essential also to take into account the interactions between the various systems in the plant. The primary system motion, for example, is governed by that of the points at which it is anchored to the containment and not directly by ground motion; if the fundamental frequency of the primary system is higher than that of the containment, it can be shown that the containment motion will significantly affect that of the primary system. The worst case appears when two natural frequencies are so close together that a resonance amplification between the two appears.

Because ground shaking affects all the systems and components in a nuclear plant, failure mechanisms and criteria are varied and complex, often taking into account many phenomena that are not immediately apparent from the dynamic analysis techniques outlined previously. Although a comprehensive exposition on earthquake engineering is beyond the scope of this text, a sketch of a few of the factors that must be considered may be instructive in providing some indication of the nature of the problems.

It is necessary to determine that the reactor core and support structure will retain their important functions. The shaking must not cause fuel elements to drop out of the core or to deflect in such a way that either the heat transfer geometry is destroyed or, in fast reactors, a major reactivity insertion is added. Deflection also must be sufficiently small that there will be no mechanical hindrance to the reentry of the control rods into the core. The effects of shaking on piping systems must be carefully analyzed both for the primary system and emergency piping. Because the mass of piping is small relative to other components (i.e., pressure vessels, steam generator, containment), it may tend to respond as a multiply-attached appendage to the movement of other, more massive systems. This situation can be exacerbated, for example, if pipes are attached to buildings where foundations are moving independently of one another. The reactor protection system must be designed to survive the direct shaking to which it is subjected as well as possible mechanical impact damage from the movement of more massive structures. Finally, containment structures must be able to withstand the shaking either at normal pressure conditions or at their design pressure. Because of the great mass of concrete containment systems, the forces induced in them may be substantially higher than those for steel systems of the same volume.

PROBLEMS

9-1. A gas-cooled reactor has a primary system coolant volume of 50×10^3 ft^3. The average coolant conditions are $T = 700°F$ and $p = 1000$ psia. The primary system is surrounded by a containment volume in which the air is at 80°F and atmospheric pressure.

A loss-of-coolant accident causes the primary system to undergo a isothermal blowdown until the primary system and containment pressure are equalized. No air, however, enters the primary system volume. Using the following data, estimate what the free volume of the containment must be if the containment pressure is not to exceed 2 atm: (a) if the reactor coolant is helium, (b) if the coolant is carbon dioxide.

A		R ft lb$_f$/(lb °R)	C_p Btu/(lb °R)	C_v Btu/(lb °R)
He	4	38.6	1.25	1.66
Air	29	53.3	0.240	1.40
CO$_2$	44	35.12	0.158	1.30

9-2. The coolant in a pressurized-water reactor has an enthalpy of $600\,\text{Btu}/\text{lb}_m$ and occupies a volume of 12×10^3 ft. The reactor is enclosed in a containment structure with a free volume of $2 \times 10^6\,\text{ft}^3$ and an effective heat transfer surface area of $0.2 \times 10^6\,\text{ft}^2$. Suppose the reactor suffers a loss-of-coolant accident that causes the primary system to blow down in 20 sec. During the course of the blowdown 2% of the coolant energy is absorbed from the coolant passing through the steam generations before exhausting to the containment. Estimate the maximum containment pressure.

9-3. In the safety analysis of the Clinch River breeder reactor an accident is postulated in which $9100\,\text{lb}_m$ of fuel occupy $63.85\,\text{ft}^3$ and have a pressure of $978\,\text{psia}$ and a temperature of $5176°\text{K}$ immediately following a nuclear excursion. The fuel vapor bubble then expands according to

$$p = p(0)\left[1.0 - 0.142\left(\frac{V(t)}{V(0)} - 1\right)^{0.52} \right]$$

a. Estimate the work done by the bubble when it expands to atmospheric pressure.

b. Estimate the work done by the bubble when the primary system comes to dynamic equilibrium at 20 atm.

9-4. Suppose that in the preceding problem the reactor is covered by 35×10^4 lb of sodium. At the top of the vessel there is a cover gas volume of $800\,\text{ft}^3$. What is the maximum kinetic energy that can be imparted to the sodium slug above the core? If the slug makes a perfectly inelastic collision with a vessel head weighing 1.0×10^6 lb, what is the maximum energy imparted to the head? If the head were not anchored, how high would it rise?

9-5. A $2440\,\text{MW(t)}$ PWR consists of 180×10^3 lb of uranium dioxide fuel and 36×10^3 lb of zircaloy cladding. The coolant volume in the primary system is $10.5 \times 10^3\,\text{ft}^3$. Twenty-four hours following the normal shutdown of the reactor the primary system has been cooled to a temperature of $120°\text{F}$ and is at atmospheric pressure. At this time a total power failure completely incapacitates the decay heat removal systems. If power is not recovered the coolant first boils off at atmospheric pressure. Suppose that the entire core then heats to the melting temperature of the cladding. The cladding falls from the fuel and the fuel continues to heat to its melting point. When 80% of the fuel has melted the molten uranium dioxide drops into the reactor vessel. Estimate the time required for the coolant to boil off. Then, using the following data, estimate the time from accident initiation until (a) the cladding begins to melt, (b) the fuel begins to melt, and (c) the molten fuel collapses into the vessel. (Assume that no heat transfer away from the core takes place during the heat-up.)

	(Fuel/UO_2)	(Cladding/Zr)
Specific Heat	0.1 cal/g-°C	0.07 cal/g-°C
Density	10.2 g/cm^3	6.49 g/cm^3
Heat of fission	67 cal/g	54 cal/g
Melting temperature	2850°C	1850°C

9-6. Verify Eq. (9-44).

9-7. For seismic analysis a structure in a nuclear power plant is modeled by taking $M_1 = 4.5 \times 10^3$ lb$_f$ sec^2/in., $M_2 = 2.5 \times 10^3$ lb$_f$ sec^2/in., $K_{11} = 10^4$ lb$_f$/in., $K_{22} = 5 \times 10^3$ lb$_f$/in., $K_{12} = K_{21} = -3 \times 10^3$ lb$_f$/in., $\beta_1 = \beta_2 = 5\%$.

a. Determine the natural frequencies of the structure.

b. Estimate the maximum modal displacement for an earthquake with a maximum ground acceleration of 0.1 g and a response spectrum given by Fig. 9-24.

RELEASES OF RADIOACTIVE MATERIALS

Chapter 10

10-1 INTRODUCTION

In the foregoing chapters the nature of the various classes of potential reactor accidents is discussed. Such analysis of real and hypothesized events should lead both to an understanding of the likelihood of their occurrence and to the ability to estimate the resulting damage to fission product barriers. In this final chapter we assume that the damage to fission product barriers and safety systems is known, and examine the paths by which radioactive-materials may escape from the reactor, disperse in the environment, and affect the health and safety of the public.

The task of determining the radiological consequences of reactor accidents is complicated by the fact that there are more than 200 radionuclides produced in fission, and as the result of neutron capture in reactor fuel. These nuclides have a wide spectrum of physical, chemical, and biological characteristics that cause large differences in the modes by which they may escape to the environment and create radiological hazards. The extent to which radioactive materials escape through the various fission product barriers to the atmosphere is most strongly dependent on the volatility and chemical characteristics of each of the radionuclides. The dispersion through the environment to populated areas then is determined primarily by the micrometeorology of the plant site and its surroundings. The determination of the radiation dose and its associated health effects then depends on the half-life, energy, and decay modes of the radionuclides, as well as on their biochemical properties. Finally, the severity of the hazard is strongly affected both by population density and by evacuation, medical treatment, and other public health measures that may be taken to mitigate the radiological consequences.

In what follows we first discuss the mechanisms by which radioactive materials may be released from the reactor fuel and penetrate the primary system envelope to reach the containment. In Section 10-3 we examine the

517

behavior of radioactive materials within the containment structures, placing emphasis on fission product removal systems whose purpose it is to minimize the escape of radionuclides to the environment. Section 10-4 deals with the dispersal of the radioactive cloud emitted from the nuclear plant. In Section 10-5 we discuss the determination of radiological dose, and the correlations between dose and health effects. Finally, we conclude the book with a brief discussion of public health measures that may be employed to reduce the radiological consequences of reactor accidents.

10-2 BEHAVIOR OF FISSION PRODUCTS WITHIN THE PRIMARY SYSTEM

The extent to which fission products and actinides are released from reactor fuel and escape from the primary system is determined primarily by their volatility, with their chemical properties also playing a significant role. Therefore it is convenient to group the elements with significant fission yields by volatility and chemical properties. This is done in Table 10-1, by listing the elements in roughly descending order of volatility. Since the release of actinides may cause a significant health hazard, they are included along with the fission products. For purposes of discussion the elements are broadly separated into three categories: inert, volatile, and nonvolatile. These categories are relevant to the behavior of the elements in oxide or carbide fuels. Group I elements, consisting of the xenon and krypton, remain gaseous even at room temperature. In Group II, consisting of halogens, alkali metals, and the tellurium group, are listed elements that are

TABLE 10-1 Groupings of Radionuclides That May Contribute Significantly to Radioactive Releases From Nuclear Power Plants

Xe, Kr	Noble gases	I	Inert
I, Br	Halogens		
Cs, Rb	Alkali metals	II	Volatile
Te, Se, Sb	Telluriam group		
Ba, Sr	Alkaline earths		
Ru, Rb, Pd, Mo, Tc	Noble metals		
Y, La, Ce, Pr, Nd, Pm, Sm, Eu, Np, Pu	Rare earths	III	Nonvolatile
Zr, Nb	Refractory oxides		

partially volatile at reactor operating temperatures and are likely to be released in significant quantities in the event of fuel melting. When yields, half-lives, and other characteristics are accounted for, iodine and cesium tend to be the most important contributions to the volatile group. In Group III are those elements that are relatively involatile even at fuel-melting temperatures. These would be dispersed significantly only in the event of fuel vaporization or in some cases by chemical reactions. In the event of fuel vaporization the nonvolatile elements would condense in aerosol particles.

A more specific description of the escape of radionuclides from the fuel and primary system is difficult without referring to particular reactor types and to particular classes of accidents. We therefore distinguish between liquid- and gas-cooled reactors and between the two most likely causes of large radioactive releases: loss-of-coolant accidents and destructive nuclear excursions. The loss of coolant most often sets the severest criteria for thermal reactors, whereas destructive power excursions tend to dominate fast reactor release considerations. With this in mind we illustrate the mechanisms for fission product release by considering loss-of-coolant accidents in water-cooled reactors, and destructive nuclear excursions in sodium-cooled fast reactors. We then address these two accident types in fast and thermal gas-cooled reactors.

Loss-of-Coolant Accidents

Fuel elements consisting of pellets of uranium dioxide, or mixed oxides of uranium and plutonium, in tubes of metal cladding are widely used in water-cooled reactors, as well as in several other reactor types. Their thermal-mechanical properties under normal operating conditions are discussed in Section 6-4. Although some of the properties of radionuclide releases from oxide fuel vary significantly, depending on cladding materials and reactor coolant, many of the characteristics are similar. We specifically examine the fission product transport out of the zircaloy-clad oxide fuel presently used in water-cooled reactors under the severe conditions that may follow a loss-of-coolant accident in which emergency core-cooling also is impaired. As a part of the U.S. Atomic Energy Commission Reactor Safety Study (1974b) a detailed study was made of fission product release from fuel elements for severe water-cooled reactor accidents. In what follows we draw heavily on this study to summarize the transport of fission products from oxide fuel.

When progressively more severe water reactor accidents are considered, the fuel may undergo cladding failure, melting, mixing with concrete or metals, and/or vapor explosions, as described, for example, in Section 9-4. As in the Reactor Safety Study (U.S. Atomic Energy Commission, 1974b), it is useful to classify fission product releases into four groups and discuss them separately:

1. *Gap release*—fission product release that takes place when the cladding is ruptured but the oxide fuel remains solid.
2. *Meltdown release*—fission product release when the oxide fuel becomes molten.
3. *Vaporization release*—fission product release that occurs when molten fuel causes vaporization of more volatile materials, such as concrete, causing internal convection and gas sparging in the molten fuel mass to enhance the release.
4. *Oxidation release*—fission product release as the immediate result of a steam explosion in which the fuel becomes finely divided and, in an oxidizing atmosphere, undergoes extensive oxidation.

The gap release component consists of that part of the fission product inventory that is free to escape as a gas or vapor from the fuel element in the event of a cladding rupture. It consists primarily of the activity that has collected in the fuel voids and fuel-cladding gap during normal operation. Because of random variation in cladding strength, core power distribution, and other parameters, a small fraction of the thousands of fuel pins may be expected to fail during normal operation. Continuous decontamination of the coolant steam, however, is an effective means of disposing of the gap release component from these failed fuel elements. Under accident conditions the temperature at which cladding rupture takes place is dependent on the internal gas pressure and mechanical properties of the cladding, as discussed in Section 6-4. For the zircaloy cladding used in water-cooled reactors the rupture temperatures are likely to be between 1400 and 2000°F. The rate at which the fission products escape through such cladding ruptures depends both on the size of the cladding rupture and on the pressure differential across the cladding. High fuel element internal pressures, caused by fission gas buildup and elevated temperatures, and primary system depressurization both increase the pressure differential and accelerate the release rate.

To determine the gap component of fission product release as a result of a failed fuel element, it is necessary to account both for the release of fission products from the oxide fuel into the fuel-cladding gap and for the fission product escape through the cladding rupture. In the absence of melting the release fractions tend to be small, and they are difficult to calculate with precision because they depend strongly on the temperature profile and exposure history of the fuel. Moreover, for short half-life nuclides, the rate of release becomes important, since the fission product may undergo substantial radioactive decay before escaping in the event that the release requires times larger than a half-life or so. Of the fission products that reach the gap, the noble gases, Kr and Xe, flow freely to the rupture point and escape, since they do not interact with other substances. The halogens, I and Br, and to a lesser extent the alkali metals, Cs and Rb, are volatile and form gases in elemental form at failure temperatures. These elements may undergo chemical reactions with cladding and other fission products, however, such as those between iodine and zirconium or cesium, and therefore may be prevented from migrating from their point of origin to the cladding rupture point. It is estimated that only about one-third of these elements present in the gap might escape through the rupture. The volatility of the remaining fission products is small, whether they exist in elemental or oxide form, and therefore their escape fraction from the gap will be substantially smaller, as in the case of the tellurium group, or altogether negligible, as in the case of the nonvolatile fission products. Typical best estimates for the total gap release fraction are given in Table 10-2 for water-cooled reactor fuel elements. The uncertainties in such numbers are large and tend to increase with decreasing volatility; these range from about a factor of 4 for the noble gases to greater than two orders of magnitude for Te, Se, and Sb.

TABLE 10-2 Fission Product Release Source Summary—Best Estimate Total Core Release Fractions

Fission Product	Gap Release Fraction	Meltdown Release Fraction	Vaporization Release Fraction[4]	Steam Explosion Fraction[5]
Xe, Kr	0.030	0.870	0.100	$(X)(Y)$ 0.90
I, Br	0.017	0.883	0.100	$(X)(Y)$ 0.90
Cs, Rb	0.050	0.760	0.190	—
Te[1]	0.0001	0.150	0.850	$(X)(Y)$ (0.60)
Sr, Ba	0.000001	0.100	0.010	—
Ru[2]	—	0.030	0.050	$(X)(Y)$ (0.90)
La[3]	—	0.003	0.010	—

From U.S. Atomic Energy Commission, 1974.
[1] Includes Se, Sb.
[2] Includes Mo, Pd, Rb, Tc.
[3] Includes Nd, Eu, Y, Ce, Pr, Pm, Sm, Np, Pu, Zr, Nb.
[4] Exponential loss over 2 hours with half-time of 30 minutes. If a steam explosion occurs prior to this, only the core fraction not involved in the steam explosion can experience vaporization.
[5] $X =$ fraction of core involved in the steam explosion. $Y =$ fraction of inventory remaining for release by oxidation.

In the event that the fuel melts, the release of fission products becomes much larger. Moreover, the release process may be complex because it is strongly dependent on the motion of the fuel as the melting progresses. Fuel melting first takes place only in the highest power density locations in the core almost on a fuel pellet to pellet scale. As a result, a relatively large surface-to-mass ratio is offered by the melting fuel for the release of fission products. Melting of individual pellets continues as the molten region of the core spreads, but molten fuel elements may then coalesce into a larger mass with a correspondingly smaller surface-to-mass ratio. If melting is taking place within the interior of such a congealed mass, the release of fission products may be inhibited by the time required for them to migrate to the more distant free surface. At the same time, the release is enhanced by the presence of gaseous fission products, since these form bubbles, causing the melt to froth and permit fission products to escape by rising to the surface of the molten mass.

In Table 10-2 are presented the best estimates for the release rates from molten UO_2 fuel. Nearly all the noble gases are predicted to be released as the fuel first becomes molten and while the surface-to-mass ratio remains large. A variable fraction of these gases is likely to be trapped in the molten mass as it enlarges during the latter stage of meltdown, with 10% being considered most probable. The high volatility of the halogens and alkali metals will also cause nearly total release to take place, while substantially smaller fractions of Te, Se, Sb, and the alkaline earths escape the molten mass.

Although Te and Sb are also quite volatile, experimental evidence indicates that in water-cooled reactors extensive reaction with unoxidized zircaloy cladding would tend to retain the tellurium in the melt, even though much of the cladding may oxidize during the meltdown process. The release fraction of the tellurium group is therefore of the same order as the less volatile alkaline earths. The noble metals may volatize as oxides. In an oxygen-deficient environment, however, the release rates are likely to be of the order of only a few percent. The rare earths generally exist in the core as various oxides of low volatility. Experimental evidence indicates that release fractions would be less than a percent and may be as little as 0.01%. Refractory metals readily form stable oxides that are of low volatility, leading to small release fractions for these elements.

Further fission product release from the molten fuel following the collapse of fuel pins into a larger molten mass depends specifically on the conditions encountered in water reactor meltdown accidents both within the primary vessel and then on the containment floor. As the melting progresses, the molten mass of fuel and cladding is likely to make contact with additional materials: coolant—if any remains in the primary system—and structural material. If the core collapses, contact will be made with the reactor pressure vessel, and if pressure vessel melt-through takes place, the molten core may come into contact with the concrete of the containment structure as well as with the containment atmosphere.

The molten core consists of a ceramic. Therefore when it makes contact with more volatile materials, they tend to decompose, and in some cases vaporize and mix with the molten fuel mass. This may result in gas sparging and internal convection. For example, concrete will violently decompose, yielding steam and carbon dioxide, which are likely to pass through the molten core mass, causing the sparging effect. Such processes increase the access of the fission products to the free surface of the molten mass and may also lead to the formation of aerosols containing fission products. Thus the potential for additional fission product release increases. Fission product release taking place under these ill-defined conditions is lumped under the heading of vaporization release; best estimates of these conditions for a water-cooled reactor meltdown followed by pressure vessel melt-through are given in Table 10-2. Of the fission products remaining in the molten fuel debris, gas sparging will remove the highly volatile ones, with the rate of release following an exponential behavior. The 30 minutes value associated with Table 10-2 is a rough average over all fission products. For Xe, Kr, I, Br, Cs, Rb, Te, Se, and Sb, total release may be expected during the vaporization phase. The remaining fission products are of low volatility, and total release should not occur. Although dense aerosols containing these products may be formed, most of these would settle back into the melt. Considering the other limiting conditions on solubility, oxidation, and so on, it is doubtful that more than 1% of the low-volatility elements would be released.

In the event that the reactor meltdown sequence gives rise to a vapor explosion, the UO_2 fuel involved in the explosion will be fragmented and scattered in a finely divided form. If this should take place in an oxidizing atmosphere, as would be the case following a vessel melt-through in a water reactor reactor accident, the uranium dioxide particles will cool and undergo reactions with oxygen to form U_3O_8 at temperatures below 1500°C. This exothermic reaction results in the release of those fission products in the uranium dioxide that are volatile under such conditions. The release of noble gases, halogens, the tellurium group, and the noble metals from that fraction of the fuel involved in the vapor explosion may be considered essentially instantaneous. The estimated release fractions are derived from the last column in Fig. 10-2.

To estimate the fraction of those fission products released from the fuel that will also escape from the primary system, the situation within the reactor vessel at the times the various release components listed in Table

10-2 take place must be known. The gap release takes place at or shortly after the initial blowdown. At this time the core is vapor blanketed and there are relatively high mass flow rates of steam escaping the primary system. Even in the event that the emergency core-cooling system works adequately, most of the gap release will have been completed before the fuel is resubmerged in water.

In the event that the emergency core-cooling system fails to the extent that a core meltdown takes place, the molten fuel will again be exposed to a steam atmosphere, albeit at a higher temperature than during the initial gap release. In these circumstances the flow of steam from the primary system may vary, depending on the size of the break and the extent of the emergency cooling system failure. If a large break is followed by a complete cooling failure, a relatively stagnant steam will exist at the time of meltdown. In this case only about two-thirds of the vapor content of the vessel would expand into the containment following the meltdown. However, for a small-break accident, or one in which a partial emergency cooling system failure results in water remaining in the lower vessel head, even as the core becomes vapor bound and melts, much higher rates of steam boil-off to the primary system will take place.

From the preceding we see that both gap and melt fission product releases are made to steam atmospheres that are in turn vented directly to the containment atmosphere. Therefore the plate-out of fission products from the flowing steam into the internal surfaces of the primary system is the primary mechanism that might prevent their escape to the containment. When the molten core debris drops into water remaining in the lower vessel head, further fission product release may be caused by the steam generated in the event of a vapor explosion. Vapor explosions or sparging of the molten core following vessel melt-through leads to a release directly to the containment atmosphere.

In all the releases, the noble gases will escape the primary system. Likewise, it is found that very little plate-out of the halogens will take place above 500°F, even for quite low steam flow rates; and pessimistically, all these elements must be assumed to escape the primary system. The situation is somewhat more complex for the volatile elements in the alkali metal and tellurium groups. Some plate-out from the gap release is likely to occur during the vessel blowdown when the wall temperatures are below 1000°F. Likewise, if a meltdown takes place, these elements may initially plate out on cooler surfaces. But within a few minutes the surfaces will heat to above 1000°F and the condensed elements will once again vaporize and be swept from the primary system. The behavior of the alkaline earths is similar during meltdown, but with the sweep-out delayed by the requirement of higher surface temperatures before the plated materials will vaporize.

The low volatility of the noble metals and rare earths should cause them to condense within the primary system during the meltdown period. However, there is a strong competition between surface plate-out and the formation of aerosol particles, particularly in the presence of the condensing vapor of other core materials. Because large fractions of these aerosols are likely to be swept from the primary system, it may be questionable to assume significant plate-out of these elements. In summary, we see that if the reactor core melts, the fission products that are released from the molten fuel debris will for the most part also escape from the primary system, regardless of their volatility.

Nuclear Excursions

For liquid-cooled reactors the fission product releases that result from overpower transients, particularly superprompt-critical power bursts, may exhibit characteristics that are quite different from loss-of-coolant accidents. Most importantly, at the time that fuel melting or vaporization is initiated, the reactor vessel is still filled with liquid, and the primary system envelope is intact. The mode of fuel element destruction is also likely to differ from a loss-of-coolant accident. As discussed in Section 6-4, fuel failure following loss-of-coolant events in which the reactor is brought subcritical progresses by cladding melting, then structural failure and fuel melting. In contrast, a sharp power burst may cause fuel to melt or even vaporize before the cladding has reached its melting temperature. The pressure of the molten or vaporized fuel then causes it to rupture the cladding and enter the coolant stream.

For most destructive power burst scenarios in liquid-cooled reactors, the coolant channels are filled either with the coolant or its vapor at the time of fuel rupture. The molten fuel thus is either ejected directly into a liquid-coolant stream or comes into contact with liquid after being swept to the outlet plenum by the expansion and/or flow of the coolant vapor. Experimental evidence indicates that when uranium dioxide is injected into pools of either water or sodium, the molten fuel fragments and freezes in the form of particulate matter (Johnson et al., 1974). With such fragmentation it may be expected that all the noble gases and halogens and large fractions of the alkali metals and tellurium group will escape to the coolant, and only the more nonvolatile elements will remain with the fuel particulate to any great extent.

Following escape to the liquid coolant, the possibility for escape of the fission products to the containment depends on the integrity of the primary system envelope and the partitioning of the radioactive materials between liquid and gas phase within the vessel. If the primary system envelope can withstand the pressure generated by the fuel vapor and fuel–coolant interactions, and if the damage to the core structure is not so extensive as to prevent decay heat removal and lead to a melt-through of the reactor vessel, the radionuclides will be contained within the primary system. If leaks develop in the primary system, the noble gases will escape to the containment. The nonvolatile fission products will remain with the fuel debris, and the escape of the halogens and other volatile fission products will depend on their partitioning between the coolant and the gas phase. For example, Table 10-3 shows representative values for the partitioning coefficients, defined as the ratio of the volumetric nuclide concentration in the coolant to that in the gas phase. Essentially all iodine forms NaI, which has a very large partition coefficient. As a result of the smaller partition coefficient and the relatively large yield of cesium per fission, this radionuclide tends to dominate the volatile release to the gas phase. Even then, with the exception of the noble gases, a large retention of all the fission products is expected in the sodium. They would then become dispersed only with aerosol formation that might accompany a sodium fire.

In present-day power reactors, nuclear excursions that are likely to be of such magnitude as to cause wholesale fuel destruction and primary system leaks, either by blast damage or by reactor vessel meltdown, are thought to be credible only for fast reactors. For these massive core disassembly accidents the mechanism for fission product release may differ significantly from the releases from individual fuel pins discussed earlier. To illustrate, we draw heavily on the design basis disassembly accident for the sodium-cooled Clinch River breeder reactor (Project Management Corp., 1975).

TABLE 10-3 Volume Partition Coefficients of Fission Products in Sodium at 900°K

Element	Partition Coefficient
Cs	6.4×10^2
Rb	1.3×10^3
Sr	2.6×10^6
Ba	6.4×10^7
NaI	1.8×10^5
Te	1.6×10^6

From Abbey, 1972.

If the reactor core is first voided of sodium and then subsequent fuel motion causes a superprompt-critical mass to form and cause an energetic excursion, a high-pressure bubble consisting of a two-phase mixture of liquid and vapor fuel may be created. In considering the dissipation of the radionuclides, the paths to the outer containment shell must be taken into account. In terms of the sodium reactor layout shown in Fig. 9-7 there are two paths. First, the radioactive material may be released to the cover gas above the sodium coolant and then vented to the outer containment through possible leak paths in the vessel head. Second, if a vessel melt-through takes place, the radioactive material will escape to the reactor cavity. If this material then becomes airborne, it may leak from the cavity into the outer containment.

The escape through the first path depends on the dynamics of the fuel vapor bubble, and as these are not well understood, significant uncertainty must be taken into account by employing pessimistic assumption. With the formation of the two-phase bubble, all the fission products vaporize with the fuel, whereas only nonvolatile radionuclides remain in the liquid droplets. The bubble is assumed to expand adiabatically until it comes to equilibrium with the cover gas pressure. It then rises buoyantly and merges with the cover gas. The noble gases, halogens, alkali metals, and tellurium group mix with the cover gas. Adiabatic cooling of the fuel vapor accompanies bubble expansion, causing a substantial fraction of the fuel and nonvolatile fission products to condense and settle into the sodium pool. After mixing with the cover gas, further convective cooling takes place, causing fuel and fission product condensation. The radioactivity in the cover gas then consists of noble gas and a fraction of the volatile fission products—primarily cesium on a mass basis—which remain in gaseous form. The halogens, along with the remaining volatile and nonvolatile fission products, condense with the fuel, forming aerosol particulates. Depending on the particulate size, some of the particles may remain suspended in the cover gas while the remainder sink into the sodium pool.

The foregoing set of assumptions is highly pessimistic, since in reality the fuel vapor bubble is likely to lose a substantial amount of heat by radiation and convection to the sodium and upper core structures as it expands, causing more of the fuel to condense and sink into the sodium pool before it can become suspended as an aerosol in the cover gas region. Likewise, any entrainment of sodium within the bubble, as is likely during the expansion and buoyant rise, will cause further fuel condensation, and more important, the affinity of the halogens and other fission products for the surrounding liquid will tend to trap them in the sodium pool.

The leakage of radionuclides through the vessel head to the outer containment may be expected to be quite small, provided the accident does not exceed the structural design basis. However, if the vessel head is lifted or penetrations through it are opened, essentially all the radioactive material suspended in the cover gas may be expected to escape to the outer containment. If such a large leak should occur, it is likely that it would be accompanied by sodium, which would then burn in the outer containment atmosphere.

As indicated in Section 9-4, if, following the excursion, the decay heat cannot be successfully removed from the core debris within the reactor vessel, molten fuel may melt through the reactor vessel and guard vessel into the reactor cavity. A second possible path for fission product release is then established. There would be no immediate sodium fire because such cavities contain a nitrogen atmosphere. Moreover, if the volume of the cavity is sufficiently small, the molten core debris would still be immersed in a pool of sodium. However, if the molten core debris should melt through the cavity liner and attack the concrete foundation, sparging releases to the sodium pool similar to those described in the preceding subsection may take place. The integrity of the barrier between the cavity and the outer containment would then have to be evaluated in terms of its ability to withstand pressure from the carbon dioxide and steam generated, and from the ensuing sodium combustion that results from the concrete decomposition. The partitioning of the radionuclides between the sodium pool and the cavity atmosphere would then determine the fission product release.

The danger associated with cavity overpressure and sodium fire resulting from concrete decomposition has been eliminated in some reactor designs by placing a suitable core catcher on the cavity floor. For in the absence of concrete decomposition, the fission product inventory available for leaking from cavity to outer containment is greatly reduced. Although some of the noble gases may enter the cavity, it is likely that a preponderance of them will be trapped within the upper reaches of the reactor vessel. The remaining fission products are largely trapped in the sodium pool in the bottom of the cavity. Even if the partition coefficients in Table 10-3 are extrapolated to higher temperatures, less than 1% of any volatile radionuclides will enter the cavity atmosphere. Small oxygen impurities present in the nitrogen atmosphere may cause some sodium combustion. As a result, some aerosol particulate of sodium oxides may contain fission products. The escape of these through small leak paths in the cavity wall would then complete leakage through the second path to the outer containment.

Gas-Cooled Reactors

The fission product transport away from the primary systems of either fast or thermal gas-cooled reactors may be dominated by mechanisms quite different from those in their liquid-cooled ccounterparts. This is due not only to the inability of gas coolants to immobilize large quantities of radionuclides, but to the use of prestressed concrete reactor vessels, and in the case of the high-temperature gas-cooled thermal reactors, to the use of all-ceramic core configurations.

The fuel elements in proposed helium-cooled fast reactors resemble those used in liquid-cooled systems. They consist of oxide pellets in metal cladding. Following cooling failures, the fission product migration from this fuel may be expected to resemble that of water reactors, with a major exception: The helium coolant will not support chemical reactions; thus there is no equivalent to the exothermic metal–water reactions in water-cooled systems. With overheating of the fuel sufficient to cause cladding failures, gap releases to the primary system similar to those listed in the first column of Table 10-2 may be expected. Similarly, if the accident causes gross melting of the fuel, leading to a larger coalesced fuel mass, the melt release fractions of the second column of Table 10-2 present a reasonable estimate. With a fast reactor, however, the meltdown of the core presents the possibility of fuel motion into a superprompt-critical configuration. As described in Section 9-3, the molten or vaporized fuel would be dispersed, creating aerosols of plutonium and fission products. In the absence of a liquid coolant, particulate matter would not be expected to form; rather, the fuel debris would more likely settle into molten pools.

Proposed gas-cooled reactors are housed in chambers of a prestressed concrete vessel. Thus in the absence of a within-vessel core catcher, a general meltdown would eventually cause the molten fuel mass to fall to the floor of the vessel, melt through the thermal shields and insulated liners, and come into contact with the concrete. At this point fission product releases comparable to the sparging releases listed in Table 10-2 are likely to result from the violent spalling and decomposition of the concrete. Further, unlike liquid-cooled systems, the gas coolant provides no trapping mechanisms for the radionuclides. On the contrary, there is concern that such fission products as iodine and strontium that leak from fuel elements during normal operation and plate out on the internal surfaces of the primary system may be desorbed to the coolant gas in the higher temperatures encountered during accidents. Of the fission products released to the coolant, the fraction that escape the primary system is largely determined by the ability of the concrete vessel, its seals, and other fittings to withstand the excessive internal pressure generated under accident conditions. This pressure is likely to become particularly intense if the molten fuel debris attacks the concrete floor of the vessel because of the generation of carbon dioxide and steam, and the resulting possibility of metal–water reactions.

The behavior of fission products in the all-ceramic cores of gas-cooled thermal reactors cannot be inferred by analogy to other reactor types, because the coated fuel particles embedded in the graphite moderator result in quite different thermal behavior and fission product transport characteristics. Even if a primary system depressurization accident is combined with a total loss of forced convection, the release of fission products to the coolant stream would be very slow (Moffette, 1975). The core configuration provides an extremely large heat capacity; the graphite moderator in which the coated fuel particles are encapsulated results in a very slow core heat-up rate from decay heat, following a total loss of flow. The fuel and graphite do not reach the initial fuel temperature until 20 minutes or more following the accident. Moreover, the graphite is not susceptible to melting, as metal cladding structures are. Rather, it initially gains strength and then does not melt, but instead sublimates slowly when temperatures in excess of 6000°F are reached.

In coated fuel particles shown in Fig. 6-13, the fission products are maintained in the carbide or oxide fuel particles by pyrolitic carbon and silicon carbide coatings. The silicone carbide coatings are particularly effective in stopping the diffusion of metal fission products through the graphite. Under normal operating conditions some small fraction of these coatings may be expected to crack, allowing the fission products to diffuse into the surrounding graphite matrix. Under accident conditions, increased fractions of the coated particles may be expected to crack, because of the mechanisms discussed in Section 6-4, and make their radionuclides available for diffusion through the graphite. With the slow heat-up and high-temperature strength of the carbide coatings, however, the release is a gradual process. The behavior shown for iodine-131 in Fig. 10-1a is indicative of the noble gas and halogen release. The strontium-90 curves shown in Fig. 10-1b are indicative of the less volatile fission products. For the metals there is an added delay because the large masses of graphite retain the solid fission products for a substantial period of time.

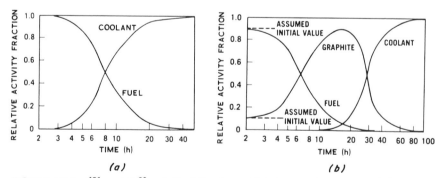

FIGURE 10-1 I^{131} and Sr^{90} activity following a high-temperature gas-cooled reactor accident. (a) I^{131}. (b) Sr^{90}. From T. R. Moffette, "Safety Evaluations of the HTGR as Related to 10 CFR 100 Siting Requirements," *Nucl. Tech.* **25**, 630–635 (1975). Used by permission.

In the event of a primary system breach, the fission products may be assumed to escape to the containment once they have reached the coolant. A particular problem that may arise in graphite-moderated systems involves the ingress of water or air into the primary system when the graphite is at high temperature. The resulting graphite combustion could then lead to increased rates of heat-up and pressure generation. In any event, the entire mass of the reactor core—typically in excess of 3 million lb—would have to reach very high temperatures before the core would sink to the foundation floor. As in the case of the gas-cooled fast reactor, the possibility then exists that the violent decomposition of the floor of the concrete vessel would generate high primary system pressures.

10-3 CONTAINMENT RELEASE

Radionuclides that escape the primary system as gases or suspended particulates enter the containment atmosphere. The fraction of each nuclide that escapes to the environment is then determined by the competition between the rate of containment leakage and the other processes that may remove the radionuclides from the containment atmosphere. As indicated in Section 9-2, the leakage is minimized both by the construction of the walls of the containment shell and by the use of active isolation systems to close nonessential shell penetrations in the event of an accident. Provided the leak rate is small—less than 0.1% volume/day if commonly required at design pressure—much of the radioactive material may be removed from the atmosphere before it can leak out of the containment. The removal mechanisms include radioactive decay and a number of natural depletion processes, including aerosol settling and adsorption to cool surfaces. Active safety features, such as filtered air recirculation or spray systems, often are included within the structures to remove radionuclides more rapidly from the containment atmosphere. Finally, in some multicompartment containments other scrubbing techniques also may be employed.

As in the preceding section, the grouping of radionuclides into a manageable number of categories according to similarity in physical and chemical properties greatly simplifies the modeling of their behavior without resulting in an undue loss of accuracy. To this end, the radioactive materials airborne in the containment atmosphere are often classified into four groups:

1. Noble gases.
2. Elemental iodine.
3. Methyl iodine.
4. Aerosol particles.

The noble gases are unique in that they are inert. The halogens, predominantly iodine, remain volatile under containment conditions, but with a behavior that is strongly dependent on whether they are in elemental form or appear as an alkaline halide. All remaining fission products and the actinides condense at temperatures much higher than those expected in containment atmospheres. They therefore remain airborne only in the form of aerosol particles; consequently, their leakage rate from the containment is largely determined by the size distribution of the aerosol. For a specific accident sequence, an estimate of the primary system release fractions may be used to determine the contribution made by the various radionuclides to each of the preceding four categories. For licensing procedures, however, it often has been found useful to evaluate containment performance and siting conditions in terms of a set of standard primary system release fractions that are large enough to envelope accidents considered to be credible.

For example, in the United States the well-known formula has been established for primary system release following a loss-of-coolant accident in a water-cooled reactor (U.S. Atomic Energy Commission, 1974*a*):

1. One hundred percent of the equilibrium radioactive noble gas inventory developed from full-power operation of the core should be assumed to be immediately available for leakage from the reactor containment.
2. Twenty-five percent of the equilibrium radioactive iodine inventory developed from full-power operation of the core should be assumed to be immediately available for leakage from the primary reactor containment. Ninety-one percent of this 25% is to be assumed to be in the form of elemental iodine, 5% of this 25% in the form of particulate iodine, and 4% of this 25% in the form of organic iodides.
3. One percent of the remaining fission products will be made available in aerosol form.

The relative magnitudes of these release factors are not necessarily appropriate for other reactor types. In sodium systems, for example, very little elemental iodine would be expected to appear, because of the formation of sodium iodide in the primary system. In this situation sodium iodide aerosols as well as those containing cesium and plutonium would tend to become relatively more significant.

The Competitive Rate Equation

Once the radionuclide source terms are specified, the effect of the containment in attenuating the radioactive release to the environment can be addressed. The modeling of multiple-compartmental containment structures can become quite complex, particularly where vapor suppression or other scrubbing techniques are employed. Most of the more important phenomena involving radionuclide behavior, however, can be illustrated by considering a single-volume containment.

We assume that the containment atmosphere consists of a single volume, which may be dry, as in the case of gas- or sodium-cooled reactors, or wet, consisting of a steam–air mixture in the event of a water reactor loss-of-coolant accident. To eliminate spatial effects, we also take the atmosphere to be well mixed. This is a reasonable approximation under accident conditions, since the primary system discharge will be at an elevated temperature, inducing convection currents into the containment atmosphere. With these assumptions the behavior of each of the radionuclide species can be described by a simple competitive rate equation of the form

$$\frac{dC}{dt} = \frac{1}{V_{ct}}\tilde{Q}'(t) - (\lambda^* + \lambda + \tilde{\lambda})C, \tag{10-1}$$

where $\lambda^* = F_{ct}/V_{ct}$,

$C(t) =$ radionuclide concentration (Ci/m^3),

$\tilde{Q}'(t) =$ primary system release rate (Ci/sec),

$V_{ct} =$ containment volume (m^3),

$F_{ct} =$ containment volumetric leak rate (m^3/sec),

$\lambda =$ radioactive decay constant (sec^{-1}),

$\tilde{\lambda} =$ removal coefficient (sec^{-1}).

The removal coefficient may be expressed as a sum of natural depletion and engineered removal mechanisms:

$$\tilde{\lambda} = \sum_j \lambda_j, \tag{10-2}$$

where λ_j is the rate constant for the species resulting from mechanism j.

An exact treatment of the fission product behavior in the containment would require that a large number of radionuclides by tracked by Eq. (10-1). However, as stated previously, the number of species that need to be treated can normally be substantially reduced by grouping them according to the physical and chemical form in which they become airborne. The characteristics of the λ_j that govern depletion from the containment atmosphere can be lumped according to the previously stated four categories: noble gases, elemental iodine, methyl iodine, and aerosol particles.

Because the noble gases are inert, there are no depletion mechanisms, and therefore $\tilde{\lambda} = 0$. The iodine may appear in elemental form, as methyl iodine or as aerosol particles. Although methyl iodine rarely constitutes more than a few percent of a total iodine release, it is considered separately because this alkyline halide is resistant to most removal mechanisms and therefore has a value of $\tilde{\lambda}$ that is much smaller than that of elemental iodine. The depletion characteristics of the iodine attached to aerosol particles, as well as those of the remaining less volatile radionuclides airborne in aerosol form, are determined predominantly by the size distribution of the aerosol.

Although a single depletion coefficient may suffice for each of the four preceding categories, each category contains several radionuclides that contribute significantly. Thus it may be necessary to treat separately several isotopes for the noble gases and for iodine. In many instances, however, the half-lives may be substantially longer than the residence time in the containment. It then may be assumed that the radioactive decay does not contribute significantly to the reduction of C, therefore λ may be equal to zero. Even where the radioactive decay does somewhat affect the result, this approximation provides a pessimistic simplification because it leads to an overestimate of the leakage. With this approximation we are able to sum all the radionuclides within a group into a single composite concentration for purposes of estimating containment leakage. At the other extreme, isotopes with very short half-lives may be neglected because they will decay before they can escape the containment; care must be taken, however, to treat properly any long half-lived daughter products of these isotopes.

The coefficients in Eq. (10-2) often may be expected to vary with time, since they frequently depend on the conditions within the containment atmosphere as well as the operation of safety systems; the parametric dependences of these coefficients are taken up in the following subsections. The containment leak rate F_{ct} is a function of the pressure drop Δp between containment atmosphere and the environment. If $\mathscr{K}$ is an appropriate orifice coefficient, the viscous pressure drop relations from Section 7-2 may be employed to obtain the rough result that

$$F_{ct} = A_l \sqrt{\frac{2\,\Delta p}{m\mathscr{K}}}, \tag{10-3}$$

where A_l is the leak area and m is the density of the containment atmosphere. Finally, the primary system release rate may vary substantially, ranging from a nearly instantaneous puff to a slow leaking of the primary system.

Although refined calculations often require that the implicit time dependence of the rate equation coefficient be taken into account, escape estimates often can be performed by using values that are pessimistically averaged over the time interval of interest. For example, the design leak rate—typically required to be less than 0.1% per day of the containment volume—can be used as a conservative overestimate of F_{ct}. Similarly, the time dependence of the primary system release may be replaced by an instantaneous release of magnitude $\tilde{Q}$ or by a constant release $\tilde{Q}'$ over a time span t_0 such that $\tilde{Q}'t_0 = \tilde{Q}$.

With the assumption of constant rate coefficients, the solution to Eq. (10-1) can be written in simple form. If we assume an instantaneous release of magnitude $\tilde{Q}$ at $t = 0$, then

$$C(t) = \frac{\tilde{Q}}{V_{ct}} e^{-(\lambda^* + \lambda + \tilde{\lambda})t}. \tag{10-4}$$

The leakage $Q(t)$ from the containment then is

$$Q'(t) = F_{ct}C(t), \tag{10-5}$$

the cumulative release to the environment between the accident initiation and time t is then

$$Q(t) \equiv \int_0^t Q'(t)\, dt, \tag{10-6}$$

and therefore for the instantaneous primary system release, Eqs. (10-4) through (10-6) yield

$$Q(t) = \frac{\lambda^*[1 - e^{-(\lambda^* + \lambda + \tilde{\lambda})t}]}{(\lambda^* + \lambda + \tilde{\lambda})} \tilde{Q}. \tag{10-7}$$

If instead we take a constant release rate $\tilde{Q}'$ between $t = 0$ and t_0, we obtain

$$C(t) = \frac{\tilde{Q}'}{V_{ct}} \frac{[1 - e^{-(\lambda^* + \lambda + \tilde{\lambda})t}]}{(\lambda^* + \lambda + \tilde{\lambda})}, \qquad \leq t \leq t_0 \tag{10-8}$$

and

$$C(t) = \frac{\tilde{Q}'}{V_{ct}} \frac{[1 - e^{-(\lambda^* + \lambda + \tilde{\lambda})t_0}]}{(\lambda^* + \lambda + \tilde{\lambda})} e^{-(\lambda^* + \lambda + \tilde{\lambda})(t - t_0)}, \qquad t > t_0. \tag{10-9}$$

These expressions can then be combined with Eqs. (10-5) and (10-6) to estimate $Q(t)$, the cumulative release to the environment.

Natural Depletion Processes

Even in the absence of engineered safety features, the concentration of radionuclides in a containment atmosphere will be depleted by natural processes (Knudson and Hilliard, 1969). For example, if there is an instantaneous release of fission products to the containment of a large pressurized-water reactor, the time dependence of the radionuclide concentrations is likely to appear similar to Fig. 10-2. In this figure no credit is taken for radioactive decay, and the leakage is assumed to be small enough that its effects on the concentration can be neglected. The curves initially follow an exponential behavior that is adequately described by taking the $\lambda_{j's}$ for natural depletion processes as constants. In the case of elemental iodine, saturation effects eventually set in and greatly slow the removal rate. This behavior may be understood by first estimating the initial decay constant and then examining the saturation effects.

Following a loss-of-coolant accident the walls of the containment become covered with a film of water as the result of steam condensation. The

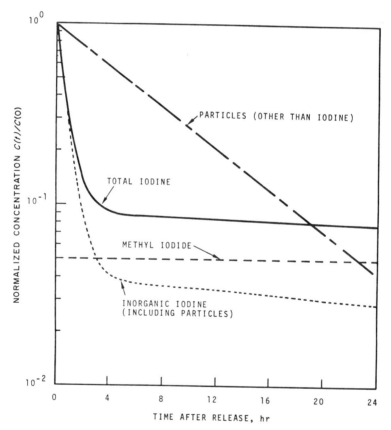

FIGURE 10-2 Prediction of iodine and particle mass concentration in the containment atmosphere of a typical large PWR. From R. K. Hillard and L. F. Coleman, "Natural Transport Effects on Fission Product Behavior in the Containment Systems Experiment," Battelle Northwest Laboratory Report, BNWL-1457, 1970.

elemental iodine is then removed from the atmosphere primarily by absorption into this liquid film, and a smaller amount is also condensed along with the condensing steam. Both of these contributions may be lumped into a single effective mass transfer coefficient;

$$k_e = \frac{\text{curies deposited/m}^2 \text{ of wall/sec}}{\text{curies/m}^3 \text{ of atmosphere}}, \qquad (10\text{-}10)$$

which is the rate at which the iodine is absorbed per unit surface area per second for a unit volumetric concentration in the atmosphere. It therefore has the units of speed, and heuristically it can be viewed as the speed at which iodine crosses the boundary layer to be absorbed on the surface. The parametric dependence of k_e on the convection and steam condensation within the atmosphere may be found elsewhere (Knudson and Hilliard, 1969). For water reactor containments it will have values of the order of 3 m/sec.

The rate constant λ_n for removal by natural processes may be considered to be just the fraction of the containment volume that is swept free of a particular radionuclide per unit time. Thus viewing k_e as a depletion velocity, the volume swept free per unit time is $k_e A_{ct}$, where A_{ct} is the containment surface area. The rate constant then must be

$$\lambda_n = \frac{k_e A_{ct}}{V_{ct}}. \qquad (10\text{-}11)$$

A similar argument can be applied to methyl iodide. Because this compound is unreactive and has very little solubility in water, however, the rate at which it is depleted is much slower. As a result, after several hours it becomes the dominant form of iodine in the containment atmosphere.

Figure 10-2 also indicates that after elemental iodine is depleted by a factor of 10 or more, the depletion rate becomes very small. This occurs when the concentration of iodine in the wall film approaches an equilibrium with the gas phase concentration. After this has occurred, the depletion rate is controlled by much less effective processes, namely, steam condensation and removal by irreversible reactions with the wall surface. Empirical correlations have been used to describe this phenomenon that are too lengthy to report here (Knudson and Hilliard, 1969). Figure 10-2, however, provides the right order of magnitude for this equilibrium effect.

The mechanisms by which radionuclide containing aerosols are eliminated naturally from the containment atmosphere are complex, depending on the temperature and density gradients in the atmosphere as well as agglomerational and gravitational settling. The gravitational settling, however, tends to be the predominant mechanism (U.S. Atomic Energy

Commission, 1974). This may be approximated by assuming that the particles quickly reach a downward terminal velocity given by

$$v_t = (m_p - m)\frac{d^2 g}{18\mu},$$ (10-12)

where

m_p = particle density,

m = atmosphere density,

μ = atmosphere viscosity,

d = particle diameter.

The volume cleared of particles per unit time is then $v_t A_f$, where A_f is the floor area of the containment and the fraction of the total containment volume cleared per unit time is

$$\lambda_n = v_t \frac{A_f}{V_{ct}}.$$ (10-13)

It should be noted that the larger particles settle more rapidly than the smaller ones. Thus if a distribution of particular sizes is to be approximated by an average particle diameter d, this value will decrease with time as the larger particles settle, thus slowing the depletion rate at long times. Typical particle diameters might vary from 15 microns at short times to 5 microns after several hours (U.S. Atomic Energy Commission, 1974b). If dense aerosol releases are considered, however, the depletion rate increases because high aerosol concentrations will favor the formation of larger, faster-settling particles.

Because the predominant mechanism for aerosol removal is gravitational settling, the foregoing arguments are also applicable in the dry containment atmosphere of gas- or sodium-cooled systems. The behavior of iodine in dry atmospheres, however, is not similar to that in steam–air mixtures for the following reason. The absence of a liquid film on the containment walls to dissolve the iodine results in a much smaller value of k_e, the effective mass transfer coefficient. Moreover, any iodine that does adhere to the wall will not be removed by a flowing film of water, but will remain in place, where it may become airborne again, if conditions in the containment atmosphere change adversely. For this reason little reliance can be placed on natural depletion mechanisms in dry containment for those fission products that are not trapped in aerosol particles.

Engineered Removal Systems

The escape of radionuclides to the atmosphere following a major accident can be greatly reduced through the use of active safety systems that remove them from the containment atmosphere. To this end recirculating air filtration systems have been included in the containment systems of reactors of several types and spray systems have been used extensively in water-cooled reactors.

Filtration. The rate constant for removal of a radionuclide by a filtration system can be written as

$$\lambda_f = \frac{F_f E_f}{V_{ct}}, \tag{10-14}$$

where F_f is the volumetric flow rate through the filtration system and E_f is the collection efficiency for the nuclide up to a designated maximum flow rate. High-efficiency filters have been rated to remove more than 99.97% of aerosols and impregnated charcoal filters to remove more than 99.9% of elemental iodine and 85% of methyl iodine (U.S. Atomic Energy Commission, 1974b). Because large quantities of radioactive material are deposited in the filters, decay heat sources may develop that are sufficient to cause excessive filter temperature if adequate cooling is not provided. Overheating must be avoided because it results in deterioration of the efficiency of the particulate filter, and if it results in charcoal filter combustion it may even cause the iodine to become airborne again.

As discussed in Section 9-2, spray systems are frequently placed in the containment of water-cooled reactors to reduce the pressure in the event of a loss-of-coolant accident. The sprays also reduce the concentration of fission products by washing them from the atmosphere. Moreover, the fission product removal can be enhanced by adding chemicals to the spray. We consider separately the effects of spray systems on iodine in gaseous form and on aerosol particles (Hilliard et al., 1971).

Spray Removal of Gases. To formulate a simple model for predicting the removal rate of fission product gases resulting from a spray system, we ignore the effects of leakage, radioactive decay, and natural depletion on the radionuclide concentration in the containment atmosphere. Following an instantaneous release, the rate of elemental iodine removed then may be written as

$$\frac{dC(t)}{dt} = -\frac{F_s}{V_{ct}}[C_f - C_0], \tag{10-15}$$

where F_s is the volumetric flow rate of the spray, C_0 is the concentration of the iodine in the spray droplets as they are emitted from the nozzles, and C_f is the iodine concentration in the droplets as they land on the containment floor.

Assuming that fresh containment spray water is used, we have $C_0 = 0$. Now, if the spray droplets are small enough, and if they remain in the containment atmosphere long enough before falling to the sump, they will eventually come to chemical equilibrium with the atmosphere. The concentration C_f is then equal to $HC(t)$, where H is the volumetric partitioning coefficient, or ratio of liquid to the gas phase volume concentration. This represents the highest rate at which iodine can be removed, however, and is not obtained in practice. Hence we define the spray efficiency E_s as the fraction of this removal rate that is obtained. Thus we have

$$\frac{dC(t)}{dt} = -\lambda_s C(t), \qquad (10\text{-}16)$$

where

$$\lambda_s = \frac{F_s H E_s}{V_{ct}}. \qquad (10\text{-}17)$$

A correlation for estimating the efficiency is given by Parsley (1970):

$$E_s = 1 - \exp\left(-\frac{6k_g t_e}{Hd}\right), \qquad (10\text{-}18)$$

where

H = volumetric partitioning coefficient

k_g = gas phase mass transfer coefficient,

t_e = drop fall time,

d = drop diameter.

The drop fall time may be written approximately as

$$t_e = \frac{z_{ct}}{v_t}, \qquad (10\text{-}19)$$

where z_{ct} is the containment height and v_t is the terminal velocity.

For small values of the efficiency, Eq. (10-18) may be expanded and replaced by the argument of the exponential. We then have

$$\lambda_s = \frac{6k_g z_{ct} F_s}{v_t d V_{ct}}. \qquad (10\text{-}20)$$

Thus for small efficiencies the dependence on the partitioning coefficient vanishes because if the iodine concentration in the droplets is small they act as a perfect sink. The form of the equation may be simplified further. Since the total volume of the spray droplets within the containment at any time is

$t_e F_s$, the number of spherical droplets is

$$N_d = \frac{6 t_e F_s}{\pi d^3},$$ (10-21)

and the total area of the droplets is $\pi d^2 N_d$ or

$$A_d = \frac{6 t_e F_s}{d}.$$ (10-22)

Thus combining Eqns. (10-19) through (10-22), we have

$$\lambda_s = \frac{k_g A_d}{V_{ct}}.$$ (10-23)

Experimentally, it is found that when caustic–borate sprays are used, the exponential decay law breaks down when the concentration is reduced to less than about 1% of its initial value, and the concentration then decays very slowly. At concentrations below this level, back diffusion from liquid films on the containment walls and floor apparently acts to establish an equilibrium between gas and liquid phases.

This equilibrium concentration is estimated through the equilibrium partitioning coefficient

$$H = \frac{C_l}{C_g},$$ (10-24)

where C_l and C_g are the equilibrium liquid and gas phase iodine concentrations. At the initiation of the washout we assume that the entire containment volume is occupied by gas phase iodine at an initial concentration $C(0)$. At equilibrium, films have formed whose volume V_l is much smaller than that of the containment. If the liquid concentration is C_l, we have

$$V_{ct} C(0) \simeq C_g V_{ct} + V_l C_l,$$ (10-25)

or using Eq. (10-24) to eliminate C_l,

$$C_g = \frac{C(0) V_{ct}}{V_{ct} + H V_l}.$$ (10-26)

In practice, C_g represents a concentration below which further iodine removal by the spray will be very slow. If the pessimistic assumption is made that no further reduction below C_g takes place, the time dependence of the elemental iodine in the gas phase may then be approximated by (Hilliard et al., 1971),

$$C(t) = \frac{C(0)}{V_{ct} + H V_l} [V_{ct} + H V_l e^{-\lambda_s t}].$$ (10-27)

This equation indicates that the fraction of elemental iodine that can be rapidly removed from the containment is strongly dependent on the partitioning coefficient. For caustic–boric acid sprays, applicable values are (Postma and Pasedag, 1973) caustic (pH 9.5), $H = 5000$; boric acid (pH = 5), $H = 200$. However, if about 1% boric sodium thiosulfate is added to the spray, a value of $H = 100,000$ may be obtained. With this additive, equilibrium concentrations of less than 0.1% of the initial value result.

Methyl iodine is removed only very slowly when caustic or boric acid sprays are employed, with removal half-lives of many hours. When one weight percent sodium thiosulfate is added to the spray, however, an appreciable increase in performance takes place. For example, in a large containment an increase in rate constants from 0.058 per hour to 0.5 per hour has been observed with the addition of the sodium thiosulfate to a boric acid spray (Hilliard et al., 1971).

Spray Removal of Particulates. Sprays may also be effective in removing particulate matter from the containment atmosphere. The dominant removal mechanism for particles of more than a few microns in diameter is by impaction. Hence for particulate spray removal the rate constant λ_p may be estimated as follows. The fraction of the particulate matter in the containment atmosphere removed per second is proportional to the volume swept out by the falling droplets. A droplet of diameter d falling at terminal velocity v_t will sweep out a volume $v_t \pi d^2/4$ per second, and the total volume swept out per second will be $N_d v_t \pi d^2/4$, where N_d is the number of droplets present. If E_p, the impaction efficiency, is defined as the fraction of particulate matter lying in the path of a falling droplet that is removed from the atmosphere, the removal rate is just E_p times the fraction of the containment volume swept out per second,

$$\lambda_p = \frac{E_p}{4 V_{ct}} N_d v_t \pi d^2, \tag{10-28}$$

or combining with Eq. (10-21),

$$\lambda_p = \frac{3 E_p F_s}{2 d V_{ct}} z_{ct}. \tag{10-29}$$

The collection efficiency of the spray depends in general on both spray drop and particle diameter. For particles with diameters greater than a few microns, where the predominant removal is by impaction, the collection efficiency is nearly independent of droplet size but decreases with decreasing particle diameter as indicated in Fig. 10-3. Thus larger particles are removed preferentially causing the average diameter of the particulate to decrease with time. For particles of less than a micron, Brownian diffusion and

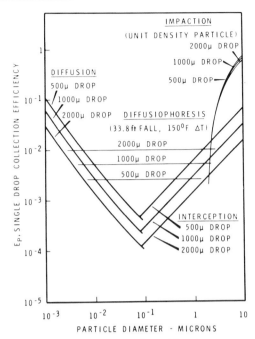

FIGURE 10-3 Aerosol washout calculated for cold-water sprays entering a containment atmosphere saturated at 250°F. From R. K. Hilliard, A. K. Postman, J. P. McCormick, and F. L. Coleman, "Removal of Iodine and Particulates by Sprays in the Containment Systems Experiment," *Nucl. Tech.* **10**, 499–514 (1971). Used by permission.

diffusiophoresis become important contributions to particle removal. Equation (10-29) may still be used to estimate the removal rate provided these effects are taken into account in the collection efficiency, as shown in Fig. 10-3.

In general, the more serious the aerosol release, the more effective the decontamination of particulates by the spray can be expected to be. For with more dense aerosol releases, the distribution of particle diameters may be expected to become larger, thus increasing the removal rate constant.

Multiple-Compartment Containments

A number of the containment configurations discussed in Section 9-2 consist of multiple compartments and therefore are not well represented as a single well-mixed volume. In such situations the competitive rate equation (10-1) must be replaced by a set of such equations that are coupled by inter-compartment flow rates. Examples of such analysis may be found elsewhere

(U.S. Atomic Energy Commission, 1974). Although a wide variety of multiple-compartment configurations is possible, there are three in use that illustrate some of the more important phenomena.

Some containment systems consist of an inner primary containment shell capable of resisting the pressures and temperatures encountered in design basis accidents and a surrounding airtight but non-pressure-resistant building. The secondary containment building is kept slightly below atmospheric pressure so that it does not leak even when the primary containment is leaking at its design basis rate. Rather, the pressure is controlled by exhausting the secondary containment atmosphere to the environment after it has been passed through a highly efficient filter train for fission product removal. The leakage to the environment may be estimated by first computing the leakage from the primary containment as if it were a well-mixed single volume. The resulting effluent then becomes the source term for calculating the radionuclide concentration in the secondary containment building. The escape rate to the environment for each radionuclide species is then

$$Q'(t) = (1 - E_f)F_f C_s, \qquad (10\text{-}30)$$

where E_f is the filter efficiency, F_f the volumetric flow through the filter train, and C_s the concentration in the secondary containment.

In water-cooled reactors, vapor suppression, and ice condenser systems are often used to condense steam and therefore reduce the necessary volume of the containment. These systems—described in Section 9-2—also result in a scrubbing of the primary system release as it passes through the vapor suppression pool or ice beds. For a typical boiling-water reactor it is estimated that the steam–air mixture passing through the vapor suppression pool will be decontaminated by a factor of 100 for elemental iodine, by 2 for methyl iodine, and by 50 to 100 for particulate matter (U.S. Atomic Energy Commission, 1974b). In ice condenser systems, more than 50% of the elemental iodine is removed as the steam–air mixture passes through the ice beds (Malinowski and Picone, 1971).

10-4 DISPERSAL OF RADIOACTIVE MATERIALS

From the discussion of the foregoing sections, we conclude that in the event of a severe accident, noble gases and halogens as well as aerosols of the less volatile radionuclides may escape to the containment atmosphere, and that some fraction of these may leak through the containment and enter the environment. In contrast, the bulk of the less volatile fission products and actinides will remain with the fuel. If, however, there is meltdown of the reactor core, the molten fuel mass may carry some of these radionuclides downward through the containment foundation to be dispersed in the earth underlying the nuclear power plant.

Our task in this section is to describe the mechanisms by which the escaping radionuclides disperse and thereby cause the public outside the reactor site to be exposed to radiation. To determine the doses incurred from the contaminant leakage the key parameter that must be calculated is χ, the atmospheric concentration (curies/m^3) of each radioisotope at the site boundary and at other points where exposure of the general public may take place. With χ known, the whole-body dose from the gamma rays emanating from the passing cloud, the skin dose from the beta particles emitted in the cloud, and the internal doses from inhalation of radionuclides can be estimated. In addition to χ, it often is necessary to determine ω, the ground concentration (in curies/m^2) of each radionuclide deposited by the passing cloud. For this quantity is needed to determine the long-term gamma exposure after the cloud has passed, and in particular to estimate the effects of the doses from ingestion of contaminated food and water.

Nonatmospheric Contamination

For major reactor accidents the atmospheric and ground surface contamination measured by the values of χ and ω represent the dominant contributions to the public radiation exposure. We therefore devote the following subsections to methods for determining these quantities. Two other—albeit less important—mechanisms by which exposure of the public may take place deserve brief discussion. First, in the event of a core meltdown some of the radionuclides associated with the molten fuel debris that has dispersed in the earth under the containment foundation may eventually reach the atmosphere or be leached from the core debris by groundwater movement to cause eventual contamination of water supplies. Second, the quantity $V_{ct}C$ of each radionuclide suspended in the containment atmosphere may cause direct gamma radiation of the surrounding area.

Meltdown accident studies indicate that the radiological hazard caused by the penetration of the molten core debris into the soil beneath the containment foundation would be small if not negligible compared to hazards likely to result from the atmospheric releases of radionuclides (U.S. Atomic Energy Commission, 1974b). All, or nearly all, of the noble gases and halogens will be separated from the core debris before the containment foundation is penetrated. Thus only the less volatile fission products and actinides may be expected to remain with the molten fuel mass as it sinks into the earth.

Although small quantities of noble gases or other volatile fission products may enter the soil with the molten fuel, there is little likelihood that they would cause a significant hazard by escaping to the atmosphere. To cause a hazard they must pass through from tens to hundreds of feet of ground before they reach the surface, and even under pressure this would require several hours even for the radioactive gases in the highest permeability sand soils, and years in lower permeability formations. Moreover, the subsurface groundwater that the fission products would be required to pass through acts as a natural filter bed for elemental iodine and other water-soluble compounds, and aerosols of the less volatile fission products would be effectively trapped by the dry upper layers of the soil (U.S. Atomic Energy Commission, 1974b).

Like the escape of radionuclides through the soil to the atmosphere, the contamination of groundwater appears to result in no significant hazard. Following core debris penetration into the earth, the high heat fluxes emanating from the molten debris tend to dry out the surrounding ground; consequently, there is no immediate contact with the groundwater. As discussed in Section 9-4, the debris will eventually come to rest 50 to 100 ft below the surface and gradually solidify as a glasslike mass. If this glassy mass undergoes fracture, as it cools toward the ambient temperature, then radioactive debris may be leached away with the return of groundwater to the surrounding soil. It is estimated, however, that it would require more than a year before the radionuclides remaining in the glassy debris would begin to be leached away. Even then, modeling of the sorption processes and the subsequent hydrology indicates that the leach rates would be relatively small.

After leaching begins, delays between months and hundreds of years—depending on the radioisotopes—would occur before the contamination could reach off-site bodies of water. Moreover, with the possible exception of strontium-90, the radioisotope concentrations in the contaminated water would be well below the maximum permissible concentrations as specified in 10CFR20 for discharges to uncontrolled waters. In the case of strontium-90, dilution in large water bodies would lead to lower concentrations before a point of human consumption was reached. Moreover, the several years required for the strontium to leach away in significant quantities would allow adequate time for the drilling of wells to monitor the strontium movement, and, if need be, to control the groundwater motion and/or extract and treat contaminated water (U.S. Nuclear Regulatory Commission, 1975).

A second potential source of exposure to the public is the direct gamma ray radiation emanating from the radionuclides suspended in the containment atmosphere. The source strength is the superposition of the number of curies $V_{ct}C$ of each of the gamma-emitting radionuclides. The distance from the containment to the closest uncontrolled boundary of the reactor site normally is enough greater than the containment dimensions that the source may be approximated as a point in estimating direct radiation exposure to the public. The gamma flux intensity (in photons/m^2/sec) received from a point source is a sharply decreasing function of distance. To a first approximation it is given by

$$I = 3.7 \times 10^{10} V_{ct} C \frac{(1 + K\mu r) e^{-\mu r}}{4\pi r^2}, \tag{10-31}$$

where r is the distance from the source. The values of μ, the attenuation coefficient, and K, the empirical buildup factor, are given as a function of gamma ray energy in Fig. 10-4. The relaxation length μ^{-1} or distance required to reduce the intensity by a factor e is about 100 m for typical gamma rays in the energy of about 0.7 MeV.

If fission products are released to the containment, but there is no containment leakage, the dose received by the public comes only as a result

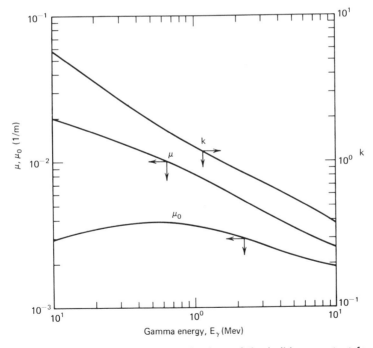

FIGURE 10-4 Absorption coefficients and values of the buildup constant for air at S.T.P. From D. L. Slade (Ed.), "Meteorology and Atomic Energy," U.S. Atomic Energy Commission Report, TID-24190, 1968.

of this direct gamma radiation. If reasonable leakage rates for the containment are assumed, however, the direct exposure tends to be minor compared to the doses received from the resulting radioactive cloud downwind from the reactor site. Because the direct dose drops off much more sharply with distance than the cloud dose, this effect is particularly pronounced for large power reactors whose site boundaries are located at correspondingly larger distances from the containment. As a result, the direct dose is of interest primarily from the standpoint of protecting plant personnel, whereas the exposure to the public is determined primarily by χ, the atmosphere concentration, and ω, the ground surface concentration, of the various radionuclides outside the plant boundary.

Dispersion in an Ideal Atmosphere

When an aerosol or gaseous contaminant is released to the atmosphere, turbulent diffusion caused by atmospheric eddy currents is the dominant mechanism for dispersal. The problem of quantitatively describing such turbulent dispersal in the lower atmosphere in such a way that the concentration of containment can be described as a function of space and time is not amenable to solution by purely analytic methods. Rather, predictions must be made from the semiempirical Gaussian plume model (Slade, 1968). In the Gaussian plume model, functional relationships for the atmospheric contamination are first derived by assuming that the dispersal is governed— as in heat conduction or molecular diffusion—by the classical Fick's law diffusion equation. To obtain tractable relationships these equations are solved by assuming that the coefficients of diffusivity are constants. In the resulting expressions the diffusivity coefficients are then allowed to become dependent on both space and atmospheric stability conditions in order to fit the plume model to experimental data over a variety of weather conditions.

In this subsection we derive the plume model results in an ideal (and nonexistent) atmosphere for which the diffusivity coefficients are constant. In the following subsection the empirical modifications in the resulting expressions are carried out so that estimates can be made of the concentration of radioactive contaminants in the atmosphere and on the ground.

We begin by assuming that the atmospheric concentration of each radionuclide obeys the diffusion equation

$$\frac{\partial \chi}{\partial t} = K_x \frac{\partial^2 \chi}{\partial \tilde{x}^2} + K_y \frac{\partial^2 \chi}{\partial \tilde{y}^2} + K_z \frac{\partial^2 \chi}{\partial \tilde{z}^2}, \tag{10-32}$$

where $\chi(\tilde{x}, \tilde{y}, \tilde{z}, t)$ is measured in curies per cubic meter of a particular radionuclide. We further assume that the eddy coefficients of diffusivity K_x,

K_y, and K_z are independent of space and time and take $\tilde{z}$ as the vertical direction. Now suppose that an instantaneous release of Q curies is made at $\tilde{x} = \tilde{y} = \tilde{z} = 0$, with $t = 0$. To find the resulting distribution χ, we must solve Eq. (10-32) with the initial condition

$$\chi(\tilde{x}, \tilde{y}, \tilde{z}, 0) = Q\delta(\tilde{x})\delta(\tilde{y})\delta(\tilde{z}), \qquad (10\text{-}33)$$

where $\delta(\tilde{x})$, $\delta(\tilde{y})$, and $\delta(\tilde{z})$ are Dirac delta functions. The solution can be found by standard methods (Carslaw and Jaeger, 1959) if we separate variables by requiring the solution to be a product of functions of $(\tilde{x}, t)$, $(\tilde{y}, t)$ and $(\tilde{z}, t)$. The result is

$$\chi(\tilde{x}, \tilde{y}, \tilde{z}, t) = Q\frac{e^{-(\tilde{x}^2/(4K_xt))}}{(4\pi K_xt)^{1/2}} \frac{e^{-(\tilde{y}^2/4K_yt)}}{(4\pi K_yt)^{1/2}} \frac{e^{-(\tilde{z}^2/4K_zt)}}{(4\pi K_zt)^{1/2}} \qquad (10\text{-}34)$$

The coordinate system in this expression has its origin at the point of release, and moves with the net motion of the atmosphere. If we define x as the downwind distance from the point of release, then a constant wind speed $\bar{u}$ can be accommodated in Eq. (10-34) by making the replacements $x = \tilde{x} + \bar{u}t$ and $y = \tilde{y}$. Similarly, if z is defined as the vertical distance from ground level, we may write $z = \tilde{z} + h$, where h is the elevation of the release. Incorporation of these substitutions in Eq. (10-34) yields

$$\chi(x, y, z, t) = Q\frac{e^{-[(x-\bar{u}t)^2/4K_xt]}}{(4\pi K_xt)^{1/2}} \frac{e^{-(y^2/4K_yt)}}{(4\pi K_yt)^{1/2}} \frac{e^{-[(z-h)^2/4K_zt]}}{(4\pi K_zt)^{1/2}}. \qquad (10\text{-}35)$$

To obtain the Gaussian plume model, we make an additional simplification. For purposes of exposure calculations, we assume that the diffusion parallel to the wind direction can be neglected. To do this we take the limit of $\chi(x, y, z, t)$ as $K_x \to 0$; using the properties of Dirac delta functions, we have

$$\lim_{K_xt \to 0} \frac{e^{-[(x-\bar{u}t)^2/4K_xt]}}{(4\pi K_xt)^{1/2}} = \delta(x - \bar{u}t) = \frac{1}{\bar{u}}\delta\left(\frac{x}{\bar{u}} - t\right), \qquad (10\text{-}36)$$

and therefore the remaining factors of Eq. (10-35) can be evaluated at $t = x/\bar{u}$ to yield

$$\chi(x, y, z, t) = \frac{Q}{2\pi\bar{u}\sigma_y\sigma_z} \exp\left[-\frac{y^2}{2\sigma_y^2} - \frac{(z-h)^2}{2\sigma_z^2}\right]\delta\left(\frac{x}{\bar{u}} - t\right); \qquad (10\text{-}37)$$

for brevity the results are expressed in terms of the standard deviations

$$\sigma_y^2 \equiv \frac{2K_yx}{\bar{u}}, \qquad (10\text{-}38)$$

and

$$\sigma_z^2 \equiv \frac{2K_z x}{\bar{u}}. \tag{10-39}$$

This has been referred to as the spreading-disk plume model, since the diffusion now takes place only in the crosswind and vertical directions and the cloud moves as an undispersed unit in the downwind direction (Slade, 1968).

The foregoing relationships do not take into account the presence of the earth's surface and its effects on the cloud concentration. A pessimistic overestimate of the concentration near ground level can be made by using a device from heat conduction problems. The ground is assumed to be a perfect reflector of the contaminant by placing an image source at $(0, 0, -h)$. We then have

$$\chi(x, y, z, t) = \frac{Q\, e^{-(y^2/2\sigma_y^2)}}{2\pi\bar{u}\sigma_y\sigma_z}\left\{\exp\left[-\left(\frac{(z-h)^2}{2\sigma_z^2}\right)\right] + \exp\left(\frac{(z+h)^2}{2\sigma_z^2}\right)\right\}\delta\left(\frac{x}{\bar{u}} - t\right). \tag{10-40}$$

Or, since we normally are interested in the concentration at ground level,

$$\chi(x, y, 0, t) = \frac{Q}{\pi\bar{u}\sigma_y\sigma_z}\exp\left[-\left(\frac{y^2}{2\sigma_y^2} + \frac{h^2}{2\sigma_z^2}\right)\right]\delta\left(\frac{x}{\bar{u}} - t\right). \tag{10-41}$$

Equation (10-40) is useful because it can be easily integrated over time to obtain the quantities of direct interest in making dose calculations at little cost in decreased accuracy. For example, the value of t in the equation is the time elapsed since the release. Thus if a source continuously emits contaminant at a rate $Q'(t)$ curies per second, the instantaneous concentration at (x, y, z) is

$$\bar{\chi}(x, y, z, t) = \int_{-\infty}^{t} \chi(x, y, z, t - t_0)Q'(t_0)\, dt_0. \tag{10-42}$$

For purposes of making experimental studies of plume behavior, a time-independent source Q' is utilized. Thus with $z = 0$ in Eqs. (10-40) and (10-42), we have

$$\bar{\chi}(x, y, 0) = \frac{Q'}{\pi\bar{u}\sigma_y\sigma_z}\exp\left[-\left(\frac{y^2}{2\sigma_y^2} + \frac{h^2}{2\sigma_z^2}\right)\right], \tag{10-43}$$

and in particular along the cloud center line, we may take $y = 0$ to yield at ground level

$$\bar{\chi}_\phi(x) = \frac{Q'}{\pi\bar{u}\sigma_y\sigma_z}\exp\left[-\left(\frac{h^2}{2\sigma_z^2}\right)\right]. \tag{10-44}$$

As a second example, for a puff or instantaneous release it is often desirable to know the cumulative exposure

$$\psi(x, y, z) = \int_0^\infty \chi(x, y, z, t)\, dt \qquad (10\text{-}45)$$

received by an observer at (x, y, z). From Eqs. (10-40) and (10-45) we obtain

$$\psi(x, y, 0) = \frac{Q}{\pi \bar{u} \sigma_y \sigma_z} \exp\left[-\left(\frac{y^2}{2\sigma_y^2} + \frac{h^2}{2\sigma_z^2} \right) \right]. \qquad (10\text{-}46)$$

This equation may also be shown to give the total exposure caused by a continuously emitting source whose time-integrated strength is

$$Q = \int_0^\infty Q'(t)\, dt. \qquad (10\text{-}47)$$

Dispersion in Real Atmospheres

The experimental data from real atmospheres is not well represented by the foregoing expressions so long as the lateral and vertical diffusion parameters σ_y and σ_z are forced to be related to constant diffusivity coefficients by Eqs. (10-38) and (10-39). However, if σ_y and σ_z are viewed as empirical quantities that can be determined by fitting experimental data as a function of x, $\bar{u}$, and the atmospheric stability, the resulting Gaussian plume formulation provides a useful means for estimating the dispersal of radioactive contaminants. In what follows, we first discuss the determination of these quantities and then examine plume rise, building wake deposition, and other corrections that must be made to gain an adequate representation of atmospheric dispersal.

Empirical Diffusion Parameters. The diffusion parameters are governed to a great extent by the stability or mixing characteristics of the atmosphere. The mixing characteristics in turn are closely correlated to the relationship between the vertical temperature gradient or lapse rate in the atmosphere and the adiabatic lapse rate. The adiabatic lapse rate is the rate of temperature decrease with increased altitude that results from the adiabatic cooling of air as it expands with the decreasing pressure of increased altitude. For dry air the adiabatic lapse rate is about $-1.0°C/100$ m.

The effects of various temperature gradients on the stability of the atmosphere are illustrated in Fig. 10-5. In Fig. 10-5a is shown a superadiabatic lapse rate. Under these conditions a parcel of air that rises

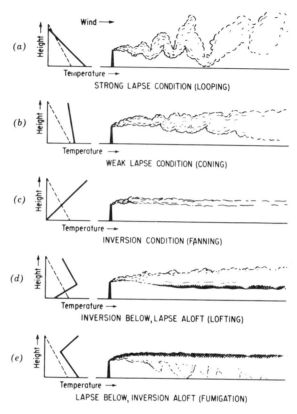

FIGURE 10-5 Schematic representation of stack-gas behavior under various conditions of vertical stability. Actual temperature (solid line) and dry adiabatic lapse rate (dashed line) are shown. From U.S. Weather Bureau, "Meteorology and Atomic Energy," 1955.

and expands adiabatically does not cool as quickly as the surrounding atmosphere. Therefore, with increased altitude, it becomes progressively hotter and less dense than the surrounding atmosphere, increasing the buoyant force driving it upward. Similarly, if the air is sinking, the adiabatic compression caused by the increasing pressure is not sufficient to increase its temperature as much as that of the surrounding atmosphere. Under these conditions the air parcel becomes colder and progressively more dense than its surroundings, tending to accelerate its downward motion. Thus, under superadiabatic conditions all vertical atmospheric motions are accelerated, creating an unstable atmosphere. This instability leads to the looping behavior of effluents illustrated in Fig. 10-5a.

In Figs. 10-5*b* and 10-5*c* are shown progressively more stable atmospheric conditions that arise when the lapse rate becomes subadiabatic (or weak) and when the atmospheric temperature gradient becomes positive. Under the latter conditions an inversion is said to exist. In these situations vertical mixing is hindered, since a parcel of air that expands adiabatically as it rises becomes cooler and more dense than the surrounding atmosphere. Thus the vertical motion is decelerated. Downward moving air parcels also come to rest, since they become progressively hotter and more buoyant than the surrounding atmosphere.

Conditions often arise in which the lapse rate is not constant. Two such situations are shown in Fig. 10-5. In Fig. 10-5*e* a low-altitude inversion causes fumigation. The latter is the most severe atmospheric condition from the standpoint of radiological safety since it results in contaminants being concentrated at ground level some distance from the source even though they may have been released at a substantial height.

The temperature lapse rate in the atmosphere typically undergoes a daily cycle caused by the solar heat absorbed by the earth's crust that then warms the lower atmosphere. During the day the lapse rate tends to be superadiabatic because of this heating; at night the surface cools faster than the atmosphere and therefore creates inversion conditions. The extent to which inversions occur also depends on many other local conditions. Low

TABLE 10-4 Relation of Turbulence Types to Weather Conditions

A—Extremely unstable conditions   D—Neutral conditions[1]
B—Moderately unstable conditions   E—Slightly stable conditions
C—Slightly unstable conditions   F—Moderately stable conditions

Surface Wind Speed, m/sec	Daytime Insolation			Nighttime Conditions	
	Strong	Moderate	Slight	Thin Overcast or $\geq \frac{4}{8}$ Cloudiness[2]	$\leq \frac{3}{8}$ Cloudiness
<2	A	A−B	B		
2	A−B	B	C	E	F
4	B	B−C	C	D	E
6	C	C−D	D	D	D
>6	C	D	D	D	D

From Slade, 1968.
[1] Applicable to heavy overcast, day or night.
[2] The degree of cloudiness is defined as that fraction of the sky above the local apparent horizon which is covered by clouds.

wind speeds and overcast days tend to increase the extent to which inversions occur. Persistent inversions also tend to be more prominent during the winter months. As a result of these factors and others, it is usually necessary to rely on experimental correlations to predict the probability and duration of inversion conditions at a particular site.

For purposes of dispersal calculations (Pasquill, 1962) stability conditions have been correlated to insulation (i.e., solar radiation intensity), wind speed, and cloud cover as shown in Table 10-4. For each of the six stability categories empirical relationships for σ_y and σ_z are given as a function of the downwind distance x as shown in Figs. 10-6 and 10-7. Note that although σ_y and σ_z are no longer proportional to $x/\bar{u}$ as in Eqs. (10-38) and (10-39), they do increase monotonically with x. For making dose rate calculations, it is

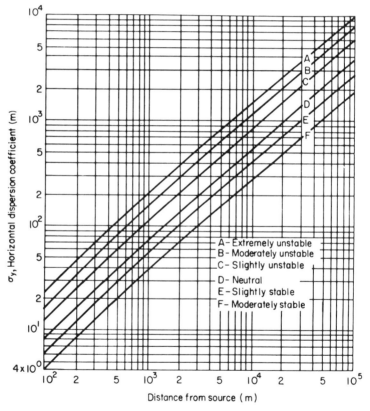

Distance from source (m)

FIGURE 10-6 Horizontal dispersion coefficient σ_y versus downwind distance from source for different turbulence categories. From D. L. Slade, (Ed.), "Meteorology and Atomic Energy, U.S. Atomic Energy Commission Report, TID-24190, 1968.

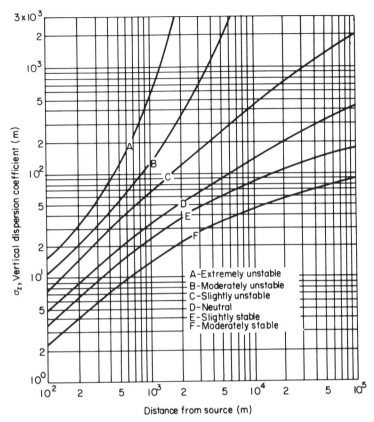

FIGURE 10-7 Vertical dispersion coefficient σ_z versus downwind distance from source for different turbulence categories. From D. L. Slade, (Ed.), "Meteorology and Atomic Energy," U.S. Atomic Energy Commission Report, TID-24190, 1968.

often the normalized ground level concentration along the cloud centerline

$$\frac{\bar{\chi}_\notin}{Q'} = \frac{e^{-[h^2/(2\sigma_z^2(x))]}}{\pi \bar{u} \sigma_y(x)\sigma_z(x)}, \tag{10-48}$$

that is of interest. In Fig. 10-8 this quantity is plotted for the various stability categories for a release height of $h = 30$ m.

Correction Terms. With the semiempirical correlations for $\sigma_y(x)$ and $\sigma_z(x)$ known, the other relationships from the preceding subsection may be useful in evaluating the dispersal of radioactive contaminants. Corrections for several other features of the atmosphere need to be included, however,

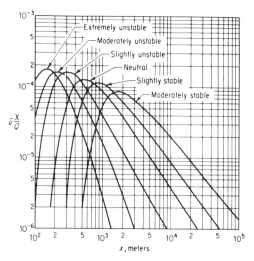

FIGURE 10-8 Normalized ground-level average concentration for an effective source height of 30 m as a function of distance from the source for different turbulence categories. From D. L. Slade, (Ed.), "Meteorology and Atomic Energy," U.S. Atomic Energy Commission Report, TID-24190, 1968.

to extend the applicability of the formulas and in some cases to eliminate unnecessary pessimism from the calculations.

Equation (10-48), which is often used to obtain the maximum off-site dose rates, may yield unrealistically high results if it is multiplied by the time interval to estimate the total exposure. The overestimate occurs because in reality the wind direction fluctuates even over relatively short time spans, and therefore a stationary observer would not remain continually at the cloud centerline. This effect can be roughly taken into account by replacing the Gaussian crosswind distribution by a rectangular or "top hat" distribution (U.S. Nuclear Regulatory Commission, 1975). Equation (10-48) is then replaced by

$$\frac{\bar{\chi}(x, y, 0)}{Q'} = \left(\frac{2}{\pi}\right)^{1/2} \frac{e^{-(h^2/2\sigma_z^2)}}{3\bar{u}\sigma_y\sigma_z}; \qquad |y| < 1.5\sigma_y(x), \qquad (10\text{-}49)$$

This replaces the Gaussian crosswind shape with a square function that has an amplitude within 20% of the peak given by Eq. (10-48).

At short distances from the nuclear power plant, the containment is not approximated well as a point source, since it may leak from a variety of places. Moreover, the presence of the containment and other buildings tends to enhance atmospheric dispersal. Detailed account of these effects require modeling of the reactor site. However, to a first approximation the foregoing relationships can be corrected for this effect by modifying σ_y and σ_z (Slade, 1968):

$$\sigma_y \rightarrow \left(\sigma_y^2 + \frac{cA}{\pi}\right)^{1/2} \tag{10-50}$$

$$\sigma_z \rightarrow \left(\sigma_z^2 + \frac{cA}{\pi}\right)^{1/2} \tag{10-51}$$

where A is the cross-section area of the building normal to the wind and c is a constant between 0.5 and 0.67.

In the event of a reactor accident giving rise to a sudden puff of radioactivity, the escaping gas is likely to be quite hot relative to the surrounding atmosphere. It will then rise buoyantly, having the effect of adding additional elevation to h, the release height. Briggs (1969) has reviewed plume rise correlations and recommended corrections to h for the various stability conditions.

In addition to the foregoing source corrections, there are two depletion effects that tend to reduce the downwind concentration. For short-half-lived radionuclides, the radioactive decay during flight may cause a significant reduction in the concentration. This may be accounted for by multiplying the source in the foregoing relationships by

$$f_r = \exp\left\{\frac{-\lambda x}{\bar{u}}\right\}. \tag{10-52}$$

The second effect, ground deposition, is more difficult to describe quantititatively. At the same time it is often substantially more significant, for deposition can significantly decrease the atmospheric concentration of the contaminant downwind from the reactor, on the one hand, but result in a residual ground contamination of radioactive materials, on the other. With the exception of the noble gases, radionuclides may be removed from the atmosphere by natural processes that are classified as either wet or dry deposition (U.S. Nuclear Regulatory Commission, 1975).

Dry deposition takes place either as a result of sedimentation under the influence of gravity or by impaction with the earth's surface or other obstacles. Sedimentation is not significant for particulate matter smaller than about 15 microns, since the vertical displacements in the turbulent atmosphere are then large compared to the small settling velocity. The mechanisms of dry deposition of gases and particulates are thought to be sensitive to many of the characteristics of the ground surface as well as the meteorological conditions. There has been insufficient experimental study, however, to quantify the deposition rates precisely, and therefore rather simple models are used to envelope the effects.

The rate at which dry deposition takes place is most often represented in terms of a deposition velocity v_d. The contaminant is deposited at a rate $v_d\chi(x, y, 0)$, and the total concentration left on the ground by the passing cloud is then

$$\omega(x, y) = v_d\psi(x, y, 0). \qquad (10\text{-}53)$$

The deposition velocity varies substantially with the character of the surface and atmospheric conditions. A value of 0.001 m/sec has been estimated as a reasonable mean value, but with an uncertainty range from 0.01 to 0.001 m/sec. Obviously, at significant distances downwind, the formula must be modified to take into account the loss of the deposited contaminant from the cloud. It has been suggested that this effect may be accounted for approximately by multiplying the source terms for χ and ψ by

$$f_d = \exp\left[\frac{-v_d x}{(\bar{u}\bar{z})}\right], \qquad (10\text{-}54)$$

where $\bar{z}$, the effective height of the plume, is given by

$$\bar{z} = \left(\frac{\pi}{2}\right)^{1/2} \sigma_z \exp\left(\frac{h^2}{2\sigma_z^2}\right). \qquad (10\text{-}55)$$

Wet deposition or washout from rain also is difficult to quantify in a precise manner. It has been handled rather simply by just assuming that the rate at which radionuclides are removed is $\Lambda_d\chi$, where the wet removal coefficient Λ_d lies between 10^{-2} and 10^{-5} m/sec, depending on the nature of the precipitation. The cloud depletion is then estimated by multiplying the source term by

$$f_\omega = \exp\left[-\Lambda_d(t - t_0)\right], \qquad (10\text{-}56)$$

where $(t - t_0)$ is the time elapsed since the onset of the precipitation (U.S. Nuclear Regulatory Commission, 1975).

The foregoing relationships are meant only to approximate atmospheric dispersion and ground deposition on relatively flat terrain. The proximity to the reactor site of mountains, bodies of water, or other prominent geographical features may preclude the direct application of the Gaussian plume methods to dispersal estimates. In such situations the dispersal properties of the site must be correlated more closely to experimental observation, and indeed in extreme cases adverse meteorological effects may cause a potential site to be eliminated from further consideration.

10-5 RADIOLOGICAL CONSEQUENCES

With the methods described in the preceding section, estimates can be made of the atmospheric and ground surface concentration, χ and ω, for each radionuclide as a function of space and time following severe accidents. Once this is done, the radiation doses to the surrounding population can be determined and the radiological consequences of the accident evaluated.

The radiation doses can be classified as early and chronic exposure depending on the time span over which they are incurred (U.S. Nuclear Regulatory Commission, 1975). The early exposure is due to the doses received from the radioactive cloud as it passes over populated areas, as well as any gamma doses received directly from the radionuclides suspended in the containment atmosphere. The cloud dose includes the external dose received over the whole body from the gamma-emitting nuclides in the cloud, the skin dose received from the beta emitters in the cloud, and the internal dose received from inhaling radionuclides as the cloud passes.

As indicated in the preceding section, the passage of the cloud will lead to ground contamination caused by the settling of radioactive materials from the air. This contamination gives rise to a chronic exposure that may last many years in the case of long-lived isotopes. This exposure may result from a combination of mechanisms that includes an external whole-body dose from the gamma-emitting nuclides on the earth's surface, the dose that results from the ingestion of contaminated food and water, and possibly an inhalation dose from resuspended aerosol particles. It is the early exposure dose that is dominant in determining the amount of radiation illness and the number of fatalities, if any, that will result from an accident. The extent of this exposure is determined not only by the magnitude of the accident and the density of the surrounding population, but also by the emergency measures that are taken to mitigate its effects. In particular, an effective evacuation plan and adequate medical care for the exposed population can greatly reduce the health hazard.

Except for the part of the gamma dose delivered from ground contamination immediately following the passage of the radioactive cloud, the chronic dose to the populace surrounding the reactor site can be reduced to levels not exceeding those recommended by the Federal Radiation Council (1964, 1965), or other appropriate organization, for exposure to the general public. Because the dose is delivered over a long period of time, land can be interdicted, decontamination procedure can be carried out, and contaminated crops and milk can be confiscated to prevent excessive chronic exposures. All this however can be very expensive, and therefore it is primarily the chronic exposure potential that determines the amount of property damage and other monetary expense that is likely to result from a reactor accident.

In what follows we outline the methods for calculating the radiation doses once the air and ground contamination is known. We then examine the health effects of exposure to radiation and consider the consequences that a large—albeit highly improbable—release of radioactivity is likely to have.

Dose Determination

The distribution of radiation damage over the body is strongly dependent on whether the exposure results from uncharged gamma rays or charged beta or alpha particles. As discussed in Section 1-2, gamma rays are penetrating radiation that are attentuated exponentially as they pass through matter. Their relaxation length in tissue is typically the order of tens of centimeters. Therefore whether exposure results from an external gamma source or from an inhaled or ingested radionuclide, it is likely to be fairly uniformly distributed over the entire body. Moreover, in most cases external exposure predominates the total dose from gamma rays.

In contrast, charged particles deposit their energy over very short tracts in tissue. Thus unless beta or alpha particles are inhaled or ingested, they are capable only of causing skin burns or cataracts. If they enter the respiratory or gastrointestinal tract, the biological damage that results will depend on their effects on the individual organs into which they are incorporated. These effects, moreover, are strongly dependent on the biochemical form in which the radionuclides are inhaled or ingested.

We first consider the external whole-body doses that may originate from the gamma-emitting fission products in the containment atmosphere, from those in the radioactive cloud released from the containment, and from those deposited on the ground from the cloud. We then examine the external dose to the skin from the beta emitters in the radioactive cloud, and, more important, the doses to critical organs from inhaled or ingested radioisotopes.

External Dose. The flux of gamma rays at a distance r from a point source is given by Eq. (10-31). To convert this quantity to a dose rate in air we first note that if $\bar{E}_\gamma$ is the energy (in MeV) of the gammas, then $\bar{E}_\gamma I$ is the energy flux (i.e., MeV/sec/m^2). Then, if μ_a represents the energy absorption coefficient for air, given in Fig. 10-4, the energy absorption density is $\mu_a \bar{E}_\gamma I / m_a$ (MeV/g), where $m_a = 1293$ g/m^3 is the density of air (at standard temperature and pressure). Now, noting that there are 1.6×10^{-6} ergs/MeV, and that 1 rad = 100 erg/g, we obtain from Eq. (10-31) for the dose rate in air[*]

$$_\gamma D_r' = \frac{\mu_a Q(3.7\times10^{10})\bar{E}_\gamma(1.6\times10^{-6})(1+K\mu r)\,e^{-\mu r}}{(1293)(100)4\pi r^2}, \qquad (10\text{-}57)$$

where Q is the number of curies in the point source located r meters away. Because of the slightly higher electron density in tissue than in air, this number is multiplied by 1.11 to obtain the tissue dose rate:

$$_\gamma D_r' = 0.0404\mu_a Q\bar{E}_\gamma \frac{(1+K\mu r)}{r^2}e^{-\mu r}. \qquad (10\text{-}58)$$

This expression may be used as a point kernel for determining either cloud or surface contamination doses. We set $r = |\mathbf{r}-\mathbf{r}'|$, the distance between the source location at $\mathbf{r}'$ and the receptor located at $\mathbf{r}$. To calculate the cloud dose rate we replace Q by the concentration $\chi(\mathbf{r}')$ and integrate Eq. (10-58) over the volume V_c' of the cloud:

$$_\gamma D' = 0.0404\mu_a\bar{E}_\gamma \int_{V_c'} \frac{(1+K\mu|\mathbf{r}-\mathbf{r}'|)\,e^{-\mu|\mathbf{r}-\mathbf{r}'|}}{|\mathbf{r}-\mathbf{r}'|^2}\chi(\mathbf{r}')\,dV'; \qquad (10\text{-}59)$$

similarly to determine the dose rate for ground contamination, we integrate the surface source $\omega(\mathbf{r}')$ over the ground area $\mathcal{S}_g$ that is contaminated,

$$_\gamma D' = 0.0404\mu_a\bar{E}_\gamma \int_{\mathcal{S}_g} \frac{(1+K\mu|\mathbf{r}-\mathbf{r}'|)\,e^{-\mu|\mathbf{r}-\mathbf{r}'|}}{|\mathbf{r}-\mathbf{r}'|^2}\omega(\mathbf{r}')\,d\mathcal{S}_g'. \qquad (10\text{-}60)$$

The evaluation of the foregoing integrals over the cloud or surface contamination geometry is a prodigious task, and one which for most cases is not justified because of the lack of precision with which the spatial dependence of χ and ω can be estimated. As a result, most dose determinations are based on the so-called infinite cloud and infinite surface source approximations. These provide suitably pessimistic approximations when proper care is taken in their application. Where additional accuracy is required, correction factors have been derived to approximate the finite size of the radioactive cloud or surface area (U.S. Nuclear Regulatory Commission, 1975; Slade, 1968).

In the infinite cloud approximation the concentration χ is assumed to have the same value everywhere in space as it does at the point $\mathbf{r}$ where the receptor is located. This provides a reasonable approximation to a finite cloud provided the cloud dimensions, as represented by $\sigma_y(x)$ and $\sigma_z(x)$, are large compared to the relaxation length μ^{-1} of the gamma rays in air. For in

---

[*] Hereafter doses in rads are denoted by D and dose rates in rads/sec by D'.

this case most of the gammas passing near point $\mathbf{r}$ originate at points where the contamination concentration is comparable to that at $\mathbf{r}$. Assuming an infinite cloud, the dose may be determined by setting $\chi(\mathbf{r}) = \chi$ in Eq. (10-59) and performing the integration over all space. The dose rate may be obtained more simply, however, from the following energy argument. In an infinite cloud the amount of energy produced per unit volume must be equal to the amount absorbed in air per unit volume. Thus the absorption rate is $\chi\bar{E}_\gamma \times 3.7 \times 10^{10}(\text{MeV}/\text{m}^3)$. Or with $m_a = 1293\ \text{g}/\text{m}^3$ as the air density under standard conditions,

$$\gamma D' = \frac{\chi(3.7 \times 10^{10})\bar{E}_\gamma(1.6 \times 10^{-6})}{(1293)(100)}. \qquad (10\text{-}61)$$

At ground level the dose rate to air amounts to one-half of this value because the cloud only occupies the infinite half-space above the ground level. To obtain the dose rate to tissue we must again multiply the air dose by 1.11 to account for the higher electron densities. Thus the tissue dose at ground level is obtained by multiplying Eq. (10-61) by 0.5×1.11:

$$\gamma D'_\infty = 0.25 \bar{E}_\gamma \chi. \qquad (10\text{-}62)$$

To obtain the total gamma dose for the passing cloud we integrate this expression over time. Thus from Eq. (10-45) we have

$$\gamma D_\infty = 0.25 \bar{E}_\gamma \psi. \qquad (10\text{-}63)$$

Simplifications can also be made in determining the dose resulting from ground contamination by gamma emitters. The simple energy balance argument, however, is no longer applicable. Rather, the assumption is made that the ground concentration ω (curies/m^2) is uniform everywhere over an infinite plane. This approximation is reasonable provided that ω does not change significantly within a few relaxation lengths μ^{-1} in air of the point over which the dose rate is to be determined. With this simplification, the dose rate at a distance b above ground level can be determined by integrating Eq. (10-60) over the infinite plane. We have for tissue

$$\gamma \tilde{D}'_\infty = 0.0404 \mu_a \bar{E}_\gamma \pi \left(\int_b^\infty \frac{e^{-\mu r}\, dr}{r} + K e^{-\mu b} \right) \omega. \qquad (10\text{-}64)$$

The dose rate is obtained by numerically evaluating the exponential integral (Janke and Emde, 1945) and utilizing the expressions for the values of K, μ, μ_a, given in Figs. 10-4. The results are plotted in Fig. 10-9 as a function of $\bar{E}_\gamma$, for $\omega = 1$ (Ci/m^2) at several distances b above ground level. The dose is then obtained by integrating Eq. (10-64) over time. If ω remains about constant for t_e, the time of exposure,

$$\gamma \tilde{D}_\infty = \gamma \tilde{D}'_\infty t. \qquad (10\text{-}65)$$

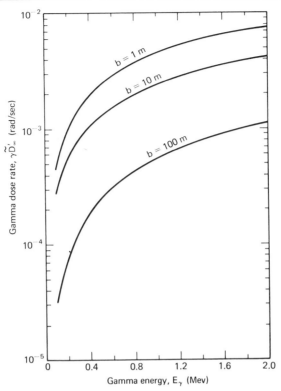

FIGURE 10-9 Gamma dose rate at heights of 1, 10, and 100 m above ground as given for infinite plane source of 1.0 Ci/m². From D. L. Slade, (Ed.), "Meteorology and Atomic Energy," U.S. Atomic Energy Commission Report, TID-24190, 1968.

The foregoing expressions for whole-body gamma doses must be multiplied by the factors f_r and f_ω given in the preceding section to account for the depletion caused by radioactive decay during flight and by upwind settling. Further attenuation of the doses will occur because of partial shielding of the receptor from the cloud or ground concentrations (U.S. Nuclear Regulatory Commission, 1975). For example, the dose rates from clouds are attenuated by buildings by a factor ranging from 0.9 to 0.2, depending on whether light wooden structures or large office or industrial buildings provide the shielding. Likewise, surface roughness and other line-of-sight obstructions can significantly attenuate the ground contamination dose rate given by Eq. (10-64). For example, the results may be multiplied by 0.7 for an open field, whereas for an urban area, buildings and other nearby structures are likely to yield values in the 0.4 to 0.6 range. Remaining indoors provides a further reduction in the ground contamination dose, ranging from 0.04 to 0.5, depending on the nature of the residence.

The range over which the energy from beta particles is deposited is of the order of only a few meters in air. Thus calculating the dose for the atmospheric contamination of beta emitters, little error is introduced by assuming that the cloud dimensions are infinite. The same energy argument applied to the infinite cloud gamma dose thus is applicable, and the beta dose in air is obtained simply by replacing $\bar{E}_\gamma$ by $\bar{E}_\beta$ in Eq. (10-61),

$$_\beta D'_\infty = 0.457 \bar{E}_\beta \chi. \tag{10-66}$$

This expression is for the beta dose in air where media of other absorption characteristics are not present. Except for protruding organs, however, the body will effectively shield the skin surface from the betas emitted by one-half of the infinite cloud. Therefore we have

$$_\beta D'_\infty = 0.23 \bar{E}_\beta \chi. \tag{10-67}$$

In the lower extremities the presence of the earth's surface blocks a larger solid angle of the infinite cloud, and exactly at ground level this dose rate would be reduced by another factor of 2. This refinement most often is neglected, since some cancellation of errors occurs between the overestimate of the cloud dose to the lower extremities and the skin dose to the feet and legs that may result from ground contamination of beta emitters. More accurate treatments of the beta dose to the skin can be carried out (Slade, 1968). However, for reactor accident calculations these beta doses generally do not constitute a significant part of the health hazard.

Internal Dose. Internal radiation doses result from inhalation of radioactive materials contained in the passing cloud, from ingestion of foodstuffs, milk, or water that have been contaminated by the deposition of radioactive materials, or in some cases by inhalation of resuspended particulate matter following the passage of the cloud. Of these the cloud inhalation dose dominates the early internal exposure. Inhalation doses result from radionuclides deposited in the lungs and passages of the upper respiratory tract. The gastrointestinal tract may also be exposed from isotopes removed from the respiratory tract by ciliary motion and then swallowed. Some of the radionuclides in the respiratory or gastrointestinal tracts are absorbed into the bloodstream and deposited in other organs. The deposition of the inhaled contaminant is determined by its biochemical properties and particularly by its solubility in body fluids. Most often there is one critical organ that suffers the most radiological damage as a result of its selective absorption of the nuclide—the thyroid for iodine, the bone for strontium, and so on.

The following model may be used to estimate the dose received by any organ from internal irradiation. Suppose that at some time $t = 0$, Q curies of a radioisotope are inhaled, and a fraction f of the inhaled nuclide is absorbed in the organ under consideration. Then the rate at which energy is dissipated in the organ immediately thereafter is $3.7 \times 10^{10} Qf\bar{E}$ (MeV/sec), where $\bar{E}$ is the average energy per disintegration that is deposited in the organ. Immediately following the deposition of the isotope in the organ, the dose rate is

$$_lD'(0) = \frac{Qf(3.7 \times 10^{10})\bar{E}(1.6 \times 10^{-6})}{100 \, M_o}, \qquad (10\text{-}68)$$

where M_o is the organ mass in grams.

The value of $\bar{E}$ can be taken as equal to the average energy per disintegration for alpha and beta emitters, since the ranges in tissue over which their energies are deposited are small compared to the dimensions of most organs. For gamma rays, however, the exponential relaxation length is of the order of 50 cm in tissue, and therefore only a small part of the emitted energy is absorbed in the organ. As a result, if mixed gamma-charged particle radiation is emitted, the gamma contribution to the dose can be ignored for many small organs. The organ dose rate does not continue at the rate given by Eq. (10-68) for two reasons. First, the radioactive decay of the isotope causes $_lD'(t)$ to decrease as $\exp(-\lambda t)$ where $0.693/\lambda$ is the half-life of the radionuclide. Second, the isotope eventually is eliminated from the organ by biological processes. The biological elimination is most often approximated by exponential decay, $\exp(-\lambda_b t)$, where $T_b = 0.639/\lambda_b$ is the so-called biological half-life of the nuclide in the organ. Multiplying Eq. (10-68) by these factors for radioactive decay and biological elimination, we have

$$_lD'(t) = 592\frac{\bar{E}Qf}{M_o}e^{-(\lambda+\lambda_b)t}. \qquad (10\text{-}69)$$

Thus the dose received over a time interval τ following the inhalation is

$$_lD_\tau = 592\frac{\bar{E}Qf}{M_o}\frac{[1 - e^{-(\lambda+\lambda_b)\tau}]}{\lambda+\lambda_b}; \qquad (10\text{-}70)$$

the nature of the dose received depends on the effective half-life

$$T_{1/2} = \frac{0.693}{\lambda+\lambda_b}. \qquad (10\text{-}71)$$

If the $T_{1/2}$ is of the order of days or weeks, the organ dose is delivered over a relatively short period of time, and therefore it is

$$_lD_\infty = \frac{592\bar{E}Qf}{M_o(\lambda+\lambda_b)} \qquad (10\text{-}72)$$

that is most closely correlated to health effects. On the other hand, if the value of $T_{1/2}$ is long, the dose may be delivered over many years at a nearly uniform rate:

$$_ID' = \frac{592\bar{E}Qf}{M_o}. \tag{10-73}$$

As examples of these two extremes, iodine-131, taken up by the thyroid, has a radioactive half-life of 8.05 days, and therefore the dose to the thyroid is computed from Eq. (10-72). In contrast, strontium-90, absorbed in bone, has a radioactive half-life of 28 years and remains essentially in place over a period of years. Therefore $\lambda + \lambda_b$ is small, and the body burden of strontium-90 results in a chronic exposure over a period of years.

To calculate the organ dose received by an organ as a result of a particular radionuclide in a passing cloud, Q, the number of curies inhaled, must be evaluated. The rate at which the nuclide is inhaled is just equal to

$$Q'(t) = B'\chi, \tag{10-74}$$

where the volumetric breathing rate B' for an average person is about 0.833 m^3/hr averaged over a 24 hr period or 1.25 m^3/hr averaged over an 8 hr workday. The total number of curies inhaled is obtained by integrating this expression over time. If the breathing rate can be approximated by $\bar{B}'$, an appropriate time-averaged rate, we have

$$Q = \bar{B}'\psi \tag{10-75}$$

for the number of curies inhaled. The total dose in Eq. (10-72) is a function only of the number of curies inhaled and not of the time elapsed since inhalation. Thus Eqs. (10-72) and (10-75) can be combined to yield

$$_ID_\infty = 592\frac{\bar{B}'\bar{E}f\psi}{M_o(\lambda + \lambda_b)}. \tag{10-76}$$

This expression is useful if the total dose to the organ is delivered over a relatively short period of time. However, if the effective half-life is long, so that the dose following exposure is delivered over a period of several years, the value of $_ID_\tau$ is required to evaluate the acute health effects. The definition of $_ID_\tau$ in Eq. (10-70) presumes that the entire exposure takes place at $t = 0$. However, if the inhalation takes place over some period of time as indicated by Eq. (10-74) then the dose for the first τ years following the initiation of exposure at $t = 0$ may be shown to be given by

$$_ID_\tau = \int_0^\tau 592\frac{\bar{E}'Q'(t)f}{M_o(\lambda + \lambda_b)}[1 - e^{-(\lambda + \lambda_b)(\tau - t)}]\, dt. \tag{10-77}$$

For long-half-lived isotopes the duration of the exposure is likely to be small compared to the effective half-life. In this situation the foregoing expression is adequately approximated by Eq. (10-70) following passage of the cloud. Therefore combining Eq. (10-70) and (10-75) we have

$$_ID_\tau = 592 \frac{\bar{B}'\bar{E}'f\psi}{M_o(\lambda + \lambda_b)}[1 - e^{-(\lambda + \lambda_b)\tau}].$$ (10-78)

Internal organ doses from ingestion of contaminated crops, milk, or water are estimated from Eqs. (10-68) through (10-73) in a manner similar to those for inhalation. The factor f must be determined by uptake from the gastrointestinal tract. The estimate of the number of curies ingested, however, is more difficult since it depends on the whole ecological chain by which ground contamination is incorporated into foodstuffs (Eisenbud, 1973).

The foregoing expressions permit the evaluation of the dose to any organ from a particular radionuclide in terms of χ, the cloud concentration, ω, the ground concentration, and the amount of the radionuclide ingested. There are, however, some 54 different radionuclides produced by neutron fission and capture that may contribute significantly to one or more of the dose components. Thus the preceding expressions must be summed over these radionuclides. Fortunately, health dosimetry models have been developed in which the organ masses, deposition fractions, and other parameters already have been evaluated for a "standard" man, and the relationships between dose, χ, and, ψ, or Q numerically evaluated for the radionuclides.

In Appendix A data are tabulated for the dosimetry models used in the Rasmussen Study of Reactor Safety (U.S. Nuclear Regulatory Commission, 1975). The radionuclides, their inventory in a 3200 MW(t) water-cooled reactor, and their half-lives are listed in Table A-1. In the subsequent tables it is assumed that cloud depletion by radioactive decay and settling as well as shielding effects have already been accounted for in the computation of χ, ψ, ω. The results, in rads, have been multiplied by the appropriate quality factor to place the biological damage from alpha, beta, and gamma emitters on an equal footing. Therefore the results are given in terms of the dose equivalent or rem.

In Table A-2 are given the values of $_\gamma D_\infty/\psi$ from Eq. (10-63) for the calculation of the gamma cloud dose. In Table A-3 are given values of $_\gamma \tilde{D}'_\infty t/\omega$ from Eq. (10-64) for calculating the external gamma dose from ground contamination over varying time periods τ. In Table A-4 are tabulated values of $_ID_\tau/Q$ from Eq. (10-70), for the doses per curie inhaled received over the time period τ following inhalations. When used in conjunction with Eqs. (10-74) and (10-75) these values permit the inhalation cloud doses to be evaluated. Finally, in Table A-5 are given the doses per curie ingested for those radionuclides that are most likely to enter the body as a result of contaminated food or water.

Health Effects

The health effects of radiation exposure can be divided into three categories: early somatic effects, late somatic effects, and genetic effects. The early somatic effects encompass the illness and possible early fatalities that result from large doses of radiation. As outlined for large whole-body doses in Chapter 1, these effects appear soon after, if not immediately following, the radiation exposure and last for a period of from a few days to months, depending on the severity of the exposure. Following recovery from such radiation sickness, and even for exposures that are too small to produce detectable early somatic effects, the recipient of radiation dose is subject to late somatic effects. These take the form of an increased risk of contracting various forms of cancer over a prolonged period. The cancer risk rises with the radiation exposure. Finally, radiation exposure results in an increased number of mutations that may be passed on as genetic defects in future generations.

Early Somatic Effects. In situations where only external gamma radiation is involved, it is appropriate to estimate somatic effects on the basis of the whole-body dose, since all organs are then subjected to approximately the same exposure. Following reactor accidents, however, this assumption may not be valid since each organ is likely to be subjected not only to an external whole-body dose but also to an internal dose from inhalation of radionuclides. The internal dose, moreover, is likely to vary greatly from organ to organ and to depend strongly on the combination of inhaled radionuclides. For this reason, often it is necessary to evaluate somatic effects by superimposing the inhalation and whole-body doses to each of the radiosensitive organs.

An extensive analysis of dose–mortality relationships for the major somatic effects has been carried out in conjunction with the Reactor Safety Study (U.S. Nuclear Regulatory Commission, 1975), and we draw heavily from the results in our discussion of these effects. Following a catastrophic reactor accident, there are doses to three organs that are likely to contribute significantly to the fatalities from early somatic effects: the bone marrow, the lungs, and the gastrointestinal tract. With each of these, there is a lower dose below which the probability of death is negligible. The likelihood of fatality then increases with dose until an upper value is reached above which death is certain. Often quoted is the LD_{50}, which is the dose that is expected to be lethal to 50% of the exposed population. Doses to the bone marrow and other blood-forming organs are the most likely to cause fatalities. Without medical attention some fatalities begin to occur at 150 rems, and above about 600 rems death is certain. The LD_{50} dose is approximately 300 rems for the bone marrow. Fortunately, blood transfusions, antibiotics, and other supportive medical treatments substantially raise the doses required to produce fatalities. For example, the LD_{50} bone marrow dose is estimated to rise to 510 rems if supportive treatment is received. If "heroic" treatment involving transplants of bone marrow is available, the LD_{50} dose is estimated to be 1050 rems.

Substantially higher doses to the lungs or gastrointestinal tract are required to produce fatalities. For the lungs, the likelihood of fatalities at doses below about 400 rems is small; the LD_{50} is 2000 rems, and death becomes certain only above 2500 rems. For the gastrointestinal tract the lower limit for fatalities is at about 2000 rems, the LD_{50} is 3500 rems, and the death of all the exposed population receiving more than 5000 rems is expected.

Comparing the LD_{50} doses, it is seen that fatalities as a result of whole-body doses are predominantly the result of damage to the bone marrow and blood-forming organs. This is true even with the availability of excellent medical follow-up, since bone marrow LD_{50} is still at least a factor of 2 lower than for the lung or gastrointestinal tract.

Under accident conditions both external and internal doses may contribute to mortalities. For the catastrophic events considered in the Reactor Safety Study, it was found that about two-thirds of the early exposure of bone marrow is caused by external gamma radiation, with the remainder resulting from the inhalation dose. In contrast, the largest part of the exposure to the gastrointestinal tract is contributed by the internal dose, as is the overwhelming part of the dose to the lungs. It is nevertheless found that the bone marrow dose always leads to the dominant risk of early mortalities. In particular, it is found that irradiation of the thyroid or other organs is insignificant in predicting the fatalities, since the LD_{50} doses are so high that it would be impossible to receive a lethal dose to the thyroid or other organs as the result of a reactor accident without also being exposed to a lethal dose to the bone marrow, lung, or gastrointestinal tract.

A population exposed to sublethal doses of whole-body radiation is likely to suffer a number of temporary effects as outlined in Section 1-2. Among these may be temporary sterility, prolonged vomiting, and the like. If large doses to a single organ are received a number of morbundities requiring medical attention may result. These include cataracts, hypothyroid, gastrointestinal problems, and respiratory impairment. However, it is unlikely that an exposure high enough to induce these effects would be received as a result of a reactor accident without the individual also receiving a lethal whole-body dose. The possible exception to this is respiratory impairment occurring at lung doses greater than 1000 rems. Fetuses are particularly sensitive to radioactivity. It is estimated that the LD_{50} dose to a fetus increases with time during pregnancy. After two weeks of pregnancy the LD_{50} is estimated to be 100 rems; after one month, 200 rems. For sublethal doses in this range, however, during the early stages of pregnancy there is an increased risk of congenital malformations and growth retardation (U.S. Nuclear Regulatory Commission, 1975).

Late Somatic Effects.   Following irradiation, the exposed population will be subject to an increased risk of cancer and the formation of thyroid nodules. Because these effects are most likely observed between 2 and 30 years following the irradiation, they are referred to as late somatic effects. The thyroid nodules and latent cancer are random phenomena whose probability for an individual is a function of the dose received. However, the cases that occur cannot be directly differentiated from those occurring spontaneously. Therefore late somatic effects appear only as increases in the normal incidence of cancer and nodules for an extended period of time following a large release of radioactive material.

Estimates made for the increased incidence of cancer caused by chronic radiation exposure in the BEIR report (National Academy of Sciences, 1972) have been revised to represent better the situation stemming from a single large release of radioactive materials (U.S. Nuclear Regulatory Commission, 1975). Upper-bound estimates are made on the basis of the linear hypothesis. That is to say, the probability of contracting cancer is assumed to be proportional to the dose. With this assumption the results are thought to represent quite pessimistic overestimates when populations exposed to fewer than tens of rems are included. With the linear hypothesis the number of cancers of a given type resulting from an accident may be calculated for the population dose, that is, from the total number of man-rems resulting from the accident. In Table 10-5 are given the number of cancer fatalities expected per million man-rems of exposure of the general population.

Estimates such as those in Table 10-5 are based primarily on doses in the 10 to 300 rems range that have resulted from the detonation of nuclear weapons. Evidence from animal experiments indicates that if either the dose rate or the total dose is lower, the risk per man-rem may decrease significantly and may possibly vanish below some threshold. In the event of a catastrophic reactor accident only a small fraction of the total exposed population would be expected to receive doses of several tens of rems or more, while a much larger number of people would receive substantially smaller doses. Thus the use of the linear hypothesis with no threshold provides a quite pessimistic overestimate of the number of cancer fatalities resulting from a reactor accident. Multiplication corrections for the results in Table 10-5 have been proposed to provide what is thought to be a more

TABLE 10-5 Expected Latent Cancer (Excluding Thyroid) Deaths per Million Man-Rem of External Exposure

Type of Cancer	Expected Deaths per 10^6 Man-rem
Leukemia	28.4
Lung	22.2
Stomach	10.2
Alimentary canal	3.4
Pancreas	3.4
Breast	25.6
Bone	6.9
All other	21.6
Total (excluding thyroid)	121.6

From U.S. Nuclear Regulatory Commission, 1975.

realistic picture of the cancer risk where lower doses and/or dose rates are predominant. These are listed in Table 10.6.

The latent effects of thyroid exposure deserve special attention because of both the large inventory of radioiodine in power reactors and the characteristics of thyroid cancer. The linear hypothesis may also be pessimistically applied to the thyroid dose using the numbers in Table 10-7. Note that the formation of nodules is more likely in children than in adults. However, the fraction of nodules that become malignant is smaller for children, and as a result the incidence of cancer is roughly the same per man-rem for children and adults.

TABLE 10-6 Dose–Effectiveness Factors

Total Dose (rem)	Dose Rate (rem per day)		
	<1	1–10	>10
<10	0.2	0.2	0.2
10–25	0.2	0.4	0.4
25–300	0.2	0.4	1.0

From U.S. Nuclear Regulatory Commission, 1975.

TABLE 10-7   Thyroid Nodules per 10^6 Persons per Rem per
                Year

	Benign	Cancerous	Total
Children (<20)	8.1	4.3	12.4
Adults	4.0	4.3	8.3

From U.S. Nuclear Regulatory Commission, 1975.

The results in Table 10-7 are for external radiation. Experimental evidence indicates that for internal radioiodine doses, the incidence of both benign and malignant nodules is at least an order of magnitude smaller. Thus a pessimistically conservative assumption is to divide the results of Table 10-7 by 10 for inhalation doses. The results only hold for thyroid doses below 1500 rems. Above this value, the incidence of both benign and malignant nodules declines until above 5000 rems no nodules appear. At these high doses the thyroid tissue is ablated, thus precluding the formation of benign or malignant nodules. It is also important to point out that survival rate for thyroid malignancies is well in excess of 90% because they are well differentiated and slow growing, and thereby amenable to surgical removal or other therapy. Moreover, periodic screening for early nodule detection and removal may be expected to further reduce the likelihood of mortalities from thyroid cancer in the exposed population.

Genetic Effects.   In addition to the early and late somatic effects discussed above, radiation damage to the reproductive cells may produce adverse genetic effects in future generations. The number of mutations that occur spontaneously in the genetic material is increased by exposure to ionizing radiation. Although ionizing radiation increases the frequency of mutations it does not induce new kinds of effects. Thus the genetic disorders that would arise would not differ in kind from those already occurring naturally. The effect of radiation in causing genetic change is expressed in terms of the doubling dose, or the dose that produces as many mutations as those already occurring naturally. It has been estimated (National Academy of Sciences, 1972) that the doubling dose for humans lies between 20 and 200 rems.

Consequence Evaluation

In the preceding sections the possible modes for the leakage of radioactive materials out of a nuclear power plant under accident conditions are traced.

Meteorological models are outlined by which the results of such leakage can be expressed in terms of cloud and ground surface concentrations of the radionuclides in populated areas. Finally, the contamination concentrations are related to radiation doses and to health effects. With the paths between radiation release and health effects completed, it is possible to evaluate the extent of health hazards, if any, caused by major reactor accidents. To place the results of such analyses in perspective it is well to recall from Section 2-2 that the likelihood that an accident will occur and the accident consequences are equally important in determining the risk incurred by the public because of the presence of the nuclear plant.

The radiological evaluations of hypothesized reactor accidents can be divided conveniently into two categories: those done in conjunction with licensing proceedings to demonstrate compliance with safety criteria and those done to estimate the magnitude and nature of the consequences of potentially catastrophic reactor accidents. The first type of determination is typified by the radiological evaluation of the design basis accidents carried out within the framework of the licensing procedure in the United States. The second type of calculation has been carried out in the Reactor Safety Study directed by Rasmussen (U.S. Nuclear Regulatory Commission, 1975).

Licensing Evaluation. Nuclear power plant licensing in the United States is discussed in some detail in Section 2-3. As a part of this process, a radiological evaluation must be made to demonstrate that the site criteria set out in 10CFR100 are met for each of a series of design basis accidents. Within the design bases are included all sequences of events that are thought to have frequency of occurrence exceeding once per million reactor-years of operation (U.S. Atomic Energy Commission, 1973). Presuming that there may some day be 1000 operating reactors, an accident exceeding the design basis should not occur more than once per thousand years. Typically the severest of the design basis accidents involve loss-of-primary-system cool-ant events for thermal reactors and severe nuclear excursions for fast reactors.

In 10CFR100 two doses are specified: 25 rems (whole-body), and 300 rems (thyroid). Two additional organ doses more recently have been used as guideline values (Project Management Corp., 1975): 150 rems (bone), and 75 rems (lung). These doses are not to be exceeded at the exclusion area boundary, within the first two hours following the accident nor at the low population zone boundary for 30 days. According to the definitions given in Section 2-3, members of the public are not allowed in the exclusion area, and it should be possible to evacuate the low population zone in two hours. Thus no member of the public should receive a dose exceeding the aforestated values in the event of a design basis accident, the presumption being that the dose following the first 30 days will be insignificant. The whole-body dose level was originally set at 25 rems, because this was the value recommended by the National Council on Radiation Protection as the maximum once-in-a-lifetime emergency or accidental dose for radiation workers that may be disregarded in determining their radiation exposure status (Di Nunno et al., 1962). Comparable dose levels are set for the thyroid and other organs. With these dose levels early somatic effects would be barely detectable, and then only for that small number of people located on or near the exclusion or low population zone boundaries. It is highly unlikely that even a single death would occur from acute radiation exposure for any accident meeting these criteria.

These criteria, however, say nothing of the number of people who may be exposed to smaller doses, and therefore they do not directly address the question of the number of latent cancer fatalities that might occur if a large population were exposed to smaller doses. There is an additional stipulation, however, that no population center of more than 25,000 people be located less than one and one-third times the distance to the boundary of the low population zone. In practice this is interpreted so that substantially larger population centers are required to be located even further from the plant than this. Therefore, although the criteria do not explicitly set an upper limit on the permissible number of man-rems, it is difficult to conceive of a set of meteorological conditions and a population distribution that would cause a significant number of latent cancers to occur if the maximum dose and population center distance criteria were being met.

Catastrophic Accidents. The foregoing criteria must be violated grossly before situations arise in which significant numbers of fatalities will result from radiation exposure. However, accidents so severe as to result in the release of a substantial fraction of the inventory of radioactive materials fom large power plants could cause large numbers of fatalities. Such events have been examined both abroad (Beattie and Bell, 1973) and in the United States (U.S. Nuclear Regulatory Commission, 1975). We conclude this text by drawing on the results from the U.S. Nuclear Regulatory Commission Study to examine the nature of the effects on the health and safety of the public that a truly catastrophic reactor accident would be likely to have.

The radiological consequences are based on the accidents that may befall a large, 3200 MW(t), pressurized-water reactor that has been operating at full power for a year. Before proceeding with the discussion of the consequences for the worst possible release of radioactive material to the environment, we must put these accidents into perspective: They are extremely unlikely, as indicated by the probability consequence diagram in Fig. 2-2. A core meltdown must occur before a substantial fraction of the fission product inventory can be released from the fuel elements, and for a pressurized water-reactor the probability of such an event is estimated to be of the order of 6×10^{-5} per reactor-year. If 1000 such reactors were operating this would amount to one meltdown event in 170 years. Several additional events must take place before a large release to the atmosphere results. We consider here the consequences for the worst release conditions for the pressurized-water reactor. For this to take place, the core meltdown must be followed by a steam explosion on contact of the molten fuel with water remaining in the reactor vessel. Failure of the containment spray and heat removal systems must also occur and cause the containment to be well above ambient pressure at the time of the explosion. Furthermore, the steam explosion must be of such magnitude as to rupture the upper portion of the reactor vessel and breach the containment shell. A substantial amount of radioactivity then may be released from the containment in a puff over a period of roughly 10 minutes; the release of radionuclides would continue thereafter at a relatively low rate.

The foregoing sequence is estimated to cause the atmospheric release of 90% of the noble gases, 70% of the halogens, 40% of the alkali metals, tellurium group, and noble metals, 5% of the alkali earths, and 0.3% of the remaining low-volatility nuclides. The release is expected to take place about 2.5 hours after the accident initiation. The release of stored heat from the containment shell would be expected to accompany the radionuclide escape, thus causing the heated discharge to rise substantially into the atmosphere before significant dispersion takes place. However, in some sequences of events the containment may be breached before the steam explosion, causing the escaping cloud to be at a lower temperature and therefore to remain near ground level. It is the latter release that is estimated to have the most dire consequence, and it is the basis for the radiological evaluation that follows. The accident sequence leading to this large ground level release is estimated to have a probability of 4×10^{-7} per reactor year! With 1000 operating reactors, less than one such release would be expected every 2500 years.

Even in the event of such a release of radioactive material, the number of fatalities is not necessarily great. Rather, the consequences vary widely, depending on the meteorological conditions following the release and the population distribution surrounding the plant. An elaborate statistical model based on weather and population data from the site of present commercial nuclear power plants in the United States is used to take into account the variety in weather and population distribution. Using this model it is found that the fatalities for releases such as that described above increase with decreasing probability. The numbers that we quote correspond to events under the adverse weather conditions that cause large numbers of casualties. The number quoted below corresponds to the worst accident (i.e., release and meteorological conditions) evaluated. This event has a probability of 10^{-9} per reactor-year. Thus even if there are 1000 reactors of comparable safety features in operation, we are talking about a once in a million years catastrophe!

In Table 10-8 are given the early and late somatic effects for the worst combination of radioactive release and meteorological conditions calculated (the accident having a probability of 10^{-9} per reactor-year). The early fatalities and illnesses occur predominantly within a few miles of the reactor where the doses are very high. Since there is estimated to be over two hours between accident initiation and the release, an effective evacuation plan can reduce the number of early fatalities by more than an order of magnitude. The 1.2 and 7 mph are the most probably and mean speeds at which the population is likely to be moved radially away from the plant site. In calculating the average values, it is assumed that the 0, 1.2 and 7 mph evacuations have probabilities of 30, 40, and 30% respectively. The number of latent cancer fatalities is insensitive to the success of evacuation because most of these are due to low doses spread over the large population living outside the 25 mile limit assumed for the evacuation area.

It is informative to analyze the dose contributions that most significantly contribute to the fatalities. Immediately downwind from the reactor, where the early fatalities are concentrated, the thyroid, lung, gastrointestinal tract, and bone marrow receive doses in descending magnitude, as indicated in Fig. 10-10. Moreover, the ratio of doses to these organs is relatively independent of distance. If the doses are compared to the dose–mortality relationships, given earlier, their order of importance is seen to be reversed.

TABLE 10-8 Sensitivity of Maximum Health Consequences to Effective Evacuation Speed

Effective Evacuation Speed (mph)	Early Fatalities	Early Illness	Latent Cancer Fatalities per Year
0	6200	80,000	1400
1.2	2300	30,000	1400
7	350	4,500	1400
Average	3300	45,000	1400

From U.S. Nuclear Regulatory Commission, 1975.

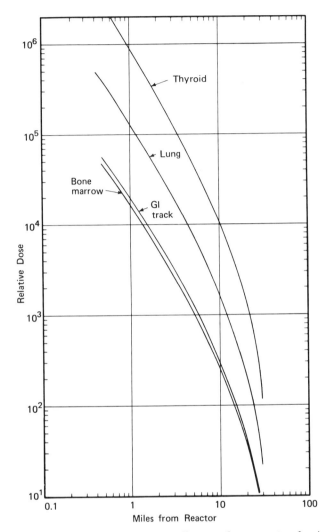

FIGURE 10-10 Total organ doses versus distance from reactor for hypothetical weather; stability F, wind speed = 2.0 m/sec. [Thyroid dose = 1 day ground + external cloud dose + 30 day inhalation dose; lung dose = 1 day ground + external cloud dose + 1 year inhalation dose; GI tract dose = 1 day ground + external cloud dose + 7 day inhalation dose (the GI tract dose is the dose to the regenerative cells of the lower large intestine); bone marrow dose = 1 day ground + external cloud dose + $\frac{1}{2}$(7 day inhalation + 30 day inhalation dose).] From U.S. Nuclear Regulatory Commission, "Reactor Safety Study," WASH-1400 (NUREG-75/104), Appendix VI, 1975.

The deaths are dominated by the bone marrow dose, which has an LD_{50} value with supportive medical care at 8 miles from the reactor. In contrast, the LD_{50} values for the lung and gastrointestinal tract occur at 1.5 and 3.0 miles from the reactor. At these distances, however, the bone marrow dose will cause certain death. Likewise, the thyroid dose is insignificant in its acute effects when compared to the bone marrow dose at comparable distances.

In calculating the doses in Fig. 10-10, external dose contributions from the passing cloud and from one day of ground contamination are included. It is assumed that in areas where doses of this level are incurred, the population must be resettled within the first day and hence is not exposed to further ground contamination. Likewise the inhalation dose is cut off or depreciated after some time to account for the fact that doses received at lower rates over large periods of time following inhalation do not contribute as efficiently to early somatic effects as do doses received over short times.

The contributions to the different organ doses from the external cloud dose, the inhalation dose, and the external ground contamination dose are quite different. All three exposure modes contribute roughly equal amounts to the bone marrow dose. However, the lung dose is dominated by inhalation, where rubidium-106 is the most important contributor, whereas both inhalation and external ground contamination doses contribute to the gastrointestinal tract dose, with tellerium-132 being the most important radionuclide in both of these contributions. The thyroid dose is dominated by the inhalation of radioiodine, with iodine-131 being the most important contributor. It is interesting to note that the iodines, tellerium, and rubidium play dominant roles in determining the acute radiological effects, and the noble gases contribute very little, and then only to the external cloud dose.

It should be pointed out that the curves in Fig. 10-10 are indicative of the downwind doses for a ground level release under adverse weather conditions. If one averages over all directions from the reactor, the probability of death is never more than 0.1 even at small distances. Moreover, if instead of a ground level release, an elevated or hot release were considered, this probability of death would decrease by more than an order of magnitude, and by nearly an additional order of magnitude in the event that an effective evacuation plan is executed.

The latent cancer deaths indicated in Table 10-8 are spread over a 30 year period. Thus the total number of fatalities for the worst calculated accident is 42,000. These fatalities are due predominantly to small doses distributed over a population of about 10 million people. For the general population the risk of death from spontaneous cancer is 1700 per million people per year. Thus the accident results in roughly a 9% increase in the number of cancer fatalities over the 30 year period for the population of 10 million. The added risk of cancer death for any individual is relatively constant out to about 100 miles from the reactor and then decreases rapidly. The estimated contributions of the different exposure modes to the latent cancer fatalities in different organs are given in Table 10-9. The predominant risk is that of lung cancer due to inhalation dose from the cloud. In contrast, the use of the 10% mortality estimate from thyroid malignancies results in their making a rather small contribution to the total risk that is included in the "all other" column of Table 10-9.

In estimating the radiation risk, two groups of people must be considered: those living from 10 to 30 miles downwind of the reactor and those living at greater distances. For the worst accident (10^{-9} per reactor-year) the population in the 10 to 30 mile ranges is likely to undergo radiation illness but then recover. However, the land that they occupy will be contaminated sufficiently that they must be relocated. Thus in addition to the initial acute dose that they have received from the passing cloud, they will be subjected to chronic exposure only from those long-lived isotopes—primarily strontium-90—that they have inhaled. Their removal precludes any additional radiation dose from ground contamination. In contrast, the population located further downwind from the reactor, say 30 to 100 miles will have received much lower doses from the passing cloud. However, the land they inhabit

TABLE 10-9 Contribution of Different Exposure Modes to Latent Cancer Fatalities

	Percentages							Whole Body[1]
	Leukemia	Lung	Breast	Bone	GI Tract	All Other	Total	
External cloud	0.2	0.5	0.5	0.1	0.1	0.3	1	3
Inhalation from cloud	0.5	59.0	10.0	0.2	1.0	0.2	71	15
External ground (<7 days)	4.0	8.0	8.0	1.0	1.0	3.0	25	47
External ground (>7 days)	2.0	2.0	6.0	1.0	1.0	2.0	13	30
Inhalation of resuspended contamination	0.1	3.0	0.1	0.1	0.1	0	3	2
Ingestion of contaminated foods	0.2	0.2	0.5	0.1	0.1	0.2	1	4
Subtotals	7	66	16	2	3	6	100	100

From U.S. Nuclear Regulatory Commission, 1975.
[1] Whole-body values are proportional to 50-year whole-body man-rem.

will be somewhat contaminated, but probably not enough to require relocation. Therefore they will be subjected to additional external chronic radiation from ground contamination and internal irradiation from the ingestion of contaminated foodstuffs.

The chronic dose received by the population may be controlled to a great extent by administration action. A criterion is set specifying the maximum permissible dose that may be received as a result of living on contaminated land, and then interdictive actions are taken to ensure that this dose is not exceeded. For purposes of analysis in the Reactor Safety Study, the criteria is that the 30 year whole-body dose be less than 10 rems in rural areas and less than 24 rems in urban areas. Three levels in interdiction may then be required to meet such specifications, as indicated in Fig. 10-11. Closest to the reactor, the population is relocated altogether. Next there is a zone in which scrubbing and other decontamination procedures can reduce the contamination to a low enough level to allow the population to return in a relatively short time. Further out, crop impoundment for one or more years is required to prevent the ingestion of contaminated foodstuffs, and finally in the most distant region only milk impoundment is necessary to minimize the uptake of radioiodine because of it reconcentration in milk. In the outer two regions the decision on crop and milk impoundment may be made on the basis of the maximum contamination of foodstuffs and milk alloted under 10CFR20. It should be pointed out that although the health effects are dominated by the early exposure from the cloud and ground contamination, the major part of the economic cost of the accident is derived from the land interdiction procedure required to limit chronic exposure.

We conclude by emphasizing that if more probable accidents are considered the health hazards and economic cost decrease sharply. Moreover, the relative importance of the different radionuclides, exposure modes, and

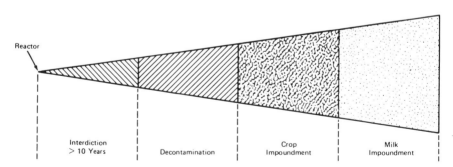

FIGURE 10-11   Simplified interdiction criteria model. From U.S. Nuclear Regulatory Commission, "Reactor Safety Study," WASH-1400 (NUREG-75/104), Appendix VI, 1975.

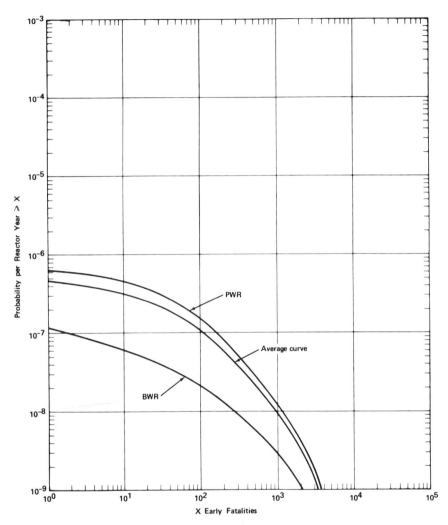

Note: Approximate uncertainties are estimated to be represented
by factors of 1/4 and 4 on consequence magnitudes and
by factors of 1/5 and 5 on probabilities.

FIGURE 10-12a Probability of reactor accidents with early fatalities greater than X. From U.S. Nuclear Regulatory Commission, "Reactor Safety Study," WASH-1400 (NUREG-75/104), Appendix VI, 1975.

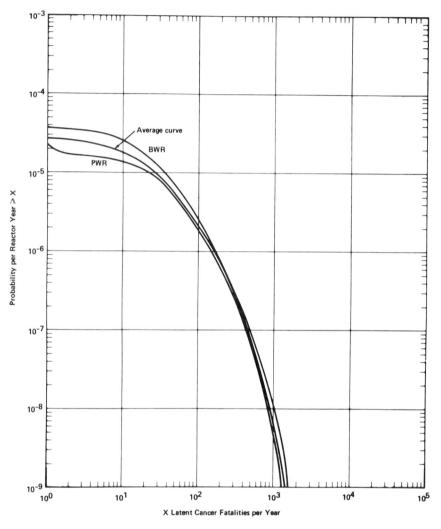

Note: Approximate uncertainties are estimated to be represented
by factors of 1/6 and 3 on consequence magnitudes and by
factors of 1/5 and 5 on probabilities.

FIGURE 10-12b Probability of reactor accidents with latent cancer fatalities per year greater than X. From U.S. Regulatory Commission, "Reactor Safety Study," WASH-1400 (NUREG-75/104), Appendix VI, 1975.

health effects also may change considerably. This is indicated by Figs. 10-12a and b, which give the number of early and latent cancer fatalities expected as a function of the probability of the reactor accident. Suppose, for example, we consider an accident with a probability of 10^{-5} instead of 10^{-9} per reactor-year. Even with 1000 operating reactors accidents of this magnitude would still be expected only once per 100 years. From Fig. 10-12a we see that there would most likely be no fatalities from early somatic effects. However, because latent cancer can be induced even by the low doses associated with smaller releases or more favorable meteorological conditions, there would still be latent cancer fatalities. Now, however, the number is estimated from Fig. 10-12b to be about 30 per year or about 900 total. Finally, according to the Reactor Safety Study only a small fraction of water reactor core meltdown accidents lead to an accident even of this magnitude. Since the probability of a pressurized-water reactor meltdown is estimated at 6×10^{-5} per reactor-year, only one in six meltdowns would lead to this accident.

PROBLEMS

10-1. After consulting the Safety Analysis Report of a fast reactor and Table A-1, compare the actinide activities for a fast and a thermal reactor of the same power rating.

10-2. A gas-cooled reactor containment has a volume of $1 \times 10^6 \, ft^3$ and a design leak rate of 0.2 volume percent per day. Radioiodine filters with removal efficiencies of 90% are to be installed. What must the volumetric flow rate through the filter be to guarantee that no more than 0.1% of the gaseous radioiodine released to the containment will escape to the environment.

10-3. A measure of containment spray performance is the dose reduction factor, $(DRF)_t$. The $(DRF)_t$ is defined as the reduction in fission product leakage from the containment vessel over some period of time t resulting from the effect of the spray.

a. Show that if the elemental iodine concentration is given by Eq. (10-27),

$$(DRF)_t = (V_{ct} + HV_l) \left[V_{ct} + HV_l \frac{1}{\lambda_s t} (1 - e^{-\lambda_s t}) \right]^{-1}.$$

b. Calculate the 2-hour DRF for a caustic–borate spray system with the following properties:

$$V_{ct} = 3 \times 10^6 \text{ ft}^3, \qquad V_l = 3 \times 10^4 \text{ ft}^3,$$

$$F_s = 400 \text{ ft}^3/\text{min}, \qquad d = 0.05 \text{ in.},$$

$$H = 5000, \qquad k_g = 18 \text{ ft/min},$$

$$z_{ct} = 100 \text{ ft}, \qquad v_t = 700 \text{ ft/min}.$$

c. Plot the concentration reduction $C(t)/C(0)$ and (DRF)$_t$ as a function of t for 2 hr.

10-4. Repeat the preceding problem when the spray system is used to remove particulate matter with an average diameter of 1 micron.

10-5. Use the properties of Dirac delta functions to verify Eq. (10-36) and to show that Eq. (10-37) follows from Eqs. (10-35) and (10-36).

10-6. Starting with Eq. (10-41), show that Eqs. (10-46) and (10-47) can be used to estimate the total exposure from a continuously emitting source.

10-7. Estimate the ratio of thyroid to whole-body dose received as a result of being exposed to an infinite cloud of iodine-131.

10-8. A 3000 MW(t) reactor has been operating at full power for several months when a major accident occurs, and 10% of the iodine-131 inventory is released to the containment. The containment leak rate is 0.1% of volume per day. At the site boundary 200 m downwind from the containment calculate (**a**) the direct dose, (**b**) the external cloud dose and (**c**) the internal cloud dose received from the iodine-131 over a 2-hour and over a 24-hour period. In making your calculations assume stability condition F with a wind speed of 2 m/sec and an effective release height of 10 m. Neglect natural iodine removal mechanisms and shielding provided by the containment.

10-9. Fit the natural depletion curve for total iodine in Fig. 10-2 by the sum of two exponential then carry out the cloud dose calculations of the preceding problem taking natural depletion in the containment atmosphere into account.

10-10. Following an accident it is desired to find the maximum cloud concentration at ground level. Suppose that σ_y and σ_z in Eq.

(10-48) can be approximated by Eq. (10-38) and (10-39):

a. Show that

$$\bar{\chi}_\phi\big|_{max} = \frac{\sqrt{2}Q}{e\pi\bar{u}h\sigma_y(x)}$$

where $\sigma_y(x)$ is evaluated at the distance x for which $\sqrt{2}\sigma_z(x) = h$.
b. For class C and class F stability conditions with $h = 30$ m, calculate $Q^{-1}\bar{u}\bar{\chi}_\phi\big|_{max}$ from the foregoing expressions and Figs. 10-6 and 10-7. Compare your results to the exact maximum values in Fig. 10-8.

10-11. After surveying the relevant articles in *Nuclear Safety* compare and contrast the radiological hazards following a major accident in a nuclear plant located

   a. At a conventional site.

   b. Offshore.

   c. Underground.

REFERENCES

F. Abbey, "Fission Product Behavior in Liquid Cooled Reactors, "Manual of Lecture Notes, Reactor Safety Course No. 13, United Kingdom Atomic Energy Authority, September 1972.

M. Abramowitz and I. A. Stegun, *Handbook of Mathematical Functions*, Natl. Bur. Stand., Washington, D.C., 1964.

Advisory Committee on Reactor Safeguards, "Report on the Integrity of Reactor Vessels for Light Water Power Reactors," *Nucl. Eng. Des.*, **28**, 147–195 (1974).

American Society of Mechanical Engineers, "ASME Boiler and Pressure Code, Section III, Nuclear Power Plant Components," 1971.

American Society for Testing and Materials, *Fracture Toughness Testing and Its Applications*, ASTM-STP 381, Philadelphia Pa., 1965.

Atomic Industrial Forum, "World Power Survey," 1976.

C. A. Anderson, Jr., R. A. Markely, and J. F. P. Henson, "Fuel Rod Design Basis," *Proc. Fast Reactor Fuel Technology*, Am. Nucl. Soc., New Orleans, La., April 1971.

L. R. Baker, Jr., and L. C. Just, "Studies of Metal–Water Reactions at High Temperatures III, Experimental and Theoretical Studies of the Zirconium–Water Reaction," Argonne National Laboratory Report, ANL-6548, 1962.

L. R. Baker, Jr. and R. C. Liimatainen, "Chemical Reactions," in *The Technology of Nuclear Reactor Safety*, Vol. 2, T. J. Thompson and J. G. Beckerley (Eds.), M.I.T. Press, Cambridge, Mass., 1973.

L. Baurmash, C. Nelson, R. Johnson, R. Koontz, and H. Morewitz, "One-Cell and Two-Cell Sodium Pool Fire Tests," *Proc. Fast Reactor Safety Meeting*, Beverly Hills, Calif., CONF-740401, 488–497, April 1974.

J. R. Beattie, "An Assessment of Environmental Hazards from Fission Product Releases," U.K. Atomic Energy Authority Report, AHSB (5), R–64, 1961.

J. R. Beattie and G. D. Bell, "A Possible Standard of Risk for Large Accident Releases," *Proc. Symposium and Principles and Standards of Reactor Safety*, Jülich, February 1973, International Atomic Energy Agency, Vienna, 1973.

A. M. Baxter and L. L. Swanson, "HTGR Plant Dynamic Analysis Under Accident Conditions," *Nucl. Eng. Des.*, **26**, 103–117 (1974).

G. I. Bell and S. Glasstone, *Nuclear Reactor Theory*, Van Nostrand Reinhold, New York, 1970.

M. Bellon, "Thermal and Hydraulic Time Constants for Different Types of Reactors," unpublished M.S. project report, Nuclear Engineering Program, Northwestern University, 1973.

H. A. Bethe and J. H. Tait, "An Estimate of the Order of Magnitude of the Explosion When the Core of a Fast Reactor Collapses," UKAE-RHM(56) 113, 1956.

J. M. Biggs, "Structural Response to Seismic Input," in *Seismic Design for Nuclear Power Plants*, R. H. Hansen (ed.), M.I.T. Press, Cambridge, Mass.,1970.

R. B. Bird, W. E. Stewart, and E. N. Lightfoot, *Transport Phenomena*, Wiley, New York, 1960.

J. O. Blomeke and M. F. Todd, "Uranium-235 Fission Product Production as a Function of Thermal Neutron Flux, Irradiation Time, and Decay Time," Oak Ridge National Laboratory Report, ORNL-2127, 1958.

J. E. Boudreau and J. F. Jackson, "Recriticality Considerations in LMFBR Accidents," *Proc. Fast Reactor Safety Meeting*, Beverly Hills, Calif., CONF-740401, April 1974.

J. A. Boure, A. B. Bergles, and L. S. Tong, "Review of Two-Phase Flow Instability," *Nucl. Eng. Des.*, **25**, 165–192 (1973).

G. A. Briggs, *Plume Rise*, U.S. Atomic Energy Commission, Critical Review Series, 1969.

G. F. Brockett, R. W. Shumway, J. O. Zane, and R. W. Griebe, "Loss of Coolant: Control of Consequences by Emergency Core Cooling," Appendix A, *Proc. 1972 Int. Conf. on Nuclear Solutions to World Energy Problems*, Washington, D.C., American Nuclear Society, November 1972.

D. Burgreen, "Flow Coastdown in a Loop After Pumping Power Cutoff," *Nucl. Sci. Eng.* **6**, 306–312 (1959).

J. C. Burnell, "Flow of Boiling Water Through Nozzles, Orifices, and Pipes," *Engineering*, **164**, 572 (1947).

D. R. Buttemer and J. A. Larrimore, "Response of Gas-Cooled Fast Breeder Reactors to Depressurization Accidents," *Nucl. Eng. Des.*, **26**, 195–200 (1974).

F. F. Caked, D. P. Dominicis, and R. H. Leyse, "PWR-FLECHT (Full Length Emergency Cooling Heat Transfer) Final Report," Westinghouse report WCAP-7665, 1971.

C. F. Carmichael and S. A. Marko, "CONTEMPT-PS—A Digital Computer Code for Predicting the Pressure-Temperature History within the Pressure-Suppression Containment Vessel in Response to a Loss-of-Coolant Accident," Phillips Petroleum Company Report IDO-17252, 1969.

H. S. Carslaw and J. C. Jaeger, *Conduction of Heat in Solids*, 2nd ed. Oxford University Press, 1959.

W. R. Castro, "Pipe Cracking in Boiling-Water Reactors," *Nucl. Saf.*, **17**, 475–482 (1976).

H. Cember, *Introduction to Health Physics*, Pergamon Press, New York, 1969.

I. Charak, "The Emergency Core-Cooling Problem in LMFBR's," *React. Technol.*, **13**, (1970).

C. V. Chester and R. O. Chester, "Civil Defense Implications of a Pressurized Water Reactor in a Thermonuclear Target Area," *Nucl. Appl. Technol.*, **9**, 788–795 (1970).

C. V. Chester and R. O. Chester, "Civil Defense Implications of an LMFBR in a Thermonuclear Target Area," *Nucl. Technol.*, **21**, 190–200 (1974).

D. H. Cho and M. Epstein, private communications, Argonne National Laboratory, 1974.

D. H. Cho, M. Epstein, and H. K. Fauske, "Work Potential Resulting from a Voided Core Disassembly," *Trans. Am. Nucl. Soc.*, **18**, 220 (1974).

W. B. Cottrell, "The ECCS Rule-Making Hearing," *Nucl. Saf.*, **15**, 30–55 (1974).

W. B. Cottrell, "Protection of Nuclear Power Plants against External Disasters," Oak Ridge National Laboratory Report, ORNL-NSIC-117, 1975.

Commonwealth Edison Company, "Final Safety Analysis Report—Zion Station," USAEC Docket No. 50-295, 1970.

Crane Co., "Flow of Fluids through Valves, Fittings, and Pipes," Technical Paper No. 410, Crane Co., Chicago, 1957.

R. C. Dahlberg, "Physics of Gas-Cooled Reactors," *Proc. Nat. Topical Meeting in New Developments in Reactor Physics and Shielding*, Kiamesha Lake, N.Y., CONF-270901, September 1972.

Department of Water and Power of the City of Los Angeles, "Preliminary Hazards Summary Report, Malibu Nuclear Plant No. 1," U.S. Atomic Energy Commission Docket No. 50-214, 1963.

J. R. Dietrich, "The Reactor Core," in *The Technology of Nuclear Reactor Safety*, Vol. 1, T. J. Thompson and J. G. Berkeley (Eds.), M.I.T. Press, Cambridge, Mass., 1964.

J. J. DiNunno, F. D. Anderson, R. E. Baker, and R. L. Waterfield, "Calculation of Distance Factors for Power and Test Reactor Sites," U.S. Atomic Energy Commission Report, TID-14844, 1962.

F. W. Dittus and L. M. K. Boelter, University California Publ. Eng., Vol. 2, p. 443, 1930.

P. L. Doan, "Tornado Considerations for Nuclear Power Plant Structures," *Nucl. Saf.*, **11**, 296–308 (1970).

R. B. Duffy and D. T. C. Porthouse, "The Physics of Rewetting in Water Reactor Emergency Core Cooling," *Nucl. Eng. Des.*, **25**, 379–394 (1973).

J. D. Duncan and J. E. Leonard, "Emergency Cooling in BWR's under Simulated Loss-of-Coolant Conditions (BWR-FLECT Final Report)," General Electric report GEAP-13197 (1971).

B. M. Dunn, C. E. Parks, and W. J. Schermer, "Multinode Analysis of B & W's 2568 Mwt Nuclear Plants During a Loss-of-Coolant Accident," Babcock and Wilcox Report, BAW-10034, 1971.

M. Eisenbud, *Environmental Radioactivity*, 2nd ed., Academic, New York, 1973.

M. M. El-Wakil; *Nuclear Heat Transport*, International Publishers, New York, 1971a.

M. M. El-Wakil, *Nuclear Energy Conversion*, Intext, New York, 1971b.

M. Epstein and D. H. Cho, "Fuel Vaporization and Quenching by Cold Sodium; Interpretation of TREAT Test S-11, *Proc. Fast Reactor Safety Meeting* Beverly Hills, Calif. CONF-740401, 268–278, April 1974.

W. H. Esselman, R. L. Ramp, and G. L. Hohmann, "Control Rod Drives and Indicating Systems," *Nuclear Power Reactor Instrumentation Systems Handbook*, Vol. 1, H. Harrer and J. G. Beckerley (Eds.), U.S. Atomic Energy Commission, TID-25952-P1, 1973.

F. R. Farmer, "Reactor Safety and Siting: A Proposed Risk Criterion," *Nucl. Saf.*, **8**, 539–548 (1967).

H. K. Fauske, "Contribution to the Theory of Two-Phase, One Component Critical Flow," Argonne National Laboratory Report, ANL-6633, 1962.

H. K. Fauske, "Some Aspects of Liquid–Liquid Heat Transfer and Explosive Boiling," *Proc. Fast Reactor Safety Meeting*, Beverly Hills, Calif., CONF-740401, April 1974.

Federal Radiation Council, "Background Material for the Development of Radiation Protection Standards," FRC Staff Report, No. 5, 1964.

Federal Radiation Council, "Background Material for the Development of Protective Guides for Strontium-89, Strontium-90 and Cesium-137," FRC Staff Report, No. 7, 1965.

S. H. Fistedis, "Structural Dynamics in Fast Reactor Accident Analysis," *Trans 3rd Int. Conf. on Structural Mechanics in Reactor Technology*, Vol. 2, E311, London, September 1975.

A. L. Florence and G. R. Abrahamson, "Simulation of a Hypothetical Core Disruptive Accident in a Fast Flux Test Facility," Stanford Research Institute Final Report, Contract AT(45-1)-2170, 1973.

M. H. Fontana (Ed.), "Core Melt-Through as a Consequence of Failure of Emergency Core Cooling," *Nucl. Saf.*, **9**, 14–24 (1968).

M. H. Fontana, "Core Melt Through in LMFBR's—A Condensed Review" Oak Ridge National Laboratory Report ORNL-TM-3504, 1971.

M. H. Fontana, D. G. Thompson, T. S. Kress and J. L. Wantland, "Thermal-Hydraulic Effects of Partial Blockages on Simulated LMFBR Fuel Assemblies with Application to CRBR," Oak Ridge National Laboratory Report, ORNL-TM-4779, 1975.

G. L. Fox, "Fast Breeder Reactor Vessel Analysis," American Society of Mechanical Engineering, 69-WA/NE-23, 1969.

A. F. Freudenthal, "Thermal-Stress Analysis and Mechanical Design," *Nuclear Engineering*, C. F. Bonella (Ed.), McGraw-Hill, New York, 1957.

A. M. Freudenthal, "Reliability of Reactor Components and Systems Subject to Fatigue and Creep," *Nucl. Eng. Des.*, **28**, 198–217 (1974).

R. Froehlich, "Current Problems in Multidimensional Reactor Calculations," *Proc. Am. Nucl. Soc. Conf. on Mathematical Models and Computational Techniques for Analysis of Nuclear Systems*, Ann Arbor, Mich., CONF-730414-P2, VII–66, April 1973.

K. Fuchs. "Efficiency for a Very Slow Assembly," Los Alamos Scientific Laboratory Report, LA-596, 1946.

J. D. Gabor, E. S. Sowa, L. Baker, Jr., and J. C. Cassulo, "Studies and Experiments on Heat Removal from Fuel Debris in Sodium," *Proc. Fast Reactor Safety Meeting*, Beverly Hills, Calif. CONF-740401, 823–844, April 1974.

S. Glasstone and A. Sesonske, *Nuclear Reactor Engineering*, Van Nostrand Reinhold, New York, 1963.

J. N. Grace, "Reactor Systems Kinetics," *Naval Reactor Physics Handbook*, Vol. 1, A. Radkowsky (Ed.), U.S. Atomic Energy Commission, 1964.

J. Graham, *Fast Reactor Safety*, Academic, New York, 1971.

A. E. Green and A. J. Bourne, *Reliability Technology*, Wiley, New York, 1972.

A. A. Griffith, "The Phenomena of Rupture and Flow in Solids, *Royal Soc.* (London) *Phil. Trans.*, *Series A*, **221**, 163 (1920).

Gulf General Atomic, "Gas-Cooled Fast Breeder Reactor Preliminary Safety Information Document," GA-10298, 1971.

G. E. Gulley, J. E. Hanson, R. D. Leggell, and F. E. Bard, "Response of an EBR-II Irradiated Mixed-Oxide Fuel Pin to an Overpower Transient in TREAT," *Proc. Fast Reactor Fuel Element Technology*, New Orleans, La, Am. Nucl. Soc., April 1971.

D. R. Gurinsky and S. Isserow, "Nuclear Fuels," in *The Technology of Nuclear Reactor Safety*, Vol. 2., T. J. Thompson and J. G. Beckerley (Eds), M.I.T. Press, Cambridge, Mass., 1973.

E. W. Hagan (Ed.), "Reactor Protection Systems, Philosophies and Instrumentation Reviews from Nuclear Safety," Oak Ridge National Laboratory Report, ORNL-NSIC-111, 1973.

Hanford Engineering Development Laboratory "Fast Flux Test Facility Design Safety Assessment," HEDL-TME, 79–92, 1972.

G. E. Hansen, "Burst Characteristics Associated with the Slow Assembly of Fissionable Materials, Los Alamos Scientific Laboratory Report, LA-1441, 1952.

B. E. Harper and R. S. Hart (Eds.), "10000-Mwe Liquid Metal Fast Breeder Reactor Follow-on Study Conceptual Design Report," Atomics International Report, AI-AEC-12792, Vol. 111, 1969.

A. F. Henry, "The Application of Reactor Kinetics to the Analysis of Experiments," *Nucl. Sci. Eng.*, **3**, 52–70 (1958).

A. Henry, *Nuclear-Reactor Analysis*, M.I.T. Press, Cambridge, Mass., 1975.

R. E. Henry, J. D. Gabor, I. O. Winsch, E. A. Spleha, D. J. Quinn, E. G. Erickson, J. J. Heibergerand, G. T. Goldfass, "Large Scale Vapor Explosions," *Proc. Fast Reactor Safety Meeting*, Beverly Hills, Calif., CONF-740401, 922–934, 1974.

D. L. Hetrick, *Dynamics of Nuclear Reactors*, University of Chicago Press, Chicago 1971.

E. P. Hicks and D. C. Menzies, "Theoretical Studies on the Fast Reactor Maximum Accident," *Proc. Conf. on Safety, Fuels and Core Design in Large Fast Power Reactors*, Argonne National Laboratory Report, ANL-7120, 1965.

R. K. Hilliard and L. F. Coleman, "Natural Transport Effects on Fission Product Behavior in the Containment Systems Experiment," Battelle Northwest Laboratory Report BNWL-1457, 1970.

R. K. Hilliard, A. K. Postma, J. D. McCormark, and L. F. Coleman, "Removal of Iodine and Particles by Sprays in the Containment Systems Experiment," *Nucl. Tech.*, **10**, 499–519 (1971).

R. S. Holcomb, "Preliminary Evaluation of an Assumed Core Meltdown Accident in a 1000 MW(e) Gas-Cooled Fast Breeder Reactor," Oak Ridge National Laboratory Report, ORNL-TM-3546, 1971.

Y. Y. Hsu, "On the Size Range of Active Nucleation Cavities on a Heated Surface," *J. Heat Trans. Trans. ASME*, Series C, **84**, 207–216 (1962).

D. E. Hudson, "Destructive Earthquake Ground Motions," in *Applied Mechanics in Earthquake Engineering*, W. D. Iwan (Ed.)., Am. Soc. Mech. Eng., AMD, Vol. 8, New York, 1974.

D. J. Hughes and R. B. Schwartz, "Neutron Cross Sections," 2nd ed., Brookhaven National Laboratory Report, BNL-325, 1958.

A. P. Hull, "Radiation in Perspective: Some Comparisons of the Environmental Risks from Nuclear and Fossil-Fueled Power Plants," *Nucl. Saf.*, **12**, 185–195 (1971).

H. Hummel and D. Okrent, *Reactivity Coefficients in Large Fast Power Reactors*, American Nuclear Society, Hinsdale, Ill., 1970.

International Atomic Energy Agency, *Proc. Symposium on Principles and Standards of Reactor Safety*, STI/PUB/392, Julich, IAEA, Vienna, February, 1973.

W. H. Irvin, A. Quirk, and E. Beritt, "Fast Fracture of Pressure Vessels," *J. Br. Nucl. Energy Soc.*, **31**, (1964).

R. O. Ivins and J. C. Hesson, "Post Accident Heat Removal, Reactor Development Program Progress Report," Argonne National Lagoratory Report, ANL-7640, 1969.

J. F. Jackson and R. B. Nicholson, "VENUS-II An LMFBR Disassembly Program," Argonne National Laboratory Report, ANL-7951, 1972.

J. F. Jackson, M. G. Stevenson, J. F. Marchaterre, R. H. Sevy, R. Avery, and K. O. Ott, "Trends in LMFBR Hypothetical-Accident Analysis," *Proc. Fast Reactor Safety Meeting*, Beverly Hills, Calif. CONF-740401, 1241–1264, April 1974.

J. F. Jackson and D. P. Weber, "Hydrodynamic Methods in Fast Reactor Safety," *Proc. Conf. or Computational Methods in Nuclear Engineering*, Vol. 1, Charleston, S.C., CONF-750413, April 1975.

E. Jahnke and F. Emde, *Table of Functions*, Dover, New York, 1945.

D. Jakeman, *Physics of Nuclear Reactors*, American Elsevier, New York, 1966.

V. Z. Jankus, "A Theoretical Study of Destructive Nuclear Power Bursts in Fast Power Reactors," Argonne National Laboratory Report, ANL-6512, 1962.

T. R. Johnson, L. Baker, Jr., and J. R. Pavlik, "Large-Scale Molten Fuel–Sodium Interaction Experiments," *Proc. Fast Reactor Safety Meeting*, Beverly Hills, Calif. CCNF-740401, 876–883, April 1974.

J. L. Kaae, D. W. Stevens, and C. S. Luby, "Prediction of the Irradiation Performance of Coated Particle Fuels by Means of Stress-Analysis Models," *Nucl. Appl.*, **10**, 44–53 (1971).

I. Kaplan, *Nuclear Physics*, 2nd ed., Addison-Wesley, Reading, Mass., 1963.

J. M. Kay, *An Introduction to Fluid Mechanics and Heat Transfer*, Cambridge, University Press, London, 1963.

J. H. Keenan and F. G. Keyes, *Thermodynamic Properties of Steam Including Data for Liquid and Solid Phases*, Wiley, New York, 1959.

G. R. Keepin, *Physics of Nuclear Kinetics*, Addison-Wesley, Reading, Mass., 1965.

C. D. G. King, *Nuclear Power Systems*, Macmillan, New York, 1964.

E. E. Kintner, R. V. Laney, and W. H. Esselman, "Status and Experimental Progress in the LMFBR Program," *Proc. Fourth Int. Conf. on the Peaceful Uses of Atomic Energy*, Vol. 5, Geneva 1971; Int. Atomic Energy Agency, Vienna, 1972.

J. Kirk and R. S. Taylor, "Design for Safety of Gas-Cooled Reactors," *Proc. Fourth Int. Conf. on Peaceful Uses of Atomic Energy*, Vol. 3, Geneva, September 1971; Int. Atomic Energy Agency, Vienna, 1972.

"Know Your Reactors," *Nucl. Eng. Int.*, **18**, 203, (1973).

J. G. Knudsen and R. K. Hilliard, "Fission Product Transport by Natural Processes in Containment Vessels," Battelle Pacific Northwest Laboratory Report, BNWL-943, January 1969.

L. J. Koch, H. O. Monson, D. Okrent, M. Levenson, W. R. Simmons, J. R. Humphreys, J. Hangsnes, V. C. Jankus, and W. B. Lowenstein, "Experimental Breeder Reactor-II (EBR-II) Hazards Summary Report," Argonne National Laboratory Report, ANL-5719, 1957.

H. Kramers, *Physische Transportverschijnselen*, Hogeschool, Delft, Holland, 1958.

F. Kreith, *Principles of Heat Transfer*, International Publishers, New York, 1965.

J. R. Lamarsh, *Introduction to Nuclear Reactor Theory*, Addison-Wesley, Reading, Mass., 1966.

J. W. Landis and J. F. Watson, "Materials and Design Concepts of Gas-Cooled Reactor Systems," *Nucl. Eng. Des.*, **26**, 38–57 (1974).

L. B. Lave and C. L. Freeburg, "Health Effects of Electricity Generation from Coal, Oil and Nuclear Fuel," *Nucl. Saf.*, **14**, 409–428 (1973).

G. G. Lawson, "Emergency Core-Cooling Systems for Light-Water-Cooled Power Reactors," Oak Ridge National Laboratory Report, ORNL-NSIC-24, October 1968.

E. E. Lewis, "A Transient, Heat Conduction Model for Reactor Fuel Elements," *Nucl. Eng. Des.*, **15**, 33–40 (1971).

D. Linehan, "Geological and Seismological Factors Influencing the Assessment of a Seismic Treat to Nuclear Reactors," in *Seismic Design for Nuclear Power Plants*, R. H. Hansen (Ed.), M.I.T. Press, Cambridge, Mass.; 1970.

W. W. Little, Jr., in *Lecture Series on Fast Reactor Safety Technology and Practices*, Vol. I, D. L. Hoffman (Ed.), Battelle Northwest Laboratory Report, BNWL-SA-3093, 1970.

T. F. Lomenick and NSIC Staff, "Earthquakes and Nuclear Power Plant Design," Oak Ridge National Laboratory Report, ORNL-NSIC-28, 1970.

H. Ludewig and L. G. Epel, "Preliminary CGFR Plant Transient Analysis Using the System Code HELAP," Brookhaven National Laboratory, BNL-FRS-74-1, 1974.

T. R. Mager, P. C. Riccardella, and E. T. Wessel, "Fracture Mechanics For Heavy Section Steel Nuclear Pressure Vessels," *Nucl. Eng. Des.*, **20**, 181–200 (1972).

D. D. Malinowski and L. F. Picone, "Iodine Removal in the Ice Condenser System," *Nucl. Tech.*, **10**, 428–443 (1971).

L. G. Marquis and I. M. Jacobs, "Quality Assurance and Reliability," in *Nuclear Power Reactor Instrumentation and Systems Handbook*, Vol. 1, J. M. Harrer and J. G. Beckerley (Eds.), U.S. Atomic Energy Commission Report, TID-25952-P1, 1974.

I. J. Martel, "Corrosion Damage in Nuclear Steam Generators," *EPRI J.*, 20–23, June 1976.

R. C. Martinelli, "Heat Transfer to Molten Metals," *Trans. ASME*, **69**, 947–959 (1947).

R. C. Martinelli and D. B. Nelson, "The Prediction of Pressure Drop During Forced Circulation Boiling Water," *Trans. ASME*, **70**, 695–702 (1948).

W. J. McAfee, W. K. Sartory, Z. P. Bazant, and P. A. Stancampiano, "Evaluation of the Structural Integrity of LMFBR Equipment Cell Liners—Result of Preliminary Investigations," Oak Ridge National Laboratory Report, ORNL-TM-5145, NRC-7, 1976.

J. R. McDonald, K. C. Mehta, and J. E. Minor, "Tornado-Resistant Design of Nuclear Power Plant Structure," *Nucl. Saf.*, **15**, 432–439 (1974).

J. E. Meyer, "Conservation Laws in One-Dimensional Hydrodynamics," Bettis Technical Review, WAPD-BT-20, 61–72, 1960.

J. E. Meyer, "Hydrodynamic Models for the Treatment of Reactor Thermal Transients," *Nucl. Sci. Eng.*, **10**, 269–277 (1961).

R. A. Meyer, B. Wolf, N. F. Friedman, and R. Seifert, "Fast Reactor Meltdown Accidents Using Bethe–Tait Analysis," General Electric Report, GEAP-4809, 1967.

T. R. Moffette, "Safety Evaluations of the HTGR as Related to 10CFR 100 Siting Requirements," *Nucl. Tech.*, **25**, 630–634 (1975).

F. J. Moody, "Maximum Flow Rate of a Single Component, Two-Phase Mixture," *J. Heat Trans.*, **87**, 134 (1965).

L. F. Moody, "Friction Factors for Pipe Flow," *Trans. ASME*, **66**, 671 (1944).

K. V. Moore and W. H. Retting, "RELAP4—A Computer Program for Transient Thermal-Hydraulic Analysis," Aerojet General Company Report, ANCR-1127, 1973.

G. E. Myers, *Analytical Methods in Conduction Heat Transfer*," McGraw-Hill, New York, 1971.

A. N. Nahavandi, "The Loss-of-Coolant Accident Analysis in Pressurized Water Reactors," *Nucl. Sci. Eng.*, **36**, 159–188 (1969).

National Academy of Sciences, *The Effects on Populations of Exposure to Low Levels of Ionizing Radiation*, Report of the Committee on the Biological Effects of Ionizing Radiation (BEIR), National Research Council, Washington, D.C., 1972.

National Bureau of Standards, "Nuclear Data," Circular 449, 1950.

N. M. Newmark, "Fifth Rankine Lecture: Effects of Earthquakes on Dams and Embankments," *Geotechnique*, **15**, 139–160 (1965).

N. M. Newmark, "Design Criteria for Nuclear Reactors Subjected to Earthquake Hazards," *Proc. IAEA Panel on Aseismic Design and Testing of Nuclear Facilities*, Tokyo, Japan Earthquake Eng. Soc., 1967.

L. W. Nordheim, "Physics Section II," University of Chicago Metallurgical Laboratories Report CP-2589, 1945.

W. E. Nyer, "Mathematical Models of Fast Transients," in *The Technology of Nuclear Reactor Safety*, Vol. 1, T. J. Thompson and J. G. Beckerley (Eds.), M.I.T. Press, Cambridge, Mass., 1964.

H. G. O'Brien and C. S. Walker, "Protection System," in *Nuclear Power Reactor Instrumentation Systems Handbook*, J. M. Harrer and J. G. Beckerley (Eds.), U.S. Atomic Energy Commission Report, TID-25952-P2, 1974.

D. Okrent, M. V. Davis, B. R. T. Frost, J. D. B. Lamerts, R. O. Meyer, R. B. Poeppel, J. T. A. Roberts, and R. W. Weeks, "Lecture Background Notes on Fast Reactor Fuel Element Behavior," Argonne National Laboratory, 1972.

D. Okrent and H. K. Fauske, "Lecture Background Notes on Transient Sodium Boiling and Voiding in Fast Reactors," Argonne National Laboratory Report, ANL/ACEA-101, 1972.

H. J. Otway and R. C. Erdmann, "Reactor Siting from a Risk Viewpoint," *Nucl. Eng. Des.*, **13**, 365–376 (1970).

A. Padilla, Jr., "Analysis of Mechanical Work Energy for LMFBR Maximum Accidents," *Nucl. Tech.*, **12**, 348–355 (1971).

P. J. Pahl, "Model Response of Containment Structures," in *Seismic Design for Nuclear Power Plants*, R. H. Hansen (Ed.), M.I.T. Press, Cambridge, Mass., 1970.

N. J. Palladino, "Mechanical Design of Components for Reactor Systems," in *The Technology of Nuclear Reactor Safety*, Vol. 2, T. J. Thompson and J. G. Beckerley (Eds.), M.I.T. Press, Cambridge, Mass., 1973.

L. F. Parsly, "Design Considerations of Reactor Containment Spray Systems—Part VII; A Method for Calculating Iodine Removal by Sprays," Oak Ridge National Laboratory Report, ORNL-TM-2412, Part VII, 1970.

F. Pasquill, *Atmospheric Diffusion*, Van Nostrand, New York, 1962.

T. Pasternak (Ed.), "HTGR Accident Initiation on Progression Analysis Status Report," General Atomic Corp, Report, GA-13617, 1975.

A. Pearson and C. G. Lennox, "Sensing and Control Instrumentation," in *The Technology of Nuclear Reactor Safety*, Vol. 1, T. J. Thompson and J. C. Beckerley (Eds.), M.I.T. Press, Cambridge, Mass., 1964.

N. T. Peters and J. G. Yevick, "Plant Structures, Containment Design, and Site Criteria," in *Fast Reactor Technology and Plant Design*, J. C. Yevick (Ed.), M.I.T. Press, Cambridge, Mass., 1966.

A. K. Postma and W. F. Pasedag, "A Review of Mathematical Models for Predicting Spray Removal of Fission Products in Reactor Containment Vessels," Battelle Northwest Laboratory Report, BNWL B-268, 1973.

Power Reactor Development Company, "Enrico Fermi Reactor Preliminary Safety Analysis Report," 1961.

Project Management Corporation, "Clinch River Breeder Reactor Plant Preliminary Safety Analysis Report," Vols. 14, 15, 1975.

J. A. Redfield and J. H. Murphy, "The FLASH-2 Method For Loss-of-Coolant Analysis," *Nucl. Appl.*, **6**, 127–136 (1969).

J. A. Richardson "Summary Comparison of West European and U.S. Licensing Regulations for LWR's," *Nucl. Eng. Int.*, **21**, 32–41 (1976).

L. C. Richardson, L. J. Finnegan, R. J. Wagner, and J. M. Waage, "CONTEMPT-A Computer Program for Predicting the Containment Pressure–Temperature Response to a Loss-of-Coolant Accident," Phillips Petroleum Company Report, IDO-17220, 1967.

C. F. Ricter, *Elementary Seismology*, Freeman, San Francisco, 1958.

P. L. Rittenhouse, "Fuel Rod Failure and Its Effects in Light-Water Reactor Accidents," *Nucl. Saf.*, **12**, 487–495 (1971).

W. Rockenhauser, "Structural Design Criteria for Primary Containment Structures (Prestressed Concrete Reactor Vessels)," *Nucl. Eng. Des.*, **9**, 449–466 (1969).

D. J. Rose, "Nuclear Electric Power," *Science*, **184**, 35–359 (1974).

R. Salvatori, "Containment and Containment Structures Loading Due to Postulated Pipe Ruptures," unpublished notes, M.I.T. Summer Course on Reactor Safety, July 1974.

M. A. Schultz, *Control of Nuclear Reactors and Power Plants*, McGraw-Hill, New York, 1961.

D. S. Schweitzer, "HTGR Safety Evaluation Division Quarterly Report," Brookhaven National Laboratory Report, BNL-50450, 1975.

J. H. Scott, G. E. Gulley, C. W. Hunter, and J. E. Hanson, "Microstructural Dependence of Failure of Oxide LMFBR Fuel Pins," *Proc. Fast Reactor Safety Meeting*, Beverly Hills, Calif., CONF-740401, April 1974.

W. T. Sha and T. H. Hughes, "VENUS—A Two Dimensional Coupled Neutronics–Hydrodynamics Fast Reactor Excursion Composite Program," Argonne National Laboratory Report, ANL-7701, 1970.

P. R. Shire, in *Lecture Series on Fast Reactor Safety Technology and Practices*, Vol. II, P. L. Hofmann (Ed.), Battelle North West Laboratory Report BNWL-SA-3093, 1970.

D. H. Slade (Ed.), "Meteorology and Atomic Energy," U.S. Atomic Energy Commission Report, TID-24190, 1968.

D. C. Slaughterbeck, "Correlations to Predict the Maximum Containment Pressure Following a Loss-of-Coolant Accident in Large Pressurized Water Reactors with Dry Containments," Idaho Nuclear Corporation Report, IN-1468, May 1971.

C. B. Smith, "Power Plant Safety and Earthquakes," *Nucl. Saf.*, **14**, 568–569 (1970).

R. L. Smith, J. L. Rowlands, A. R. Baker, D. C. Smith, E. P. Hicks, J. E. Mann, and J. Weale, "Fast Reactor Physics, Including Results from the U.K. Zero Power Reactors," *Proc. Third United Nations Conference on Peaceful Uses of Atomic Energy*, Vol. 6, United Nations, Geneva, 1964, New York, 1965.

W. M. Stacey, Jr., *Space Time Nuclear Reactor Kinetics*, Academic, New York, 1969.

O. M. Stansfield, C. B. Scott, and J. Chin, "Kernel Migration in Coated Carbide Fuel Particles," *Nucl. Technol.* **25**, 517–430 (1975).

C. Starr, "Benefit-Cost Studies in Sociotechnical Systems," *Proc. Colloquium on Perspectives on Benefit–Risk Decision-Making*, National Academy of Engineering, Washington, D.C., 1972.

C. Starr and M. A. Greenfield, "Public-Health Risks of Thermal Power Plants," *Nucl. Saf.*, 267–274 (1973).

A. J. Stepanoff, *Centrifugal and Axial Flow Pumps*, Wiley, New York, 1957.

H. B. Stewart and M. M. Merril, "Kinetics of Solid-Moderator Reactors," in *The Technology of Nuclear Reactor Safety*, Vol. 1, T. J. Thompson and J. G. Beckerley (Eds.), M.I.T Press, Cambridge, Mass., 1964.

K. Shure, "Fission Product Decay Energy," *Bettis Technol. Rev.*, Bettis Atomic Power Laboratory Report, WAPD-BJ-24, 1–17, 1961.

H. J. Teague, "Cooling Failure in a Sub-Assembly," *An Appreciation of Fast Reactor Safety*, UKAEA Report NP–18487, 1970.

Tennessee Valley Authority, "Browns Ferry Nuclear Power Station Design and Analysis Report," U.S. Atomic Energy Commission Docket No. 50-259, 1963.

T. J. Thompson, "Accidents and Destructive Tests," in *The Technology of Nuclear Reactor Safety*, Vol. 1, T. J. Thompson and J. G. Beckerley (Eds.), M.I.T. Press, Cambridge, Mass., 1964.

T. J. Thompson and C. R. McCullough, "The Concepts of Reactor Containment," in *The Technology of Nuclear Reactor Safety*, Vol. 2, T. J. Thompson and J. G. Beckerley (Eds.), M.I.T. Press, Cambridge, Mass., 1973.

L. S. Tong, *Boiling Crisis and Critical Heat Flux*, U.S. Atomic Energy Commission TID-25887, 1972.

L. S. Tong and J. Weisman, *Thermal Analysis of Pressurized Water Reactors*, American Nuclear Society, Hinsdale, Ill., 1970.

S. E. Turner, C. R. McCullough, and R. L. Lyerly, "Industrial Sabotage in Nuclear Power Plants," *Nucl. Saf.*, **11**, 107–114 (1970).

U.S. Atomic Energy Commission, "Theoretical Possibilities and Consequences of Major Accidents in Nuclear Power Plants," WASH-740, 1957.

U.S. Atomic Energy Commission, "The Safety of Nuclear Power Reactors (Light Water-Cooled) and Related Facilities," WASH-1250, 1973a.

U.S. Atomic Energy Commission, "Anticipated Transients without Scram for Water-Cooled Power Reactors," WASH-1270, 1973b.

U.S. Atomic Energy Commission, "Seismic and Geological Criteria for Nuclear Power Plants," 10CFR 100 Proposed Appendix A, 1973c.

U.S. Atomic Energy Commission "Design Response Spectra for Seismic Design of Nuclear Power Plants," Regulatory Guide 1.6, 1973d.

U.S. Atomic Energy Commission, "Damping Values for Seismic Design of Nuclear Power Plants," Regulatory Guide 1.61, 1973e.

U.S. Atomic Energy Commission, "Assumptions Used for Evaluating the Potential Radiological Consequences of a Loss of Coolant Accident for Pressurized Water Reactors," Regulatory Guide 1.4 Riv. 2, 1974a.

U.S. Atomic Energy Commission, "Reactor Safety Study," WASH-1400 (Draft), 1974b.

U.S. Atomic Energy Commission, "Preapplication Safety Evaluation of the Gas Cooled Fast Breeder Reactor," Directorate of Licensing Project No. 456, 1974c.

U.S. Nuclear Regulatory Commission, "Reactor Safety Study," WASH-1400 (NUREG-75/104), Appendix VI, 1975.

U.S. Weather Bureau, "Meteorology and Atomic Energy," 1955.

S. Untermyer and J. T. Weills, "Heat Generation in Irradiated Uranium," Argonne National Laboratory Report, ANL-4790, 1962.

J. P. van Erp, T. C. Chawla, and H. K. Fauske, "An Evaluation of Pint-to-Pin Failure Propagation in LMFBR Fuel Assemblies," *Proc. Fast Reactor Safety Meeting*, Beverly Hills, Calif., CONF-740401, April 1974.

G. E. Wade, "Evolution and Current Status of BWR Containment System," *Nucl. Saf.*, **15**, 163–179 (1974).

I. B. Wall, "Probabilities Assessment of Flooding Hazard for Nuclear Power Plants," *Nucl. Saf.*, **15**, 399–408 (1974a).

I. B. Wall, "Probabilities Assessment of Aircraft Risk for Nuclear Power Plants," *Nucl. Saf.*, **15**, 276–284 (1974).

A. E. Waltar, in *Lecture Series in Fast Reactor Safety Technology and Practices*, Vol. 2, P. L. Hoffman (Ed.), Batelle Northwest Laboratory Report, BNWL-SA-3093, 1970.

T. N. Washburn and J. L. Scott, "Performance Capability of Advanced Fuels for Fast Breeder Reactors," *Proc. Fast Reactor Fuel Technology*, Am. Nucl. Soc., New Orleans, La., April, 1971.

J. B. Waters and V. Walker, "Fuel Elements," in *The Design of Gas-Cooled Graphite Moderated Reactors*, D. R. Poulter (Ed.), Oxford University Press, London, 1963.

K. Way and E. P. Wigner, "The Rate of Decay of Fission Products," *Phys. Rev.*, **73**, 1318–1330 (1948).

E. T. Weber, O. D. Slagle, and C. A. Hinman, "Laboratory Studies on Melting and Gas Release Behavior of Irradiated Fuel," *Proc. Fast Reactor Safety Meeting*, Beverly Hills, Calif., CONF-740401-P2, April 1974.

S. J. Weems, W. G. Lyman, and R. B. Haga, "The Ice-Condenser Reactor Containment System," *Nucl. Saf.*, **11**, 215–222 (1970).

A. M. Weinberg and E. P. Wigner, *The Physical Theory of Neutron Chain Reactors*, University of Chicago Press, Chicago 1958.

R. U. Whitman, "Basic Concepts and Important Problems," in *Seismic Design for Nuclear Power Plants*, R. H. Hansen (Ed.), M.I.T. Press, Cambridge, Mass., 1970.

G. D. Whitman, G. C. Robinson, Jr., and A. W. Savolainen, "Technology for Steel Pressure Vessels for Water-Cooled Reactors," Oak Ridge National Laboratory Report, ORNL-NSIC-21, 1967.

H. U. Wider, J. F. Jackson, L. L. Smith, and D. T. Eggen, "An Improved Analysis of Fuel Motion During an Overpower Excursion," *Proc. Fast Reactor Safety Meeting*, Beverly Hills, Calif., CONF-740401, April 1974.

J. H. Wilkenson, *The Algebraic Eigenvalue Problem*, Clarendon, Oxford, 1965.

W. R. Wise and J. F. Proctor, "Explosion Containment Laws for Nuclear Reactor Vessels," U.S. Naval Ordnance Laboratory Report, NOLTR-63-140, 1965.

W. R. Wise, et al., "Response of Enrico Fermi Reactor to TNT Simulated Nuclear Accidents," U.S. Naval Ordnance Laboratory Report, NOLTR 62-207, 1962.

A. Yamanouchi, "Effect of Core Spray Cooling in Transient State after Loss-of-Coolant Accident, *J. Nucl. Sci. Technol.*, **5**, 547 (1968).

J. B. Yasinsky and A. F. Henry, "Some Numerical Experiments Concerning Space–Time Reactor Kinetics Behavior," *Nucl. Sci. Eng.*, **22**, 171–181 (1965).

L. J. Ybarranto, C. W. Solbrig, and H. S. Isbin, "The Calculated Loss-of-Coolant Accident: A Review," AICHE Monograph Series No. 7, Am. Inst. Chem. Eng., New York, 1972.

F. C. Zapp, "Testing of Containment Systems Used with Light-Water Cooled Power Reactors," Oak Ridge National Laboratory Report, ORNL-NSIC-26, 1968.

Appendixes

A DOSE CONVERSION FACTORS

APPENDIX A   DOSE CONVERSION FACTORS*

TABLE A-1   Activity of Radionuclides in a 3200 MW(t) Water-Cooled Reactor Core Following One Year of Operation[1]

No.	Radionuclide	Radioactive Inventory Source (curies $\times 10^{-8}$)	Half-life (days)
1	Cobalt-58	0.0078	71.0
2	Cobalt-60	0.0029	1,920
3	Krypton-85	0.0056	3,950
4	Krypton-85m	0.24	0.183
5	Krypton-87	0.47	0.0528
6	Krypton-88	0.68	0.117
7	Rubidium-86	0.00026	18.7
8	Strontium-89	0.94	52.1
9	Strontium-90	0.037	11,030
10	Strontium-91	1.1	0.403
11	Yttrium-90	0.039	2.67
12	Yttrium-91	1.2	59.0
13	Zirconium-95	1.5	65.2
14	Zirconium-97	1.5	0.71
15	Niobium-95	1.5	35.0
16	Molybdenum-99	1.6	2.8
17	Technetium-99m	1.4	0.25
18	Ruthenium-103	1.1	39.5
19	Ruthenium-105	0.72	0.185
20	Ruthenium-106	0.25	366
21	Rhodium-105	0.49	1.50
22	Tellurium-127	0.059	0.391

No.	Radionuclide	Radioactive Inventory Source (curies $\times 10^{-8}$)	Half-life (days)
23	Tellurium-127m	0.011	109
24	Tellurium-129	0.31	0.048
25	Tellurium-129m	0.053	0.340
26	Tellurium-131m	0.13	1.25
27	Tellurium-132	1.2	3.25
28	Antimony-127	0.061	3.88
29	Antimony-129	0.33	0.179
30	Iodine-131	0.85	8.05
31	Iodine-132	1.2	0.0958
32	Iodine-133	1.7	0.875
33	Iodine-134	1.9	0.0366
34	Iodine-135	1.5	0.280
35	Xenon-133	1.7	5.28
36	Xenon-135	0.34	0.384
37	Cesium-134	0.075	750
38	Cesium-136	0.030	13.0
39	Cesium-137	0.047	11,000
40	Barium-140	1.6	12.8
41	Lanthanum-140	1.6	1.67
42	Cerium-141	1.5	32.3
43	Cerium-143	1.3	1.38
44	Cerium-144	0.85	284
45	Praseodymium-143	1.3	13.7
46	Neodymium-147	0.60	11.1
47	Neptunium-239	16.4	2.35
48	Plutonium-238	0.00057	32,500
49	Plutonium-239	0.00021	8.9×10^6
50	Plutonium-240	0.00021	2.4×10^6
51	Plutonium-241	0.034	5,350
52	Americium-241	0.000017	1.5×10^5
53	Curium-242	0.0050	163
54	Curium-244	0.00023	6,630

[1] From U.S. Nuclear Regulatory Commission, 1975.

TABLE A-2 Photon Dose-Conversion Factors for Immersion in Contaminated Air (rem per Ci-sec/m³)[1]

Radionuclide	Whole Body[2]	Total Marrow	Lung
CO-58	2.16E−01	2.40E−01	2.01E−01
CO-60	6.00E−01	6.31E−01	5.67E−01
KR-85	4.75E−04	5.78E−04	4.47E−04
KR-85M	3.64E−02	5.50E−02	3.22E−02
KR-87	1.81E−01	1.02E−01	1.72E−01
KR-88	4.67E−01	4.83E−01	4.47E−01
RB-86	2.07E−02	2.27E−02	1.94E−02
SR-89	0.0	0.0	0.0
SR-90	0.0	0.0	0.0
SR-91	1.69E−01	1.93E−01	1.60E−01
Y-90	0.0	0.0	0.0
Y-91	6.25E−04	6.39E−04	5.94E−04
ZR-95	1.62E−01	1.87E−01	1.52E−01
ZR-97	4.22E−02	4.72E−02	4.00E−02
MB-95	1.66E−01	1.83E−01	1.56E−01
MO-99	3.64E−02	4.44E−02	3.42E−02
TC-99M	3.06E−02	5.42E−02	2.54E−02
RU-103	1.11E−01	1.36E−01	1.05E−01
RU-105	1.79E−01	2.21E−01	1.67E−01
RU-106	4.31E−02	5.22E−02	4.06E−02
RH-105	1.82E−02	2.74E−02	1.61E−02
TE-127	9.36E−04	1.16E−03	8.78E−04
TE-127M	1.10E−03	1.79E−03	5.61E−04
TE-129	1.47E−02	1.81E−02	1.35E−02
TE-129M	7.83E−03	9.92E−03	6.97E−03
TE-131M	3.14E−01	3.56E−01	2.84E−01
TE-132	4.75E−02	7.31E−02	4.19E−02
SB-127	1.51E−01	1.84E−01	1.43E−01
SB-129	2.68E−01	2.97E−01	2.53E−01
I-131	8.72E−02	1.08E−01	8.22E−02
I-132	5.11E−01	5.89E−01	4.83E−01
I-133	1.54E−01	1.83E−01	1.46E−01
I-134	5.33E−01	5.80E−01	5.00E−01
I-135	4.19E−01	4.42E−01	4.00E−01
XE-133	9.06E−03	1.59E−02	6.97E−03
XE-135	5.67E−02	8.47E−02	5.06E−02
CS-134	3.50E−01	4.03E−01	3.28E−01
CS-136	4.78E−01	5.42E−01	4.44E−01
CS-137	1.22E−01	1.49E−01	1.15E−01
BA-140	4.44E−02	5.61E−02	4.14E−02

Radionuclide	Whole Body[2]	Total Marrow	Lung
LA-140	5.67E−01	6.06E−01	5.39E−01
CE-141	1.83E−02	3.22E−02	1.50E−02
CE-143	6.81E−02	9.36E−02	6.08E−02
CE-144	4.31E−03	7.61E−03	3.44E−03
PR-143	0.0	0.0	0.0
ND-147	3.14E−02	4.39E−02	2.78E−02
NP-239	3.08E−02	4.07E−02	2.65E−02
PU-238	5.25E−05	4.25E−05	9.58E−06
PU-239	2.30E−05	2.17E−05	5.42E−06
PU-240	4.64E−05	3.80E−05	9.17E−06
PU-241	4.17E−10	8.53E−10	2.94E−10
AM-241	4.56E−03	9.33E−03	3.22E−03
CM-242	5.00E−05	3.89E−05	8.31E−06
CM-244	1.42E−03	2.81E−03	1.07E−03

[1] From U.S. Nuclear Regulatory Commission, 1975.
[2] For organs not listed the dose-conversion factors are similar to those of the whole body.

TABLE A-3 Gamma Dose-Conversion Factors for Exposure to Contaminated Ground (rem per Ci/m²) (Time-Integral Dose to N Days)[1,2]

Radio nuclide	Whole Body[3]		Total Marrow		Lung	
	1 Day	7 Days	1 Day	7 Days	1 Day	7 Days
CO-58	3.29E+02	2.24E+03	3.67E+02	2.50E+03	3.08E+02	2.10E+03
CO-60	8.48E+02	5.88E+03	8.89E+02	6.22E+03	7.99E+02	5.58E+03
KR-85	0.0	0.0	0.0	0.0	0.0	0.0
KR-85M	0.0	0.0	0.0	0.0	0.0	0.0
KR-87	0.0	0.0	0.0	0.0	0.0	0.0
KR-88	0.0	0.0	0.0	0.0	0.0	0.0
RB-86	2.93E+01	1.85E+02	3.22E+01	2.02E+02	2.75E+01	1.73E+02
SR-89	0.0	0.0	0.0	0.0	0.0	0.0
SR-90	0.0	0.0	0.0	0.0	0.0	0.0
SR-91	1.68E+02	2.05E+02	1.96E+02	2.39E+02	1.59E+02	1.83E+02
Y-90	0.0	0.0	0.0	0.0	0.0	0.0
Y-91	8.82E−01	5.91E+00	8.98E−01	6.06E+00	8.39E−01	5.66E+00
ZR-95	2.47E+02	1.77E+03	2.85E+02	2.04E+03	2.32E+02	1.67E+03
ZR-97	3.33E+02	5.38E+02	4.02E+02	6.52E+02	3.14E+02	5.10E+02
NB-95	2.49E+02	1.64E+03	2.73E+02	1.80E+03	2.33E+02	1.54E+03
MO-99	7.73E+01	3.25E+02	1.07E+02	4.65E+02	6.95E+01	2.91E+02
TC-99M	1.52E+01	1.62E+91	2.70E+01	2.88E+01	1.26E+01	1.35E+01

Radio-nuclide	Whole Body[3]		Total Marrow		Lung	
	1 Day	7 Days	1 Day	7 Days	1 Day	7 Days
RU-103	1.75E+02	1.16E+03	2.14E+02	1.42E+03	1.65E+02	1.09E+03
RU-105	7.23E+01	7.94E+01	9.04E+01	9.99E+01	6.77E+01	7.37E+01
RU-106	6.51E+01	4.56E+02	7.96E+01	5.54E+02	6.17E+01	4.30E+02
RH-105	2.20E+01	5.67E+01	3.32E+01	8.55E+01	1.95E+01	5.01E+01
TE-127	6.77E−01	8.13E−01	8.43E−01	1.01E+00	6.38E−01	7.67E−01
TE-127M	7.90E+00	5.84E+01	1.27E+01	9.20E+01	4.40E+00	3.37E+01
TE-129	1.98E+00	1.98E+00	2.44E+00	2.44E+00	1.83E+00	1.83E+00
TE-129M	3.63E+01	2.46E+02	4.54E+01	3.07E+02	3.28E+01	2.22E+02
TE-131M	3.81E+02	9.60E+02	4.36E+02	1.10E+03	3.56E+02	8.90E+02
TE-132	6.77E+02	3.08E+03	8.04E+02	3.63E+03	6.36E+02	2.88E+03
SB-127	2.12E+02	9.20E+02	2.57E+02	1.11E+03	2.00E+02	8.65E+02
SB-129	1.00E+02	1.04E+02	1.12E+02	1.16E+02	9.47E+01	9.78E+01
I-131	1.28E+02	7.08E+02	1.59E+02	8.73E+02	1.21E+02	6.63E+02
I-132	1.07E+02	1.07E+02	1.23E+02	1.23E+02	1.01E+02	1.01E+02
I-133	1.63E+02	3.11E+02	1.93E+02	3.75E+02	1.54E+02	2.91E+02
I-134	4.14E+01	4.14E+01	4.56E+01	4.56E+01	3.88E+01	3.88E+01
I-135	2.52E+02	2.85E+02	2.77E+02	3.18E+02	2.38E+02	2.69E+02
XE-133	0.0	0.0	0.0	0.0	0.0	0.0
XE-135	0.0	0.0	0.0	0.0	0.0	0.0
CS-134	5.30E+02	3.60E+03	6.10E+02	4.26E+03	4.07E+02	3.47E+03
CS-136	6.84E+02	4.10E+03	7.79E+02	4.68E+03	6.37E+02	3.32E+03
CS-137	1.86E+02	1.31E+03	2.28E+02	1.60E+03	1.76E+02	1.24E+03
BA-140	2.13E+02	3.65E+03	2.42E+02	3.98E+03	2.02E+02	3.46E+03
LA-140	6.49E+02	1.80E+03	6.88E+02	1.92E+03	6.12E+02	1.71E+03
CE-141	2.77E+01	1.82E+02	4.93E+01	3.24E+02	2.28E+01	1.50E+02
CE-143	9.00E+01	2.24E+02	1.24E+02	3.98E+02	3.06E+01	2.00E+02
CE-144	1.72E+01	1.20E+02	2.39E+01	1.67E+02	1.52E+01	1.07E+02
PR-143	0.0	0.0	0.0	0.0	0.0	0.0
ND-147	5.20E+01	3.05E+02	7.28E+01	4.26E+02	4.63E+01	2.70E+02
NP-239	5.89E+01	2.02E+02	9.53E+01	3.26E+02	5.09E+01	1.74E+02
PU-238	8.95E−01	6.20E+00	7.17E−01	5.92E+00	1.82E−01	1.14E+00
PU-239	3.76E−01	2.63E+00	3.56E−01	2.49E+00	8.88E−02	6.22E−01
PU-240	7.84E−01	5.47E+00	6.53E−01	4.57E+00	1.55E−01	1.08E+00
PU-241	4.52E−05	2.21E−03	9.23E−05	4.52E−03	3.18E−05	1.56E−03
AM-241	2.06E+01	1.43E+02	4.21E+01	2.95E+02	1.45E+01	1.02E+02
CM-242	7.92E−01	5.46E+00	6.14E−01	4.25E+00	1.31E−01	9.05E−01
CM-244	4.96E+00	3.46E+01	9.73E+00	6.81E+01	3.73E+00	2.61E+01

[1] From U.S. Nuclear Regulatory Commission, 1975.

[2] Note: noble gases do not deposit on the ground.

[3] For organs not listed the dose-conversion factors are similar to those of the whole body.

TABLE A-4 Dose-Conversion Factors for Inhaled Radionuclides (rem/Ci Inhaled) Body Organ: Total Marrow[1]

Radionuclide	0–2 Days	0–30 Days	0–30 Years
CO-58	2.3E+02	1.1E+03	3.1E+03
CO-60	5.5E+02	2.8E+03	5.8E+04
KR-85	6.0E−01	6.1E−01	6.1E−01
KR-85M	3.9E−01	3.9E−01	3.9E−01
KR-87	1.3E+00	1.3E+00	1.3E+00
KR-88	3.1E+00	3.1E+00	3.1E+00
RB-86	5.0E+02	4.8E+03	6.5E+03
SR-89	5.2E+02	5.0E+03	1.3E+04
SR-90	3.6E+02	1.0E+04	6.0E+05
SR-91	1.9E+02	2.3E+02	3.2E+02
Y-90	2.0E+02	5.1E+02	5.1E+02
Y-91	1.6E+02	2.3E+03	9.3E+03
ZR-95	1.8E+02	9.4E+02	3.6E+03
ZR-97	1.7E+02	1.9E+02	1.9E+02
NB-95	1.8E+02	7.7E+02	1.4E+03
MO-99	7.5E+01	1.3E+02	1.3E+02
TC-99M	1.1E+01	1.1E+01	1.1E+01
RU-103	1.3E+02	5.5E+02	1.1E+03
RU-105	2.3E+01	2.4E+01	2.4E+01
RU-106	7.6E+01	6.6E+02	6.2E+03
RH-105	1.6E+01	2.3E+01	2.3E+01
TE-127	3.7E+00	3.9E+00	3.9E+00
TE-127M	2.6E+01	2.7E+02	8.0E+02
TE-129	1.1E+00	1.1E+00	1.1E+00
TE-129M	6.7E+01	5.3E+02	8.4E+02
TE-131M	2.1E+02	3.1E+02	3.1E+02
TE-132	4.8E+02	1.0E+03	1.0E+03
SB-127	1.7E+02	3.3E+02	3.3E+02
SB-129	4.5E+01	4.6E+01	4.6E+01
I-131	7.7E+01	1.8E+02	1.9E+02
I-132	5.0E+01	5.0E+01	5.0E+01
I-133	8.8E+01	9.4E+01	9.4E+01
I-134	2.0E+01	2.0E+01	2.0E+01
I-135	9.1E+01	9.1E+01	9.1E+01
XE-133	1.5E+00	1.6E+00	1.6E+00
XE-135	2.1E+00	2.1E+00	2.1E+00
CS-134	5.8E+02	7.0E+03	4.8E+04
CS-136	6.9E+02	5.0E+03	6.0E+03
CS-137	3.6E+02	5.2E+03	3.7E+04
BA-140	4.8E+02	2.8E+02	3.4E+03
LA-140	4.4E+02	6.8E+02	6.8E+02

Radionuclide	0–2 Days	0–30 Days	0–30 Years
CE-141	3.7E+01	1.5E+02	2.7E+02
CE-143	6.6E+01	1.0E+02	1.1E+02
CE-144	6.2E+01	3.5E+02	9.2E+03
PR-143	3.0E+00	2.6E+01	3.4E+01
ND-147	5.5E+01	1.7E+02	2.0E+02
NP-239	4.0E+01	6.4E+01	6.4E+01
PU-238	1.8E+01	2.8E+02	8.7E+05
PU-239	1.6E+01	2.6E+02	9.2E+05
PU-240	1.6E+01	2.6E+02	9.2E+05
PU-241	3.3E−03	7.2E−02	1.5E+04
AM-241	4.0E+01	4.2E+02	9.9E+05
CM-242	2.1E+01	3.3E+02	8.5E+03
CM-244	2.0E+01	3.3E+02	5.8E+05

TABLE A-4 (Continued) Body Organ: Lower Large Intestine Wall

Radionuclide	0–2 Days	0–7 Days	0–30 Days
CO-58	3.2E+03	6.9E+03	7.1E+03
CO-60	8.0E+03	1.8E+04	1.8E+04
KR-85	1.8E−01	1.8E−01	1.8E−01
KR-85M	2.2E−01	2.2E−01	2.2E−01
KR-87	1.0E+00	1.0E+00	1.0E+00
KR-88	2.3E+00	2.3E+00	2.3E+00
RB-86	1.2E+03	2.7E+03	5.8E+03
SR-89	9.2E+03	1.4E+04	1.4E+04
SR-90	7.3E+03	1.4E+04	1.6E+04
SR-91	2.4E+03	2.6E+03	2.6E+03
Y-90	2.2E+04	4.0E+04	4.0E+04
Y-91	2.0E+04	4.6E+04	4.9E+04
ZR-95	6.1E+03	1.4E+04	1.4E+04
ZR-97	1.5E+04	1.8E+04	1.9E+04
NB-95	3.2E+03	6.9E+03	7.0E+03
MO-99	1.2E+04	2.0E+04	2.1E+04
TC-99M	1.1E+01	1.1E+01	1.1E+01
RU-103	4.8E+03	1.0E+04	1.1E+04
RU-105	9.1E+02	1.2E+03	1.2E+03
RU-106	5.5E+04	1.2E+05	1.3E+05
RH-105	3.3E+03	5.1E+03	5.1E+03
TE-127	7.3E+02	7.8E+02	7.8E+02
TE-127M	6.9E+03	1.7E+04	2.0E+04
TE-129	5.7E+00	5.7E+00	5.7E+00

TABLE A-4—continued

Radionuclide	0–2 Days	0–7 Days	0–30 Days
TE-129M	1.5E+04	3.4E+04	3.6E+04
TE-131M	5.8E+03	8.5E+03	8.5E+03
TE-132	3.3E+03	6.0E+03	6.1E+03
SB-127	1.3E+04	2.6E+04	2.6E+04
SB-129	7.9E+02	8.1E+02	8.1E+02
I-131	2.8E+02	3.3E+02	3.6E+02
I-132	6.0E+01	6.0E+01	6.0E+01
I-133	3.0E+02	3.3E+02	3.3E+02
I-134	2.0E+01	2.0E+01	2.0E+01
I-135	2.2E+02	2.2E+02	2.2E+02
XE-133	4.2E−01	4.2E−01	4.2E−01
XE-135	9.9E−01	9.9E−01	9.9E−01
CS-134	9.0E+02	2.8E+03	1.0E+04
CS-136	9.9E+02	2.8E+03	6.4E+03
CS-137	6.8E+02	1.9E+03	6.3E+03
BA-140	9.4E+03	1.6E+04	1.6E+04
LA-140	1.3E+04	2.1E+04	2.1E+04
CE-141	6.7E+03	1.5E+04	1.5E+04
CE-143	9.8E+03	1.6E+04	1.6E+04
CE-144	5.2E+04	1.2E+05	1.2E+05
PR-143	1.2E+04	2.5E+04	2.5E+04
ND-147	1.0E+04	2.1E+04	2.2E+04
NP-239	7.4E+03	1.3E+04	1.3E+04
PU-238	2.2E+04	5.0E+04	5.1E+04
PU-239	2.0E+04	4.6E+04	4.7E+04
PU-240	2.1E+04	4.7E+04	4.8E+04
PU-241	2.1E+02	4.7E+02	4.8E+02
AM-241	2.3E+04	5.2E+04	5.4E+04
CM-242	2.4E+04	5.5E+04	5.6E+04
CM-244	2.3E+04	5.2E+04	5.4E+04

TABLE A-4 (Continued) Body Organ: Thyroid

Radionuclide	0–2 Days	0–7 Days	0–30 Days
CO-58	1.1E+02	2.0E+02	9.5E+02
CO-60	2.4E+02	6.5E+02	2.3E+03
KR-85	1.8E−01	1.8E−01	1.8E−01
KR-85M	2.0E−01	2.0E−01	2.0E−01
KR-87	9.7E−01	9.7E−01	9.7E−01
KR-88	2.0E+00	2.0E+00	2.0E+00

Radionuclide	0–2 Days	0–7 Days	0–30 Days
RB-86	5.0E+02	1.7E+03	4.8E+03
SR-89	2.0E+02	5.0E+02	9.2E+02
SR-90	1.8E+02	7.0E+02	1.0E+03
SR-91	1.3E+02	1.4E+02	1.6E+02
Y-90	8.2E+00	1.7E+01	2.1E+01
Y-91	7.1E+00	2.5E+01	9.8E+01
ZR-95	7.9E+01	2.2E+02	8.0E+02
ZR-97	7.7E+01	8.6E+01	8.6E+01
MB-95	8.1E+01	2.1E+02	6.4E+02
MO-99	9.4E+01	1.4E+02	1.5E+02
TC-99M	4.6E+01	4.6E+01	4.6E+01
RU-103	5.2E+01	1.4E+02	4.3E+02
RU-105	1.4E+01	1.5E+01	1.5E+01
RU-106	4.8E+01	1.7E+02	6.2E+02
RH-105	6.4E+00	9.6E+00	9.9E+00
TE-127	2.9E+00	3.0E+00	3.0E+00
TE-127M	1.6E+01	6.3E+01	1.6E+02
TE-129	8.1E−01	8.1E−01	8.1E−01
TE-129M	4.3E+01	1.4E+02	3.2E+02
TE-131M	4.5E+03	3.4E+04	8.7E+04
TE-132	4.8E+04	8.8E+04	9.7E+04
SB-127	1.0E+02	1.8E+02	2.2E+02
SB-129	3.7E+01	3.7E+01	3.8E+01
I-131	1.3E+05	4.8E+05	1.0E+06
I-132	6.6E+03	6.6E+03	6.6E+03
I-133	1.2E+05	1.8E+05	1.8E+05
I-134	1.1E+03	1.1E+03	1.1E+03
I-135	4.3E+04	4.4E+04	4.4E+04
XE-133	3.9E−01	4.0E−01	4.0E−01
XE-135	9.1E−01	9.1E−01	9.1E−01
CS-134	5.8E+02	2.0E+03	7.9E+03
CS-136	6.9E+02	2.1E+03	5.0E+03
CS-137	3.6E+02	1.3E+03	5.1E+03
BA-140	2.2E+02	5.9E+02	1.0E+03
LA-140	1.5E+02	2.2E+02	2.3E+02
CE-141	6.0E+00	1.6E+01	4.6E+01
CE-143	1.8E+01	2.4E+01	2.5E+01
CE-144	5.1E+00	1.3E+01	4.5E+01
PR-143	9.3E−02	2.9E−01	7.9E−01
ND-147	1.2E+01	2.8E+01	6.0E+01
NP-239	8.2E+00	1.4E+01	1.5E+01
PU-238	1.8E+01	6.2E+01	2.8E+02

TABLE A-4—continued

Radionuclide	0–2 Days	0–7 Days	0–30 Days
PU-239	1.6E+01	5.8E+01	2.6E+02
PU-240	1.6E+01	5.8E+01	2.6E+02
PU-241	1.7E-03	6.9E-03	4.5E-02
AM-241	2.0E+01	7.3E+01	3.3E+02
CM-242	2.1E+01	7.5E+01	3.3E+02
CM-244	2.0E+01	7.3E+01	3.3E+02

TABLE A-4 (Continued) Body Organ: Lung

Radionuclide	0–2 Days	0–30 Days	0–30 Years
CO-58	1.9E+03	1.7E+04	6.1E+04
CO-60	4.8E+03	4.9E+04	1.3E+06
KR-85	1.8E-01	1.8E-01	1.8E-01
KR-85M	2.1E-01	2.1E-01	2.1E-01
KR-87	9.6E-01	9.6E-01	9.6E-01
KR-88	2.0E+00	2.0E+00	2.0E+00
RB-86	7.2E+03	1.2E+04	1.4E+04
SR-89	6.0E+03	7.2E+03	7.8E+03
SR-90	4.6E+03	1.2E+04	1.8E+04
SR-91	3.4E+03	3.8E+03	4.3E+03
Y-90	1.7E+04	3.3E+04	3.3E+04
Y-91	1.3E+04	1.0E+05	2.0E+05
ZR-95	3.5E+03	3.4E+04	1.3E+05
ZR-97	1.4E+04	1.5E+04	1.5E+04
NB-95	1.8E+03	1.5E+04	3.1E+04
MO-99	7.7E+03	1.6E+04	1.6E+04
TC-99M	8.9E+01	8.9E+01	8.9E+01
RU-103	2.9E+03	2.4E+04	5.4E+04
RU-105	2.1E+03	2.2E+03	2.2E+03
RU-106	3.3E+04	3.3E+05	3.9E+06
RH-105	2.5E+03	3.6E+03	3.6E+03
TE-127	1.5E+03	1.6E+03	1.6E+03
TE-127M	5.2E+03	5.2E+04	1.2E+05
TE-129	5.6E+02	5.6E+02	5.6E+02
TE-129M	1.3E+04	9.5E+04	1.5E+05
TE-131M	6.6E+03	1.1E+04	1.1E+04
TE-132	1.3E+04	3.0E+04	3.0E+04
SB-127	9.1E+03	2.4E+04	2.5E+04
SB-129	3.1E+03	3.1E+03	3.2E+03
I-131	2.1E+03	2.4E+03	2.4E+03
I-132	1.0E+03	1.0E+03	1.0E+03

TABLE A-4—continued

Radionuclide	0–2 Days	0–30 Days	0–30 Years
I-133	3.1E+03	3.1E+03	3.1E+03
I-134	5.6E+02	5.6E+02	5.6E+02
I-135	2.5E+03	2.5E+03	2.5E+03
XE-133	4.1E−01	4.1E−01	4.1E−01
XE-135	9.4E−01	9.4E−01	9.4E−01
CS-134	3.0E+03	1.1E+04	5.1E+04
CS-136	2.9E+03	7.3E+03	8.2E+03
CS-137	3.2E+03	8.3E+03	4.0E+04
BA-140	4.6E+03	6.1E+03	6.3E+03
LA-140	1.0E+04	1.6E+04	1.6E+04
CE-141	4.0E+03	3.1E+04	6.2E+04
CE-143	7.0E+03	1.2E+04	1.3E+04
CE-144	2.9E+04	3.0E+05	2.9E+06
PR-143	6.9E+03	4.0E+04	4.9E+04
ND-147	6.1E+03	3.1E+04	3.8E+04
NP-239	5.0E+03	9.2E+03	9.3E+03
PU-238	1.3E+06	1.3E+07	3.1E+08
PU-239	1.2E+06	1.2E+07	2.0E+08
PU-240	1.2E+06	1.2E+07	2.9E+08
PU-241	1.3E+02	1.8E+03	5.7E+05
AM-241	1.3E+06	1.3E+07	3.1E+08
CM-242	1.4E+06	1.4E+07	8.7E+07
CM-244	1.4E+06	1.4E+07	3.1E+08

TABLE A-4 (Continued) Body Organ: Skeleton

Radionuclide	0–2 Days	0–7 Days	0–30 Days	0–1 Year	0–30 Years
CO-58	1.5E+02	3.3E+02	8.4E+02	2.5E+03	2.6E+03
CO-60	3.5E+02	8.0E+02	2.2E+03	1.8E+04	5.0E+04
KR-85	1.5E−01	1.5E−01	1.5E−01	1.5E−01	1.5E−01
KR-85M	1.9E−01	1.9E−01	1.9E−01	1.9E−01	1.9E−01
KR-87	8.3E−01	8.3E−01	8.3E−01	8.3E−01	8.3E−01
KR-88	1.8E+00	1.8E+00	1.8E+00	1.8E+00	1.8E+00
RB-86	5.0E+02	1.7E+03	4.8E+03	6.5E+03	6.5E+03
SR-89	7.2E+02	3.8E+03	1.9E+04	3.0E+04	3.0E+04
SR-90	5.3E+02	3.8E+03	2.1E+04	2.8E+05	2.4E+05
SR-91	1.9E+02	2.0E+02	2.3E+02	3.4E+02	3.4E+02
Y-90	4.1E+02	8.6E+02	1.0E+03	1.0E+03	1.0E+03
Y-91	3.4E+02	1.2E+03	4.7E+03	1.0E+04	1.9E+04
ZR-95	1.2E+02	2.0E+02	7.9E+02	3.3E+03	3.4E+03
ZR-97	1.2E+02	1.3E+02	1.3E+02	1.3E+02	1.3E+02

Radionuclide	0–2 Days	0–7 Days	0–30 Days	0–1 Year	0–30 Years
NB-95	1.2E+02	2.6E+02	6.2E+02	1.2E+03	1.2E+03
MO-99	6.2E+01	1.0E+02	1.1E+02	1.1E+02	1.1E+02
TC-99M	1.0E+01	1.0E+01	1.0E+01	1.0E+01	1.0E+01
RU-103	7.7E+01	1.8E+02	4.3E+02	8.8E+02	8.8E+02
RU-105	1.6E+01	1.7E+01	1.7E+01	1.7E+01	1.7E+01
RU-106	5.8E+01	1.9E+02	6.0E+02	3.4E+03	5.9E+03
RH-105	1.1E+01	1.6E+01	1.6E+01	1.6E+01	1.6E+01
TE-127	5.0E+00	5.2E+00	5.2E+00	5.2E+00	5.2E+00
TE-127M	3.1E+01	1.3E+02	4.4E+02	1.8E+03	2.0E+03
TE-129	1.2E+00	1.2E+00	1.2E+00	1.2E+00	1.2E+00
TE-129M	7.5E+01	2.6E+02	7.5E+02	1.4E+03	1.4E+03
TE-131M	1.6E+02	2.3E+02	2.5E+02	2.5E+02	2.5E+02
TE-132	4.4E+02	8.0E+02	9.1E+02	9.1E+02	9.1E+02
SB-127	1.2E+02	2.1E+02	2.5E+02	2.6E+02	2.6E+02
SB-129	3.8E+01	3.8E+01	3.8E+01	3.9E+01	3.9E+01
I-131	7.6E+01	1.2E+02	2.0E+02	2.1E+02	2.1E+02
I-132	4.7E+01	4.7E+01	4.7E+01	4.7E+01	4.7E+01
I-133	8.7E+01	9.2E+01	9.2E+01	9.2E+01	9.2E+01
I-134	1.9E+01	1.9E+01	1.9E+01	1.9E+01	1.9E+01
I-135	8.7E+01	8.7E+01	8.7E+01	8.7E+01	8.7E+01
XE-133	3.5E−01	3.6E−01	3.6E−01	3.6E−01	3.6E−01
XE-135	7.2E−01	7.2E−01	7.2E−01	7.2E−01	7.2E−01
CS-134	5.7E+02	2.0E+03	7.8E+03	4.2E+04	4.7E+04
CS-136	6.6E+02	2.0E+03	4.9E+03	5.9E+03	5.9E+03
CS-137	3.6E+02	1.3E+03	5.1E+03	3.1E+04	3.6E+04
BA-140	6.1E+02	2.0E+03	4.2E+03	5.2E+03	5.2E+03
LA-140	4.2E+02	6.7E+02	7.0E+02	7.0E+02	7.0E+02
CE-141	2.5E+01	6.1E+01	1.6E+02	3.2E+02	3.2E+02
CE-143	4.7E+01	7.1E+01	9.4E+01	1.1E+02	1.1E+02
CE-144	1.0E+02	2.0E+02	6.3E+02	7.2E+03	1.0E+04
PR-143	7.6E+00	2.4E+01	6.5E+01	8.6E+01	8.6E+01
ND-147	3.8E+01	8.5E+01	1.6E+02	1.9E+02	2.4E+02
NP-239	3.0E+01	4.9E+01	5.4E+01	5.4E+01	1.6E+02
PU-238	9.9E+03	3.5E+04	1.6E+05	3.4E+06	4.9E+08
PU-239	9.2E+03	3.3E+04	1.5E+05	3.1E+06	5.2E+08
PU-240	9.2E+03	3.3E+04	1.5E+05	3.1E+06	5.2E+08
PU-241	2.4E−01	1.2E+00	1.3E+01	2.7E+03	8.3E+06
AM-241	1.1E+04	3.9E+04	1.8E+05	3.6E+06	5.5E+08
CM-242	1.2E+04	4.3E+04	1.8E+05	1.8E+06	4.8E+06
CM-244	1.1E+04	4.1E+04	1.9E+05	3.7E+06	3.3E+08

[1] From U.S. Nuclear Regulatory Commission, 1975.

TABLE A-5 Dose-Conversion Factors for Ingestion of Radionuclides (rem per curie Ingested)[1]

	0–10 Years	0–30 Years
Whole Body		
Cs-134	7.14×10^4	7.14×10^4
Cs-136	8.96×10^3	8.96×10^3
Cs-137	5.49×10^4	5.49×10^4
Sr-89	1.91×10^3	1.91×10^3
Sr-90	5.52×10^4	8.29×10^4
I-131	8.79×10^2	8.79×10^2
I-133	2.70×10^2	2.70×10^2
Total Marrow		
Cs-134	7.34×10^4	7.34×10^4
Cs-136	9.29×10^3	9.29×10^3
Cs-137	5.61×10^4	5.61×10^4
Sr-89	5.26×10^3	5.26×10^3
Sr-90	2.08×10^5	2.74×10^5
I-131	2.87×10^2	2.87×10^2
I-133	1.48×10^2	1.48×10^2
Bone (Mineral)		
Cs-134	7.24×10^4	7.24×10^4
Cs-136	9.10×10^3	9.10×10^3
Cs-137	5.56×10^4	5.56×10^4
Sr-89	1.19×10^4	1.19×10^4
Sr-90	6.15×10^5	9.70×10^5
I-131	3.10×10^2	3.10×10^2
I-133	1.46×10^2	1.46×10^2
Thyroid		
Cs-134	7.33×10^4	7.33×10^4
Cs-136	9.33×10^3	9.23×10^3
Cs-137	5.55×10^4	5.55×10^4
Sr-89	5.81×10^2	5.81×10^2
Sr-90	3.18×10^3	3.26×10^3
I-131	1.68×10^6	1.68×10^6
I-133	3.21×10^5	3.21×10^5

TABLE A-5—continued

	0–10 Years	0–30 Years
	Lung	
Cs-134	7.31×10^4	7.31×10^4
Cs-136	8.82×10^3	8.82×10^3
Cs-137	5.59×10^4	5.59×10^4
Sr-89	5.81×10^2	5.81×10^2
Sr-90	3.18×10^3	3.74×10^3
I-131	3.56×10^2	3.56×10^2
I-133	1.58×10^2	1.58×10^2
	Lower Large Intestine Wall	
Cs-134	9.33×10^4	9.33×10^4
Cs-136	1.35×10^4	1.35×10^4
Cs-137	6.64×10^4	6.64×10^4
Sr-89	8.53×10^4	8.53×10^4
Sr-90	8.12×10^4	8.12×10^4
I-131	1.91×10^3	1.91×10^3
I-133	1.82×10^2	1.82×10^2

[1] From U.S. Nuclear Regulatory Commission, 1975.

B BESSEL FUNCTIONS

APPENDIX B   BESSEL FUNCTIONS

The solution of Bessel's equation is often required in dealing with neutronic problems in cylindrical geometry. The Bessel equation of order n is

$$\frac{d^2\Phi}{dr^2}+\frac{1}{r}\frac{d\Phi}{dr}+\left(\alpha^2-\frac{n^2}{r^2}\right)\Phi=0.$$

The solution to this equation is

$$\Phi(r)=AJ_n(\alpha r)+BY_n(\alpha r),$$

where A and B are arbitrary constants and $J_n(\alpha r)$ and $Y_n(\alpha r)$ are known as ordinary Bessel functions of the first and second kind, respectively.

The second form of the Bessel equation results when the sign of the parameter α^2 is changed:

$$\frac{d^2\Phi}{dr^2}+\frac{1}{r}\frac{d\Phi}{dr}-\left(\alpha^2+\frac{n^2}{r^2}\right)\Phi=0.$$

The solution to this equation is

$$\Phi(r)=A'I_n(\alpha r)+B'K_n(\alpha r),$$

where A' and B' are arbitrary constants and $I_n(\alpha r)$ and $K_n(\alpha r)$ are known as the modified Bessel functions of the first and second kinds, respectively.

In neutronics problems only the Bessel equations of order zero usually appear. The ordinary and modified Bessel functions of order zero are plotted in Fig. B-1. To apply boundary conditions and to determine integral parameters it is often necessary to differentiate or integrate the zero order Bessel functions. The required relationships are

$$\frac{dJ_0(\alpha r)}{dr} = -\alpha J_1(\alpha r); \qquad \frac{dY_0(\alpha r)}{dr} = -\alpha Y_1(\alpha r)$$

$$\frac{dI_0(\alpha r)}{dr} = \alpha I_1(\alpha r); \qquad \frac{dK_0(\alpha r)}{dr} = -\alpha K_1(\alpha r)$$

$$\int J_0(\alpha r)r\, dr = \frac{r}{\alpha} J_1(\alpha r); \qquad \int Y_0(\alpha r)r\, dr = \frac{r}{\alpha} Y_1(\alpha r)$$

$$\int I_0(\alpha r)r\, dr = \frac{r}{\alpha} I_1(\alpha r); \qquad \int K_0(\alpha r)r\, dr = -\frac{r}{\alpha} K_1(\alpha r)$$

Values of the ordinary and modified Bessel functions of order zero and one are tabulated in Table B-1. The values of I_n become infinite as the argument goes to infinity:

$$I_n(\alpha r) \xrightarrow[\alpha r \to \infty]{} \frac{1}{\sqrt{2\pi\alpha r}} e^{\alpha r}$$

The values of Y_n and K_n become infinite as the argument goes to zero. Specifically:

$$Y_0(\alpha r) \xrightarrow[\alpha r \to 0]{} \frac{2}{\pi} [\ln(\alpha r) - 0.11593]$$

$$Y_1(\alpha r) \xrightarrow[\alpha r \to 0]{} -\frac{2}{\pi \alpha r}$$

$$K_0(\alpha r) \xrightarrow[\alpha r \to 0]{} -[\ln(\alpha r) - 0.11593]$$

$$K_1(\alpha r) \xrightarrow[\alpha r \to 0]{} \frac{1}{\alpha r}$$

More thorough treatments of the properties of Bessel functions may be found in standard texts on mathematical physics (Abramowitz and Stegun, 1964).

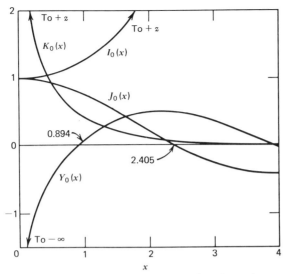

FIGURE B-1 Ordinary and modified Bessel functions of zero order.

TABLE B-1 Bessel Functions of Order Zero and One

x	$J_0(x)$	$J_1(x)$	$Y_0(x)$	$Y_1(x)$	$I_0(x)$	$I_1(x)$	$K_0(x)$	$K_1(x)$
0	1.0000	0.0000	$-\infty$	$-\infty$	1.000	0.0000	∞	∞
0.05	0.9994	0.0250	-1.979	-12.79	1.001	0.0250	3.114	19.91
0.10	0.9975	0.0499	-1.534	-6.459	1.003	0.0501	2.427	9.854
0.15	0.9944	0.0748	-1.271	-4.364	1.006	0.0752	2.030	6.477
0.20	0.9900	0.0995	-1.081	-3.324	1.010	0.1005	1.753	4.776
0.25	0.9844	0.1240	-0.9316	-2.704	1.016	0.1260	1.542	3.747
0.30	0.9776	0.1483	-0.8073	-2.293	1.023	0.1517	1.372	3.056
0.35	0.9696	0.1723	-0.7003	-2.000	1.031	0.1777	1.233	2.559
0.40	0.9604	0.1960	-0.6060	-1.781	1.040	0.2040	1.115	2.184
0.45	0.9500	0.2194	-0.5214	-1.610	1.051	0.2307	1.013	1.892
0.50	0.9385	0.2423	-0.4445	-1.471	1.063	0.2579	0.9244	1.656
0.55	0.9258	0.2647	-0.3739	-1.357	1.077	0.2855	0.8466	1.464
0.60	0.9120	0.2867	-0.3085	-1.260	1.092	0.3137	0.7775	1.303
0.65	0.8971	0.3081	-0.2476	-1.177	1.108	0.3425	0.7159	1.167
0.70	0.8812	0.3290	-0.1907	-1.103	1.126	0.3719	0.6605	1.050
0.75	0.8642	0.3492	-0.1372	-1.038	1.146	0.4020	0.6106	0.9496
0.80	0.8463	0.3688	-0.0868	-0.9781	1.167	0.4329	0.5653	0.8618
0.85	0.8274	0.3878	-0.0393	-0.9236	1.189	0.4646	0.5242	0.7847
0.90	0.8075	0.4059	-0.0056	-0.8731	1.213	0.4971	0.4867	0.7165
0.95	0.7868	0.4234	0.0481	-0.8258	1.239	0.5306	0.4524	0.6560

TABLE B-1—continued

x	$J_0(x)$	$J_1(x)$	$Y_0(x)$	$Y_1(x)$	$I_0(x)$	$I_1(x)$	$K_0(x)$	$K_1(x)$
1.0	0.7652	0.4401	0.0883	−0.7812	1.266	0.5652	0.4210	0.6019
1.1	0.6957	0.4850	0.1622	−0.6981	1.326	0.6375	0.3656	0.5098
1.2	0.6711	0.4983	0.2281	−0.6211	1.394	0.7147	0.3185	0.4346
1.3	0.5937	0.5325	0.2865	−0.5485	1.469	0.7973	0.2782	0.3725
1.4	0.5669	0.5419	0.3379	−0.4791	1.553	0.8861	0.2437	0.3208
1.5	0.4838	0.5644	0.3824	−0.4123	1.647	0.9817	0.2138	0.2774
1.6	0.4554	0.5699	0.4204	−0.3476	1.750	1.085	0.1880	0.2406
1.7	0.3690	0.5802	0.4520	−0.2847	1.864	1.196	0.1655	0.2094
1.8	0.3400	0.5815	0.4774	−0.2237	1.990	1.317	0.1459	0.1826
1.9	0.2528	0.5794	0.4968	−0.1644	2.128	1.448	0.1288	0.1597
2.0	0.2239	0.5767	0.5104	−0.1070	2.280	1.591	0.1139	0.1399
2.1	0.1383	0.5626	0.5183	−0.0517	2.446	1.745	0.1008	0.1227
2.2	0.1104	0.5560	0.5208	−0.0015	2.629	1.914	0.0893	0.1079
2.3	0.0288	0.5305	0.5181	0.0523	2.830	2.098	0.0791	0.0950
2.4	0.0025	0.5202	0.5104	0.1005	3.049	2.298	0.0702	0.0837
2.5	0.0729	0.4843	0.4981	0.1459	3.290	2.517	0.0623	0.0739
2.6	−0.0968	0.4708	0.4813	0.1884	3.553	2.755	0.0554	0.0653
2.7	−0.1641	0.4260	0.4605	0.2276	3.842	3.016	0.0493	0.0577
2.8	−0.1850	0.4097	0.4359	0.2635	4.157	3.301	0.0438	0.0511
2.9	−0.2426	0.3575	0.4079	0.2959	4.503	3.613	0.0390	0.0453
3.0	−0.2601	0.3391	0.3769	0.3247	4.881	3.953	0.0347	0.0402
3.2	−0.3202	0.2613	0.3071	0.3707	5.747	4.734	0.0276	0.0316
3.4	−0.3643	0.1792	0.2296	0.4010	6.785	5.670	0.0220	0.0250
3.6	−0.3918	0.0955	0.1477	0.4154	8.028	6.793	0.6175	0.0198
3.8	−0.4026	0.0128	0.0645	0.4141	9.517	8.140	0.0140	0.0157
4.0	−0.3971	−0.0660	−0.0169	0.3979	11.302	9.759	0.0112	0.0125

C CONVERSION FACTORS

APPENDIX C CONVERSION FACTORS

TABLE C-1 Length

Centimeters, cm	Meters, m	Inches in.	Feet, ft	Miles	Microns, μ	Angstroms, Å
1	0.01	0.3937	0.03281	6.214×10^{-6}	10^4	10^8
100	1	39.37	3.281	6.214×10^{-4}	10^6	10^{10}
2.540	0.0254	1	0.08333	1.578×10^{-5}	2.54×10^4	2.54×10^8
30.48	0.3048	12	1	1.894×10^{-4}	0.3048×10^6	0.3048×10^{10}
1.6093×10^5	1609.3	6.336×10^4	5280	1	1.6093×10^9	1.6093×10^{13}
10^{-4}	10^{-6}	3.937×10^{-5}	3.281×10^{-6}	6.2139×10^{-10}	1	10^4
10^{-8}	10^{-10}	3.937×10^{-9}	3.281×10^{-10}	6.2139×10^{-14}	10^{-4}	1

TABLE C-2 Time

Seconds, sec	Minutes, min	Hours, hr	Days	Years, yr
1	1.667×10^{-2}	2.778×10^{-4}	1.157×10^{-5}	3.169×10^{-8}
60	1	1.667×10^{-2}	6.944×10^{-4}	1.901×10^{-6}
3,600	60	1	0.04167	1.141×10^{-24}
8.64×10^4	1,440	24	1	2.737×10^{-3}
3.1557×10^7	5.259×10^5	8,766	365.24	1

TABLE C-3 Mass

Grams, g_m	Kilograms, kg_m	Pounds, lb_m	Tons (short)	Tons (metric)	Atomic mass units, amu
1	0.001	2.2046×10^{-3}	11.102×10^{-6}	10^{-6}	0.60248×10^{24}
1,000	1	2.2046	0.001102	10^{-3}	6.0248×10^{26}
453.6	0.4536	1	5.0×10^{-4}	4.536×10^{-4}	2.7328×10^{26}
9.072×10^{5}	907.2	2,000	1	0.9072	5.4657×10^{29}
10^{6}	1,000	2,204.7	1.1023	1	6.0248×10^{29}
1.6598×10^{-24}	1.6598×10^{-27}	3.6593×10^{-27}	1.8297×10^{-30}	1.6598×10^{-30}	1

TABLE C-4 Flow

cm^3/sec	ft^3/min	U.S. gal/min
1	0.002119	0.01585
472.0	1	7.481
63.09	0.1337	1

TABLE C-5 Pressure

kg/cm^2	$lb_f/in.^2$	lb_f/ft^2	cm Hg (0°C)	in. Hg (32°F)	in. H_2O (60°F)	atm
1	14.22	2,048	73.56	28.96	394.1	0.9678
0.07031	1	144	5.171	2.036	27.71	0.06805
4.882×10^{-4}	0.006944	1	0.03591	0.01414	0.1924	4.725×10^{-4}
0.01360	0.1934	27.85	1	0.3937	5.358	0.01316
0.03453	0.4912	70.73	2.540	1	13.61	0.03342
0.002538	0.03609	5.197	0.1866	0.07348	1	0.002456
1.033	14.70	2,116	76.0	29.92	407.2	1

TABLE C-6 Energy

Ergs	Joules	kwhr	gm-cal	ft.-lb_f	Btu	eV
1	10^{-7}	2.778×10^{-14}	2.388×10^{-8}	7.376×10^{-8}	9.478×10^{-11}	6.2421×10^{11}
10^{7}	1	2.778×10^{-7}	0.2388	0.7376	9.478×10^{-4}	6.2421×10^{18}
3.6×10^{13}	3.6×10^{6}	1	$8.598 + 10^{5}$	2.655×10^{6}	3412	2.25×10^{25}
4.187×10^{7}	4.187	1.163×10^{-5}	1	3.088	3.968×10^{-3}	2.616×10^{19}
1.356×10^{7}	1.356	3.766×10^{-7}	0.3238	1	1.285×10^{-3}	8.462×10^{18}
1.055×10^{10}	1055	2.931×10^{-4}	252	778.2	1	6.584×10^{21}
1.6021×10^{-12}	1.6021×10^{-19}	4.44×10^{-26}	3.826×10^{-20}	1.178×10^{-19}	1.519×10^{-22}	1

TABLE C-7 Power Density

W/cm^3	cal/sec cm^3	Btu/hr in.^{3}	Btu/hr ft^3	MeV/sec cm^3
1	0.2388	55.91	9.662×10^4	6.2420×10^{12}
4.187	1	234.1	4.045×10^5	2.613×10^{13}
0.01788	4.272×10^{-3}	1	1728	1.1164×10^{11}
1.035×10^{-5}	2.472×10^{-6}	5.787×10^{-4}	1	6.4610×10^7
1.602×10^{-13}	3.826×10^{-14}	8.9568×10^{-12}	1.5477×10^{-8}	1

TABLE C-8 Heat Flux

W/cm^2	cal/sec cm^2	Btu/hr ft^2	MeV/sec cm^2
1	0.2388	3170.2	6.2420×10^{12}
4.187	1	1.3272×10^4	2.6134×10^{13}
3.155×10^{-4}	7.535×10^{-5}	1	1.9691×10^9
1.602×10^{-13}	3.826×10^{-14}	5.0785×10^{-10}	1

TABLE C-9 Thermal Conductivity

W/cm °C	cal/sec cm °C	Btu/hr ft °F	Btu in./hr ft^2 °F	MeV/sec cm °C
1	0.2388	57.78	693.3	6.2420×10^{12}
4.187	1	241.9	2903	2.6134×10^{13}
0.01731	4.134×10^{-3}	1	12	1.0805×10^{11}
1.441×10^{-3}	3.445×10^{-4}	0.08333	1	9.004×10^9
1.602×10^{-13}	3.8264×10^{-14}	9.2551×10^{-12}	1.111×10^{-10}	1

TABLE C-10 Viscosity

cP	P	kg$_m$/sec m	lb$_m$/sec ft	lb$_m$/hr ft	lb$_f$ sec/ft^2
1	0.01	0.001	6.720×10^{-4}	2.419	2.089×10^{-5}
100	1	0.1	0.06720	241.9	2.089×10^{-3}
1,000	10	1	0.6720	2,419	0.02089
1,488	14.88	1.488	1	3,600	0.03108
0.4134	4.134×10^{-3}	4.134×10^{-4}	2.778×10^{-4}	1	8.634×10^{-6}
4.788×10^4	478.8	47.88	32.17	1.158×10^5	1

D SOME THERMODYNAMIC PROPERTIES OF DRY SATURATED STEAM

APPENDIX D SOME THERMODYNAMIC PROPERTIES OF DRY SATURATED STEAM PRESSURE

Abs. Press., psi	Temp., °F	Specific Volume		Enthalpy			Entropy		
		Sat. Liquid	Sat. Vapor	Sat. Liquid	Evap.	Sat. Vapor	Sat. Liquid	Evap.	Sat. Vapor
p	t	v_f	v_g	h_f	h_{fg}	h_g	s_f	s_{fg}	s_g
14.696	212.00	0.01672	26.80	180.07	970.3	1150.4	0.3120	1.4446	1.7566
20	227.96	0.01683	20.089	196.16	960.1	1156.3	0.3356	1.3962	1.7319
30	250.33	0.01701	13.746	218.82	945.3	1164.1	0.3680	1.3313	1.6993
40	267.25	0.01715	10.498	236.03	933.7	1169.7	0.3919	1.2844	1.6763
50	281.01	0.01727	8.515	250.09	924.0	1174.1	0.4110	1.2474	1.6585
100	327.81	0.01774	4.432	298.40	888.8	1187.2	0.4740	1.1286	1.6026
150	358.42	0.01809	3.015	330.51	863.6	1194.1	0.5138	1.0556	1.5694
200	381.79	0.01839	2.288	355.36	843.0	1198.4	0.5435	1.0018	1.5453
300	417.33	0.01890	1.5433	393.84	809.0	1202.8	0.5879	0.9225	1.5104
400	444.59	0.0193	1.1613	424.0	780.5	1204.5	0.6214	0.8630	1.4844
500	467.01	0.0197	0.9278	449.4	755.0	1204.4	0.6487	0.8147	1.4634
600	486.21	0.0201	0.7698	471.6	731.6	1203.2	0.6720	0.7734	1.4454
700	503.10	0.0205	0.6554	491.5	709.7	1201.2	0.6925	0.7371	1.4296
800	518.23	0.0209	0.5687	509.7	688.9	1198.6	0.7108	0.7045	1.4153
900	531.98	0.0212	0.5006	526.6	668.8	1195.4	0.7275	0.6744	1.4020
1000	544.61	0.0216	0.4456	542.4	649.4	1191.8	0.7430	0.6467	1.3897
1100	556.31	0.0220	0.4001	557.4	630.4	1187.8	0.7575	0.6205	1.3780
1200	567.22	0.0223	0.3619	571.7	611.7	1183.4	0.7711	0.5956	1.3667
1300	577.46	0.0227	0.3293	585.4	593.2	1178.6	0.7840	0.5719	1.3559
1400	587.10	0.0231	0.3012	598.7	574.7	1173.4	0.7963	0.5491	1.3454
1500	596.23	0.0235	0.2765	611.6	556.3	1167.9	0.8082	0.5269	1.3351
2000	635.82	0.0257	0.1878	671.7	463.4	1135.1	0.8619	0.4230	1.2849
2500	668.13	0.0287	0.1307	730.6	360.5	1091.1	0.9126	0.3197	1.2322
3000	695.36	0.0346	0.0858	802.5	217.8	1020.3	0.9731	0.1885	1.1615
3206.2	705.40	0.0503	0.0503	902.7	0	902.7	1.0580	0	1.0580

Index